Environmental Microbiology

Environmental Microbiology

Editor: Rosemary Charles

www.callistoreference.com

Callisto Reference,
118-35 Queens Blvd., Suite 400,
Forest Hills, NY 11375, USA

Visit us on the World Wide Web at:
www.callistoreference.com

ISBN: 978-1-63239-835-2 (Hardback)

The publisher's policy is to use permanent paper from mills that operate a sustainable forestry policy. Furthermore, the publisher ensures that the text paper and cover boards used have met acceptable environmental accreditation standards.

Printed in the United States of America.

Cataloging-in-publication Data

Environmental microbiology / edited by Rosemary Charles.
p. cm.
Includes bibliographical references and index.
ISBN 978-1-63239-835-2
1. Microbial ecology. 2. Ecology. 3. Microbiology. I. Charles, Rosemary.

QR100 .E58 2017
579.17--dc23

Table of Contents

Preface

Environmental microbiology also referred to as microbial ecology, is the study of microorganisms and their relation and interaction to each other and the environment. As microorganisms play a very important role in the ecology thus, environmental microbiology is an essential part of scientific study. This book provides comprehensive insights into this field. It provides interesting topics for research which readers can take up. The text presents researches and studies performed by experts across the globe. Coherent flow of topics, student-friendly language and extensive use of examples make this text an invaluable source of knowledge. It is a vital tool for all researching or studying environmental microbiology as it gives incredible insights into emerging trends and concepts.

I am honored to present to you this unique book which encompasses the most up-to-date data in the field. I was extremely pleased to get this opportunity of editing the work of experts from across the globe. I have also written papers in this field and researched the various aspects revolving around the progress of the discipline. I have tried to unify my knowledge along with that of stalwarts from every corner of the world, to produce a text which not only benefits the readers but also facilitates the growth of the field.

Finally, I would like to thank all the contributing authors for their valuable time and contributions. This book would not have been possible without their efforts. I would also like to thank my friends and family for their constant support.

Editor

1

Should We Build "Obese" or "Lean" Anaerobic Digesters?

Aurelio Briones[1]*, **Erik Coats**[2], **Cynthia Brinkman**[2]

1 Department of Plant, Soil and Entomological Sciences, University of Idaho, Moscow, Idaho, United States of America, 2 Department of Civil Engineering, University of Idaho, Moscow, Idaho, United States of America

Abstract

Conventional anaerobic digesters (ADs) treating dairy manure are fed with raw or fermented manure rich in volatile fatty acids (VFAs). In contrast, pre-fermented AD (PF-AD) is fed with the more recalcitrant, fiber-rich fraction of manure that has been pre-fermented and depleted of VFAs. Thus, the substrate of PF-AD may be likened to a lean diet rich in fibers while the pre-fermentation stage fermenter is fed a relatively rich diet containing labile organic substances. Previous results have shown that conventional and pre-fermented ADs fed with raw or pre-fermented manure, respectively, produced comparable methane yields. The primary objective of this study was to characterize, using next-generation DNA sequencing, the bacterial communities in various bioreactors (pre-fermentation stage fermenter; various operational arrangements PF-AD; conventional single-stage AD; and a full scale AD) and compare the *Firmicutes* to *Bacteroidetes* (F/B) ratios in these different systems. *Firmicutes* and *Bacteroidetes* constituted the two most abundant phyla in all AD samples analyzed, as well as most of the samples analyzed in the fermenters and manure samples. Higher relative abundance of *Bacteroidetes*, ranging from 26% to 51% of bacteria, tended to be associated with PF-AD samples, while the highest relative abundance of *Firmicutes* occurred in the fermenter (maximum of 76% of bacteria) and manure (maximum of 66% of bacteria) samples. On average, primary stage fermenters exhibited microbiological traits linked to obesity: higher F/B ratios and a 'diet' that is less fibrous and more labile compared to that fed to PF-AD. On the other hand, microbial characteristics associated with leanness (lower F/B ratios combined with fibrous substrate) were associated with PF-AD. We propose that bacterial communities in AD shift depending on the quality of substrate, which ultimately results in maintaining VFA yields in PF-AD, similar to the role of bacterial communities and a high fiber diet in lean mice.

Editor: Marie-Joelle Virolle, University Paris South, France

Funding: The authors gratefully acknowledge the Idaho Center for Advanced Energy Studies (CAES) for funding the project (Contract No. 00042246-00074). The funders had no role in study design, data collection and analysis, decision to publish, or preparation of the manuscript.

Competing Interests: The authors have declared that no competing interests exist.

* E-mail: abriones@uidaho.edu

Introduction

Obesity and or leanness are not traits commonly associated with engineered systems like anaerobic digesters (ADs). However, AD microbial consortia may be greatly influenced by substrate qualities that in the gut, elicit metabolic responses that lead to either obesity or leanness. Like gut bacteria, ADs rely on the activities of different functional groups of anaerobic microbes that work together to degrade organic matter to produce methane. Thus, ADs and many animal digestive systems are faced with similar metabolic challenges that require similar biochemical reactions involving hydrolysis, acidogenesis, acetogenesis, and methanogenesis. The first three processes ultimately lead to the production of volatile fatty acids (VFAs) that are critical intermediates in both animal and engineered systems. The fate of VFAs bifurcate depending on the system: they are mainly absorbed and converted to energy and/or body mass in animals or converted to methane in ADs. Most of the methane formed in animals is considered energetically wasteful, although hydrogen consumption by methanogens may enhance fermentation. Regardless of the system however, bacteria play similar roles in the generation of VFAs from organic matter.

Our interest in anaerobic digestion – and ultimately to the potential linkage between obesity/leanness and substrate – stems from the need to address the environmental impacts posed by manure produced from intensive dairy operations. Since ADs produce energy while treating an organic source of pollution such as excessive manure, anaerobic digestion of dairy waste appears to be a win-win situation. However, it has not gained traction in places like the United States because of unfavorable economics driven primarily by low electricity and natural gas rates [1]. One way to address the low adoption of ADs in the US and other places is to diversify and create added value to the bioproducts from ADs. Thus we have proposed that anaerobic digestion be conducted as a two-stage operation wherein the first stage operates as a high-rate fermenter to produce VFAs from the labile fraction of dairy manure [1]. VFAs are diverted to other bioreactors to produce high value bioproducts such as bioplastics and biofuels [1–3]. The thickened, pre-fermented manure is fed to the AD for biogas production. In contrast to conventional single-stage or second-stage AD that is fed with raw manure and VFA-rich supernatant from a primary fermenter respectively, our pre-fermented AD (PF-AD) is fed with the more recalcitrant, fiber-rich fraction of manure that persists through the pre-fermentation stage. Thus, the PF-AD substrate may be likened to a lean diet rich in fibers while the primary stage fermenter is fed a relatively rich diet containing labile organic substances that are more easily converted to VFAs. A single-stage AD (SS-AD) supports both the fermentation of raw manure and subsequent anaerobic digestion and receives the same feed type as a primary fermenter. However, the microbial community of a SS-AD will reflect characteristics of anaerobic digestion, since conditions such as turnover time of solids (solids retention time = SRT) are imposed to optimize methanogenesis.

Recent findings relating gut microbiota to the obese and lean states suggest that leanness is a characteristic associated with the interaction between increased abundance of *Bacteroidetes* relative to *Firmicutes* (the two most dominant phyla in the gut) and the presence of high fiber (and low fat) in the diet [4]. This combination produces high levels of VFAs derived from plant fiber that promotes leanness by inhibiting fat accumulation in adipose tissue and through other possible physiological mechanisms [5–7]. If similar bacterial processes were to occur in the PF-AD, then this may help explain the high rates of methane production that are routinely achieved from pre-fermented manure when compared to single-stage AD fed with raw manure [8]. This also suggests that the microbiota in ADs fed with a high-fiber 'diet' adjusts to the substrate and thus remains potentially productive in terms of VFA production as has been observed in mice fed with high fiber diets.

The primary objective of this study was to characterize the bacterial communities in various bioreactors (primary stage fermenter; various operational arrangements PF-AD; and conventional single-stage AD) and compare the *Firmicutes* to *Bacteroidetes* ratios in these different systems. To obtain a more complete picture of the microbiology of AD, we performed similar analysis on samples obtained from a full scale AD processing manure from 11,000 dairy cows. Our second objective was to identify the key bacterial populations that are associated with fermentation and anaerobic digestion. Identifying the average bacterial composition and core microbiomes (a core microbiome is comprised of members common to two or more microbial assemblages associated with a habitat [9]) in each process are keys to better understand the links between substrate quality, microbiology and bioreactor performance. This may also be the first step in developing useful mixtures of bacteria for bioaugmentation during times of reactor failure or other perturbations.

Materials and Methods

Description of the Bioreactors

Concurrent with the research in the present study, we have been investigating multiple issues related to PF-AD operation and optimization [1,8]. Samples from these earlier studies were included in the present study for next-generation DNA sequencing (NGS) analysis. Construction of the laboratory-scale SS-AD and PF-AD digesters (Table 1), with the exception of the 0.4L PF-AD, is described elsewhere [8]. The 0.4L PF-AD consisted of a 0.5L glass incubation bottle, capped with a rubber stopper (vented to a wet tip gas flow meters (wettipgasmeter.com), placed inside a water bath and mixed using a magnetic stir plate. Digesters SS-AD and all PF-ADs were operated as completely and continuously mixed systems under the conditions specified; each AD was manually decanted and fed once daily to maintain the SRT/HRT (solids/hydraulic retention time). Each AD received manure obtained from the University of Idaho dairy (Moscow, ID, USA), a facility that maintains around 100 milking cows. The manure substrate was supplied to the ADs either as raw and unprocessed manure, or pre-fermented in a laboratory-scale fermenter (see Coats et al. 2012 [8] and Coats et al. 2013 [1] for additional details).

Table 1. Summary of operational and performance characteristics of anaerobic digesters and fermenters that were sampled for next generation sequencing.

	Single-stage AD (SS-AD)	Pre-fermented AD PF-AD	PF-AD	PF-AD	PF-AD	PF-AD	PF-AD	Large Scale AD (LS-AD)	Fermenter
No. of samples sequenced (labeling codes)	6 (O1-O6)	6 (T1-T6)	1 (T7)	2 (T8, T9)	1 (T10)	2 (T11, T12)	1 (T13)	4 (D1-D4)	5 (F1-F5)
Received Pre-fermented Manure?	No	Yes	Yes	Yes	Yes	Yes	Yes	No	No
Substrate Characteristics	Raw manure	All solids from manure fermenter	All solids from manure fermenter	Screened solids (large fraction) from manure fermenter	All solids from manure fermenter	Screened solids (large fraction) from manure fermenter	Centrifuged solids (finer fraction) from screened fermenter effluent	Manure from 11,000 cows	Raw manure
Operating Volume, L	40	40	40	40	40	40	0.4	3.78×10^6	20
SRT = HRT, d	20	16	20	20	30	30	20	n.d.*	4
Organic Loading Rate, gVS/L-d	3.7	4.2	3.31	3.39	2.60	3.14	3.66	n.d.	8.75–9.5
Operating Temperature	35°C	35°C	35°C	35°C	35°C	35°C	35°C	37–38°C	35°C
Average Methane Yield (L CH4/L d	0.70	0.71	0.56	0.59	0.55	0.59	0.48	n.d.	n.d.

*n.d. = not determined.

Moreover, the fermented solids were recovered either through a fine-mesh screen or through centrifugation (10,000 rpm for 10 minutes). ADs fed with all pre-fermented solids (*via* centrifugation), only the coarse solids (*via* screening), and only the residual fine solids (*via* centrifugation of screened effluent) were separately investigated.

DNA Extraction and PCR

Genomic DNA was extracted from biomass obtained from each AD using the MO BIO PowerSoil DNA Isolation Kit (MO BIO Laboratories, Inc., Carlsbad, CA, USA). Biomass samples were collected on nine dates during AD operational analysis period. Samples were stored at −20°C until further use. Amplification of 16S rRNA fragments for next-generation (Ion Torrent) DNA sequencing was carried out on genomic DNA using *Bacteria*-specific primer set 338F (5′-ACT CCT ACG GGA GGC AGC AG-3′) and 533R (5′- TTA CCG CGG CTG CTG GCA C-3′) [10]. The PCR reaction was performed using 50 ng of DNA template with 5 minutes of initial denaturation at 94°C and 20 cycles of 94°C for 30 seconds, 56°C annealing for 30 seconds, 72°C extension for 1 minute, and a 72°C final extension for 7 minutes. DNA was purified using GeneJet Gel Extraction Kit (Thermo Scientific, Pittsburgh, PA, USA) and quantified using Synergy H1 micro plate reader (BioTek, Winooski, VT, USA).

Ion Torrent Sequencing and Data Analysis

End-repair, adapter ligation and nick repair of 16S rRNA gene amplicon libraries were done using Ion Plus Fragment Library Kit (Life Technologies Corp., Carlsbad, CA, USA; Cat # 4471252) according to the manufacturer's instructions. Each sample library was amplified and then purified using Agencourt AMPure XP system (Beckman Coulter Inc., Brea, CA, USA). Libraries were quantified using an Agilent Bioanalyzer (Agilent Technologies Inc., Santa Clara, CA, USA). Template preparation and enrichment on Ion Sphere Particles was done using One Touch 200 Template Kit version 2 (Cat # 4478320) on an Ion One Touch Enrichment System. Sequencing was done with the Ion PGM 200 Sequencing Kit (Cat # 4474007) using an Ion Torrent Personal Genome Machine at the Molecular Research Core Facility in Idaho State University (Pocatello, Idaho, USA).

Data processing was done using Mothur software [11] designed to process and analyze 16S rRNA gene sequence data, which was implemented within the Galaxy bioinformatics platform [12]. Sequence reads of less than 150 bp were deleted from the data sets as were sequences with average base quality score of less than 24 (a quality score of 20 corresponds to 99% base call accuracy). Chimeric sequences were removed using the UCHIME algorithm [13] and further denoised (i.e., removal of sequences that most likely arise from sequencing errors) using a pseudo-single linkage algorithm implemented in Mothur that is based on a method described by Huse et. al., 2010 [14].

Analysis of alpha diversity and richness (i.e., diversity/richness within samples), classification of sequences and rarefaction analysis were done for each sample library using a down-sampled library size of 96,000 sequences to prevent possible bias due to effects of variable library sizes. Determinations of diversity, richness and classifications were done on operational taxonomic units (OTUs) defined by clustering of sequences at 3%, 5% and 20% levels of dissimilarity, nominally corresponding to groupings at the species, genus and phylum levels. Analysis of beta diversity (compositional similarity among samples) was performed on pooled sequences representing random sub-samples of 9,600 sequences from 21 AD samples (Table 2), three manure samples and a compost sample included as an outgroup. Compositional similarity among samples was determined using principal coordinates analysis (PCoA) implemented in Mothur. The same data set used for PCoA was used to compare and determine statistical significance of differences between bacterial communities of samples using AMOVA (analysis of molecular variance) as implemented in Mothur. The core microbiomes of two major groupings were determined from the beta-diversity data set; these consisted of a group containing all of the manure and fermenter samples (FERMAN, consisting of four fermenter and 3 manure sequence libraries) and a group containing all of the AD samples (ALL-AD, consisting of 17 AD libraries; see Table 2 for details). The core microbiomes of these two major groups were determined by identifying the sequences and OTUs that are shared among all the members of each major group. The average microbiomes of each group were determined by identifying the sequences and OTUs of a random sub-sample of sequences equal in size to the respective core microbiomes of each major group, i.e., FERMAN and ALL-AD. The sizes of the core and average microbiomes of FERMAN and ALL-AD were 24,880 and 30,849 sequences, respectively.

Sequence accession numbers

Sequence data were deposited at the National Center for Biotechnology Information Sequence Read Archive (accession SRP035673).

Results

We processed a total of 28 bioreactor (Table 1) and 3 manure samples for Ion Torrent sequencing. The average sequence length after removal of chimeras and non-bacterial sequences was 181.8 nucleotides and the average sequence quality (Phred score) was 26.4, meaning that the average probability of an incorrect base call is 1 in 518.5, or the average base call accuracy was 99.7%.

Alpha Diversity Analysis

Estimates of OTU richness in all the bioreactors and manure were based on a target sample size of 96,000 sequences per sample, which in all cases provided greater than 90% coverage at the 5% dissimilarity level (nominally corresponding to grouping by genus). At both 3% and 5% (nominally species and genus) dissimilarity, the highest average numbers of OTUs observed were found in the large-scale anaerobic digesters processing manure from thousands of cows (Table S1). Rarefaction curves of the most diverse (LS-AD) and least diverse (manure) data sets suggest that these have been sampled sufficiently (Figure S1). A surprising result is the high levels of OTU diversity that were observed at the phylum level in the fermenter samples (Table S1). Even with the large sample sizes used in this study, OTUs occurring once or twice (singletons or doubletons) may constitute a dominant fraction of the total OTUs. The Chao1 estimator of OTU richness takes into account the frequency of singletons and doubletons in estimating OTU richness. When comparing the different sample types by Chao1 richness, the differences observed in OTU richness were even more apparent (Table S1), suggesting that much of the diversity in LS-AD lies in the rare phylotypes. Similarly, much of the diversity at the phylum level in the fermenters are driven by rare phylotypes detected as singletons.

Beta diversity analysis

Unless otherwise stated, similarity calculations at the 5% level of dissimilarity were used to delineate OTUs in the succeeding analysis. This level of dissimilarity, nominally grouping sequences at the genus level, was found to accurately predict the number of genera in an artificial mixture of short 16S rRNA gene sequences

Table 2. Sampling dates of samples used for next-generation DNA sequencing and subsequent beta-diversity analysis.

Sampling date	Bioreactor type	Number of samples sequenced	Samples included in beta-diversity analysis
28-Jan-11	SS-AD	2	O1
	PF-AD	2	T2
25-Feb-11	FERMAN	2	F1, M1
	SS-AD	1	O3
	PF-AD	1	T3
6-Apr-11	SS-AD	2	O4
	PF-AD	2	T4
20-Apr-11	FERMAN	2	F2, M2
	SS-AD	1	O6
	PF-AD	1	T6
10-Oct-11	FERMAN	3	F3, F4, M3
	PF-AD	2	T7, T10
9-Feb-12	PF-AD	2	T8, T11
18-Jun-12	PF-AD	3	T9, T12, T13
9-Aug-12	LS-AD	4	D1, D2
15-Jan-13	FERMAN	1	

FERMAN = fermenters and manure.
SS-AD = single-stage anaerobic digester.
PF-AD = pre-fermented anaerobic digester.
LS-AD = large-scale anaerobic digester.

[15]. The similarities between the bacterial communities of the different bioreactors were compared by principal coordinates analysis (PCoA) on 21 bioreactor samples (Table 2) plus 3 manure samples and a compost sample that were included to characterize the baseline communities and as an out-group, respectively. As expected, the bacterial communities of fermenters, AD and compost were clearly separated by axis 1, which accounted for 21.4% of the variation in the data (Figure 1). Among the AD samples, the samples from the full-scale digesters (D1, D2) were clearly separated along axis 2 from the rest of the bench-scale AD reactors. There was no significant difference ($p = 0.327$) in the bacterial communities between manure and fermenter samples. The results also showed no clear difference between the overall bacterial communities of the SS-AD and PF-AD reactors. Substrate had a stronger effect on bacterial communities as compared to AD bioreactor type: as mentioned above, large-scale AD fed with a diverse source of manure (i.e., 11,000 cows, with each individual contributing some variation in fecal microbiota) could clearly be differentiated from other ADs. Sample T13, obtained from a PF-AD fed with the fine fraction of pre-fermented manure, could also be clearly differentiated from the rest of the AD samples. In Figure 1, group I represents a relatively tight cluster of points (signifying higher compositional similarity) centering around the mean coordinates of samples collected from PF-AD fed only with the coarse or large particle sized fraction of pre-fermented manure. Similarly, group II is centered around the mean coordinates of samples collected from PF-AD fed with all solids obtained by centrifugation from the manure fermenter. Compared to group I, samples fed with all solids tended to be more compositionally diverse and could not be clearly differentiated from SS-AD samples. Statistically, the difference between the two groups (PF-AD fed with either coarse or all solids) was significant although not highly so ($p = 0.012$). Overall, the result of PCoA suggest that bacterial communities of fermenter and manure (FERMAN) are statistically indistinguishable; and different manure feed types influences bacterial communities to a moderate (differences between fine/coarse/all fractions of manure) or high (diverse manure) degree.

Analysis of Firmicutes and Bacteroidetes

Firmicutes and *Bacteroidetes* constituted the two most abundant phyla in all AD samples analyzed, as well as most of the samples analyzed in the fermenters and manure samples. The exceptions were one manure and two fermenter samples – M3, F3, and F4, respectively, in which *Actinobacteria* were more dominant that *Bacteroidetes*. In general (for all bioreactors), the relative abundance of *Firmicutes* and *Bacteroidetes* averaged ($n = 28$) 70.0% (± *s.d.* 5.0) of the bacterial community. Plotting the relative abundances of *Bacteroidetes vs. Firmicutes* in all the bioreactor samples (plus three manure samples) revealed an overall negative relationship (Figure 2), which held true for three of the four groups, i.e., PF-AD, FERMAN and LS-AD, although the latter was represented by only four points, requiring caution in interpreting this relationship for LS-AD. Higher relative abundance of *Bacteroidetes*, ranging from 25.5% to 50.7% of bacteria, tended to be associated with PF-AD samples. The lower range of relative abundance of *Bacteroidetes* occurred in the fermenter (minimum of 3.8% of bacteria) and manure (minimum of 1.4% of bacteria) samples. Thus the highest relative abundance of *Firmicutes* also occurred in the fermenter (maximum of 75.9% of bacteria) and manure (maximum of 66.0% of bacteria) samples. Conventional single-stage AD and large-scale AD, both fed with raw manure and accomplishing primary fermentation and AD activities within a single vessel were associated with relatively narrower ranges of *Firmicutes* (33.9–38.1% and 33.5−38.6%, respectively) and *Bacteroidetes* (31.3–41.3% and 24.3–27.9%, respectively).

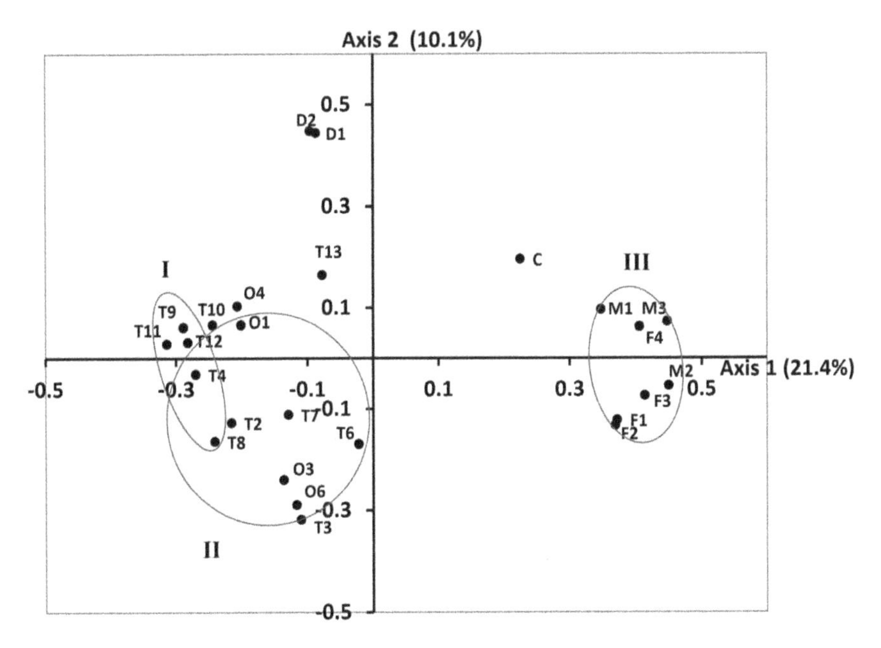

Figure 1. Principal coordinates analysis showing the compositional similarities among the bacterial communities of manure, fermenter, and AD samples as well as a compost sample (C) included as an outgroup. The 21 fermenter and AD samples included in the analysis and the sample codes are listed in Table 2; all samples represent a total of 240,000 sequences. Coordinates were derived from a Bray-Curtis distance matrix that was calculated from OTUs identified at the 5% level of dissimilarity. The centers of the ellipses correspond to the means of the clusters defined by the following groups: I, pre-fermented anaerobic digester (PF-AD) fed with large solids; II, PF-AD fed with all solids; and III, fermenter and manure samples (FERMAN). The radii of the ellipses were determined by 1.5 times the standard deviations from the mean and ellipses were oriented to encompass the maximum number of points in a group minus outliers.

Average and Core Microbiomes

For the purpose of simplifying analysis, samples were grouped based on bioreactor source type or clustering obtained from PCoA. Manure and fermenter samples were combined into a single major group which we collectively refer to as FERMAN. The second major group consisted of all anaerobic digesters (ALL-AD = SS-AD + PF-AD + LS-AD). The average microbiome is determined mostly by differences in abundances of individual

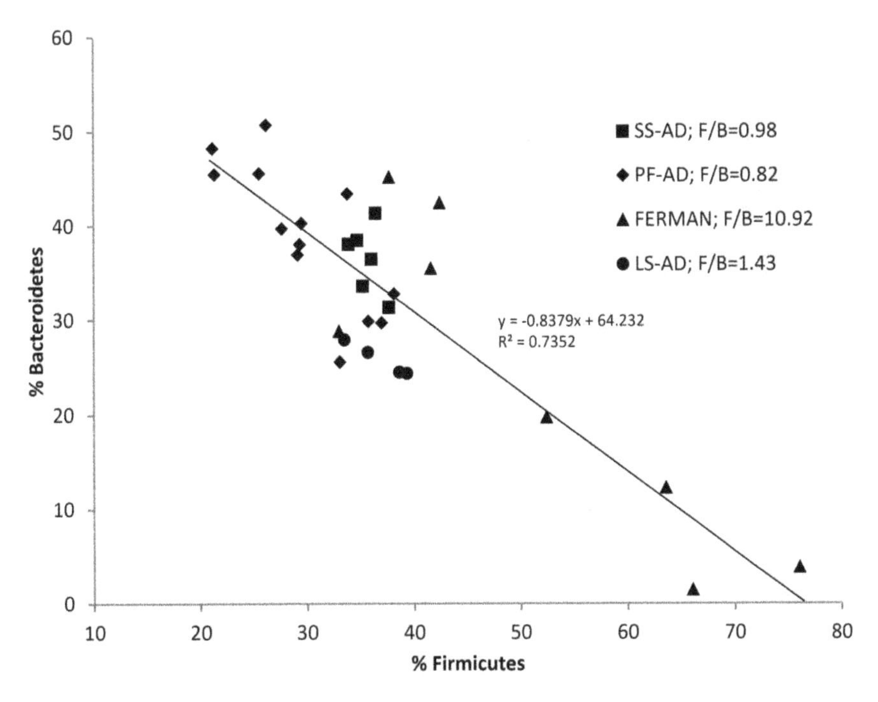

Figure 2. Relative abundances of *Bacteroidetes* vs. *Firmicutes* in FERMAN (fermenters and manure), SS-AD (single-stage anaerobic digesters), PF-AD (pre-fermented anaerobic digesters) and LS-AD (large-scale anaerobic digesters) as revealed by next-generation sequencing analysis. Target sample size was 96,000 sequences. F/B = average *Firmicutes* to *Bacteroidetes* ratio.

populations, while membership in the core microbiome connotes functional indispensability associated with a particular phylotype resulting in consistent presence throughout all the samples within a particular group. In the case of ALL-AD, the core microbiome consisted of only 36 OTUs, out of a total of 2,937 possible OTUs that can obtained by randomly sampling 30,849 sequences from a pool size of 163,200 ALL-AD sequences (Figure 3; *Firmicutes* and *Bacteroidetes* constituted on average 627 AND 723 OTUs, respectively). As expected, the majority of core OTUs were classified as *Firmicutes* and *Bacteroidetes* (14 OTUs each) while a single core OTU was classified as the *Synergistetes*. The latter finding is surprising considering that the *Synergistetes* was a rare phylum, present at only 0.42% of the bacteria in the average microbiome of ALL-AD. On the other hand, the *Verrucomicrobia*, which on average occurred at a higher frequency in ALL-AD, was not a core bacterial group in ALL-AD.

The core microbiome of the FERMAN was more diverse compared to ALL-AD and consisted of 56 OTUs, the majority of which (43 OTUs) were identified as *Firmicutes* (Figure 3). The dominance of the *Firmicutes* highlights its critical function in primary fermentation of manure. On average, *Bacteroidetes* were also a dominant phylum in FERMAN (347 *Bacteroidetes* OTUs on average) but only a single core *Bacteroidetes* OTU (consisting of 105 sequences) was consistently detected in all FERMAN samples; thus the *Bacteroidetes* was not dominant in the core microbiome of FERMAN. The exceptions to the dominance of *Bacteroidetes* occurred in one out of three manure samples and 2 out of 5 fermenter samples. In all these instances, the second most dominant phylum was the *Actinobacteria*. In summary, the core microbiome of fermenters and manure is dominated by the *Firmicutes*. Although not as abundant as the *Bacteroidetes*, more *Actinobacteria* sequences were consistently detected in all fermenter and manure samples, making this group the second most important core phylum in fermenters and manure, while the *Bacteroidetes* and *Proteobacteria* occupied less important niches in these systems.

Key Bacteria Associated with Fermentation and Anaerobic Digestion

Identifying the members of the core microbiome of primary fermenters and anaerobic digesters allowed us to analyze at greater sequencing depth: whereas beta-diversity analysis and classifications of average and core microbiomes relied on individual sample size of 9,600 sequences (pooled together totaling 240,000 sequences from all bioreactors plus manure and outgroup) – at this number of sequences, the minimum coverage value obtained was 0.81. A coverage value of 1.0 means all of the sequences were sampled more than once. At a larger sample size of 96,000, the minimum coverage value obtained was 0.93, making classifications, rarefactions, and diversity analysis more exhaustive at this larger sampling size. In the case of the *Bacteroidetes*, the two most frequently occurring orders were the *Bacteroidales* (12.4% and 17.8% of bacteria in ALL-AD and FERMAN, respectively) and *Flavobacteriales* (10.3% and 0.2% of bacteria in ALL-AD and FERMAN, respectively) (Figure S2). The distribution of these two bacteroidete orders in all the samples suggests that while *Bacteroidales* was consistently present in all bioreactors and manure, the order *Flavobacteriales* was more consistently associated with anaerobic digesters (Figure 4).

Within the *Firmicutes*, the order *Clostridiales* was dominant in all bioreactors and manure samples (Figure S3 and Figure 4). This is consistent with the known role of *Clostridia* in both fermentation and anaerobic digestion. Aside from the *Clostridia*, the orders *Erysipelotrichales* and *Lactobacillales* were detected at relative abundances reaching maxima of 4% and 30% of bacteria, respectively. Both *Erysipelotrichales* and *Lactobacillales* tended to be relatively more abundant in FERMAN – averaging ($n=8$) 8.1% (±10.4) and 2.0%

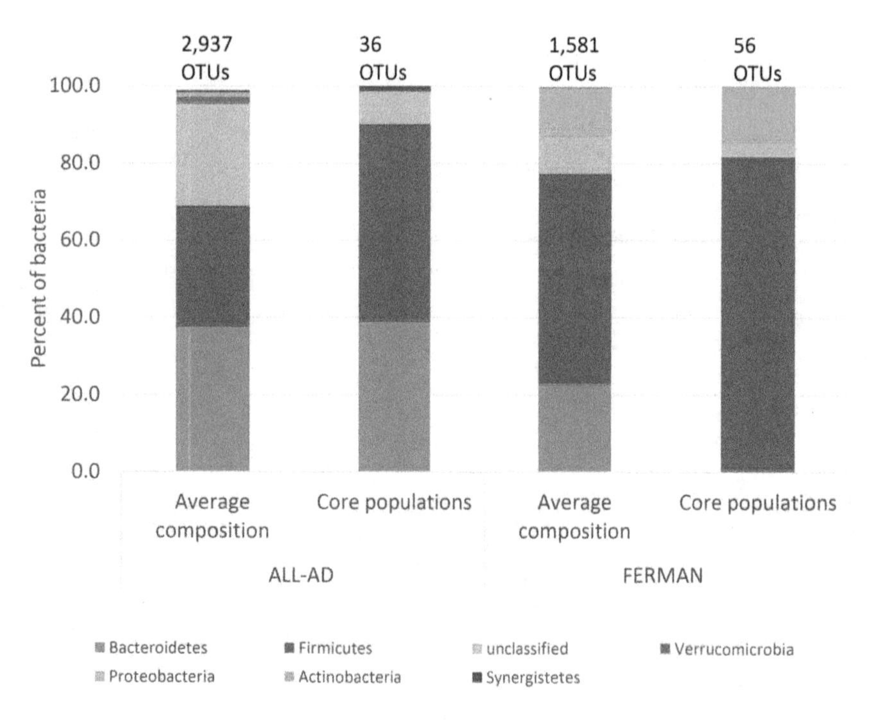

Figure 3. Average and core bacterial phyla associated with ALL-AD (all anaerobic digesters) and FERMAN (fermenters and manure). The average sample sizes for ALL-AD and FERMAN are 30,849 and 24,877 sequences, respectively. The phyla included in the figure comprise at least 99% of bacteria sequenced.

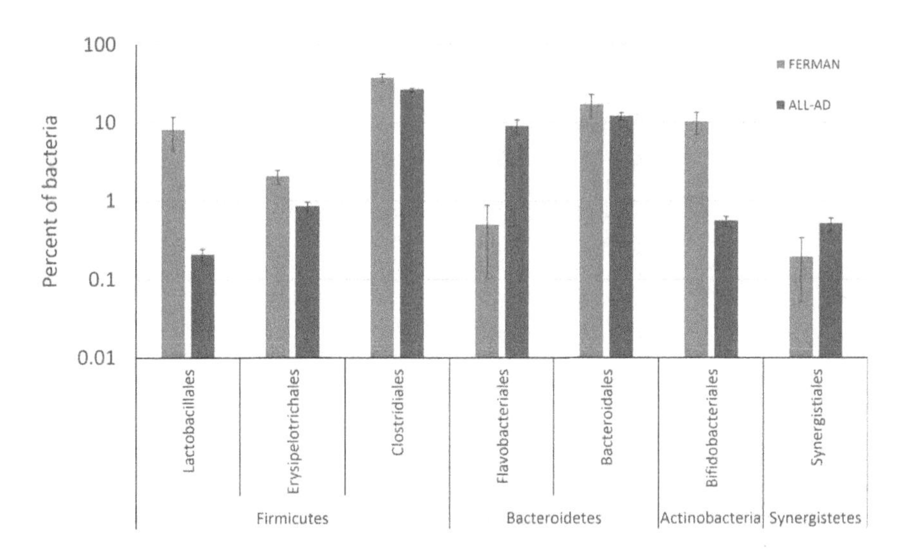

Figure 4. Key orders of bacteria belonging to *Firmicutes, Bacteroidetes, Actinobacteria,* and *Synergistetes* in FERMAN (fermenters and manure samples; $n=8$) and ALL-AD (all anaerobic digester samples; $n=23$). Error bars represent standard errors of the means.

(±1.1) of bacteria, respectively compared to averages ($n=23$) of less than 1% in ALL-AD for both groups.

Aside from the dominant phyla *Firmicutes* and *Bacteroidetes*, two minor core phyla associated with the FERMAN and ALL-AD were the *Actinobacteria* and *Synergistetes*, respectively. Within the *Actinobacteria*, the most dominant classified genus was *Bifidobacterium* (26–47% of actinobacterial sequences) while genus-level taxa within the *Synergistetes* could not be reliably classified for most sequences (11% of *Synergistetes* were classified as *Aminobacterium*).

Discussion

Animal-to-animal variation in fecal bacterial community structure at the species level has been observed in cows [17], which probably reflects the process of speciation occurring at the individual cow level. Thus it is not surprising that genus and species level OTUs are most diverse in the large scale AD facility processing manure from thousands of cows. What is unclear at this point is the ecological significance or explanation for high phylum level richness in an ecosystem, such as that observed in fermenters.

Bacteroidetes and *Firmicutes* are dominant phyla in anaerobic systems, including anaerobic digesters, where they have been shown to be also stable when fed with manure and co-digested with different substrates such as casein, starch and cream [16,18]. While there have been numerous microbial ecological studies on anaerobic digesters, to our knowledge, this study is the first to compare the bacterial communities in fermenters, SS-AD, PF-AD and LS-AD systems processing bovine manure. One study investigating the average and core microbiomes in several anaerobic digesters processing activated sludge found a core group consisting of *Chloroflexi*, *Betaproteobacteria*, *Bacteroidetes*, and *Synergistetes* [19]. Riviere et al. 2009 used a clone library-based approach and also identified the *Firmicutes* as the fourth most abundant group in activated sludge ADs. However, firmicute OTUs tended not to be shared across majority of the activated sludge ADs – and thus did not comprise a 'core' bacterial group. A similar study using next-generation 16S rRNA gene sequencing (pyrosequencing) identified similar average bacterial compositions in seven full-scale ADs processing activated sludge from municipal wastewater treatment plants (MWWTP) [20]. In our study, the core phylotypes identified were taxonomically less diverse compared to those found in WWTP ADs – specifically, *Chloroflexi*, *Proteobacteria*, *Spirochaetes*, and *Thermotoga* were relatively less abundant in AD processing dairy manure. On average, these groups were also present in dairy manure AD, but specific phylotypes were not shared across different reactor types. Differences in substrate, system configuration, scale, and analysis methods (e.g., clone library vs NGS) may account for the differences in the core microbiomes of different types of ADs. However, the consistent identification of *Synergistetes* across different studies [19, 20, 22, this study] suggests that this group of bacteria plays a consistent role in anaerobic digestion. While the role of *Synergistetes* in anaerobic systems remains largely unknown [21], there is experimental evidence to suggest that members of this group utilize acetate syntrophically with hydrogenotrophic methanogens [22]. Moreover, culture-based studies show that cultured representatives of this group are amino acid degraders with a limited repertoire of carbon sources (Godon et al. 2005 [21] and references within). This carbon source limitation dictates that while *Synergistetes* are very frequently present in anaerobic systems, they are present at low abundance – which is in agreement with our results.

Bacteroidetes and *Firmicutes* are also dominant in mammalian gut, with relative abundances that are determined by a number of host-related factors, such as obese or lean states [23] or age [24]. The intestinal microbiota may be one of many factors that can influence obesity – however the manner in which gut microbes do this is still unclear. Microbial contribution to obesity may be related to increased energy harvest which has been related to increased ratios of *Firmicutes* to *Bacteroidetes* during the obese state [25]. However, more recent findings suggest that although obesity and leanness are influenced by the gut microbial ecology, leanness is better described as the result of the interaction between microbial ecology and diet. Specifically, a higher relative abundance of *Bacteroidetes* combined with a high fiber diet are two conditions that promote leanness [4]. Moreover, energy harvest by a *Bacteroidetes*-dominated community from a high-fiber diet is greater compared to a *Firmicutes*-dominated community, which nevertheless promotes leanness, suggesting that energy harvest from substrate and obesity are not necessarily positively correlated. These findings are relevant to our study because we have consistently observed comparable methane yields from single- and two-stage digesters despite the latter being fed with manure depleted of a large fraction of its labile carbon through pre-fermentation. This can be explained by more efficient energy harvest from pre-fermented manure by a *Bacteroidetes*-dominated community comparable to energy harvest from a high-fiber diet by a *Bacteroidetes*-dominated gut community.

More efficient energy harvest may also be achieved by faster reaction kinetics, which is promoted in the primary stage fermenter that is operated at a much shorter solids retention time compared to the ADs. On average, primary stage fermenters exhibit microbiological traits linked to obesity: higher F/B ratios and a 'diet' that is less fibrous and more labile compare to that fed to PF-AD. That the *Bacteroidetes* is only a minor component of the core microbiome of manure fermenters does not suggest that this group is less functionally important in fermenters: our results suggest that hydrolysis and primary fermentation-related functions in *Bacteroidetes* are shared across *Bacteroidetes* phylotypes such as those identified in the average FERMAN microbiome (347 OTUs). Therefore, if majority of *Bacteroidetes* phylotypes carry out the same hydrolysis and primary fermentation-related functions, then any of these phylotypes can dominate different systems. Thus these functions are probably ancestral traits within the strictly anaerobic branch of the *Bacteroidetes* (*Bacteroidales*) which were consistently detected across all samples.

The *Bacteroidetes* are also known for two other attributes: this phylum is known to support frequent horizontal gene transfer events that allow the spread of novel metabolic capabilities such as the degradation of plant-derived fibers [26]. This phylum is also well known to include members specialized in biopolymer degradation [27]. Our more consistent detection of *Flavobacteriales* in the anaerobic digesters are consistent with this observation. Thus it makes great sense for this group of bacteria to be enhanced in PF-AD, which ultimately enables the system to produce comparable energy yields as SS-AD despite the re-directing of a substantial fraction of the manure's energy content to external uses (not AD) after pre-fermentation. Thus, the microbiology of animals and anaerobic digesters are intricately linked to their diet [28]. The populations of microbes that come to dominate both systems ultimately determine how much energy can be harvested from food.

While leanness is considered a healthier state, certain microbes associated with health also tended to be associated with primary fermenters. *Lactobacilli* sp. and *Bifidobacteria* sp. are typical components of probiotics and are recognized for their positive effects on human health especially in the prevention and treatment of intestinal disorders. Our observation that these groups of bacteria are mainly associated with FERMAN is evidence that certain attributes of the obese state, such as the presence of labile organic substances can promote health-associated bacteria. This may have implications on how to enhance the survival and growth of these bacteria when ingested as probiotics. In engineered systems, if the ultimate objective is to harvest more energy and value added products from manure, then our results argue in favor of operating separate "obese" and "lean" bioreactors (fermenters and PF-ADs). These facilitate the production of valuable co-products from the VFAs and comparable methane output as the less "lean" SS-AD that generates only methane and digestate. Overall, linking our findings with the science of gut microbiology demonstrates the possibilities of cross-pollination of knowledge in natural and engineered systems.

Supporting Information

Figure S1 Rarefaction curves of A) least OTU-rich manure sample and B) most OTU-rich LS-AD (large scale anaerobic digester) sample based on analysis of 96,000 sequences per sample.

Figure S2 Percentages of average and core orders of *Bacteroidetes* out of total bacteria associated with FERMAN (fermenter + manure) and ALL-AD (all anaerobic digesters).

Figure S3 Percentages of average and core orders of *Firmicutes* out of total bacteria associated with FERMAN (fermenter + manure) and ALL-AD (all anaerobic digesters).

Acknowledgments

The authors gratefully acknowledge the staff of the Molecular Research Core Facility in Idaho State University for performing Ion Torrent sequencing and Luobin Yang for technical assistance in bioinformatics; and to Jay Kesting for providing samples at the Rock Creek Digesters in Filer, Idaho.

Author Contributions

Conceived and designed the experiments: AMB ERC. Performed the experiments: CKB. Analyzed the data: AMB. Contributed reagents/materials/analysis tools: ERC AMB. Wrote the paper: AMB ERC CKB.

References

1. Coats ER, Searcy E, Feris K, Shrestha D, McDonald AG, et al. (2013) An integrated two-stage anaerobic digestion and biofuel production process to reduce life cycle GHG emissions from US dairies. Biofuels, Bioproducts and Biorefining 7: 459–473.
2. Agler MT, Wrenn BA, Zinder SH, Angenent LT (2011) Waste to bioproduct conversion with undefined mixed cultures: the carboxylate platform. Appl Microbiol 29: 70–78.
3. Angenent LT, Karim K, Al-Dahhan MH, Wrenn BA, Domiguez-Espinosa R (2004) Production of bioenergy and biochemicals from industrial and agricultural wastewater. TRENDS Biotechnol 22: 477–485.
4. Ridaura VK, Faith JJ, Rey FE, Cheng J, Duncan AE, et al. (2013) Gut Microbiota from Twins Discordant for Obesity Modulate Metabolism in Mice. Science 341.
5. Kimura I, Ozawa K, Inoue D, Imamura T, Kimura K, et al. (2013) The gut microbiota suppresses insulin-mediated fat accumulation via the short-chain fatty acid receptor GPR43. Nat Commun 4: 1829.
6. Gao Z, Yin J, Zhang J, Ward RE, Martin RJ, et al. (2009) Butyrate improves insulin sensitivity and increases energy expenditure in mice. Diabetes 58: 1509–1517.
7. Keenan MJ, Zhou J, McCutcheon KL, Raggio AM, Bateman HG, et al. (2006) Effects of resistant starch, a non-digestible fermentable fiber, on reducing body fat. Obesity (Silver Spring) 14: 1523–1534.
8. Coats ER, Ibrahim I, Briones A, Brinkman CK (2012) Methane production on thickened, pre-fermented manure. Bioresource Technology 107: 205–212.
9. Shade A, Handelsman J (2012) Beyond the Venn diagram: the hunt for a core microbiome. Environmental Microbiology 14: 4–12.
10. Huse SM, Dethlefsen L, Huber JA, Welch DM, Relman DA, et al. (2008) Exploring Microbial Diversity and Taxonomy Using SSU rRNA Hypervariable Tag Sequencing. Plos Genetics 4.
11. Schloss PD, Westcott SL, Ryabin T, Hall JR, Hartmann M, et al. (2009) Introducing mothur: open-source, platform-independent, community-supported software for describing and comparing microbial communities. Appl Environ Microbiol 75: 7537–7541.
12. Goecks J, Nekrutenko A, Taylor J, Galaxy T (2010) Galaxy: a comprehensive approach for supporting accessible, reproducible, and transparent computational research in the life sciences. Genome Biol 11: R86.
13. Edgar RC, Haas BJ, Clemente JC, Quince C, Knight R (2011) UCHIME improves sensitivity and speed of chimera detection. Bioinformatics 27: 2194–2200.

14. Huse SM, Welch DM, Morrison HG, Sogin ML (2010) Ironing out the wrinkles in the rare biosphere through improved OTU clustering. Environmental Microbiology 12: 1889–1898.
15. Roesch LF, Fulthorpe RR, Riva A, Casella G, Hadwin AKM, et al. (2007) Pyrosequencing enumerates and contrasts soil microbial diversity. Isme Journal 1: 283–290.
16. Ziganshin AM, Liebetrau J, Proter J, Kleinsteuber S (2013) Microbial community structure and dynamics during anaerobic digestion of various agricultural waste materials. Applied Microbiology and Biotechnology 97: 5161–5174.
17. Durso LM, Harhay GP, Smith TPL, Bono JL, DeSantis TZ, et al. (2010) Animal-to-Animal Variation in Fecal Microbial Diversity among Beef Cattle. Applied and Environmental Microbiology 76: 4858–4862.
18. Kampmann K, Ratering S, Kramer I, Schmidt M, Zerr W, et al. (2012) Unexpected Stability of Bacteroidetes and Firmicutes Communities in Laboratory Biogas Reactors Fed with Different Defined Substrates. Applied and Environmental Microbiology 78: 2106–2119.
19. Riviere D, Desvignes V, Pelletier E, Chaussonnerie S, Guermazi S, et al. (2009) Towards the definition of a core of microorganisms involved in anaerobic digestion of sludge. ISME J 3: 700–714.
20. Lee SH, Kang HJ, Lee YH, Lee TJ, Han K, et al. (2012) Monitoring bacterial community structure and variability in time scale in full-scale anaerobic digesters. Journal of Environmental Monitoring 14: 1893–1905.
21. Godon JJ, Moriniere J, Moletta M, Gaillac M, Bru V, et al. (2005) Rarity associated with specific ecological niches in the bacterial world: the 'Synergistes' example. Environmental Microbiology 7: 213–224.
22. Ito T, Yoshiguchi K, Ariesyady HD, Okabe S (2011) Identification of a novel acetate-utilizing bacterium belonging to Synergistes group 4 in anaerobic digester sludge. Isme Journal 5: 1844–1856.
23. Ley RE, Backhed F, Turnbaugh P, Lozupone CA, Knight RD, et al. (2005) Obesity alters gut microbial ecology. PNAS 102: 11070–11075.
24. Mariat D, Firmesse O, Levenez F, Guimaraes VD, Sokol H, et al. (2009) The Firmicutes/Bacteroidetes ratio of the human microbiota changes with age. Bmc Microbiology 9.
25. Turnbaugh PJ, Ley RE, Mahowald MA, Magrini V, Mardis ER, et al. (2006) An obesity-associated gut microbiome with increased capacity for energy harvest. Nature 444: 1027–1031.
26. Xu J, Mahowald MA, Ley RE, Lozupone CA, Hamady M, et al. (2007) Evolution of symbiotic bacteria in the distal human intestine. Plos Biology 5: e156.
27. Kirchman DL (2002) The ecology of Cytophaga-Flavobacteria in aquatic environments. Fems Microbiology Ecology 39: 91–100.
28. Thomas F, Hehemann JH, Rebuffet E, Czjzek M, Michel G (2011) Environmental and gut bacteroidetes: the food connection. Front Microbiol 2: 93.

2

Sequence Depth, Not PCR Replication, Improves Ecological Inference from Next Generation DNA Sequencing

Dylan P. Smith, Kabir G. Peay*

Department of Biology, Stanford University, Stanford, California, United States of America

Abstract

Recent advances in molecular approaches and DNA sequencing have greatly progressed the field of ecology and allowed for the study of complex communities in unprecedented detail. Next generation sequencing (NGS) can reveal powerful insights into the diversity, composition, and dynamics of cryptic organisms, but results may be sensitive to a number of technical factors, including molecular practices used to generate amplicons, sequencing technology, and data processing. Despite the popularity of some techniques over others, explicit tests of the relative benefits they convey in molecular ecology studies remain scarce. Here we tested the effects of PCR replication, sequencing depth, and sequencing platform on ecological inference drawn from environmental samples of soil fungi. We sequenced replicates of three soil samples taken from pine biomes in North America represented by pools of either one, two, four, eight, or sixteen PCR replicates with both 454 pyrosequencing and Illumina MiSeq. Increasing the number of pooled PCR replicates had no detectable effect on measures of α- and β-diversity. Pseudo-β-diversity – which we define as dissimilarity between re-sequenced replicates of the same sample – decreased markedly with increasing sampling depth. The total richness recovered with Illumina was significantly higher than with 454, but measures of α- and β-diversity between a larger set of fungal samples sequenced on both platforms were highly correlated. Our results suggest that molecular ecology studies will benefit more from investing in robust sequencing technologies than from replicating PCRs. This study also demonstrates the potential for continuous integration of older datasets with newer technology.

Editor: Christina A. Kellogg, U.S. Geological Survey, United States of America

Funding: Financial support for this work was provided by National Science Foundation (NSF) Dimensions of Biodiversity grant (DBI 1249341) to KGP. The funders had no role in study design, data collection and analysis, decision to publish, or preparation of the manuscript.

Competing Interests: The authors have declared that no competing interests exist.

* E-mail: kpeay@stanford.edu

Introduction

Next generation DNA sequencing (NGS) has changed the face of microbial ecology in the space of a few years. As a result, we have gained unprecedented insight into the community dynamics of morphologically cryptic organisms such as fungi, bacteria and viruses [1][2][3]. However, the outcome of NGS based ecological inquiry may be sensitive to technical practices that in many cases have not been adequately tested. For instance, the assumption that NGS read counts accurately reflect absolute abundance in ecological analyses may not be appropriate due to taxon specific PCR and sequencing biases [4]. These technical practices have important effects on our view of underlying biological reality, but also on the allocation of resources (time, money, reagents) that often define the scope of ecological inquiry.

Early optimization of NGS methods has focused on correcting platform specific sequencing issues, such as the known homopolymer error rates in 454 pyrosequencing [5], often with bioinformatic solutions. However, potential distortions may also arise prior to DNA sequencing during sample collection [6], DNA extraction [7][8], or PCR amplification [9][10]. Recognition of these problems has led to a loosely knit collection of best lab and bioinformatics practices that have emerged in the microbial ecology literature and that are aimed at increasing the robustness of whole community amplification. Among other things, such practices include the use of hot-start Taq polymerase, reducing the number of amplification cycles, and the pooling of multiple PCR replicates per sample [11]. The necessity of pooling PCR replicates is thought to arise from stochasticity in the PCR process that results in variable composition of DNA fragments across individual PCR reactions. Possible causes for this may be sampling effects that lead to variation in the initial population of DNA template used to start the reaction, slight variation in initial conditions, or priority effects of amplification in the early rounds of PCR. The few studies that have actually reported results from replicate NGS of the same sample (e.g. [8][12]) have found sample-to-sample variance in sequence composition that seem to support the importance of stochastic PCR effects. However, these studies have focused on the comparison of individual PCR replicates that were sequenced separately, and therefore the extent to which their conclusions rely on PCR or sequencing stochasticity is still unknown.

Though many studies have suggested pooling PCR replicates prior to sequencing, to our knowledge no study has directly tested whether samples comprised of multiple, pooled PCR replicates capture a more robust sample of the true diversity within the sample. Because there have been no explicit tests of PCR pooling,

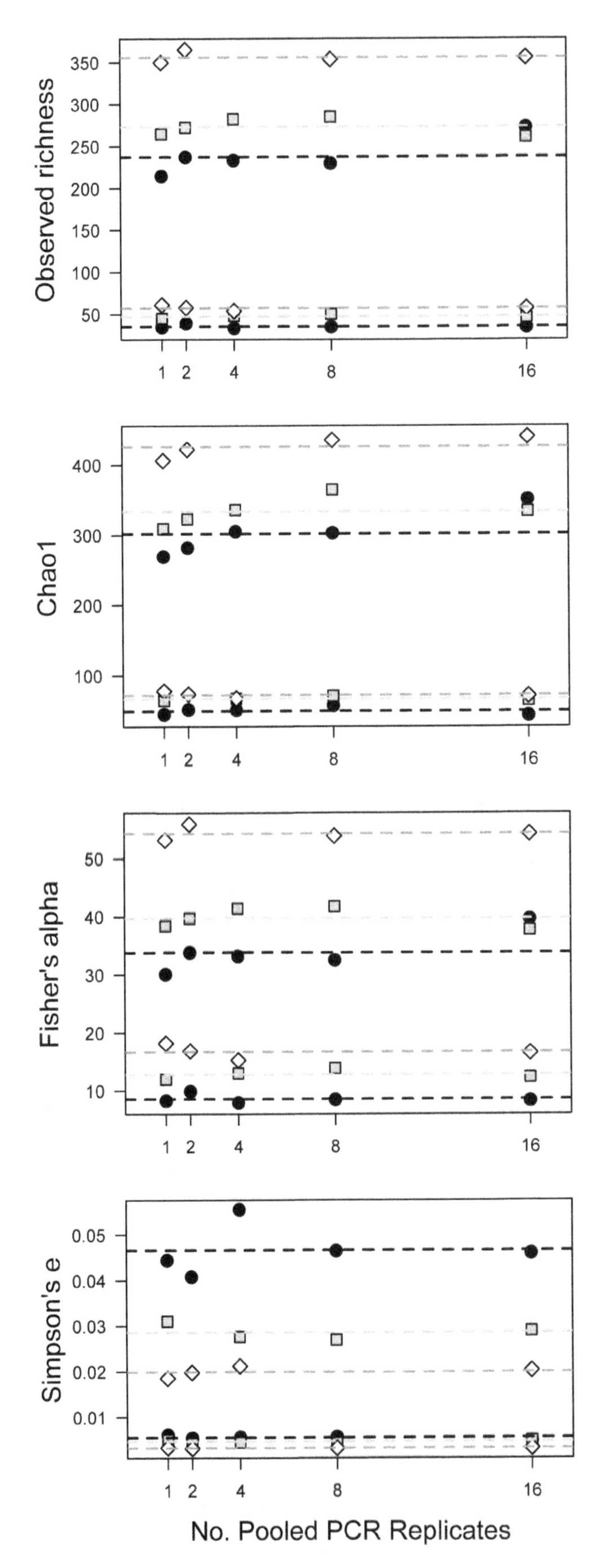

Figure 1. Estimated species richness does not depend on the number of PCR reactions pooled prior to sequencing. Plots of independent replicates representing different levels of PCR pooling for samples CT2, OR1, and OR4 against four different diversity indicators. Points are colored by sample ID. Dotted lines represent the average between different replicates of the same sample. The top three lines represent samples sequenced with Illumina MiSeq and the bottoms three lines represent the same samples sequenced with 454.

papers vary wildly in the number of pooled PCR replicates they use or recommend. While three appears to be a somewhat canonical number (e.g. it has been adopted by the Earth Microbiome Project [13]) other studies have used a single PCR replicate [14], five [15] or even ten [16].

The rapid rise and fall of NGS sequencing platforms also raises major concerns about portability of data across studies. One of the major advantages of DNA based community profiling is the collection of standardized data that can be compared or combined across studies. However, if the observed structure of a community is platform dependent it would seriously weaken the additive nature of sequencing efforts in microbial ecology [17].

In this study our primary goals were (1) to determine quantitatively the number of pooled PCR replicates that maximizes ecological inference in NGS studies and (2) to test the robustness of ecological inferences about community structure using two different NGS sequencing platforms. We did this using a two-pronged experimental approach where the same DNA sample was sequenced from a pool of 1, 2, 4, 8 or 16 separate PCR reactions using both Roche's 454 Pyrosequencing (454) and Illumina's MiSeq (MiSeq). We then compared patterns of α- and β-diversity (the primary response variables in most community ecology studies) among samples and replicates. In addition, we used both NGS platforms to sequence a larger set of soil samples taken from Pine forests in geographically distinct parts of North America where we expected to see differences in community composition.

Based on previous studies, we hypothesized that increasing the number of PCR replicates prior to sequencing would increase α-diversity, reduce β-diversity and increase reproducibility by averaging out PCR noise. Surprisingly, we found that increasing PCR replication did not meaningfully change any of our ecological response variables and may be a poor investment of resources in molecular ecology studies. By contrast, increased sequencing depth markedly improved estimates of β-diversity using both NGS platforms. In addition, we found that community sequencing results from 454 and MiSeq provide largely similar results, suggesting that data from the two platforms can be combined in a meaningful way.

Materials and Methods

Experimental design

We investigated the effects of PCR replication and sequencing platforms on ecological inference using samples from an ongoing project to characterize ectomycorrhizal fungal communities across North America pine forests. To test the hypothesis that pooling PCR replicates improves ecological inference, we selected three soil samples taken from two sites in Oregon (42.79° E –121.62° N) and one in Connecticut (41.82° E, –72.96° N). To test whether NGS sequencing platform affects ecological inference, we selected 60 samples from two sites in North Carolina (NC1 36.01° E, –78.97° N, NC2 35.99° E, –79.10° N), two sites in California (CA1 37.84° E, –119.94° N, CA2 37.81° E, –119.91° N), and two sites in

Table 1. Analysis of variance tests for an effect of PCR replicate number, sample identity, and sequencing method on fungal richness from the CT2, OR1, and OR4 samples.

	No. PCR Replicates		Sample ID		Method		Replicates × Sample ID		Replicates × Method	
	$F_{1,22}$	P	$F_{2,22}$	P	$F_{1,22}$	P	$F_{2,22}$	P	$F_{1,22}$	P
Observed	0.372	0.548	18.748	<0.001*	646.450	<0.001*	0.320	0.730	0.261	0.615
Chao1	2.380	0.137	15.480	<0.001*	717.970	<0.001*	0.171	0.844	2.428	0.134
Fisher's Alpha	0.415	0.526	38.256	<0.001*	490.635	<0.001*	0.439	0.650	0.430	0.519
Simpson	0.060	0.809	42.190	<0.001*	17.250	<0.001*	0.185	0.832	0.000	0.991
Simpson's E	0.001	0.977	13.502	<0.001*	137.701	<0.001*	0.054	0.947	0.001	0.979

Samples were sequenced with both 454 and Illumina MiSeq.

Alaska (AK1 64.77° E, −148.27° N, AK2 64.76° E, −148.25° N) (for all sites n = 10).

All sample sites in Oregon and California were located on United States Department of Agriculture (USDA) National Forest land. The Alaskan sites were located at the Bonanza Creek Long Term Ecological Research (LTER) site. Sites in North Carolina were located on private land owned by Duke University, and sites in Connecticut were located in the Gold's Pine State Forest. No permits were required for any site, and all necessary permissions were obtained prior to sampling. Soil samples were taken from either the homogenized organic or mineral layer of a core approximately 7.5 cm diameter ×14 cm deep (**File S1**). Soils were stored cool until ~0.25 g were extracted using the Powersoil DNA extraction kit (MoBio, Carlsbad CA). DNA extracts were diluted 1:20 and then 1 µl used for PCR.

Molecular methods

For sequencing using the 454 platform, PCR was carried out using modified versions of the fungal specific primer set ITS1F [18] and ITS4 [19]. The 5′ end of the ITS1F primer was modified to include the 454 Lib-L A adapter plus a 10-bp molecular identification (MID) tag to allow for sample multiplexing as in [14]. The 5′ end of the ITS4 primer was modified to include the 454 Lib-L B adapter.

For sequencing using the Illumina MiSeq platform, we designed modified versions of the primer set ITS1F and ITS2 [19]. This primer set targets a shorter section of the fungal ITS region because of the shorter read lengths possible with MiSeq. The 5′ end of the ITS1F primer was modified to include the forward Illumina Nextera adapter and a two basepair "linker" sequence designed to mismatch against all major fungal lineages immediately upstream of the gene primer (**Fig S1**). The induced mismatch is designed to decrease potential taxon-specific PCR bias from downstream matches to the adapter or barcode. The 5′ end of the ITS2 primer was modified with the appropriate reverse Illumina Nextera adapter, linker sequence, and a 12-bp error-correcting Golay barcode as in [17]. Using the program NetPrimer (Premier Biosoft, Palo Alto CA) we designed three custom sequencing primers that demonstrated low dimerization potential and high thermodynamic compatibility with each other and with the Illumina-specific PhiX sequencing primer. The Read 1 and Read 2 sequencing primers were designed to anneal to the gene priming regions of the amplicons and extend further into the conserved 18S portion of the amplified region, thereby maximizing the amount of ITS sequence returned by the reads. The Index sequencing primer was designed to sequence only the 12 bp barcode of each amplicon.

PCR was carried out in 25 µl reactions including 1 µl genomic DNA, 0.5 µl of each 10 µM primer, 5 µl of 5× OneTaq Standard Reaction Buffer (New England BioLabs, Ipswitch MA), 0.5 µl of 10 mM dNTPs (New England BioLabs, Ipswitch MA), and 0.63 units Taq polymerase. All PCR reactions were set up on ice and using Fusion hot start Taq polymerase (New England Biolabs, Ipswitch MA) to minimize non-specific amplification and primer dimerization. PCR conditions were: denaturation at 94°C for 1 min; 30 amplification cycles of 30 sec at 94°C, 30 sec at 52°C and 30 sec at 68°C; followed by a 7 min final extension at 68°C. PCR products were visualized using gel electrophoresis and successful samples cleaned using the Agencourt Ampure XP kit (Beckman Coulter, Brea CA). For the replication experiment, the three samples were each amplified 1, 2, 4, 8 or 16 times using a separate MID tag or barcode for each replication treatment (N = 3 samples ×5 replication levels = 15). Individual PCR reactions for a given sample × replication treatment were pooled and then 20 µl of each pool cleaned using the Ampure Kit as above.

Cleaned PCR products were quantified using the Qubit hs-DS-DNA kit (Invitrogen, Carlsbad CA) on a Tecan Infinite F200 Pro plate reader reading at 485 nm excitation and 530 nm emission. PCR products to be sequenced with 454 were then pooled in equimolar concentration and sent to the Duke University Institute for Genome Sciences & Policy core and sequenced on a ¼ plate partition using Titanium FLX chemistry. PCR products generated for Illumina sequencing were pooled at equimolar concentration and then multiplexed with 44 additional bacterial samples containing 16S rDNA amplicons used for an unrelated study. The final pool containing both loci was sent to the Stanford Functional Genomics Facility for 250 bp paired-end sequencing on an Illumina MiSeq. Bacterial and fungal sequencing primers were also pooled for each read before submission to the sequencing facility. A spike of 30% PhiX was included in the amplicon library in order to achieve sufficient sample heterogeneity. Raw sequence data are deposited at NCBI's Short Read Archive under study accession SRP035367. Sample metadata information is provided as File S1.

Bioinformatics

Sequence de-multiplexing and bioinformatic processing of the 454 and Illumina datasets were performed using aspects of the QIIME [20] and the UPARSE [21] pipelines. Initial quality filtering of 454 sequences excluded all sequences <350 or >1200 bp, with any primer mismatches, with a homopolymer run >10 bp, or with a mean quality score below 25. The remaining sequences were denoised using flowgram clustering [22]. Pre-filtered forward and reverse reads from the Illumina

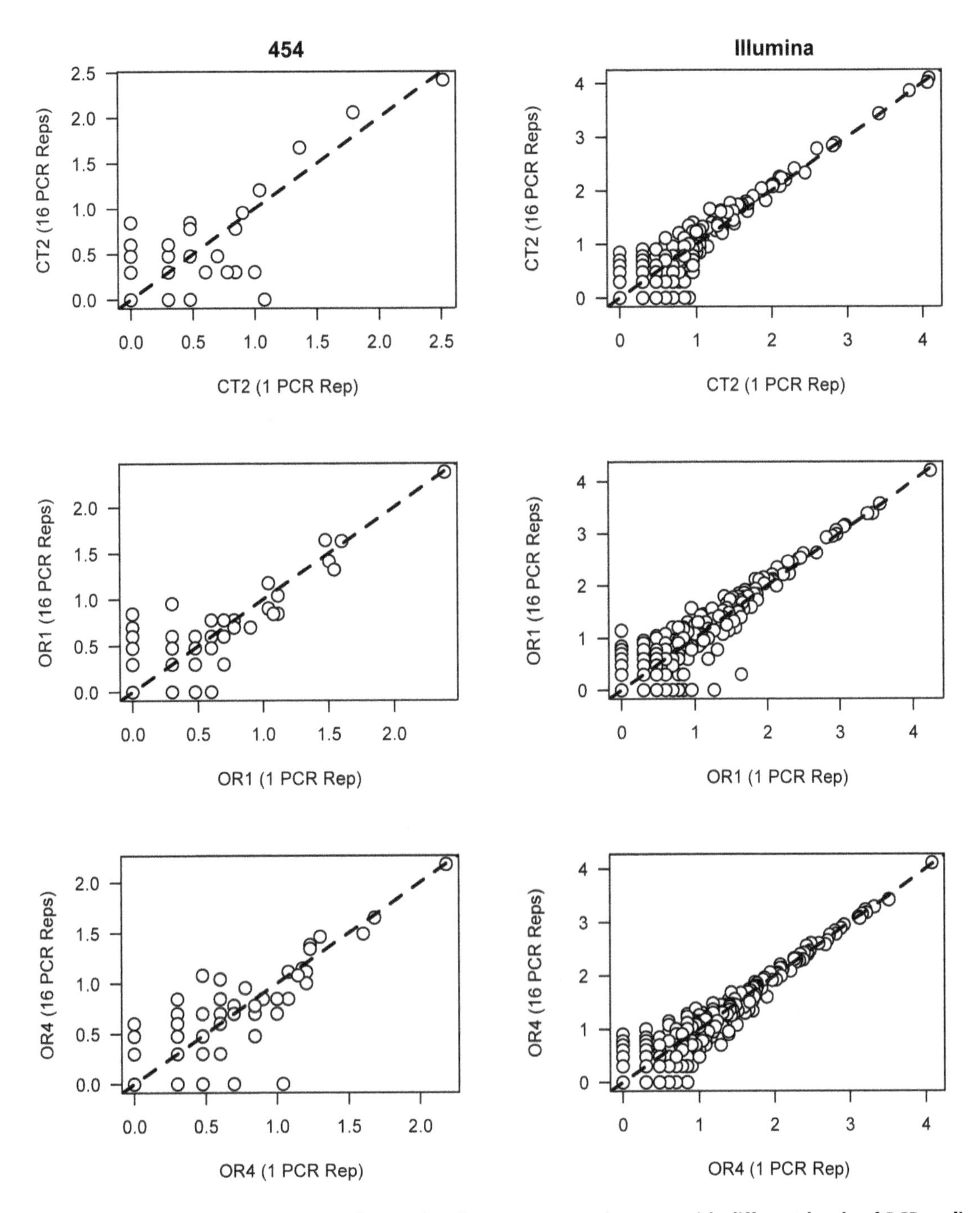

Figure 2. Taxon abundance is strongly correlated across sequencing runs with different levels of PCR replication. Circles represent individual taxa, and the relationship between the $\log_{10}$ of their abundance in samples comprised of 16 PCR replicates (y axis) and one PCR replicate (x axis) for CT2 (top two panels), OR1 (middle two panels), and OR4 (bottom two panels). Dashed lines represent a 1:1 relationship. The left three panels show samples sequenced with 454 and the right three panels show samples sequenced with Illumina MiSeq.

dataset were 238 bp long and our multiplexing strategy resulted in high quality sequences for both fungal and bacterial samples. For the fungal samples analyzed in this study, reads were trimmed with CutAdapt [23] to the point where the sequence met the distal priming site, and further trimmed using Trimmomatic [24] to remove any additional low quality end regions. After quality trimming forward reads averaged 208 bp and reverse reads averaged 185 bp. Reads were paired using USEARCH v. 7.0.1001 with a minimum Phred score sequence cutoff threshold of 3 and a minimum sequence length of 75 bp. Paired reads averaged 230 bp and were discarded if they contained >0.25 expected errors. The final fasta file containing all sequences used for analysis is available from the authors upon request.

All final, high-quality sequences from both the 454 and Illumina datasets were combined and grouped into operational taxonomic units (OTUs) in USEARCH using the UPARSE-OTU and UPARSE-OTUref algorithms (which included chimaera detection and filtering and dropped all global singleton reads) at a 97% sequence similarity cutoff. OTUs were given taxonomic assignments in QIIME based on a previously published sequence

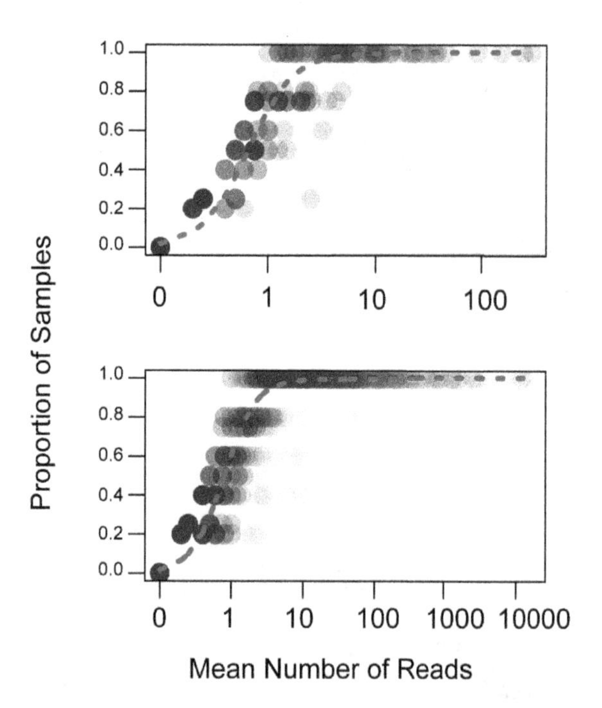

Figure 3. Taxa represented by more sequences are observed most consistently across replicates of the same sample. Plots of the proportion of samples in which an OTU is observed against the average number of reads representing that OTU. Zeros were assigned a value of 0.1 prior to log transformation. Dashed lines represent the logistic model predicting the relationship. The top panel shows results from 454 and the bottom panel shows results from Illumina.

database [8] modified for QIIME compatibility as in [25]. To compare samples on an equal basis all samples were rarefied to even sampling depths prior to statistical analysis. Rarefaction depths were determined ad-hoc to maximize the number of samples included while still maintaining a reasonable number of sequences. For the replication experiment, the 454 samples were rarefied to 500 sequences and the Illumina samples to 38,000 sequences. For the larger, cross platform-comparison dataset, 454 samples were rarefied to 1000 sequences and the Illumina samples to 40,000 sequences.

Statistical analysis

To see how replication affects ecological inference we calculated a number of common α-diversity (observed richness, Fisher's Alpha, Chao 1 and Simpson, Simpson Evenness) and β-diversity (Jaccard, Bray-Curtis, β-sim) metrics used in community ecology. We used a linear model to test whether or not the number of PCR replicates and sequencing platform affected different richness estimators (S = Replication × Sample ID × Platform). We used a similar approach to test whether average β-diversity changed in any predictable way with the number of replicates used to generate each sample. This was done by calculating β-diversity (Bray Curtis or Jaccard) for each replication level compared with all other samples sequenced from those plots.

This dataset also allowed us a unique look at data reproducibility with repeated sequencing of the same sample. To see how sequencing depth affects estimates of sample β-diversity, we calculated within sample β-diversity (that is, β-diversity between independent replicates of the same sample – hereafter termed pseudo-β-diversity) at a range of sequencing depths, from 50–1000 (454) and 100 to 80,000 (Illumina). Because there has been much debate about the handling and validity of low abundance OTUs, we tested the effect of within sample sequence abundance on the repeatability with which an OTU is detected across replicate sequencing of the same sample. We used logistic regression to model the relationship between $\log_{10}$ transformed mean within sample abundance and frequency of detection across samples (for this analysis 0 values were assigned ½ the minimum observed value prior to log transformation). We also looked at quantitative reproducibility by comparing OTU read abundance ($\log_{10}$ X+1 transformed) between the single replication treatment and the 16 replication treatment for each DNA sample.

Finally, with the larger dataset we compared the similarity of ecological inferences made with different sequencing platforms. α-diversity estimates were generated for each sample based on rarefaction to a common sequence depth within each platform. We used a Mantel test to determine whether community similarity estimates were similar across platforms. Pairwise sample similarity across samples and across platforms was visualized using non-metric dimensional scaling (NMDS), and a perMANOVA tested for the effect of sequencing platform and geographic origin on estimates of community similarity. To compare whole community overlap we generated a Venn diagram to illustrate the proportion of shared and unique taxa generated with each platform. To look for taxonomic bias we plotted relative abundance of lineages for shared and unique OTUs across platforms. Statistics were performed using the R software package [26] and the Vegan community analysis package for NMDS and perMANOVA [27].

Several data points were left out of our analyses due to either insufficient or low-quality sequences or sample mishandling (454 data: the 2 PCRs treatment for OR1, the 8 PCRs treatment for OR4, and individual points CA1.A5.OH and CA1.A5.AH from the larger dataset. MiSeq data: the 4 PCRs treatment for OR1 and individual point NC2.0.OH from the larger dataset).

Results

After quality control, denoising, and chimera removal of the smaller dataset, sequencing depth for successful samples ranged from 573–1783 sequences in the 454 dataset and from 38,423–92,189 sequences in the Illumina dataset. Observed richness at 500 sequences (454 dataset) ranged from approximately 30 to 60 OTUs/sample and at 38,000 sequences (Illumina dataset) ranged from approximately 200 to 350 OTUs/samples. As expected for Pine soils the most commonly observed taxa belonged to lineages of Basidiomycota ectomycorrhizal fungi and saprotrophs (data not shown).

Increasing the number of PCR replicates pooled prior to sequencing had no effect on the estimated α-diversity of a sample regardless of the sequencing method used (**Table 1**). This effect was consistent regardless of the richness metric chosen (**Fig 1**) and whether or not an interaction term was included in the model. Similarly, estimates of β-diversity compared with other samples in the same site did not show any trends related to the number of pooled PCR replicates (454 Jaccard ANOVA $F_{1,7}=0.588$, $P=0.468$; Bray-Curtis ANOVA $F_{1,7}=1.682$, $P=0.236$; Illumina Jaccard ANOVA $F_{1,8}=0.118$, $P=0.741$; Bray-Curtis ANOVA $F_{1,8}=1.685$, $P=0.231$; **Fig S2**). That is to say, a sample sequenced from 1 PCR did not show higher or lower β -diversity with other samples from the same plot than the same sample sequenced from 16 PCR replicates. When ordinated the different replication levels from the same sample clustered together with little variation from the centroid (**Fig S2**).

Sequence counts for individual OTUs were highly correlated across resequencing instances of the same sample. This relation-

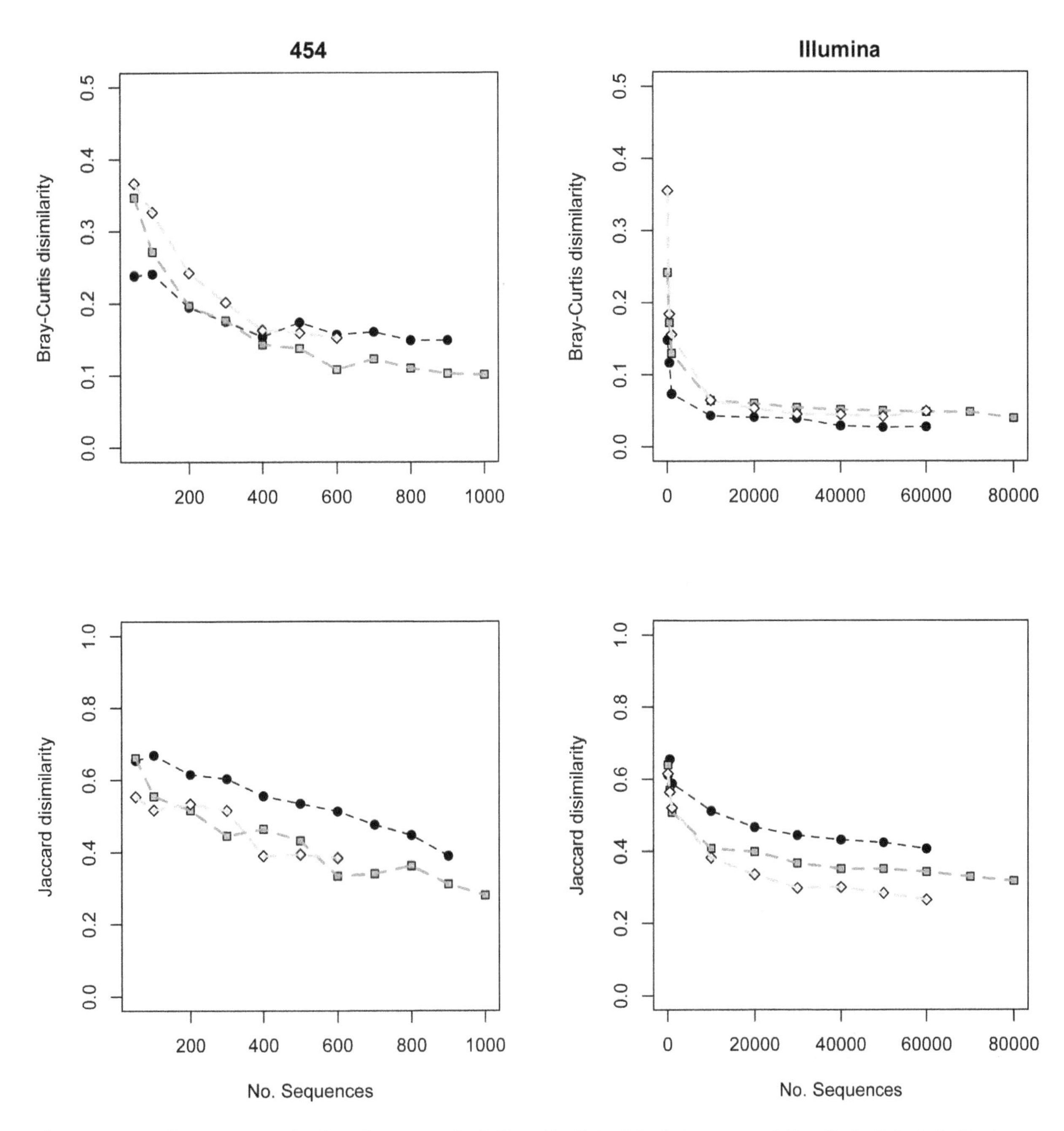

Figure 4. Increasing sequence depth reduces pseudo-β-diversity. Plots of the between-sample Bray-Curtis dissimilarity (top two panels) and Jaccard dissimilarity (bottom two panels) in CT2, OR1, and OR4 against per-sample sequencing depth. Points represent the β-diversity values between different replicates of the same sample and are colored by sample ID. Dashed lines connect each symbol within a sample. The left two panels show samples sequenced with 454 and the right two panels show samples sequenced with Illumina MiSeq.

ship was true regardless of the PCR replicate number. Plots relating the number of sequences per taxon in a sample composed of 16 PCR replicates to that of a single PCR replicate were strongly log linear and in almost all cases followed a 1:1 relationship (**Fig 2**). As a result, per OTU sequence counts were highly significantly correlated (Pearson's product moment correlation: 454 dataset CT2 $r = 0.84$, $P<0.001$; OR1 $r = 0.89$, $P<0.001$; OR4 $r = 0.87$, $P<0.001$; Illumina dataset CT2 $r = 0.92$, $P<0.001$; OR1 $r = 0.94$, $P<0.001$; OR4 $r = 0.96$, $P<0.001$).

High abundance OTUs were detected more consistently across replicate sequencing runs. The proportion of samples that an OTU was observed in increased significantly with average within sample read depth for that taxon (**Fig 3**; Overall effects tests: 454 No. reads $\chi^2_1 = 1{,}496$, $P<0.001$, Site $\chi^2_2 = 1.0$, $P = 0.60$; Illumina No. reads $\chi^2_1 = 10{,}742$, $P<0.001$, Site $\chi^2_2 = 2.3$, $P = 0.32$). There were no differences in this relationship across samples and the same patterns were seen for analyses run with median, maximum and minimum read depth (data not shown). For both 454 and Illumina OTUs with mean read abundance >10 sequences were detected nearly 100% of the time. However, many low abundance OTUs were also detected with a high degree of regularity.

Pseudo-β-diversity estimates decreased exponentially with increasing sequencing depth (**Fig 4**). In the 454 dataset,

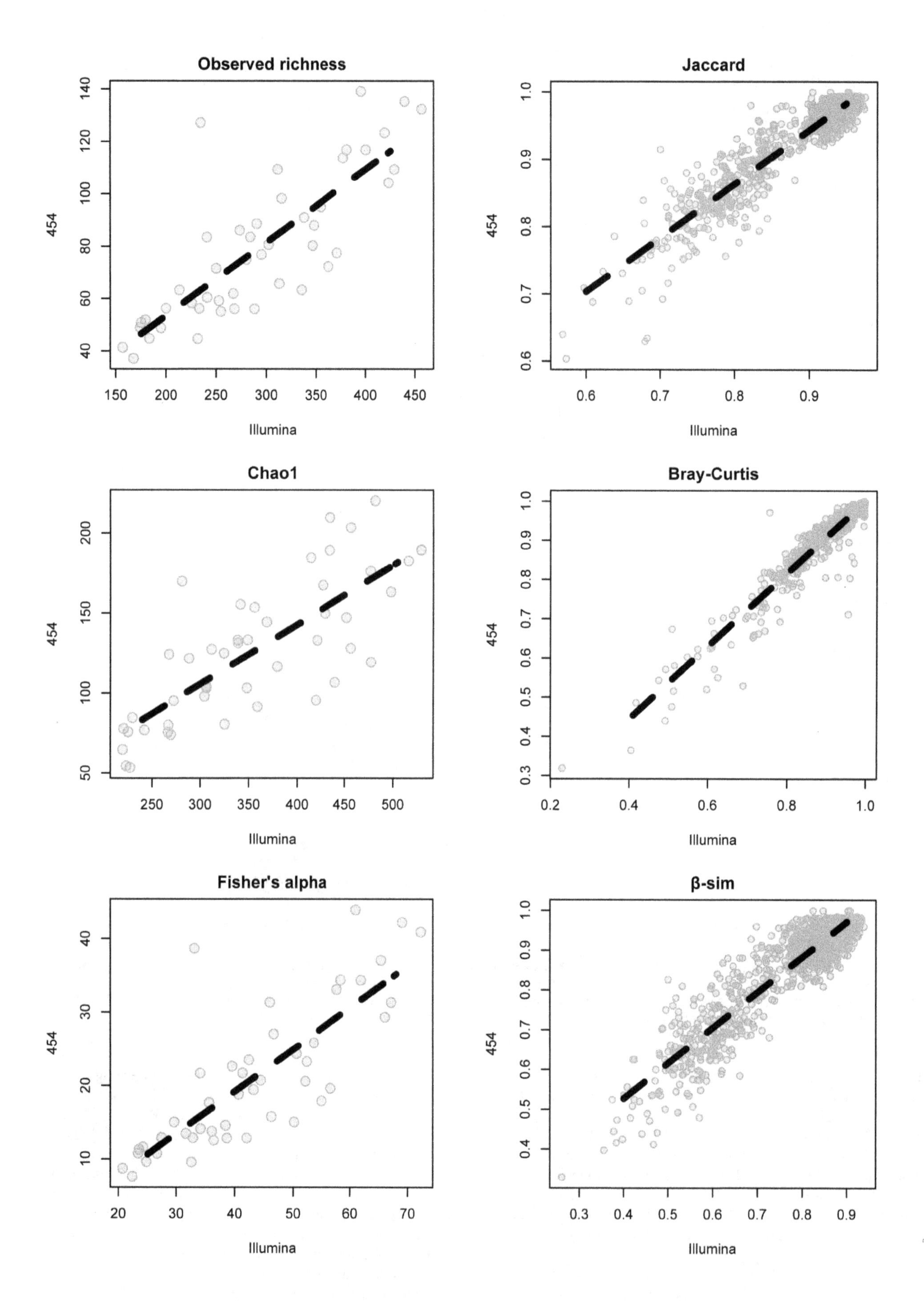
Observed richness
Jaccard
Chao1
Bray-Curtis
Fisher's alpha
β-sim
454
Illumina

Figure 5. Patterns of α- and β-diversity are highly reproducible when samples are sequenced on different platforms. Regressions of diversity found in 55 soil samples sequenced on both platforms. Left three columns: points represent individual samples, and the relationship between the total richness per sample found when sequenced with 454 (y axis) and Illumina MiSeq (x axis). Right three columns: points represent pairwise differences in between-sample community composition and the relationship between dissimilarity found with 454 (y axis) and Illumina (x axis). Dashed lines represent the linear models predicting the relationships.

dissimilarity measures consistently remained above 0.1 (Bray-Curtis) and 0.3 (Jaccard) at the maximum amount of sequences/sample recovered. In the Illumina dataset, Bray-Curtis dissimilarity approached zero above 20,000 sequences/sample while Jaccard dissimilarity remained above 0.3 at the maximum amount of sequences/sample recovered.

From the larger set of samples, sequencing on the 454 and Illumina MiSeq platforms resulted in 3,660 total OTUs. Of these 3 (0.08%) were unique to the 454 dataset, 1,798 (49.13%) were unique to the Illumina dataset, and 1,859 (50.79%) were shared (**Fig S3**). Richness was significantly higher for samples sequenced with Illumina MiSeq than for the same samples sequenced with 454 (**Table 2**). Regressions of per-sample observed species, Chao1 estimated richness, and Fisher's Alpha index between the 454 dataset and the Illumina dataset indicated that diversity recovered with either sequencing method was highly correlated (**Fig 5**: Observed slope $= 0.28 \pm 0.03$, $r^2 = 0.67$, $P<0.001$; Chao1 slope $= 0.37 \pm 0.05$, $r^2 = 0.59$, $P<0.001$, Fisher's Alpha slope $= 0.57 \pm 0.06$, $r^2 = 0.65$, $P<0.001$).

β-diversity estimates of between-sample dissimilarity were somewhat affected by sequencing platform (Jaccard perMANOVA $F_{1,98} = 30.26$, $r^2 = 0.08$ $P = 0.001$; Bray-Curtis perMANOVA $F_{1,94} = 45.40$, $r^2 = 0.09$, $P = 0.001$; β-sim perMANOVA $F_{1,94} = 29.30$, $r^2 = 0.05$, $P = 0.001$). However, regressions of β-diversity between the 454 dataset and the Illumina dataset showed that between-sample dissimilarity was highly correlated between the two sequencing platforms (**Fig 5**: Jaccard slope $= 0.80 \pm 0.01$, Mantel $r = 0.94$, $P = 0.001$; Bray-Curtis slope $= 0.93 \pm 0.01$, Mantel $r = 0.96$, $P = 0.001$, β-sim slope $= 0.92 \pm 0.01$, Mantel $r = 0.92$, $P = 0.001$). Both sequencing methods recovered the significant differences in community structure expected between sampling bioregions (Jaccard perMANOVA $F_{2,98} = 124.40$, $r^2 = 0.62$, $P = 0.001$; Bray-Curtis perMANOVA $F_{2,98} = 145.32$, $r^2 = 0.61$, $P = 0.001$; β-sim perMANOVA $F_{2,98} = 204.45$, $r^2 = 0.74$, $P = 0.001$) which explained an order of magnitude more variation than did sequencing method. All samples from both sequencing datasets ordinated primarily by sampling bioregion (**Fig 6**).

Taxonomic assignment was highly consistent between OTUs found in the 454 dataset and OTUs found in the Illumina dataset at the phylum, class, and ordinal levels (**Fig 7, Fig S4**). Taxonomic bias between the two sequencing platforms was primarily limited to low abundance taxa (e.g. relatively more species of Zygomycota and Chytridomycota taxa found in the 454 dataset), but taxonomic composition of each dataset looked nearly identical when relative abundance of taxonomic groups was considered.

Discussion

In this study we compare the effects of different lab, sequencing and bioinformatic protocols on a number of ecological metrics of α- and β-diversity. These metrics form the basis for conclusions in most microbial community studies and so our results should have important ramifications for how this work is carried out. This is particularly true because of the limited resources and options that can be explored in a single study.

Despite the popularity of PCR replication in molecular ecology studies, we find that increasing the number of PCR replicates that are pooled prior to sequencing has no meaningful effect on ecological measures of diversity or community structure and thus likely no effect on the conclusions of a given study. We present several lines of evidence to support this conclusion. First, the effect of PCR replication is highly insignificant in all statistical models predicting α- and β-diversity of replicates of three soil samples taken from different pine biomes across North America (**Table 1**, **Fig 1**, **Fig S2**). Although visually there appears to be a slight increase in species richness, Chao1 richness, and Fisher's Alpha diversity as more PCR replicates are pooled, the trend does not hold in a linear fashion and is both sample, sequencing method, and metric-specific (e.g. only seen in the Illumina dataset, only true for OR1 and OR4, and is most apparent only for Chao1 richness estimates). On the other hand, sample ID (i.e. sampling location, a more relevant ecological factor) has a comparably large effect on α- and β-diversity metrics. In all models tested, sample ID and sequencing method are the predominant drivers between differences in diversity and community structure.

Second, the number of sequences observed for each taxon between low-replicated samples (e.g. 1 PCR replicate) and high-replicated samples (16 PCR replicates) is highly correlative in a 1:1 relationship (**Fig 2**), suggesting that pooling more PCR replicates prior to sequencing does not affect the relative abundances of taxa found in each sample, and that sequence abundance per taxon in one PCR can accurately predict sequence abundance per taxon in a pool of 16 replicates. It is important to note that the relationship between sequence abundance per taxon in high vs. low replicated samples is weakest with low-abundance taxa, highlighting the

Table 2. Average per-sample fungal richness for 55 soil samples from pine forests.

	454 Avg. Richness/Sample	Illumina Avg. Richness/Sample	$F_{1,102}$	*P*
Observed	79.907	289.000	322.800	<0.001*
Chao1	125.835	349.610	269.600	<0.001*
Fisher's Alpha	21.190	42.522	82.110	<0.001*
Simpson	0.814	0.819	0.032	0.858
Simpson's E	0.019	0.005	83.970	<0.001*

Samples were sequenced with both 454 and Illumina MiSeq.

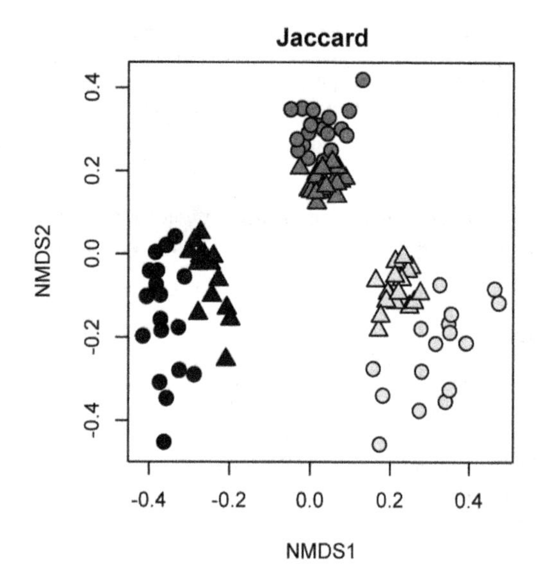

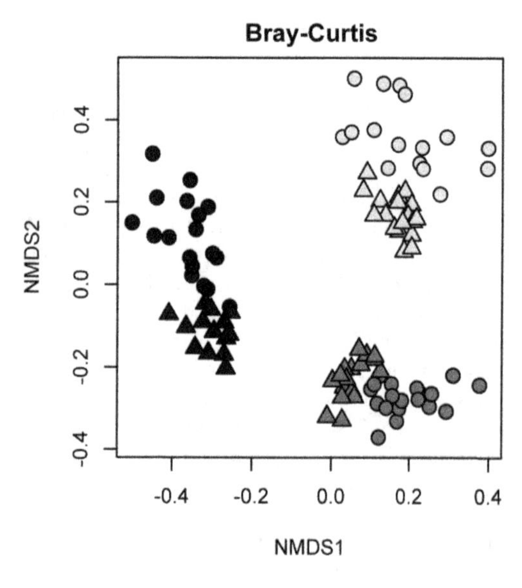

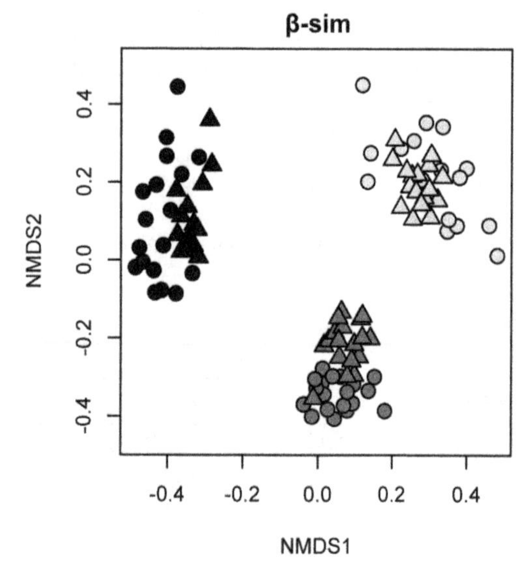

Figure 6. Broad ecological patterns of β-diversity are recovered equally well with each sequencing platform. Non-metric multidimensional scaling of fungal communities from 55 soil samples sequenced with both 454 (circles) and Illumina MiSeq (triangles). Points are colored by the three regions of sample collection. Ordinations are based on between-sample dissimilarity calculated with Jaccard (top panel), Bray-Curtis (middle panel), and β-sim (bottom panel).

importance of adequate sequencing depth for obtaining an accurate depiction of diversity within samples. Together, these results suggest that ecological studies focused on comparing diversity levels and differences between multiple samples across gradients or treatments varying in space or environmental conditions will not be improved significantly by using multiple PCR replicates.

In searching through the literature the origins of this practice are actually somewhat unclear. In general, most evidence in favor of PCR replication is based on differences in OTUs detected in repeated sequencing of single PCR replicates of the same sample [8][11][12]. However, we would argue that sampling error during PCR is a small problem compared with sampling error in the actual sequencing process, in which a few thousand molecules are down sampled from an overall population of billions. Our data show that replicates of the same sample sequenced multiple times at low sequencing depth can lead to higher levels of between-replicate dissimilarity – i.e. pseudo-β-diversity - than should ideally be the case. This is most likely due to inadequate sequencing depth or suboptimal levels of rarefaction making low-abundance taxa unlikely to be appear in all samples. We find that the proportion of replicates of the same sample in which an OTU is present increases logarithmically with the average number of reads representing that OTU (**Fig 3**). This is to say that the reproducibility of taxon coverage and composition of a sample will improve with increasing sequencing depth per taxon. For both 454 and Illumina, OTUs represented by an average of >10 sequences are detected nearly 100% of the time. While many low abundance OTUs are detected repeatedly across samples and thus likely to be real, restricting analyses to these core OTUs may restrict the influence of pseudo-β-diversity due to limited sampling when making ecological conclusions about similarity of microbial communities.

In addition, we find that pseudo-β-diversity between sequencing replicates of the same sample decreases with an increasing number of sequences per sample (**Fig 4**). Pseudo-β-diversity of abundance sensitive metrics like Bray-Curtis dissimilarity decreases exponentially as more sequences are added, approaching zero (indicating little to no difference between replicates of the same sample) at a depth of >10,000 sequences/sample. In our study, sequencing on the 454 platform was unable to capture this sequencing depth and thus dissimilarity between replicates in this dataset remain higher (>0.1) at the maximum sequencing depth recovered. Sequencing on the Illumina platform recovered approximately 70× more sequences per sample and thus easily reaches the lowest Bray-Curtis dissimilarity values possible between replicates within the recovered sequencing depth. Interestingly, values of a binary metric like Jaccard dissimilarity remain higher than might be desired for multiple replicates of the same sample even at the maximum sequencing depth recovered for both platforms. This indicates that rare taxa continue to be detected in low abundance as more sequences are recovered on either platform, regardless of the total amount of additional sequences.

The implications of this result are several. First, it could suggest that extremely low-abundance microbial taxa are always present in high diversity systems such as soils. As technology progresses to achieve orders of magnitude more sequences per sample with each new sequencing platform, microbial ecology studies will tend to detect more and more rare taxa, perhaps without ever saturating

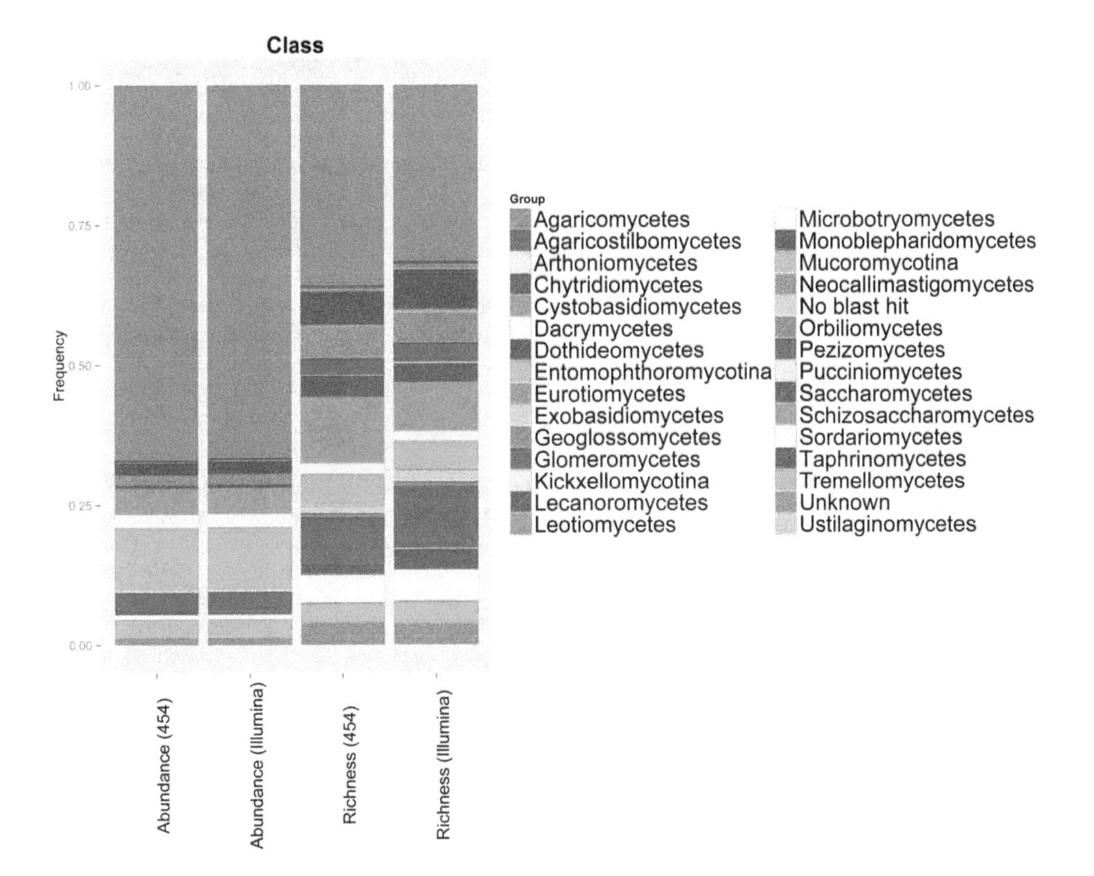

Figure 7. Taxonomic assignment to OTUs observed between both sequencing platforms is highly consistent. Bar chart indicating the proportional richness and abundance of taxa identified to the class level in 55 soil samples sequenced with both 454 and Illumina MiSeq.

taxa-accumulation curves. Using binary metrics like Jaccard could lead to artificially high estimates of between sample dissimilarity given how low-abundance and rare these new taxa are, and thus should be used with caution. Perhaps more realistically, the rare taxa that persist in detection at extremely low abundance as sequencing depth grows could also be a result of sequencing error and spurious OTU formation during sequence processing [21][28][29][30]. As a result, the ability to saturate taxa-accumulation curves by increasing sequencing depth could be somewhat confounded by subsequent increases in sequencing or processing errors. However, given that we observed many low-abundance taxa in different replicates of the same sample (**Fig 3**), it is likely that many of these taxa represent real organisms. Again, we would argue that this should encourage microbial ecologists to consider relative sequence abundance when examining β-diversity comparisons between highly diverse samples. While there is no silver bullet, the right choice of metrics will depend on the relative risk of pseudo-β-diversity vs. taxon bias in addressing the particular ecological question at hand.

Our results from the larger dataset of soil samples from three different North American pine biomes reveal interesting insights about the influence of sequencing platform on ecological conclusions and thus the adaptability of microbial ecology studies to the latest NGS platforms like Illumina MiSeq. Since MiSeq reads are at present maximum only 300 bp in either direction, adapting ecological sequencing studies to newer technology presents challenges for longer loci previously sequenced with 454. Often this means designing new primers or switching loci altogether, and the degree to which similar patterns and conclusions can be drawn from sequencing the same organisms as labs adapt their protocols for the future remains uncertain. In our study, the 454 dataset and Illumina dataset differ in the reverse PCR primers used and in the length of the amplicons.

Sequencing of the same samples on the Illumina platform vs. the 454 platform results in a ~40× increase in high quality read coverage as well as considerably more total OTUs and levels of richness per sample (**Table 2**). Encouragingly, the Illumina dataset finds nearly all OTUs that are detected with 454 but largely expands the total taxonomic coverage (**Fig S3**). The only three OTUs unique to the 454 dataset are identified as taxa in the genera Trichophaea (Ascomycota), Camarophyllopsis (Basidiomycota), and Amanita (Basidiomycota). Interestingly, eight Trichophaea taxa and 14 Amanita taxa are also present in the Illumina dataset, indicating no explicit lineage bias of the Illumina primers against members of these genera. The unique Thrichophaea and Amanita OTUs found only in the 454 dataset are thus likely due to sampling effects or error in the OTU clustering step of the sequencing processing pipeline. Additionally, since the read pairing step of Illumina processing is designed to correct sequencing errors from one read with higher-quality base calls from the other [31], it is possible that the 454-uniqe OTUs simply represent sequencing error that was not otherwise corrected.

Despite the large increase in diversity recovered with Illumina vs. 454, values of α- and β-diversity remain highly correlative between the two sequencing platforms (**Fig 5**). Sample bioregion is the predominant driving factor in the ordination of all samples from both sequencing methods (**Fig 6**), with 454 and Illumina replicates of the same samples clustering strongly by their sampling location. Within each bioregion cluster, samples cluster further by sequencing method. Variation around the centroid is greater for the 454 dataset, suggesting that β-diversity relationships between samples will depend slightly on the method used to sequence samples. This is likely due to differences in sequencing depths and species richness attainable with the two platforms. Some of the platform-specific β-diversity differences disappear in the ordination based on the β-sim metric, which controls for differences in richness between samples and thus a major difference with varied sequencing depth. It is important to note, however, that a certain degree of variation in ordination seen between the two datasets is to be expected given the different primer sets used for each platform. Still, our results strongly suggest that larger scale patterns of α- and β-diversity are as equally and consistently recoverable with newer, Illumina sequencing technology as with older 454 methods, and that ecologists should be able to transition their research with little hesitation to newer, more high-resolution sequencing technology.

Taxonomic assignment to OTUs is also consistent across the two sequencing platforms (**Fig 7, Fig S4**). At the phylum level, we observe almost complete agreement in the number and types of taxa identified. Differences between taxonomic assignment in the two datasets are primarily in the number of different Neocallimastigomycota, Glomeromycota, and Chytridiomycota taxa present. However, these groups represent a proportionally small amount of the total sequences recovered, and thus differences in the abundance of each taxonomic group recovered by the two datasets are accordingly quite small and could be due to stochasticity rather than bias. This relationship additionally holds true for the class and order groups.

Despite it's advent several years ago, amplicon sequencing with Illumina for higher order eukaryotic organisms like fungi remains scarce in the literature (but see [12] and [32]). Our results from the larger pine biome dataset present novel evidence that large scale ecological patterns of diversity, structure, and taxonomic resolution are easily attainable with an Illumina-specific fungal ITS primer set, and that ecological studies have much to gain by adopting newer NGS methods.

Supporting Information

Figure S1 Primer constructs for the amplification and sequencing of ITS1 for Illumina MiSeq. a) Sequences of PCR and sequencing primers designed to amplify and sequence ITS1, specific to the Illumina MiSeq platform. b) Partial diagram of the ITS region in fungi (not to scale), with approximate annealing locations of PCR and sequencing primers. The PCR primers are designed to generate large amplicons comprising the variable ITS1 region and conserved 18S and 5.8S regions. The Read 1 and Read 2 sequencing primers are designed to sequence a smaller region comprised mostly of ITS1, eliminating most of the conserved flanking regions. The Index sequencing primer sequences the barcode on each amplicon.

Figure S2 Pseudo-β-diversity is not significantly affected by the number of PCR replicates pooled prior to sequencing. The top four panels show the average between-replicate dissimilarity between independent replicates of CT2, OR1, and OR4 plotted against increasing PCR replication level, as determined by sequencing with 454 and Illumina MiSeq. The bottom two panels show non-metric dimensional scaling (NMDS) ordinations of the same dissimilarity values. Different colored symbols represent the different sample IDs; different shaped symbols represent the PCR replication level of each replicate.

Figure S3 Sequencing on the Illumina platform greatly expands the detectable taxonomic diversity. Venn diagram of the total fungal OTUs found from 55 soil samples sequenced with both Illumina MiSeq (light grey circle) and 454 (dark grey circle). OTUs found only when samples were sequenced with Illumina or 454 are represented by the non-overlapping regions of the circles on the left (1798 OTUs) and right (3 OTUs), respectively. OTUs present in both sequencing runs are represented by the overlapping region in the middle (1859 OTUs).

Figure S4 Taxonomic assignment to OTUs observed between both sequencing platforms is highly consistent. Bar charts indicating the proportional richness and abundance of taxa in 55 soil samples sequenced with both 454 and Illumina MiSeq at the phylum, class, and order levels.

File S1 Metadata for all samples collected and sequenced for this study.

Acknowledgments

We gratefully acknowledge Devin Leopold for helpful discussions on our sequence processing methods and for help with bioinformatic scripts.

Author Contributions

Conceived and designed the experiments: KGP. Performed the experiments: DPS. Analyzed the data: DPS KGP. Contributed reagents/materials/analysis tools: DPS KGP. Wrote the paper: DPS KGP. Performed fieldwork: DPS KGP.

References

1. Angly F, Felts B, Breitbart M, Salamon P, Edwards RA, et al. (2006) The marine viromes of four oceanic regions. PLoS Biol 4: e368.
2. Buée M, Reich M, Murat C, Morin E, Nilsson RH, et al. (2009) 454 Pyrosequencing analyses of forest soils reveals an unexpectedly high fungal diversity. New Phytol 184: 449–456.
3. Amend AS, Seifert KA, Bruns TD (2010) Quantifying microbial communities-with 454 pyrosequencing: does read abundance count? Mol Ecol 19: 5555–5565.
4. Fierer N, Strickland MS, Liptzin D, Bradford MA, Cleveland CC (2009) Global patterns in belowground communities. Ecol Lett 12: 1238–1249
5. Quince C, Lanzen A, Curtis TP, Davenport RJ, Hall N, et al. (2009) Accurate determination of microbial diversity from 454 pyrosequencing data. Nat Methods 6: 639–U627
6. Wu GD, Lewis JD, Hoffmann C, Chen Y, Knight R, et al. (2010) Sampling and pyrosequencing methods for characterizing bacterial communities in the human gut using 16S sequence tags. BMC Microbiol 10: 206.
7. Feinstein LM, Sul WJ, Blackwood CB (2009) Assessment of bias associated with incomplete extraction of microbial DNA from soil. Appl Envrion Micriobiol 75: 5428–5433.
8. Tedersoo L, Nilsson RH, Abarenkov K, Jairus T, Sadam A, et al. (2010) 454 Pyrosequencing and Sanger sequencing of tropical mycorrhizal fungi provide similar results but reveal substantial methodological biases. New Phytol 188: 291–301.
9. Bellemain E, Carlsen T, Brochmann C, Coissac E, Taberlet P, et al. (2010) ITS as an environmental DNA barcode for fungi: an in silico approach reveals potential PCR biases. BMC microbiol 10.
10. Ihrmark K, Bodeker IT, Cruz-Martinez K, Friberg H, Kubartova A, et al. (2012) New primers to amplify the fungal ITS2 region–evaluation by 454-sequencing of artificial and natural communities. FEMS Microbiol Ecol 82: 666–677.
11. Lindahl BD, Nilsson RH, Tedersoo L, Aberkenov K, Carlsen T, et al. (2013) Fungal community analysis by high-throughput sequencing of amplified markers–a user's guide. New Phytol 199: 288–299.
12. Schmidt PA, Balint M, Greshake B, Bandow C, Rombke J, et al. (2013) Illumina metabarcoding of a soil fungal community. Soil Biol Biochem 65: 128–132.
13. Gilbert JA, Meyer F, Antonopoulos D, Balaji P, Brown CT, et al. (2010) Meeting Report. The terabase Metagenomics Workshop and the Vision of an Earth Microbiome Project. Stand Genomic Sci 3:3.

14. Peay KG, Schubert MG, Nguyen NH, Bruns TD (2012) Measuring ectomycorrhizal fungal dispersal: macroecological patterns driven by microscopic propagules. Mol Ecol 16: 4122–4136.
15. Orgiazzi A, Lumini E, Nilsson RH, Girlanda M, Vizzini A, et al. (2012) Unravelling soil fungal communities from different Mediterranean land-use backgrounds. PLoS One 7: e34847.
16. Medinger R, Nolte V, Pandey RV, Jost S, Ottenwälder B, et al. (2010) Diversity in a hidden world: potential and limitation of next-generation sequencing for surveys of molecular diversity of eukaryotic microorganisms. Mol Ecol 19 Suppl 1: 32–40.
17. Caporaso JG, Lauber CL, Walters WA, Berg-Lyons D, Huntley J, et al. (2012) Ultra-high-throughput microbial community analysis on the Illumina HiSeq and MiSeq platforms. ISME J 6: 1621–1624.
18. Gardes M, Bruns T (1993) ITS primers with enhanced specificity for basidiomycetes - application to the identification of mycorrhizae and rusts. Mol Ecol 2: 113–118.
19. White TJ, Bruns TD, Lee S, Taylor J (1990) Amplification and direct sequencing of fungal ribosomal RNA genes for phylogenetics. In: Innis MA, Gefland DH, Sninsky JJ, White TJ, editors. PCR protocols: a guide to method and applications. San Diego, Academic Press. pp. 315–322.
20. Caporaso JG, Kuczynski J, Stombaugh J, Bittinger K, Bushman FD, et al. (2010) QIIME allows analysis of high-throughput community sequencing data. Nat Methods 7: 335–336.
21. Edgar RC (2013) UPARSE: highly accurate OTU sequences from microbial amplicon reads. Nat Methods 10: 996–998.
22. Knight R, Reeder J (2010) Denoising pyrosequencing reads by flowgram clustering. Nat Methods 7: 668–669.
23. Martin M (2011) Cutadapt removes adapter sequences from high-throughput sequencing reads. EMBnet.journal North America. Available: http://journal.embnet.org/index.php/embnetjournal/article/view/200/458. Accessed 2013 September 18.
24. Lohse M, Bolger AM, Nagel A, Fernie AR, Lunn JE, et al. (2012) RobiNA: a user-friendly, integrated software solution for RNA-Seq-based transcriptomics. Nucleic Acids Res 40: W622–627.
25. Peay KG, Baraloto C, Fine PV (2013) Strong coupling of plant and fungal community structure across western Amazonian rainforests. ISME J 7:1852–1861.
26. R Core Team (2013) R: A language and environment for statistical computing. R Foundation for Statistical Computing, Vienna, Austria. Available: http://www.R-project.org/. Accessed 2013 August 10.
27. Oksanen J, Kindt R, Legendre P, O'Hara B, Simpson GL, et al. (2008) vegan: Community Ecology Package. R package version 2.0-9. Available: http://CRAN.R-project.org/package = vegan. Accessed 2013 October 10.
28. Dickie IA (2010) Insidious effects of sequencing errors on perceived diversity in molecular surveys. New Phytol 188: 916–918.
29. Kunin V, Engelbrektson A, Ochman H, Hugenholtz P (2009) Wrinkles in the rare biosphere: pyrosequencing errors lead to artificial inflation of diversity estimates. Environ Microbiol 12: 118–123.
30. Quince C, Curtis TP, Sloan WT (2008) The rational exploration of microbial diversity. ISME J 2: 997–1006.
31. Masella AP, Bartram AK, Truszkowski JM, Brown DG, Neufeld JD (2012) PANDAseq: paired-end assembler for illumina sequences. BMC Bioinformatics 13: 31.
32. McGuire KL, Payne SG, Palmer MI, Gillikin CM, Keefe D, et al. (2013) Digging the New York City Skyline: soil fungal communities in green roofs and city parks. PLoS One 8: e58020.

3

Spatial Ecology of Bacteria at the Microscale in Soil

Xavier Raynaud[1]*, Naoise Nunan[2]

1 Sorbonne Universités, UPMC Univ Paris 06, Institute of Ecology and Environmental Sciences – Paris, Paris, France, 2 CNRS, Institute of Ecology and Environmental Sciences – Paris, Campus AgroParisTech, Thiverval-Grignon, France

Abstract

Despite an exceptional number of bacterial cells and species in soils, bacterial diversity seems to have little effect on soil processes, such as respiration or nitrification, that can be affected by interactions between bacterial cells. The aim of this study is to understand how bacterial cells are distributed in soil to better understand the scaling between cell-to-cell interactions and what can be measured in a few milligrams, or more, of soil. Based on the analysis of 744 images of observed bacterial distributions in soil thin sections taken at different depths, we found that the inter-cell distance was, on average 12.46 µm and that these inter-cell distances were shorter near the soil surface (10.38 µm) than at depth (>18 µm), due to changes in cell densities. These images were also used to develop a spatial statistical model, based on Log Gaussian Cox Processes, to analyse the 2D distribution of cells and construct realistic 3D bacterial distributions. Our analyses suggest that despite the very high number of cells and species in soil, bacteria only interact with a few other individuals. For example, at bacterial densities commonly found in bulk soil (10^8 cells g^{-1} soil), the number of neighbours a single bacterium has within an interaction distance of ca. 20 µm is relatively limited (120 cells on average). Making conservative assumptions about the distribution of species, we show that such neighbourhoods contain less than 100 species. This value did not change appreciably as a function of the overall diversity in soil, suggesting that the diversity of soil bacterial communities may be species-saturated. All in all, this work provides precise data on bacterial distributions, a novel way to model them at the micrometer scale as well as some new insights on the degree of interactions between individual bacterial cells in soils.

Editor: Francesco Pappalardo, University of Catania, Italy

Funding: This research was supported by the INSU EC2CO Microbiologie Environnementale and ANR Syscomm Programme MEPSOM. The funders had no role in study design, data collection and analysis, decision to publish, or preparation of the manuscript.

Competing Interests: The authors have declared that no competing interests exist.

* E-mail: xavier.raynaud@upmc.fr

Introduction

The application of novel molecular techniques (such as high throughput sequencing) during the past two decades has uncovered a phenomenal bacterial diversity in soils. For example, a single gram of soil can harbour up to 10^{10} bacterial cells and an estimated species diversity of between $4{\cdot}10^3$ [1] to $5{\cdot}10^4$ species [2]. Several studies have identified major environmental influences on soil bacterial diversity (such as soil pH [3], nitrogen [4], plant communities [5] or land use [6]) and soil bacterial biomass (soil organic carbon [7]), that vary between geographical regions and across biomes. It is intriguing however, that experiments manipulating microbial diversity have found no or only weak links between diversity and many important microbial-driven processes, such as soil carbon mineralization [8–10], nitrite oxidation [8,9] or denitrification [8,9]. This lack of relationship raises the question about the importance of microbial diversity for soil and ecosystem functioning [11] and has even lead some authors to question the value of studying the soil metagenome for understanding soil microbial functioning [12].

The diversity of biological components can affect ecosystem processes through interactions among species, such as when there is competition for resources, mutualism or predation. The extent and intensity of these interactions depend not only on the interacting species but also on their proximity to one another. The role of space in ecosystem function is widely recognised in higher plant and animal ecology: the spatial distribution of species and the spatial organisation of communities regulate the extent to which individuals interact, such as in competition for resources [13–15], mutualism [16] or predation [17,18], which, in turn, affects ecosystem properties [19]. However, compared to the vast amount of studies focusing on microbial diversity in soils, relatively little attention has been paid to spatial aspects of ecology in microbial systems at the scales at which cell-to-cell interactions occur although there have been some attempts to characterize the spatial distribution of diversity and microbial processes at the scale of aggregates [20–22]. As microbial-driven ecosystem processes are sums of the activities of microbial cells, most of which are subject to cell-to-cell interactions, such interactions are likely to have significant effects on overall processes.

In microbial systems, the scale at which individuals interact is related to the distance over which they can effect changes in the concentration of gases or solutes. This may vary depending on the gas or solute and the concentration at which it has an effect on bacterial physiology, however, two notable studies have suggested that the vast majority of interactions occur within 20 µm of bacterial cells [23,24]. Studies on microbial systems (whether they focus on microbial activity or diversity) are generally carried out at scales many orders of magnitude larger than those at which microorganisms interact with other organisms or with their surrounding environment [25]. This disparity of scale is not as prevalent in the study of higher organisms [26] and so the effects of local interactions on ecosystem processes are better understood. In soil microbial ecology, the effects of local interactions are likely to be obscured by relatively large samples that encompass environmental heterogeneity and local interactions [25].

Microbial-scale processes and local spatial organisation are known to be significant regulators of microbial community stability, function and evolution. Spatial separation has been identified as playing a major role in several microbial processes: 1. it is thought to be responsible for the emergence and maintenance of high levels of bacterial diversity observed in structured media [27]; 2. the relative importance of horizontal gene transfer in bacterial evolution is believed to depend on the proximity of bacterial neighbours, with areas of low cell density dominated by clonal reproduction and densely populated areas harbouring communities in which horizontal gene transfer can be significant [28] and 3. it has been shown that the stability of bacterial communities can depend on the distance among constituent members [29]. A common feature of these bacterial community ecology studies is that they are carried out in model or artificial systems rather than *in situ* and, as a result, the pertinence of the processes identified for real communities in their natural habitats can be questioned. An understanding of the importance of space in bacterial ecology requires knowledge of the distribution of bacterial cells in their environment. This paper aims to explore the spatial distribution of soil bacterial cells in soils at the micrometer scale. To this end, we present a method to analyse and model distributions of individual bacterial cells in soil in order to better understand how bacteria interact with one another. Measuring the distribution of bacterial cells in volumes of soil that are relevant to cell-to-cell interactions is not technically possible at present, therefore our analysis was carried out in three steps. We first studied the distributions of bacterial cells measured in 2 dimensional thin sections of soil and then extended these observations to 3 dimensions using point pattern modelling methods. Finally, we used a simple species abundance model to gain some insight into the degree to which different bacterial species may interact with each other.

Materials and Methods

Ethics statement

The Scottish samples on which this study was based were obtained from land belonging to the Scottish Crop Research Institute when one of the authors was a member of staff there and permission was granted. The French samples were given to the authors by Geneviève Grundman of the University of Lyon. However, no new sampling took place for this study. Only previously obtained data were used, most of which has already been published [30–32]. The samples were taken from agricultural fields and did not involve any endangered species. Dataset with bacterial distributions is available from the authors upon request.

Bacterial distribution data

Bacterial distributions in this study consisted of 2D point patterns (the x and y coordinates of individual cells) measured in images of soil thin sections (Fig. 1). We used 752 new or previously measured bacterial distributions taken from 94 soil thin sections sampled at different depths from a Scottish sandy loam (723 images, [30–33]) and 2 soil thin sections of surface soil (20 cm below surface) a French sandy loam (29 images; Table 1). The Scottish samples were taken from topsoil (0–30 cm, sandy loam: 71% sand, 19% silt, 10% clay, pH_{H2O} 6.2, 1.9% C and 0.07% N) and subsoil (30–80 cm, sandy loam: 72% sand, 17% silt, 11% clay, pH_{H2O}: 6.5, 0.68% C and 0.02% N) of an arable soil. The French samples were taken from topsoil of an arable soil (0–30, loam: 47.7% sand, 35.3% silt, 17.0% clay, pH_{H2O}: 7.0, 1.4% C and 0.13% N). Digital images were acquired with a Zeiss Axioplan 2 microscope fitted for epifluorescence and were equivalent to an effective area of 620×460 µm^2 for the Scottish samples and 516×410 µm^2 for the French samples (different cameras were used). Details on the image analysis procedures to extract bacterial coordinates from digital images can be found in [30]. Bacterial distributions that contained 5 cells or less (n = 8) were discarded for the analysis of the spatial distribution of cells, as low cell densities make the analysis of point patterns unreliable. Two thin sections (2×8 microscopic observations) were prepared from a single sample so that one was orthogonal to the other (i.e., XY and XZ). These observations were used to determine whether bacterial distributions at the micro- to millimetre scales were isotropic or not.

Spatial model fitting of bacterial distribution observed in thin sections

The aim of this first section is to describe the spatial structure of the observed distributions of bacteria. Each distribution was compared to two different spatial null models. The first null model, Complete Spatial Randomness (CSR) or Poisson process, assumes that the position of one point in the point pattern is independent of the position of the others. The second null model, Log Gaussian Cox Process (LGCP) is a model for aggregated point patterns where the aggregation is caused by some environmental heterogeneity [34]. As bacteria live in the soil pore network, the soil structure (Fig 1a), that determines the architecture of the pore network [35], is an important environmental heterogeneity affecting the distribution of bacteria. LGCP are processes defined in n-dimensions and are a form of inhomogeneous Poisson process where the intensity is a Gaussian random measure (Fig. 1c). In this study, LGCP with an exponential covariance function were used. These LGCP are determined by three parameters, the mean (μ), variance (σ) and scale (β) of the Gaussian random measure. The three parameters determine the intensity of the point process (the number of points in the point pattern) and the extent of aggregation. The average intensity, λ, of a LGCP is given by:

$$\lambda = e^{\mu+\sigma^2/2} \quad (1)$$

The CSR model is a special case of LGCP where $\sigma \rightarrow 0$. Details on the theory of LGCP are given in the Appendix S1.

All the point patterns used in this study (observations of bacterial distributions and simulations), were characterized by their intensity λ and 2 summary statistics: Ripley's $K(r)$ function and the nearest neighbour distance distribution function, $G(r)$ (see [36] for a mathematical definition of these two functions). The intensity λ is the number of points per unit surface and was estimated by dividing the number of points in the distribution by its surface. Ripley's $K(r)$ summary statistic is related to the number of points in a point pattern that are within distance r of an "average" point. The theoretical expression of Ripley's $K(r)$ is known for the CSR and LGCP models. For 2D point patterns, they are given by the following equations:

$$K_{CSR}(r) = \pi r^2 \quad (2)$$

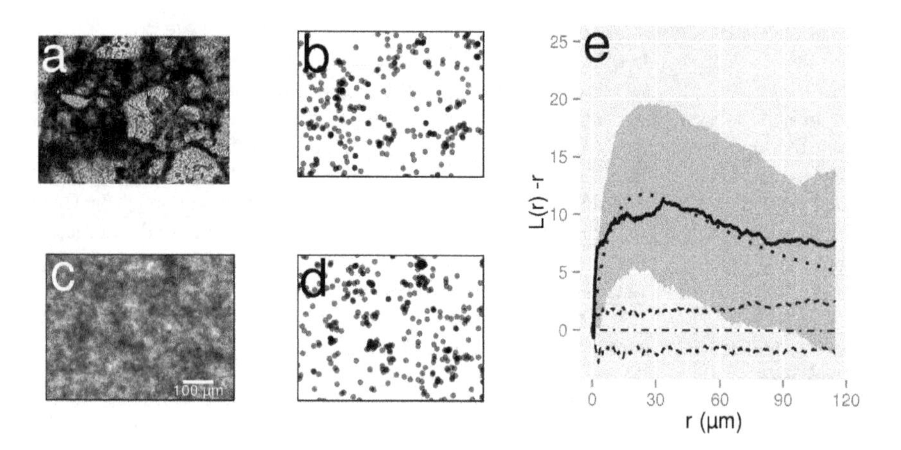

Figure 1. Bacterial habitat, observed and simulated distribution of bacteria in a soil thin section. (a) Bright field image of a soil thin section showing various soil features that characterise the soil microbial habitat and (b) bacterial distribution observed in the same thin section. (c) Random probability field generated using parameters estimated from b ($\mu = -7.64$; $\sigma^2 = 2.0$ and $\beta = 12.95$) and (d) simulated bacterial distribution using c. as random intensity (LGCP model). Colours for the random field were chosen to match those of the thin section image. Darker shades indicate higher probability of bacterial presence. The scale for all panels is identical and is indicated in c. (e) Transformed Ripley functions L(r) - r for the observed bacterial distribution shown in a. (solid line) and envelopes of 99 simulations under CSR and LGCP. The dashed lines indicate the envelope of CSR simulations of the same intensity as a. and the shaded area the envelope of simulations of an LGCP with parameters estimated from the observed point distribution. The dotted line corresponds to the theoretical functions under LGCP and the dashed-dotted line to the theoretical function under CSR. In e., L(r)-r above 0 indicates that the point pattern is more aggregated than a random process.

$$K_{LGCP}(r) = 2\pi \int_0^r s \cdot \exp\left(\sigma^2 e^{-s/\beta}\right) ds \qquad (3)$$

The theoretical expression of $K(r)$ for the LGCP model makes it possible to estimate the parameters μ, σ^2 and β for all observed bacterial distributions [37]. The nearest neighbour distance distribution function, $G(r)$, of a point pattern is the distribution function of the distance from an average point to its nearest neighbour. The theoretical expression of $G(r)$ for a 2D Poisson process is

$$G_{CSR}(r) = 1 - \exp(-\pi^{l} r^2) \qquad (4)$$

where λ is the intensity of the point process. There is no theoretical expression for $G(r)$ for a LGCP process.

Estimates of the two summary statistics and of the LGCP parameters were all carried out using R 2.15 with packages *spatstat* [38] and *Randomfield* [39]. These packages provide all the necessary functions to manipulate and analyse point patterns. We used the Ripley isotropic correction for estimates of $K(r)$ and Kaplan-Mayer estimator for $G(r)$ [38].

The goodness of fit between each observed distribution (see example in Fig 1b) and the null models (see example in Fig 1d for the LGCP null model) was carried out as in [40]. To begin with, we simulated 99 point processes of the same intensity under the two null models and calculated the estimated $K(r)$ and $G(r)$ functions for each simulation. For both the simulations and the observed distributions, we then computed the maximum absolute difference between the estimated summary statistics $K(r)$ or $G(r)$

Table 1. General properties of all observed distribution maps of bacteria at different depths in soil.

Site	Depth (cm)	# samples	Cell density(mm^{-2})	Cell density(g_{soil}^{-1})	Average distance to nearest neighbour (µm)
Scotland	0–30 cm	359	668.7±568.0	$1.03\ 10^9 \pm 8.74\ 10^8$	12.12±6.06
			(17.7–3572.8)	($2.73\ 10^7$–$5.50\ 10^9$)	(0.38–366.17)
Scotland	30–60 cm	261	297.3±289.5	$4.57\ 10^8 \pm 4.45\ 10^8$	18.37±11.05
			(17.7–1821.8)	($2.72\ 10^7$–$2.80\ 10^9$)	(0.45–366.75)
Scotland	>60 cm	103	121.2±165.5	$1.86\ 10^8 \pm 2.54\ 10^8$	28.94±23.98
			(7.1–1003.1)	($1.09\ 10^7$–$1.54\ 10^9$)	(0.66–532.36)
France	20 cm	29	3531.28±1809.7	$5.43\ 10^9 \pm 2.78\ 10^9$	4.95±0.96
			(155.6–7539.1)	($2.39\ 10^8$–$1.16\ 10^{10}$)	(0.34–221.12)
	Total	752	576.8±841.0	$8.87\ 10^8 \pm 1.29\ 10^9$	12.46±9.38
			(7.1–7539.1)	($1.09\ 10^7$–$1.16\ 10^{10}$)	(0.34–532.36)

Data for cell numbers, cell densities and average distance to nearest neighbour are given as mean±sd (range). Cell density in g_{soil}^{-1} is calculated assuming a microscope depth of field of 2 µm and a soil density of 1.3 g cm^{-3}.

and its theoretical counterparts to build a two-sided test of significance at $P=0.01$ which was used to reject the hypothesis that the observed pattern followed the null model. Because there is no theoretical expression of $G(r)$ for the LGCP model, the average of the 99 simulations was used as a theoretical counterpart. In the results section, the transformation $L(r)=\sqrt{(K(r)/\pi)}$ was plotted as it is simpler to interpret (e.g., $L(r)-r=0$ for all r in the case of a 2D CSR process).

Modelling bacterial neighbourhoods in 3 dimensions

As we found that the LGCP model was adequate for modelling bacterial cell distributions in 2D (see Results section), a similar modelling approach, based on LGCP, was used to estimate the number of neighbours a single bacterial cell had within a given distance in 3 dimensions. The expected values of the parameters μ, σ^2, β of an isotropic Gaussian random measure of a 3D LGCP are the same as those of a plane from the same Gaussian field [41,42]. Therefore, we used the 2D estimates of the LGCP parameters to simulate 3D distributions after ascertaining that bacterial distributions were isotropic at the bacterial neighbourhood scale. Simulations of LGCP 3D distributions were carried out following the method given in [34]. Average parameter values corresponding to bacterial densities of ca. 10^8 cells g^{-1} ($\mu=-10.26$, $\sigma^2=2.90$, $\beta=20$), 10^9 cells g^{-1} ($\mu=-7.52$, $\sigma^2=1.90$, $\beta=25$) and 10^{10} cells g^{-1} ($\mu=-4.91$, $\sigma^2=1.29$, $\beta=25$) were used to simulate 39 cubes ($300\times300\times300$ μm^3) for each bacterial density. The number of cells each bacterium had in its neighbourhood as a function of distance was calculated, and the average across the 39 simulations determined. This was compared with the theoretical number of neighbours for a 3D LGCP, given by the following equation (see Appendix S1 for details):

$$N(r)=e^{\mu+\sigma^2/2}\int_0^r 4\pi s^2 \exp\left(\sigma^2 e^{-s/\beta}\right)ds \tag{5}$$

where μ, σ^2, and β are the parameters of the LGCP. In our simulations the realised densities were (mean±s.e.) 1.06 $10^8\pm2.48$ 10^6, 1.04 $10^9\pm2.44$ 10^7 and 1.02 $10^{10}\pm2.47$ 10^8 cells g^{-1}, respectively.

Diversity in the bacterial neighbourhood

In order to study the number of bacterial species in the bacterial neighbourhood (i.e., the number of species with which a bacterium might interact), 3D bacterial distributions in which bacterial cells were attributed a species identity were simulated. However, the spatial structure of bacterial diversity at the micrometer scale is unknown, so a modelling approach was taken to estimate the number of species a single bacterium interacts with. As the microbial diversity in our soil samples was not known, we used published data of soil microbial diversity [1,2] to simulate the 3D bacterial communities. To do so, a simple species-abundance model, the log-series distribution of species [43], was used to calculate the number of individuals per species from a total number of individuals and species. This species abundance model has been found to fit bacterial species distributions in soils at larger spatial scales [44] and has a very simple mathematical formulation. The Fisher species abundance curve relates the number of species (S) to the number of individuals (N) as described in the following equation:

$$S=\alpha\ln(1+N/\alpha) \tag{6}$$

where α is Fisher's α index of diversity. Values for α were calculated using Eq 6 and data found in [1] ($S=4000$, $N=1.5$ 10^{10}) or [2] (Agricultural soil, Brazil: $S=3559$, $N=10^9$; Forest soil, Canada: $S=15188$, $N=10^9$; Agricultural soil, Florida, USA: $S=4477$, $N=10^9$; Agricultural soil, Illinois, USA $S=4010$, $N=10^9$). This gave values of α of 221.86, 233.04, 1107.53, 297.94 and 264.79, respectively. Because the spatial structure of species at such scales is unknown, species identities were assigned at random among individuals in the simulations. In this case, the average number of neighbouring species can be derived from the theoretical number of neighbours (Eq. 5) and the Fisher species abundance curve (Eq. 6) following the equation:

$$\begin{aligned}S(r)&=\alpha\ln(1+N(r)/\alpha)\\&=\alpha\ln\left(1+\left[e^{\mu+\sigma^2/2}\int_0^r 4\pi s^2\exp\left(\sigma^2 e^{-s/\beta}\right)ds\right]/\alpha\right)\end{aligned} \tag{7}$$

It should be noted that this type of modelling approach is rather simplistic and was only used to illustrate how interactions between different bacterial species might affect soil functioning (see Discussion).

Results

Bacterial densities in thin sections

The number of bacterial cells in the analysed images ranged from 5 to 1599. This corresponded to densities between 7 and 7539 cells mm^{-2} or ca. 1.09 10^7 to 1.16 10^{10} cells g^{-1} soil, assuming a microscope depth of field of 2 μm and a soil density of 1.3 g cm^{-3} (Table 1). On average, bacterial densities decreased with depth, with the highest densities found above 30 cm depth and the lowest below 60 cm. However, the variability in cell density was high at all depths (Table 1), indicating that bacterial cells were distributed in a heterogeneous way throughout the soil volume. Due to the decrease in density associated with depth, inter-cell distance also varied with depth, with an average of approximately 10 μm at depths of 0–30 cm (corresponding to high cell densities) and 29 μm at depths below 60 cm (Table 1). It should be noted that inter-cell distances were also highly variable at each depth.

Comparison of observed distributions with the null models

The goodness of fit between the measured distributions and the CSR model revealed clear deviations from CSR ($P<0.05$) in at least 630 of the 744 distributions (Table 2). Clustering was more pronounced in the surface layers of the soil, where bacterial cells were more abundant. Whereas in the surface layers, to a depth of 30 cm, at least 366 of the 387 distributions (94.6%) differed from CSR, only 52 of 98 distributions (53.1%) did so in soil taken from depths below 60 cm (Table 2). An example of the deviation of the Ripley's $K(r)$ function of an observed distribution from CSR is given in Fig 1e.

In contrast, the LGCP null model adequately described the observed distributions in, at worst, 80% of the cases (Table 2). Estimates of the LGCP parameters μ, σ^2 and β parameters were in the range [−13.49, −5.81], [0.59, 6.85] and [2.59 10^{-3}, 161.18], respectively. Complete spatial randomness is the limit of LGCP when σ^2 tends to 0 and therefore the values of σ^2 provide a first indication that the bacterial distributions in these soils ranged from highly aggregated (high σ^2) to near random (low σ^2)

Table 2. Total number of samples, number (proportions %) of samples deviating from CSR and number (proportions %) of samples deviating from the LGCP model for different soil depths for all bacterial distribution having more than 5 bacterial cells in the field of view.

Depth	# samples	samples deviating from CSR		samples deviating from LGCP	
		Ripley's K	G function	Ripley's K	G function
0–30 cm	387	376 (97.1%)	366 (94.6%)	37 (9.6%)	97 (25.1%)
30–60 cm	259	228 (88.0%)	212 (81.8%)	14 (5.4%)	40 (15.4%)
>60 cm	98	68 (69.4%)	52 (53.1%)	4 (4.1%)	11 (11.2%)
Total	744	672 (90.3%)	630 (84.68%)	55 (7.4%)	148 (19.9%)

Deviations are calculated based on a goodness of fit test between summary statistics for each observed distribution (Ripley's K or G function) and the corresponding statistics under the null model (CSR of LGCP). Ripley's K is a summary statistics related to the number of points in a point pattern that are within a certain distance to an average point. The nearest neighbour distance distribution G is the distribution function of the distance from an average point to its nearest neighbour.

distributions. The mean (μ) decreased ($R^2 = 0.25$, $P<0.001$) and variance (σ^2) increased ($R^2 = 0.08$, $P<0.001$) linearly with depth between 30 cm and 80 cm below-ground, but no trends were apparent for depths between 0 and 30 cm. There was no relation between β and depth (data not shown). An example of the concordance between the Ripley's function of an observed distribution and simulations of LGCP distribution is given in Fig 1e. The distributions that did not fit the LGCP null model occurred primarily in topsoil and rhizosphere samples where very high bacterial densities were observed, or in subsoil samples in which individual, isolated colonies were detected.

Isotropy of bacterial distributions

The isotropic nature of the bacterial distributions was tested using the two slides (2×8 images) that were prepared orthogonally. The number of bacterial cells in these slides ranged from 78 to 1008 for one slide and from 31 to 920 for the other. There were no statistical differences in the number of observed cells between the two slides ($P = 0.27$). The estimates of the LGCP parameters for μ, σ^2 and β were, respectively, in the ranges [−10.19, −6.51], [1.44, 3.99] and [8.94, 61.65] for one slide and [−10.17, −6.37], [1.29, 2.60] and [7.21, 60.09] for the other. Here also, there were no statistical differences between these estimates for the two slides ($P >$ 0.16 at least), suggesting that the distribution of bacteria at these micrometer to millimetre scales is isotropic.

Modelling bacterial distributions in 3D

Due to the apparent isotropic nature of the bacterial neighbourhoods at micrometer to millimetre scales, the LGCP parameters obtained in 2D space were used to simulate 39 independent 3D distributions [41]. An example of such a 3D distribution of bacteria (density equivalent to 10^9 cells g^{-1}) is given in Fig. 2a. For each simulation, the number of neighbours each bacterium had as a function of distance was computed (shaded scatterplot in Fig. 2b) and averaged across all bacteria (red line in Fig. 2b). As expected, the average number of neighbours bacteria had as a function of distance was similar to the theoretical number of neighbours derived from the theoretical expression of *K(r)* for a LGCP (Eq 2, blue line in Fig 2b). The number of neighbours that a single bacterium had within a distance of 20 µm ranged from 7 to 250 for a bacterial density of 10^9 cells g^{-1} (Fig. 2b). The minimum and maximum average number of neighbours obtained across the 39 simulations are given in Fig 3a. Overall, these results suggest that the number of cells in the neighbourhood of a typical bacterium is rather limited. For an average density of 10^9 cells g^{-1} soil, the average number of neighbours around a single cell was ca. 1043 (±250) cells within a distance of 50 µm, decreasing to ca. 120 (±40) cells at 20 µm (Fig 3a). Similarly, for an average density of 10^8 cells g^{-1} soil, the average number of neighbours was ca. 82 (±22) cells within a distance of 50 µm and ca. 12 (±4) cells at 20 µm. Finally, for bacterial densities close to what one would expect in the rhizosphere, (10^{10} cells g^{-1} soil), the average number of neighbours was ca. 5806 (±1000) cells within a distance of 50 µm decreasing to ca. 555 (±100) cells within 20 µm.

Bacterial diversity in the bacterial neighbourhood

In order to estimate the diversity of bacterial species a typical bacterium interacts with, species were distributed at random among individual cells in the 3D simulations of bacterial distributions. Using these simulations, we estimated the number of species in the neighbourhood of each cell as a function of distance (shaded scatterplot in Fig. 2c) and computed its average value (red line Fig. 2c). The average number of species (*S(r)*) in the neighbourhood of a single cell was adequately approximated by the theoretical average number of species around a single bacterium given in Eq. 7 (red and blue lines in Fig 2c).

The average number of species in the neighbourhood of bacteria was estimated from the 39 3D simulations at densities of 10^8, 10^9 and 10^{10} cells g^{-1} soil (Fig 3b). At lower bacterial densities, typical of bulk soil (10^8 cells g^{-1}), the average number of species that a bacterium can be expected to interact with (species within 20 µm, assuming a random distribution of species) was ca. 11 (±4) species. For a density of 10^9 cells g^{-1} soil, the average number of neighbouring species around a single cell increased to 97 (±24) species. Finally, when bacterial densities were set at values expected in the rhizosphere (10^{10} cells g^{-1} soil), the average number of species within an interaction distance of 20 µm was ca. 284 (±30) species. These values were higher when the interaction distance was set at 50 µm (Fig 3b). The diversity indices derived from Roesch et al (2007) [2] were similar to that derived from Torsvik et al (1990) [1] with one exception: bacterial diversity in the forest soil in Roesch et al (2007) [2] was much higher ($\alpha = 1107.53$), resulting in a greater number of species per volume of soil (Fig S1). However, despite the increase in the overall number of species, no increase in the number of neighbouring species within 20 µm was observed for bacterial densities of 10^9 cells g^{-1} soil and lower.

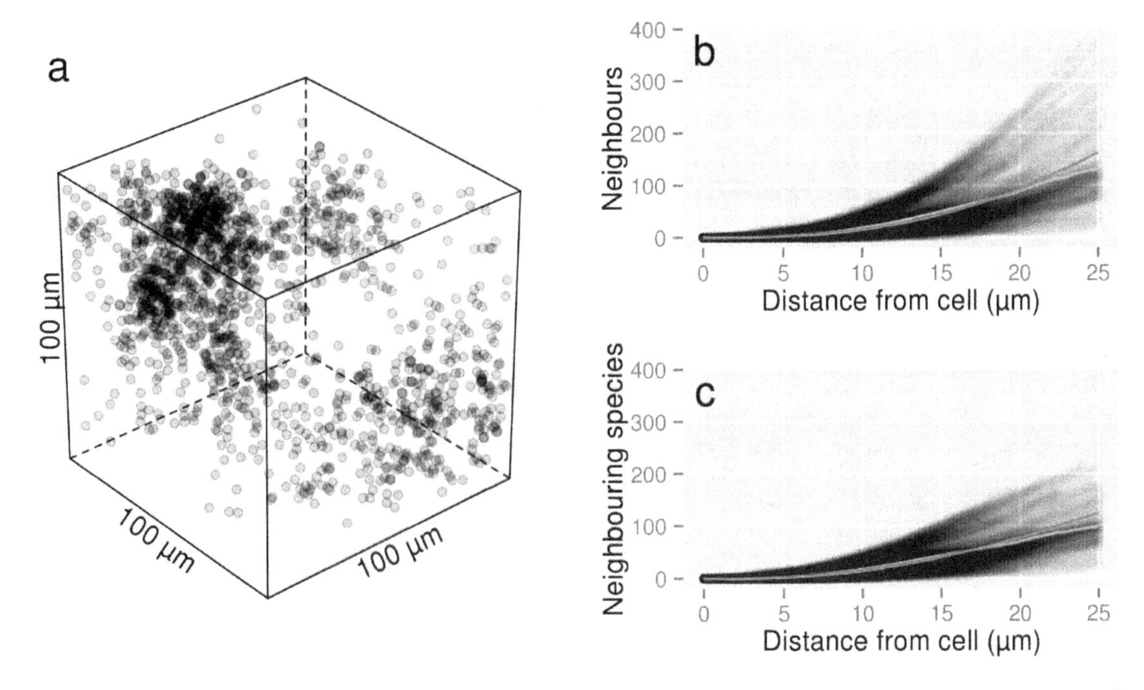

Figure 2. 3D simulation of bacterial distribution. (a) 3D representation of bacterial distribution in a $100\times100\times100\ \mu m^3$ cube as simulated by LGCP. The model parameters for the simulation were $\mu=-7.52$, $\sigma^2=1.90$ and $\beta=25$. (b) Number of neighbours as a function of distance for each bacterium (scatterplot), average number of neighbours (red line) and theoretical number of neighbours for the LGCP (blue line) and (c) Number of neighbouring species assuming a random distribution of species among individuals (line colours are the same as in b). The number of species considered in this simulation was $S=450$, corresponding to Fisher's $\alpha=221.86$ (estimated from [1]).

Discussion

The distribution of bacteria in soils

Unsurprisingly, the bacterial distributions measured in the 744 soil thin sections studied showed a high degree of aggregation at all depths, although aggregation was more frequent in the surface strata than in the subsoil. This corroborates similar observations by [30] on a sample of this dataset. We also found that bacterial distributions were isotropic at the millimetre scale. Although this was based on a limited number of observed distributions and would need to be confirmed on a larger dataset, such an observation is not surprising as the vertical gradients observed in soils generally occur at larger scales [45,46].

Log Gaussian Cox Processes (LGCP) were used to characterise the bacterial distributions and construct 3D distributions from these observations. LGCP are particularly useful for modelling aggregated spatial point patterns where aggregation is due to a stochastic environmental heterogeneity [34]; that is to say, an environmental factor that is continuously and heterogeneously distributed in space. They do not, however, account for the contribution of intrinsic processes, such as birth and death, to the

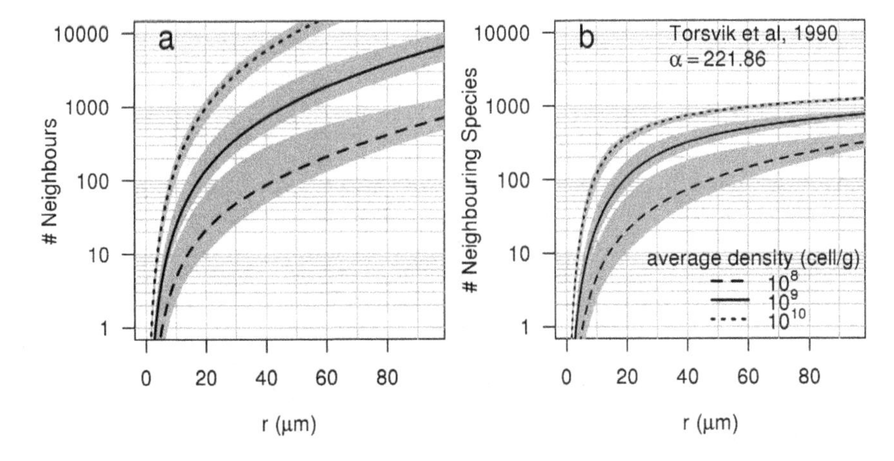

Figure 3. Number of neighbours and neighbouring species. (a) Mean number of neighbours and (b) neighbouring species around an "average" bacterium as a function of distance from the bacterium and as a function of bacterial density. The mean number of neighbours and neighbouring species are derived from Ripley's K(r) function of LGCP with parameters $\mu=-10.26$, $\sigma^2=2.90$, $\beta=20$ for the 10^8 cells g^{-1} soil density, $\mu=-7.52$, $\sigma^2=1.90$, $\beta=25$ for the 10^9 cells g^{-1} soil density and $\mu=-4.91$, $\sigma^2=1.29$, $\beta=25$ for the 10^{10} cells g^{-1} soil density. Grey envelopes surrounding curves represent the maximum and minimum of these numbers calculated from 39 simulations. Note the logarithmic scale on the y axis.

distribution of points in a given point pattern. The distribution of bacteria in soil is clearly a consequence of both extrinsic (environmental conditions, such as pore size and organic matter availability) and intrinsic (reproduction by binary fission) processes. However, it is likely that intrinsic processes such as reproduction are related to the extrinsic processes, as the probability of growth is greater where external environmental conditions are most suitable (i.e. presence of organic substrate, O_2, water…). It is worth noting that the majority of samples that were inadequately described by LGCP were from the rhizosphere or topsoil. There, intrinsic processes (e.g. cell division) would have played a significant role in the generation of these distributions as bacterial growth may be somewhat decoupled from environmental heterogeneity due to the overall high availability of resources. This may possibly be because environmental heterogeneity (with respect to bacterial growth) was reduced due to the input of organic substrate from plants. It can be concluded therefore, that the relative importance of the underlying processes contributing to the generation of the bacterial distributions (intrinsic vs. extrinsic) can change with situation.

Despite the deviation of some of the bacterial distributions from the LGCP model in some specific cases, LGCP were used to simulate bacterial distributions because they have the distinct advantage of being analytically tractable (i.e. the theoretical expression of $K(r)$ is known), making it possible to estimate model parameters from which an observed point pattern could have emerged [37].

The neighbourhood of bacteria in soils

Three properties of microbial communities emerge from our analysis. The first property is that the number of cells within interacting distances in the neighbourhood of a bacterium is, on average, rather limited compared to the number of cells commonly found in a single gram of soil. For densities equivalent to 10^9 cells g^{-1}, the number of neighbours a typical bacterium has within a distance of 20 µm is approximately 120, increasing to 1000 within a distance of 50 µm. The actual number of neighbours within interaction distances may in fact be lower as the calculations carried out here used the euclidian distance between cells rather than the geodesic distance and therefore did not account for pore geometry. For example, two cells on either side of a sand grain will not interact through the sand grain but rather around it.

The second property is that, given the high variability in the number of neighbours around bacterial cells in our dataset (Table 1) and in our simulations (Fig 2 and 3), the density of interactions is highly variable in space, even at very fine scales, with some cells interacting with few others and other cells interacting with many more. For example, some filamentous bacteria or bacteria in colonies or biofilms are likely to have completely different interaction environments to those of isolated individual bacterial cells. Such localized "pockets of interactions" might have important consequences for ecosystem processes and microbial community evolutionary dynamics [47].

The third property to emerge from this analysis, an obvious corollary of low levels of cell-to-cell interactions, is that the number of different species an individual cell interacts with is also limited. The simulations of bacterial distributions in 3D suggest that interspecific interactions in soils are orders of magnitude lower than what is possible in view of the species diversity often measured in soil (4000–50000 species g^{-1} soil; [1,2]). For densities equivalent to 10^9 cells g^{-1} soil, we found that the number of bacterial species within interaction distances (<20 µm) ranged from 1 to 120 (±40) and, even at the highest densities (equivalent to 10^{10} cells g^{-1} soil), never exceeded 1000 species (Fig 3b, Fig S1). Moreover, these data are most likely an overestimate of the actual local diversity, as we assumed that species were distributed randomly among individuals. However, as bacteria reproduce by binary fission, this assumption is almost certainly wrong. The approach taken here thus provides only an upper limit to the number of species in the bacterial neighbourhood. Furthermore, the model also assumes that there is a positive relationship between abundance and diversity as the diversity parameter (α) is constant. This assumption may also be untrue. In reality, species are aggregated and the extent of species aggregation is most probably positively related to growth intensity because bacteria grow by binary fission. This means that diversity may not increase monotonically with abundance when conditions are favourable for growth and it has indeed been found that bacterial diversity in the rhizosphere is lower than in bulk soil [48,49]. A more realistic way to model the spatial distribution of diversity should therefore account for both the environmental determinism of the spatial distribution of cells (as was done here) and cellular reproduction processes in order to yield aggregation in species distribution.

It is noteworthy that the diversity at the microscale (species within 20 µm of a bacterial cell) did not change appreciably as a function of the global diversity (species found in a gram of soil) for bacterial densities of 10^9 cells g^{-1} soil or less (i.e., diversity at 20 µm is similar in Fig 3b and Fig S1, Forest Soil, despite very different α values), suggesting that local communities may be species saturated [50]. It is now widely recognised that local diversity is not solely dependant on local interactions, but that regional processes are also important (in the case of soil bacteria, a gram of soil can be considered a region [26]). Species saturation of local communities can arise from species interactions (community membership is limited by competitive exclusion or local environmental conditions [50]) or from the physical limitations of the environment (if the local environment can only accommodate 100 individuals then there cannot be more than 100 species regardless of the overall diversity [51]). It has been suggested that the lack of relationship between ecosystem processes and diversity in soils is due to functional redundancy within soil microbial communities. Regardless of the underlying cause for the apparent local species saturation, species saturation of local microbial communities may also explain the relative insensitivity of many ecosystem processes (e.g. soil organic matter decomposition, denitrification), as well as the resistance or resilience of these processes to environmental stresses, to experimental changes in microbial diversity [52,53]. If the number of species that the average bacterium interacts with does not change as the overall diversity is changed then the functioning the bacteria are responsible for may not be affected by changes in overall diversity either, as the actual levels of diversity in communities remains the same.

Should this hypothesis be confirmed, there are a number of important consequences for soil microbial ecology. The first is that the relationship between diversity and functioning may only be understood if studied at an appropriate scale. The second is that the relationship between regional (gram of soil) and local diversity (microbial neighbourhood) must be understood if the effects of diversity on soil functioning is to be apprehended.

Conclusions

Our analysis of 744 observations of *in situ* bacterial distributions in soils indicates that bacterial cells are aggregated at the scale of a few micrometers, most likely due to soil structure and the way bacterial cells reproduce. The analysis also suggests that, because cells interact only at very small distances, the number of cells that a typical bacterial cell interact with is relatively limited, as is the number of bacterial species. Such low levels of bacterial interactions could be a reason why several soil microbial processes appear not to be affected during microbial diversity erosion experiments.

Acknowledgments

The authors wish to thank Geneviève Grundmann for a cutting remark that prompted us to initiate this work and C. Chenu and S. Boudsocq for constructive comments on an earlier version of the manuscript.

Supporting Information

Figure S1 Number of neighbouring bacterial species around a typical bacterium as a function of distance at different bacterial densities for the four species diversity levels estimated by Roesch et al (2007). Grey envelopes surrounding curves represent the maximum and minimum of these numbers calculated from 39 simulations with the same parameters values as in Fig. 3. Note the similarity in the number of neighbouring species, for all diversity levels, when bacterial density is 10^9 cells g^{-1} or less. $\alpha = 1107.53$ corresponds to a species richness of 15000 species for 10^9 cells whereas $\alpha = 264.79$ corresponds to a species richness of 4010 species for the same number of cells.

Appendix S1 Notes on the theory of Log Gaussian Cox Processes.

Author Contributions

Conceived and designed the experiments: XR NN. Performed the experiments: XR. Analyzed the data: XR NN. Contributed reagents/materials/analysis tools: NN. Wrote the paper: XR NN.

References

1. Torsvik V, Goksøyr J, Daae FL (1990) High diversity in DNA of soil bacteria. Appl Environ Microbiol 56:782–787.
2. Roesch LFW, Fulthorpe RR, Riva A, Casella G, Hadwin AKM, et al. (2007) Pyrosequencing enumerates and contrasts soil microbial diversity. ISME J 1:283–290. doi:10.1038/ismej.2007.53.
3. Lauber CL, Strickland MS, Bradford MA, Fierer N (2008) The influence of soil properties on the structure of bacterial and fungal communities across land-use types. Soil Biol Biochem 40:2407–2415. doi:10.1016/j.soilbio.2008.05.021.
4. Fierer N, Lauber CL, Ramirez KS, Zaneveld J, Bradford MA, et al. (2012) Comparative metagenomic, phylogenetic and physiological analyses of soil microbial communities across nitrogen gradients. ISME J 6:1007–1017. doi:10.1038/ismej.2011.159.
5. Kowalchuk GA, Buma DS, De Boer W, Klinkhamer PGL, van Veen JA (2002) Effects of above-ground plant species composition and diversity on the diversity of soil-borne microorganisms. A Van Leeuw J Microb 81:509–520. doi:10.1023/A:1020565523615.
6. Ranjard L, Dequiedt S, Chemidlin Prévost-Bouré N, Thioulouse J, Saby NPA, et al. (2013) Turnover of soil bacterial diversity driven by wide-scale environmental heterogeneity. Nat Commun 4:1434. doi:10.1038/ncomms2431.
7. Fierer N, Strickland MS, Liptzin D, Bradford MA, Cleveland CC (2009) Global patterns in belowground communities. Ecol Lett 12:1238–1249. doi:10.1111/j.1461-0248.2009.01360.x.
8. Wertz S, Degrange V, Prosser JI, Poly F, Commeaux C, et al. (2006) Maintenance of soil functioning following erosion of microbial diversity. Environ Microbiol 8:2162–2169. doi:10.1111/j.1462-2920.2006.01098.x.
9. Rousk J, Brookes PC, Glanville HC, Jones DL (2011) Lack of Correlation between Turnover of Low-Molecular-Weight Dissolved Organic Carbon and Differences in Microbial Community Composition or Growth across a Soil pH Gradient. Appl Environ Microbiol 77:2791–2795. doi:10.1128/AEM.02870-10.
10. Langenheder S, Bulling MT, Solan M, Prosser JI (2010) Bacterial biodiversity-ecosystem functioning relations are modified by environmental complexity. PLoS One 5:e10834. doi:10.1371/journal.pone.0010834.
11. Prosser JI (2012) Ecosystem processes and interactions in a morass of diversity. FEMS Microbiol Ecol 81:507–519. doi:10.1111/j.1574-6941.2012.01435.x.
12. Baveye PC (2009) To sequence or not to sequence the whole-soil metagenome? Nat Rev Microbiol 7:756. doi:10.1038/nrmicro2119-c2.
13. Goreaud F, Loreau M, Millier C (2002) Spatial structure and the survival of an inferior competitor: a theoretical model of neighbourhood competition in plants. Ecol Model 158:1–19. doi:10.1016/S0304-3800(02)00058-3.
14. Hodge A (2003) Plant nitrogen capture from organic matter as affected by spatial dispersion, interspecific competition and mycorrhizal colonization. New Phytol 157:303–314. doi:10.1046/j.1469-8137.2003.00662.x.
15. Raynaud X, Jaillard B, Leadley PW (2008) Plants may alter competition by modifying nutrient bioavailability in rhizosphere: a modeling approach. Am Nat 171:44–58. doi:10.1086/523951.
16. Amarasekare P (2004) Spatial dynamics of mutualistic interactions. J Anim Ecol 73:128–142. doi:10.1046/j.0021-8790.2004.00788.x.
17. Holt R (1984) Spatial heterogeneity, indirect interactions and the coexistence of prey species. Am Nat 124:377–406.
18. Jansen V (1995) Regulation of predator-prey systems through spatial interactions: a possible solution to the paradox of enrichment. Oikos 74:384–390.
19. Hooper DU, Chapin III FS, Ewel JJ, Hector A, Inchausti P, et al. (2005) Effects of biodiversity on ecosystem functioning: a consensus of current knowledge. Ecol Monogr 75:3–35. doi:10.1890/04-0922.
20. Dechesne A, Pallud C, Debouzie D, Flandrois JP, Vogel TM, et al. (2003) A novel method for characterizing the microscale 3D spatial distribution of bacteria in soil. Soil Biol Biochem 35:1537–1546. doi:10.1016/S0038-0717(03)00243-8.
21. Stefanic P, Mandic-Mulec I (2009) Social interactions and distribution of *Bacillus subtilis* pherotypes at microscale. J Bacteriol 191: 1756–1764. doi:10.1128/JB.01290-08.
22. Vieublé-Gonod L, Chenu C, Soulas G (2003) Spatial variability of 2,4-dichlorophenoxyacetic acid (2,4-D) mineralisation potential at a millimetre scale in soil. Soil Biol Biochem 35:373–382. doi:10.1016/S0038-0717(02)00287-0.
23. Franklin RB, Mills AL (2008) The importance of microbial distribution in space and spatial scale to microbial ecology. In: Franklin RB, Mills AL, editors. The spatial distribution of microbes in the environment. New York, USA: Kluwer Academic Publishers. pp. 1–30.
24. Gantner S, Schmid M, Dürr C, Schuhegger R, Steidle A, et al. (2006) In situ quantitation of the spatial scale of calling distances and population density-independent N-acylhomoserine lactone-mediated communication by rhizobacteria colonized on plant roots. FEMS Microbiol Ecol 56:188–194. doi:10.1111/j.1574-6941.2005.00037.x.
25. Vos M, Wolf AB, Jennings SJ, Kowalchuk G A (2013) Micro-scale determinants of bacterial diversity in soil. FEMS Microbiol Rev 37:936–954. doi:10.1111/1574-6976.12023.
26. Fierer N, Lennon JT (2011) The generation and maintenance of diversity in microbial communities. Am J Bot 98:439–448. doi:10.3732/ajb.1000498.
27. Dechesne A, Or D, Smets BF (2008) Limited diffusive fluxes of substrate facilitate coexistence of two competing bacterial strains. FEMS Microbiol Ecol 64:1–8. doi:10.1111/j.1574-6941.2008.00446.x.
28. Van Elsas JD, Bailey MJ (2002) The ecology of transfer of mobile genetic elements. FEMS Microbiol Ecol 42:187–197. doi:10.1111/j.1574-6941.2002.tb01008.x.
29. Kim HJ, Boedicker JQ, Choi JW, Ismagilov RF (2008) Defined spatial structure stabilizes a synthetic multispecies bacterial community. Proc Natl Acad Sci USA 105:18188–18193. doi:10.1073/pnas.0807935105.
30. Nunan N, Ritz K, Crabb D, Harris K, Wu KJ, et al. (2001) Quantification of the in situ distribution of soil bacteria by large-scale imaging of thin sections of undisturbed soil. FEMS Microbiol Ecol 36:67–77. doi:10.1111/j.1574-6941.2001.tb00854.x.
31. Nunan N, Wu K, Young IM, Crawford JW, Ritz K (2002) In situ spatial patterns of soil bacterial populations, mapped at multiple scales, in an arable soil. Microb Ecol 44:296–305. doi:10.1007/s00248-002-2021-0.
32. Nunan N, Wu KJ, Young IM, Crawford JW, Ritz K (2003) Spatial distribution of bacterial communities and their relationships with the micro-architecture of soil. FEMS Microbiol Ecol 44:203–215. doi:10.1016/S0168-6496(03)00027-8.
33. Nunan N, Young IM, Crawford JW, Ritz K (2007) Bacterial interactions at the microscale - Linking habitat to function in soil. In: Franklin RB, Mills AL, editors. The spatial distribution of microbes in the environment. New York, USA: Kluwer Academic Publishers. pp. 61–85.
34. Møller J, Waagepetersen RP, Syversveen AR (1998) Log Gaussian Cox processes. Scand J Stat 25: 451–482.
35. Ngom NF, Garnier P, Monga O, Peth S (2011) Extraction of three-dimensional soil pore space from microtomography images using a geometrical approach. Geoderma 163:127–134. doi:10.1016/j.geoderma.2011.04.013.
36. Stoyan D (2006) Fundamentals of Point Process Statistics. In: Baddeley A, Gregori P, Mateu J, Stoica R, Stoyan D, editors. Case studies in spatial point process modeling. NY, USA: Springer. pp. 3–22.

37. Møller J, Waagepetersen RP (2003) Statistical inference and simulation for spatial point processes. Boca Raton, USA: Chapman And Hall/CRC.
38. Baddeley A, Turner R (2005) Spatstat: an R package for analysing spatial point patterns. J Stat Soft 12:1–42.
39. Schlather M (2001) Simulation and analysis of random fields. R News 1:18–20.
40. Andrey P, Kiêu K, Kress C, Lehmann G, Tirichine L, et al. (2010) Statistical analysis of 3D images detects regular spatial distributions of centromeres and chromocenters in animal and plant nuclei. PLoS Comput Biol 6:e1000853. doi:10.1371/journal.pcbi.1000853.
41. Bonami A, Estrade A (2003) Anisotropic Analysis of Some Gaussian Models. J Fourier Anal Appl 9:215–236. doi:10.1007/s00041-003-0012-2.
42. VanMarcke E (1988) Random fields: analysis and synthesis. Massachusetts Institute of Technology. 382p
43. Fisher RA, Corbet A, Williams CB (1943) The relation between the number of species and the number of individuals in a random sample of an animal population. J Anim Ecol 12:42–58.
44. Hill TCJ, Walsh KA, Harris JA, Moffett BF (2003) Using ecological diversity measures with bacterial communities. FEMS Microbiol Ecol 43:1–11. doi:10.1111/j.1574-6941.2003.tb01040.x.
45. Van Groenigen JW, Zwart KB, Harris D, van Kessel C (2005) Vertical gradients of $\delta^{15}N$ and $\delta^{18}O$ in soil atmospheric N_2O–temporal dynamics in a sandy soil. Rapid Commun Mass Spectrom 19:1289–1295. doi:10.1002/rcm.1929.
46. Charley J, West N (1975) Plant-induced soil chemical patterns in some shrub-dominated semi-desert ecosystems of Utah. J Ecol 63:945–963.
47. Kinkel LL, Bakker MG, Schlatter DC (2011) A coevolutionary framework for managing disease-suppressive soils. Annu Rev Phytopathol 49:47–67. doi:10.1146/annurev-phyto-072910-095232.
48. Marilley L, Aragno M (1999) Phylogenetic diversity of bacterial communities differing in degree of proximity of Lolium perenne and Trifolium repens roots. Appl Soil Ecol 13:127–136. doi:10.1016/S0929-1393(99)00028-1.
49. Sanguin H, Remenant B, Dechesne A, Thioulouse J, Vogel TM, et al. (2006) Potential of a 16S rRNA-based taxonomic microarray for analyzing the rhizosphere effects of maize on Agrobacterium spp. and bacterial communities. Appl Environ Microbiol 72:4302–4312. doi:10.1128/AEM.02686-05.
50. Cornell H V, Lawton JH (1992) Species interactions, local and regional processes, and limits to the richness of ecological communities: a theoretical perspective. J Anim Ecol 61:1–12.
51. Loreau M (2000) Are communities saturated? On the relationship between alpha, beta and gamma diversity. Ecol Lett 3:73–76. doi:10.1046/j.1461-0248.2000.00127.x.
52. Wertz S, Degrange V, Prosser JI, Poly F, Commeaux C, et al. (2007) Decline of soil microbial diversity does not influence the resistance and resilience of key soil microbial functional groups following a model disturbance. Environ Microbiol 9:2211–2219. doi:10.1111/j.1462-2920.2007.01335.x.
53. Griffiths BS, Ritz K, Bardgett RD, Cook R, Christensen S, et al. (2000) Ecosystem response of pasture soil communities to fumigation-induced microbial diversity reductions: an examination of the biodiversity-ecosystem function relationship. Oikos 90:279–294. doi:10.1034/j.1600-0706.2000.900208.x.

4

Shift in the Microbial Ecology of a Hospital Hot Water System following the Introduction of an On-Site Monochloramine Disinfection System

Julianne L. Baron[1,2], Amit Vikram[3], Scott Duda[2], Janet E. Stout[2,3], Kyle Bibby[3,4]*

1 Department of Infectious Diseases and Microbiology, University of Pittsburgh, Graduate School of Public Health, Pittsburgh, Pennsylvania, United States of America, **2** Special Pathogens Laboratory, Pittsburgh, Pennsylvania, United States of America, **3** Department of Civil and Environmental Engineering, University of Pittsburgh, Swanson School of Engineering, Pittsburgh, Pennsylvania, United States of America, **4** Department of Computational and Systems Biology, University of Pittsburgh Medical School, Pittsburgh, Pennsylvania, United States of America

Abstract

Drinking water distribution systems, including premise plumbing, contain a diverse microbiological community that may include opportunistic pathogens. On-site supplemental disinfection systems have been proposed as a control method for opportunistic pathogens in premise plumbing. The majority of on-site disinfection systems to date have been installed in hospitals due to the high concentration of opportunistic pathogen susceptible occupants. The installation of on-site supplemental disinfection systems in hospitals allows for evaluation of the impact of on-site disinfection systems on drinking water system microbial ecology prior to widespread application. This study evaluated the impact of supplemental monochloramine on the microbial ecology of a hospital's hot water system. Samples were taken three months and immediately prior to monochloramine treatment and monthly for the first six months of treatment, and all samples were subjected to high throughput Illumina 16S rRNA region sequencing. The microbial community composition of monochloramine treated samples was dramatically different than the baseline months. There was an immediate shift towards decreased relative abundance of Betaproteobacteria, and increased relative abundance of Firmicutes, Alphaproteobacteria, Gammaproteobacteria, Cyanobacteria and Actinobacteria. Following treatment, microbial populations grouped by sampling location rather than sampling time. Over the course of treatment the relative abundance of certain genera containing opportunistic pathogens and genera containing denitrifying bacteria increased. The results demonstrate the driving influence of supplemental disinfection on premise plumbing microbial ecology and suggest the value of further investigation into the overall effects of premise plumbing disinfection strategies on microbial ecology and not solely specific target microorganisms.

Editor: Stefan Bereswill, Charité-University Medicine Berlin, Germany

Funding: This project was funded by a grant from the Alfred P. Sloan Foundation (grant B2013-12). The funders had no role in study design, data collection and analysis, decision to publish, or preparation of the manuscript.

Competing Interests: The authors have declared that no competing interests exist.

* Email: BibbyKJ@pitt.edu

Introduction

Drinking water distribution systems, including premise plumbing, contain a diverse microbiological population [1]. Once new pipes have been added to an existing system, microbial colonization begins rapidly, with microbial communities being established in as little as one year [2]. For the purposes of this study, the 'microbial community' is defined as planktonic microbes within the hospital hot water system during the study period. The microbial ecology of drinking water distribution systems varies widely, depending upon system parameters such as disinfection scheme [3], hydraulic parameters [4], location in the system, age of the system [5], and pipe materials [6]. Microbes are capable of corroding pipes within distribution systems, possibly releasing harmful chemicals such as lead [7–9]. It is largely believed that within a drinking water distribution system, the disinfection scheme is one of the primary factors controlling the abundance and make-up of microbes [3,6,10]. Additionally, the effectiveness of disinfection in removing pathogens from drinking water is mediated by the microbial ecology of the drinking water system [1]. However, the impact of on-site disinfection on premise plumbing microbial ecology is not well understood, motivating the current study.

The complex microbial ecology of premise plumbing systems can serve as a reservoir for opportunistic pathogens, such as *Legionella* spp., non-tuberculous Mycobacteria, *Pseudomonas* spp., *Acinetobacter* spp., *Stenotrophomonas* spp., *Brevundimonas* spp., *Sphingomonas* spp., and *Chryseobacterium* spp. [11–13]. Biofilms and amoeba within the water system can protect opportunistic pathogens from disinfection [1,14–16], and may even allow their regrowth and increase in pathogenicity [17–19]. As an example of the utility of microbial ecology-based approaches, a recent landmark microbial ecology-based study showed that biofilms in showerheads are actually enriched in opportunistic pathogens, creating the potential for an aerosol route of infection [20]. Additionally, antibiotic resistance genes have been detected in the biofilms of drinking water distribution systems [21,22]. Each of these points highlight the necessity for a greater understanding of premise plumbing microbial ecology.

Premise plumbing systems have an approximately ten-times greater microbial load than full-scale drinking water distribution systems, due to many factors including greater water stagnation and surface area to volume ratio [23,24]. Premise plumbing systems of hospitals are of particular concern, as hospitals may contain immunocompromised patients [25], who may not be protected by current drinking water monitoring standards [26], and who would be more susceptible to infections caused by opportunistic pathogens. To date, the majority of on-site disinfection systems have been installed in hospitals, creating a valuable testing ground to observe the impact of on-site disinfection systems on premise plumbing microbial ecology prior to more widespread application.

In addition to use in on-site systems, monochloramine as a secondary disinfectant has been advocated in the US as an effective method to reduce the production of disinfection-by-products [27,28] and control biofilm growth within water distribution systems [29]. While monochloramine is able to penetrate biofilms better than alternative disinfectants, this may not result in a reduction in biofilm growth [8]. Additionally, chloramine treatment requires the addition of an excess of ammonia, which may cause increased growth by ammonia-oxidizing bacteria [28], such as members of the genera *Nitrospira* spp. and *Nitrosomonas* spp. [30]. Bacterial nitrification is known to increase the degradation rate of monochloramine [31], thereby reducing the expected longevity and effectiveness of chloramine. Denitrifying bacteria have previously been identified in chloraminated drinking water systems [32]; however, this topic has not been fully explored in the literature.

The effectiveness of chloramination in removing opportunistic pathogens in premise plumbing remains unclear [27]. On-site monochloramine addition has been proposed as a disinfection strategy for the control of *Legionella* [33–36], but long-term studies have not yet been conducted [33,34]. Recently, a culture-based study of monochloramine on-site disinfection in a hospital's hot water system for the purpose of *Legionella* control demonstrated a significant reduction in *L. pneumophila* and no change in nitrate or nitrite levels [37]. Observed discrepancies in system performance are potentially due to differing microbial ecologies or water chemistries of the systems tested. A more holistic view of system microbial ecology, such as presented in this study, may allow more efficient application of supplemental disinfection.

Despite the obvious importance of the microbial ecology of drinking water systems in modulating disinfectant effectiveness and as a reservoir for opportunistic pathogens, there is a notable lack of studies detailing the shift in microbial diversity and composition in response to on-site disinfection. The objective of this study was to determine the effects of on-site monochloramine disinfection on the microbial ecology of a hospital hot water system. Both the microbial ecology of hot water systems and the response of premise plumbing microbial ecology to on-site disinfection are not currently well described in the literature. This study utilizes 216 samples taken from 27 sites and pooled into five composites for two time points prior to and six time points following the addition of on-site monochloramine addition. Samples were analyzed utilizing Illumina DNA sequencing of the microbial community 16S rRNA region and results demonstrate a dynamic shift of the microbial ecology of a hospital's hot water system in response to monochloramine addition.

Materials and Methods

Hospital setting

For these activities no specific permissions were required for these locations. This study took place in a 495-bed tertiary care hospital complex in Pittsburgh, PA. The building has 12 floors and receives chlorinated, municipal cold water. The hospital's hot water system was treated with the Sanikill monochloramine injection system (Sanipur, Lombardo, Flero, Italy). Monochloramine was dosed to a target concentration between 1.5 and 3.0 ppm as Cl_2. Details regarding monochloramine dosing and water chemistry are included in Text S1.

Sample collection and processing

Hot water was collected from 27 sites throughout the hospital at two time points before monochloramine injection (three months and immediately prior) and monthly for the first six months of monochloramine application. Water samples were collected from a variety of locations throughout the hospital (Table 1). Samples were taken from hot water tanks, the hot water return line, faucets in the intensive care units, rehabilitation suites including both automatic and standard faucets, and other patient rooms on the upper floors. The faucets in the intensive care units are located on the third, fourth, and fifth floors. The faucets in the rehabilitation suites are located on floors six and seven and represent both electronic sensor (automatic) faucets and standard faucets. The final grouping of sites was from short-term use patient rooms located on floors eight, nine, ten, eleven, and twelve. At each site, hot water was flushed for one minute prior to sample collection into sterile HDPE bottles with enough sodium thiosulfate to neutralize 20 ppm chlorine (Microtech Scientific, Orange, CA). For hot water tank sampling, the drain valve was opened, allowed to flush for one minute, then sampled into sterile HDPE bottles as described above. Following sampling, 100 mL of sample water was filtered through a 0.2 μm, 47 mm, polycarbonate filter membrane (Whatman, Florham Park, NJ), placed into 10 mL of the original water sample, and vortexed vigorously for 10 seconds as described in methods ISO Standards 11731:1998 and 11731:2004 for *Legionella* isolation. Five mL of each concentrated sample was frozen at −80°C until DNA extraction.

DNA extraction, PCR, and Sequencing

Frozen water samples were thawed and pooled as described in Table 1. The 27 samples were divided into five pools including the hot water tanks and hot water return line (HWT), floors 3–5 (the intensive care units, F3), floors 6 and 7 automatic faucets (the rehabilitation suites' automatic faucets, F6A), floors 6 and 7 standard faucets (the rehabilitation suites' standard faucets, F6S), and floors 8–12 (the short-term use patient rooms, F8). These samples were then filtered through 0.2 μm, 47 mm, Supor 200 Polyethersulfone membranes (Pall Corporation), housed in sterile Nalgene filter funnels (Thermo Scientific; Fisher). Filter membranes were subjected to DNA extraction using the RapidWater DNA Isolation Kit (MO-BIO Laboratories) as described by the manufacturer. PCR was performed in quadruplicate using 16S rRNA region primers 515F and 806R including sequencing and barcoding adapters as previously described [38]. These primers amplify an approximately 300 base pair region of the rRNA region spanning variable regions 3 and 4. The specificity of this primer set is considered to be well optimized and 'nearly universal' [39]; analysis of these primers against the 97% Greengenes 13.5 OTU database demonstrated a specificity of 99.9% and 98.3% for the 515f and 806r primers, respectively. Dreamtaq Mastermix (Thermo Scientific) was used and PCR product was checked on

Table 1. Sample pool description, abbreviation, and number of pooled sites.

Sample Description	Sample Abbreviation	Number of Pooled Sites
Outlets of Hot Water Tanks and Hot Water Return Line	HWT	3
Floors 3–5 Patient Room Faucets	F3	4
Floors 6 & 7 Patient Room Automatic Faucets	F6A	7
Floors 6 & 7 Patient Room Standard Faucets and Showers	F6S	7
Floors 8–12 Patient Room Faucets	F8	6
Technical Replicates of Floors 8–12 Patient Room Faucets	F8rep	6

Hot water was collected after a one-minute flush from the following locations throughout the hospital.

a 1% agarose gel. An independent negative control was run for each sample and primer set and all negative controls were negative for PCR amplification. PCR products were pooled and purified using the UltraClean PCR Clean-Up Kit (MO-BIO Laboratories). Each sample then underwent additional cleaning with the Agencourt AMPure XP PCR purification kit (Beckman Coulter) and quantified using the QuBit 2.0 Fluorometer (Invitrogen). Following quantification, 0.1 picomoles of each sample PCR product were pooled. The sample pool underwent two additional clean up steps with a 1.5:1 ratio of Agencourt AMPure XP beads followed by a 1.2:1 bead ratio (Beckman Coulter) to eliminate primer dimers. Samples were sequenced on an in-house Illumina MiSeq sequencing platform as previously described [38].

Data analysis

Data was analyzed within the MacQIIME (http://www.wernerlab.org/software/macqiime) implementation of QIIME 1.7.0 [40]. Sequences were parsed based upon sample-specific barcodes and trimmed to a minimum quality score of 20. Operational taxonomic units (OTUs) at 97% were then picked against the Greengenes 13.5 database using UCLUST [41] for taxonomic assignment. Following assignment, 7,000 successfully assigned sequences from each sample were chosen at random to allow for even downstream analyses and even cross-sample comparison. Observed OTUs were defined as observed species whereas unassigned sequences were removed from subsequent analyses (closed reference OTU picking). Alpha-diversity evenness was calculated using the 'equitability' metric within QIIME. Beta diversity analyses were conducted by UNIFRAC analysis [42]. OTUs were also open-reference picked, where unassigned sequences are placed in the taxa "other" and therefore not removed. Discussion and results from this open-reference OTU picking analysis is included in Text S1. Open-reference OTU picking did not result in a shift in any fundamental conclusions with the exception of the increase in the genus *Stenotrophomonas* spp. following monochloramine addition; closed-reference OTU picking is presented for higher-quality taxonomic assignment. Morisita-Horn indices were calculated as previously described [43,44]. Sequences are available under MG RAST accession numbers 4552832.3 to 4552878.3.

Results

Sequence Data

Sequencing reads were split by sample-specific barcodes, trimmed to a minimum quality score of 20, and placed into OTUs at 97% through comparison with the Greengenes 13.5 coreset. For each sample, 7,000 sequences with assigned taxonomy were selected to allow for even comparison across samples. Two types of OTU picking were done for this study: closed reference (sequences were compared to a reference set of sequences for OTU clustering, sequences not matching one of these pre-defined sequences were discarded) and open reference (sequences were compared to each other for OTU picking, sequences not mapping to the reference database were grouped as 'other') in Text S1.

Alpha Diversity

Alpha diversity (number of observed OTUs) of samples treated with monochloramine was significantly higher than samples from the baseline months (Figure 1). Prior to treatment, the average number of observed OTUs at 97% similarity was 151.2 ± 39.7, whereas during treatment the average number of observed OTUs was 225.2 ± 61.2 ($p<0.001$) (Figure 1). This shift was not associated with a statistically significant loss of sample evenness (Figure S1). The same statistical trends in alpha diversity were observed for open-reference picked OTUs (Figure S2).

Beta Diversity

Beta diversity (sample interrelatedness) was analyzed using weighted UNIFRAC [42]. The principal coordinate analysis (PCoA) plot from this analysis is shown in Figure 2. Samples from the first two months prior to treatment cluster together whereas those following disinfection tend to cluster by sample site more strongly than sample time (Figure 2). The same trend was observed for open-reference picked OTUs (Figure S3).

Taxonomic Comparison

Figure 3 shows the phyla-level taxonomy for each of the sample pools. Phyla <1.3% relative abundance are listed as 'minor phyla'. Prior to treatment, samples from all locations were similarly

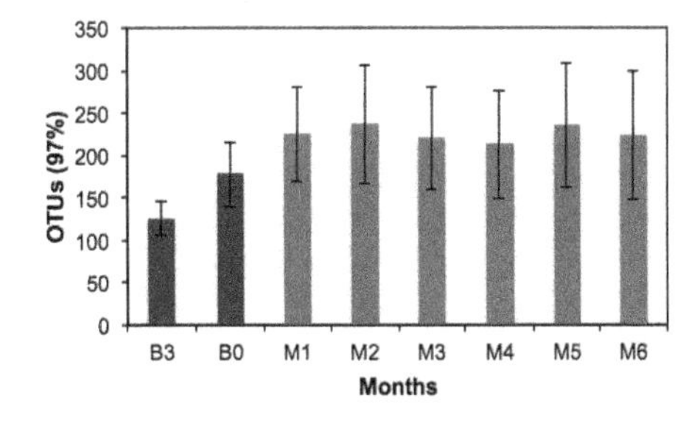

Figure 1. Comparison of the number of OTUs (97% similarity) for each month. Bars represent standard deviation. Each sample pool was normalized to 7,000 sequences. Samples from B3 and B0 represent those taken three months and immediately prior to monochloramine treatment, respectively. Samples from M1, M2, M3, M4, M5, and M6 were taken monthly during the first six months of treatment.

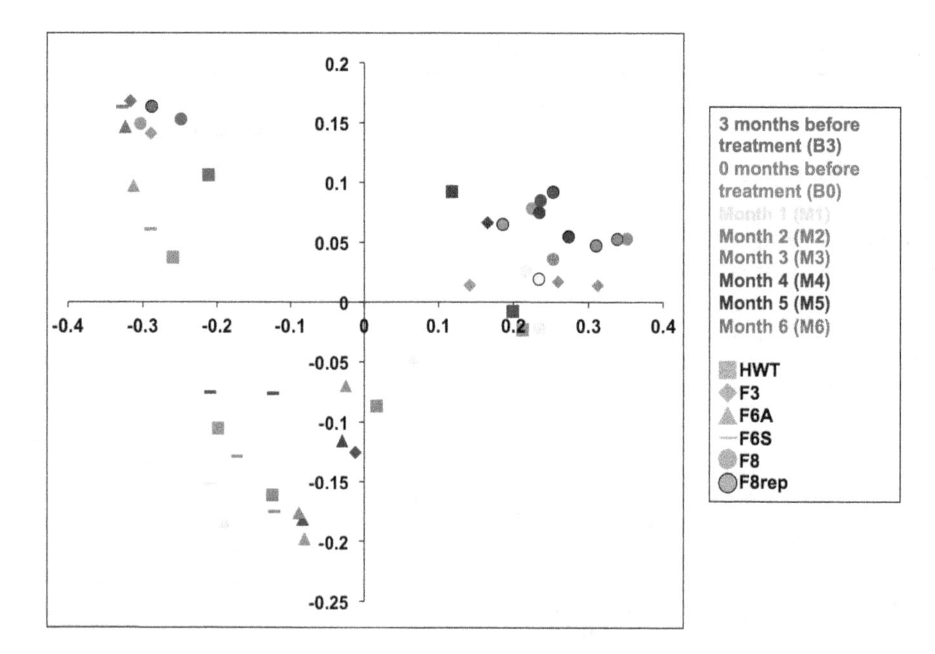

Figure 2. PCoA analysis of samples pools. Samples that cluster more closely together share a greater similarity in microbial community structure. Colors represent months sampled whereas shapes represent sample pool. Samples from B3 and B0 represent those taken three months and immediately prior to monochloramine treatment, respectively. Samples from M1, M2, M3, M4, M5, and M6 were taken monthly during the first six months of treatment.

structured, predominantly comprised of Betaproteobacteria, with lesser quantities of Firmicutes, Bacteroidetes, Alphaproteobacteria, and Gammaproteobacteria (Figure 3 Panels A–E). Following initiation of treatment (M1) there was a shift away from the predominance of Betaproteobacteria and towards a greater relative abundance of Firmicutes, Alphaproteobacteria, Gammaproteobacteria, and minor fractions of Cyanobacteria and Actinobacteria (Figure 3 Panels A–E). The same taxonomy trends were observed for open-reference picked data (Figure S4 Panels A–E).

The samples from the hot water tank (HWT) from pre-treatment months (B3 and B0) were approximately 60% Betaproteobacteria with approximately 35% Firmicutes, Bacteroidetes, Alphaproteobacteria, and Gammaproteobacteria in aggregate (Figure 3 Panel A). Following treatment the relative abundance of Betaproteobacteria was reduced to approximately 20% and Firmicutes, Alphaproteobacteria, and Gammaproteobacteria subsequently increased to comprise an average of 78% of the total relative abundance (Figure 3 Panel A).

The microbial community profile of samples from the lower floors of the hospital (intensive care units, F3) was slightly different than those of the hot water tank samples but a similar trend was observed (Figure 3 Panel B). Over 65% of pre-treatment samples were Betaproteobacteria with Firmicutes, Bacteroidetes, Alphaproteobacteria, and Gammaproteobacteria accounting for a combined 20% of community relative abundance (Figure 3 Panel B). Following treatment the amount of Betaproteobacteria and Bacteroidetes decreased to an average of 23% relative abundance, while the relative abundance of Firmicutes and Alphaproteobacteria increased sharply to approximately 68% (Figure 3 Panel B).

In spite of being from the same rooms, the taxonomic composition of samples from F6A and F6S differed after treatment (Figure 3 Panels C and D). Prior to treatment both the automatic (F6A) and standard faucets (F6S) in the rehabilitation suites contained 65–80% Betaproteobacteria, with Bacteroidetes, Alphaproteobacteria, Gammaproteobacteria, and Cyanobacteria

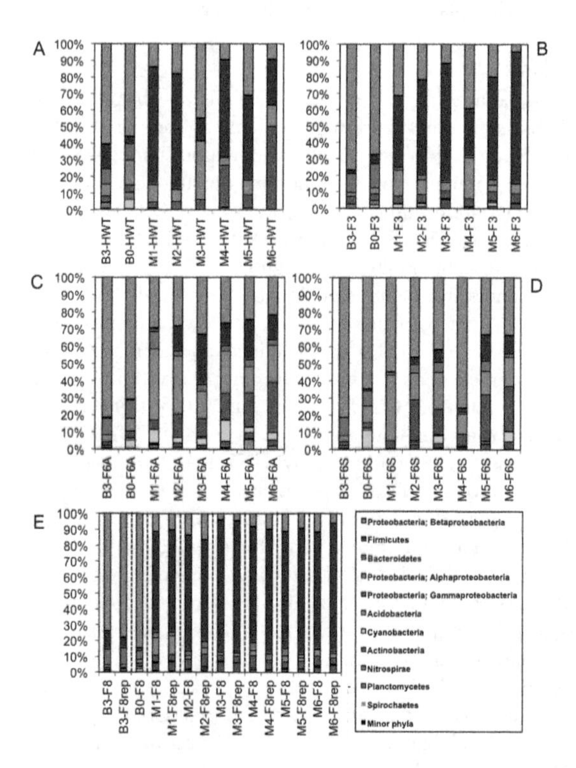

Figure 3. Taxonomic assignments of sequences from HWT (hot water tank samples) (Panel A), F3 (floors 3–5) (Panel B), F6A (floors 6 and 7 automatic faucets) (Panel C), F6S (floors 6 and 7 standard faucets) (Panel D), F8 (floors 8–12) and F8rep (replicate barcoded PCRs of samples from floors 8–12) (Panel E). Samples from B3 and B0 represent those taken three months and immediately prior to monochloramine treatment, respectively. Samples from M1, M2, M3, M4, M5, and M6 were taken monthly during the first six months of treatment. Black lines in Panel E separate pairs of replicates.

accounting for the other 20–35% of relative abundance (Figure 3 Panels C and D). However, after monochloramine application, the automatic faucets (F6A) underwent a 50% reduction in the total relative abundance of Betaproteobacteria and became enriched in Firmicutes, Alphaproteobacteria, Gammaproteobacteria, Actinobacteria, and Spirochaetes (Figure 3 Panel C). The standard faucets (F6S) lost only 26% of Betaproteobacteria, but also saw an increase in members of the Firmicutes, Alphaproteobacteria, Gammaprotobacteria, and Actinobacteria phyla from an average relative abundance of 10% before treatment to 46% after monochloramine addition (Figure 3 Panel D).

Prior to treatment, the microbial community in samples from the upper floors of the hospital (short-term use patient rooms, F8) resembled most of the other baseline samples with over 70% Betaproteobacteria and approximately 20% of Firmicutes, Bacteroidetes, Alphaproteobacteria, Gammaproteobacteria, Acidobacteria, and Cyanobacteria (Figure 3 Panel E). Following monochloramine treatment, the relative abundance of Betaproteobacteria was reduced from approximately 70% to 10% and replaced by Firmicutes, which increased from 7% of the relative abundance in the baseline months to 74% after treatment (Figure 3 Panel E). There was only a slight increase, from 2% to 9% relative abundance, in the amount of Gammaproteobacteria and Actinobacteria present (Figure 3 Panel E).

Sample Replicates

Separately amplified and barcoded technical replicates of sample pool F8 for 7 of the 8 sample pools were also sequenced to verify technical reproducibility. There is no replicate for month B0. UNIFRAC analysis demonstrated that the replicates from each month cluster very closely (Figure 2). All of the samples from F8 in samples M1–M6 and their replicates (circles and outlined circles) clustered together in the upper-right hand quadrant (Figure 2). Morisita-Horn analyses of replicates demonstrate high levels of community similarity, ranging from 0.990 (M2) to 0.9998 (M3). These results further validate the technical reproducibility of the methodology (Figure 3 Panel E) [43,44]. The open-reference picked UNIFRAC analysis and taxonomy also show replicates to have similar profiles to their original samples (Figure S3 and S4 Panel E). Morisita-Horn analyses of these samples showed similarly high levels of community similarity ranging from 0.991 (M2) to 0.9992 (M1).

Genera Containing Opportunistic Pathogens

Sequence data was further analyzed to observe the change in genera containing opportunistic pathogens of interest during treatment. Genera analyzed were: *Legionella* spp., *Pseudomonas* spp., *Acinetobacter* spp., and *Stenotrophomonas* spp. (Gammaproteobacteria group); *Brevundimonas* spp. and *Sphingomonas* spp. (Alphaproteobacteria group); *Chryseobacterium* spp. (Bacteroidetes group); and *Mycobacterium* spp. (Actinobacteria group). These genera are of special interest as some to all of the species contained within them are pathogens; however, the nature of short-read 16S rRNA region sequence analysis is such that species-level pathogens cannot be definitively identified. Trends demonstrated by this analysis could be used to direct future analyses targeting opportunistically pathogenic organisms more specifically. Analysis of the relative abundance of each of these organism groups over time shows a statistically significant increase in relative abundance for *Acinetobacter* ($p = 0.0054$), *Mycobacterium* ($p = 0.0017$), *Pseudomonas* ($p = 0.031$) and *Sphingomonas* ($p = 0.034$) as treatment progressed (Figure 4). *Brevundimonas*, *Chryseobacterium*, Legionellaceae, and *Stenotrophomonas* did not demonstrate a statistically significant increase in relative abundance following treatment (Figure 4).

The open-reference picked data demonstrated an increase in the same opportunistic pathogen containing genera as the closed-reference picked data, *Acinetobacter* ($p = 0.004$), *Mycobacterium* ($p = 0.002$), *Pseudomonas* ($p = 0.015$), and *Sphingomonas* ($p = 0.025$), but also showed a significant increase in the genera *Stenotrophomonas* ($p = 0.03$) (Figure S5).

Nitrification and Denitrification

Additionally, we investigated the shift in relative abundance of representative genera associated with nitrification and denitrification (Figure 5). There was no statistically significant difference in the relative abundance of the potential nitrifiers *Nitrospira* and Nitrosomonadaceae, before ($mean = 0.0015 \pm 0.0018$) and after treatment ($mean = 0.0005 \pm 0.0011$) ($p = 0.175$). Other nitrifier-containing genera such as *Nitrosococcus*, *Nitrobacter*, *Nitrospina*, or *Nitrococcus*, were not identified in any samples. The total relative abundance of genera containing denitrifiers (*Thiobacillus*, *Micrococcus*, and *Paracoccus*) underwent a statistically significant increase before ($mean = 0.00005 \pm 0.000074$) and after treatment with monochloramine ($mean = 0.0029 \pm 0.0029$) ($p = 0.026$). The denitrifier-containing genera *Rhizobiales* and *Rhodanobacter* were not identified in any samples. The same trends were observed in open-reference picked data (Figure S6).

Discussion

Our study objective was to examine the shift in the microbial ecology of a hospital hot water system associated with the introduction of on-site monochloramine addition. To evaluate the shift in microbial community structure we sampled 27 sites in a hospital and pooled samples into 5 groups for 8 sample time points. Sites were pooled based on their location and use in the hospital and faucet type (automatic versus standard). This study took place during the first U.S. trial of the Sanikill on-site monochloramine generation system (Sanipur, Brescia, Italy) [45–47]. These samples were subjected to DNA extraction, 16S rRNA region barcoded PCR, and Illumina sequencing to analyze the response of the microbial ecology to the addition of monochloramine.

The microbial population shift in response to monochloramine addition was immediate. The number of OTUs observed (alpha diversity) significantly increased following monochloramine treatment (Figure 1). It is possible that the overall loss of dominance of initially abundant microbial groups (e.g. Betaproteobacteria) allowed for a greater number of other bacterial species to grow, or for selected individuals to die off, thereby increasing the alpha diversity. Samples from different sites taken before monochloramine treatment were comprised of similar microbial populations and samples taken after treatment were distinct from samples taken in the baseline months (Figures 2 and 3, Figures S3 and S4). Interestingly, it appears that following monochloramine treatment the location of sampling matters more in sample similarity (beta diversity) than does the month they were taken (Figure 2, Figure S3). Microbial communities from the lower floors' intensive care units (F3) and the upper floors' short term patient rooms (F8) were more similar than to the floors 6 and 7's rehabilitation suites (F6A and F6S) automatic and standard faucet samples. These sites were located in single patient rooms in rehabilitation units and may experience as much use as some locations on the lower and upper floors, which include the trauma burn unit, the intensive care unit (ICU), the neonatal ICU, and the cardiovascular ICU. The HWT samples from earlier months of treatment closely resembled floors 6 and 7 (F6A and F6S) whereas the HWT microbial ecology from

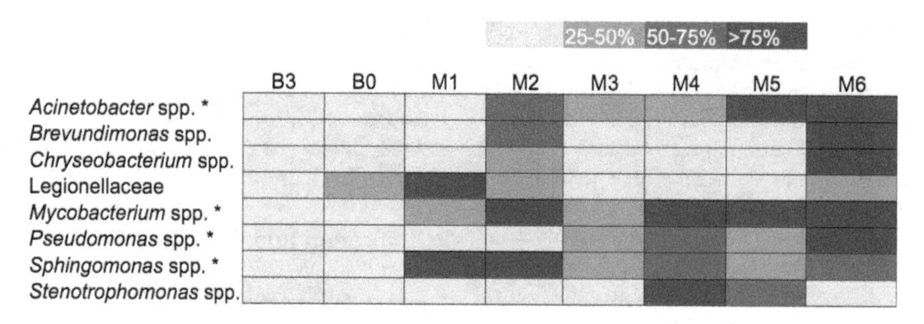

Figure 4. Relative abundance of different genera of opportunistic waterborne pathogens. Samples color coded into four groupings calculated by 25% of the maximum relative abundance for each organism. Months with the least relative abundance are lightest in color, whereas months with the highest relative abundance are darkest. *denotes a statistically significant increase in the relative abundance of this organism following treatment.

the later months was more related to the lower (F3) and upper floors (F8).

We investigated the possible differences in microbial ecology between automatic and standard faucets as it has been previously demonstrated that opportunistic pathogens, including *Legionella* [48] and *Pseudomonas aeruginosa* [49], are detected more frequently and in greater concentrations in automatic faucets. It has been suggested that the reason for the differences between automatic and standard faucets could be due to water flow, temperature, and structural issues. Automatic faucets may have diluted monochloramine concentrations due to low flow and poor flushing [48,49] and automatic faucets also contain mixing valves, which are made of materials such as rubber, polyvinylchloride, and plastic, which more easily support the growth of biofilms [48,49]. Potentially due to these biofilms, the increased colonization can persist even following disinfection with chlorine dioxide [48]. We observed a differential reduction in the relative abundance of Betaproteobacteria in standard and automatic faucets following treatment. The automatic faucets lost 50% of their relative abundance of Betaproteobacteria whereas the standard faucets only saw an average 26% reduction.

There was an overall shift towards less relative abundance of Betaproteobacteria, and more relative abundance of Firmicutes, Alphaproteobacteria, Gammaproteobacteria, Cyanobacteria and Actinobacteria after monochloramine treatment. A previous microbial ecology study of a simulated drinking water distribution system treated with monochloramine demonstrated a different trend, with an increase in specific genera within the Actinobacteria, Betaproteobacteria, and Gammaproteobacteria phyla [3]. The dissimilarity of these studies may be due to the fact that the latter occurred in a cold water system whereas our study was in a hot water supply.

Several waterborne pathogen-containing genera were examined for changes in relative abundance due to monochloramine treatment. The relative abundance of a few of the waterborne pathogen-containing genera examined, including *Acinetobacter*, *Mycobacterium*, *Pseudomonas*, and *Sphingomonas*, showed an increase after monochloramine treatment. Other studies have described an increase in some of these organisms including *Legionella*, *Mycobacterium*, and *Pseudomonas* in chloraminated water [3,6] as well as biofilms treated with monochloramine [50]. Feazel et al. previously demonstrated that *Mycobacterium* spp. can be enriched in showerhead biofilms compared to the source water [20]. An increased relative abundance of *Mycobacterium* spp. due to monochloramine treatment is of concern, specifically if this increase in relative abundance is due to the presence of more viable mycobacterial cells. These microorganisms may pose a specific threat of aerosol exposure to immunocompromised

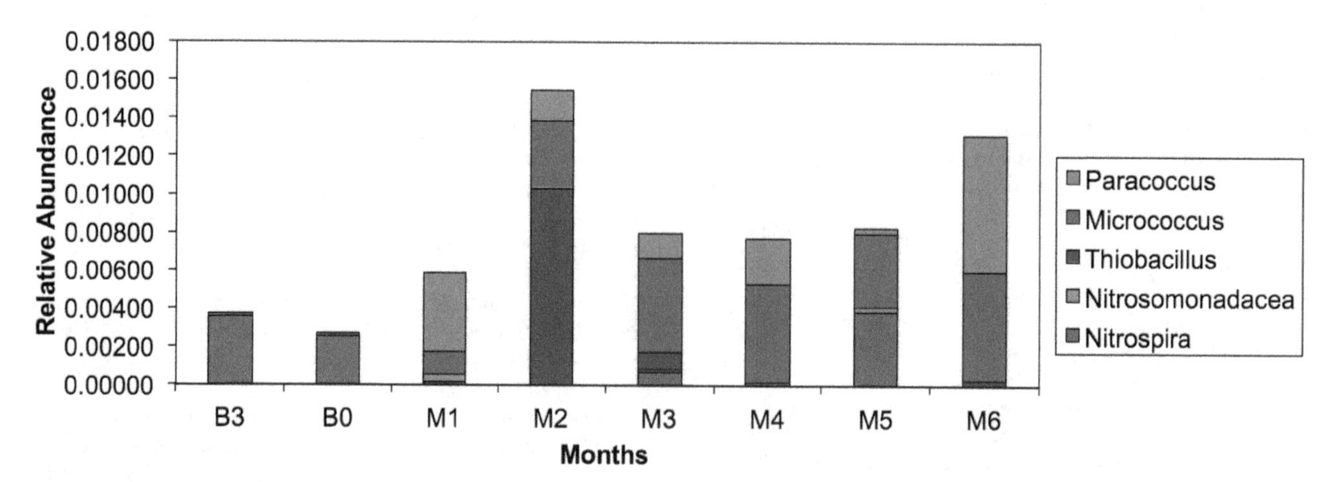

Figure 5. Relative abundance of genera containing nitrifying (*Nitrospira* and Nitrosomonadacea) and denitrifying bacteria (*Thiobacillus, Micrococcus,* and *Paracoccus*). No other genera associated with nitrification (*Nitrosococcus, Nitrobacter, Nitrospina,* or *Nitrococcus,*) or denitrification (*Rhizobiales* and *Rhodanobacter*) were found in any of our samples. The x-axis represents sampling months with months B3 and B0 being before monochloramine treatment and months M1–M6 representing the first six months of treatment. The y-axis represents the relative abundance.

patients who reside in buildings with an increased abundance of these organisms in hot water [20]. Interestingly, a recent study demonstrated that while the concentration of live bacteria is reduced after monochloramine treatment, only the viable microbial community structure is altered and genera containing opportunistic pathogens persist [51]. While we did not directly quantify microorganisms in the samples collected or verify that microorganisms detected were viable, our parallel culture-based study observed a statistically significant reduction in culturable total bacteria and *Legionella* species following monochloramine treatment (Table S1) [45–47,52].

Previous studies have found an increase in nitrification in chloraminated systems, which effectively decreased monochloramine concentration [6,31]. This chemical decay led to higher levels of *Legionella*, *Mycobacterium* spp., and *P. aeruginosa* at earlier water ages than in chlorinated simulated distribution systems [6]. A change in potentially nitrifying bacteria following monochloramine addition was not observed in the culture-based portion of this study [45–47], consistent with our molecular observations. Concentrations of nitrate and nitrite remained fairly stable throughout the study months, with the exception of a spike in nitrate levels in M6 (Table S1) [52]. We observed a statistically significant increase in the relative abundance of genera associated with denitrification in monochloramine treated samples. A previous study found a high absolute abundance, up to 200,000 cfu/mL, of potentially denitrifying bacteria in a chloraminated system even after regular flushing [32]. The highest relative abundance of bacterial genera associated with denitrification occurred during M6 when there was a spike in nitrate concentrations (Table S1) [52]. However, in months 1 and 2 there was also a large relative abundance of these bacteria present with fairly low nitrate concentrations, suggesting that some other factor might be important in their relative abundance. We do not believe that these trends were due to seasonality in our study as microbiological data were largely consistent across the study period. However, the possibility for seasonal effects cannot be excluded.

A notable increase in the relative abundance of the genus *Alicyclobacillus* spp. (Firmicutes phylum) was observed following monochloramine treatment, from an average of 4.1±4.5% of the microbial population prior to treatment to an average of 40.9±27.1% following treatment ($p<0.001$). This genera is comprised primarily of spore-formers that are of concern in food spoilage [53], and has previously been detected in drinking water [54]. The high relative abundance of *Alicyclobacillus* spp. suggests a potentially dominant role in chloraminated hot water system microbial ecology worthy of future investigation.

The incidence of reported Legionnaires' disease cases increased threefold from 2000 to 2009 [55]. This fact, coupled with an increasingly elderly and immunocompromised population [55], has lead to an increased concern about *Leginonella* and other opportunistic waterborne pathogens. Additionally, the American Society of Heating, Refrigerating, and Air-Conditioning Engineers (ASHRAE) has recently proposed Standard 188P for the prevention of legionellosis associated with premise plumbing systems [56]. This standard serves to reduce the risk of *Legionella* infections through a risk management approach [56]. For these reasons, on-site disinfection has become progressively important to protect patients in hospitals and long-term care facilities from waterborne opportunistic pathogens. An increased understanding of the influence of on-site disinfection on premise plumbing microbial ecology is necessary to maximize effectiveness and to limit undesired side effects.

This study demonstrates that there exists the potential for unwanted consequences of supplemental disinfectant addition for the removal of *Legionella* such as the potential enrichment of other waterborne pathogens, including *Acinetobacter*, *Mycobacterium*, *Pseudomonas*, and *Sphingomonas*. Understanding the impact of supplemental disinfection on water system microbial ecology, through a holistic approach, is necessary to maximize disinfectant effectiveness and to ensure that supplemental disinfectant does not select for alternative opportunistic pathogens. A recent review emphasizes not only the role of disinfectants but also other system factors that may impact microbial ecology such as temperature, pipe material, organic carbon, presence of automatic faucets, and point-of-use filtration [24]. The authors suggest a probiotic approach to opportunistic pathogen control which would either add microbes that can outcompete these pathogens, remove key species, or using engineering controls to favor benign organisms that are antagonistic to opportunistic pathogens [24]. This systematic, probiotic, approach to premise plumbing opportunistic pathogen management is an inventive concept for dealing with the diverse microbial ecology of these systems, but requires a greater understanding of the drivers of premise plumbing microbial ecology, such as provided by this study.

In conclusion, we observed a shift in the microbial ecology of a hospital's hot water system treated with on-site chloramination. This shift occurred immediately following monochloramine treatment. Prior to treatment, the bacterial ecology of all samples was dominated by Betaproteobacteria; following treatment, members of Firmicutes and Alphaproteobacteria dominated. Differences in community composition were seen in different locations within the hospital as well as between automatic and standard faucets. This suggests that water from different locations and outlet types should be sampled to get a more thorough picture of the microbiota of a system. There was an increase in the relative abundance of several genera containing opportunistic waterborne pathogens following the onset of monochloramine treatment, including *Acinetobacter*, *Mycobacterium*, *Pseudomonas*, and *Sphingomonas* and genera associated with denitrification. The benefits and risks of each supplemental disinfection strategy should be evaluated before implementation in any building, especially in hospitals, long term care facilities, and other buildings housing immunocompromised patients. This work demonstrates the effects of a supplemental monochloramine disinfection system on the microbial ecology of premise plumbing biofilms. Given the importance of premise plumbing microbial ecology on opportunistic pathogen presence and persistence, understanding the driving influence of supplemental disinfectants on microbial ecology is a crucial component of any effort to rid premise plumbing systems of opportunistic pathogens. As additional facilities turn to on-site water disinfection strategies, more long-term studies on the effects of disinfectants on microbial ecology in premise plumbing are needed as well as those evaluating a probiotic approach to opportunistic pathogen eradication.

Supporting Information

Figure S1 Sample evenness for closed-reference OTU picking. No statistically significant different was observed for samples taken prior to or following monochloramine addition.

Figure S2 Alpha diversity for open-reference OTU picking. A statistically significant difference was observed for samples taken prior to or following monochloramine addition ($p = 0.046$).

Figure S3 Beta diversity for open-reference OTU picking. Samples from before monochloramine treatment clustered together whereas following treatment samples clustered by location more so than month of treatment.

Figure S4 Taxonomic assignment of sequences from HWT (hot water tank samples) (Panel A), F3 (floors 3–5) (Panel B), F6A (floors 6 and 7 automatic faucets) (Panel C), F6S (floors 6 and 7 standard faucets) (Panel D), F8 (floors 8–12) and F8rep (replicate barcoded PCRs of samples from floors 8–12) (Panel E) for open-reference OTU picking.

Figure S5 Relative abundance of waterborne pathogen containing genera for open-reference OTU picking. A statistically significant increase in *Acinetobacter* spp., *Mycobacterium* spp., *Pseudomonas* spp., *Sphingomonas* spp., and *Stenotrophomonas* spp. was observed following treatment.

Figure S6 Relative abundance genera containing nitrifying (*Nitrospira* and Nitrosomonadacea) and denitrifying bacteria (*Thiobacillus*, *Micrococcus*, and *Paracoccus*) for open-reference OTU picking. No other genera containing nitrifying bacteria (*Nitrosococcus*, *Nitrobacter*, *Nitrospina*, or *Nitrococcus*,) or denitrifying bacteria (*Rhizobiales* and *Rhodanobacter*) were found in our samples.

Table S1 Physicochemical data obtained during the study.

Text S1 Supplementary Information. Water chemistry and monochloramine dosing methods, description of minor phyla observed, and open-reference OTU picking results.

Acknowledgments

We would like to thank the staff of Special Pathogens Laboratory for their assistance in sample collection.

Author Contributions

Conceived and designed the experiments: JLB JES KB. Performed the experiments: JLB AV SD. Analyzed the data: JLB JES KB. Wrote the paper: JLB KB.

References

1. Berry D, Xi C, Raskin L (2006) Microbial ecology of drinking water distribution systems. Curr Opin Biotechnol 17: 297–302.
2. Martiny AC, Jorgensen TM, Albrechtsen HJ, Arvin E, Molin S (2003) Long-term succession of structure and diversity of a biofilm formed in a model drinking water distribution system. Appl Environ Microbiol 69: 6899–6907.
3. Gomez-Alvarez V, Revetta RP, Santo Domingo JW (2012) Metagenomic analyses of drinking water receiving different disinfection treatments. Appl Environ Microbiol 78: 6095–6102.
4. Douterelo I, Sharpe RL, Boxall JB (2013) Influence of hydraulic regimes on bacterial community structure and composition in an experimental drinking water distribution system. Water Res 47: 503–516.
5. Henne K, Kahlisch L, Brettar I, Hofle MG (2012) Analysis of structure and composition of bacterial core communities in mature drinking water biofilms and bulk water of a citywide network in Germany. Appl Environ Microbiol 78: 3530–3538.
6. Wang H, Masters S, Edwards MA, Falkinham JO, Pruden A (2014) Effect of Disinfectant, Water Age, and Pipe Materials on Bacterial and Eukaryotic Community Structure in Drinking Water Biofilm. Environ Sci Technol 48: 1426–1435.
7. White C, Tancos M, Lytle DA (2011) Microbial community profile of a lead service line removed from a drinking water distribution system. Appl Environ Microbiol 77: 5557–5561.
8. Zhang Y, Griffin A, Rahman M, Camper A, Baribeau H, et al. (2009) Lead contamination of potable water due to nitrification. Environ Sci Technol 43: 1890–1895.
9. Zhang Y, Triantafyllidou S, Edwards M (2008) Effect of nitrification and GAC filtration on copper and lead leaching in home plumbing systems. Journal of Environmental Engineering-Asce 134: 521–530.
10. Mathieu L, Bouteleux C, Fass S, Angel E, Block JC (2009) Reversible shift in the alpha-, beta- and gamma-proteobacteria populations of drinking water biofilms during discontinuous chlorination. Water Res 43: 3375–3386.
11. Squier C, Yu VL, Stout JE (2000) Waterborne Nosocomial Infections. Curr Infect Dis Rep 2: 490–496.
12. Perola O, Nousiainen T, Suomalainen S, Aukee S, Karkkainen UM, et al. (2002) Recurrent Sphingomonas paucimobilis-bacteraemia associated with a multi-bacterial water-borne epidemic among neutropenic patients. J Hosp Infect 50: 196–201.
13. Mondello P, Ferrari L, Carnevale G (2006) Nosocomial Brevundimonas vesicularis meningitis. Infez Med 14: 235–237.
14. Buse HY, Ashbolt NJ (2011) Differential growth of Legionella pneumophila strains within a range of amoebae at various temperatures associated with in-premise plumbing. Lett Appl Microbiol 53: 217–224.
15. Emtiazi F, Schwartz T, Marten SM, Krolla-Sidenstein P, Obst U (2004) Investigation of natural biofilms formed during the production of drinking water from surface water embankment filtration. Water Res 38: 1197–1206.
16. Berry D, Horn M, Xi C, Raskin L (2010) Mycobacterium avium Infections of Acanthamoeba Strains: Host Strain Variability, Grazing-Acquired Infections, and Altered Dynamics of Inactivation with Monochloramine. Appl Environ Microbiol 76: 6685–6688.
17. Swanson MS, Hammer BK (2000) Legionella pneumophila pathogesesis: a fateful journey from amoebae to macrophages. Annu Rev Microbiol 54: 567–613.
18. Lau HY, Ashbolt NJ (2009) The role of biofilms and protozoa in Legionella pathogenesis: implications for drinking water. J Appl Microbiol 107: 368–378.
19. van der Wielen PW, van der Kooij D (2013) Nontuberculous mycobacteria, fungi, and opportunistic pathogens in unchlorinated drinking water in The Netherlands. Appl Environ Microbiol 79: 825–834.
20. Feazel LM, Baumgartner LK, Peterson KL, Frank DN, Harris JK, et al. (2009) Opportunistic pathogens enriched in showerhead biofilms. Proc Natl Acad Sci U S A 106: 16393–16399.
21. Schwartz T, Kohnen W, Jansen B, Obst U (2003) Detection of antibiotic-resistant bacteria and their resistance genes in wastewater, surface water, and drinking water biofilms. FEMS Microbiol Ecol 43: 325–335.
22. Shi P, Jia S, Zhang XX, Zhang T, Cheng S, et al. (2013) Metagenomic insights into chlorination effects on microbial antibiotic resistance in drinking water. Water Res 47: 111–120.
23. NRC (2006) Drinking Water Distribution Systems: Assessing and Reducing Risks.
24. Wang H, Edwards MA, Falkinham JO, Pruden A (2013) Probiotic Approach to Pathogen Control in Premise Plumbing Systems? A Review. Environmental Science & Technology 47: 10117–10128.
25. Williams MM, Armbruster CR, Arduino MJ (2013) Plumbing of hospital premises is a reservoir for opportunistically pathogenic microorganisms: a review. Biofouling 29: 147–162.
26. Williams MM, Braun-Howland EB (2003) Growth of Escherichia coli in model distribution system biofilms exposed to hypochlorous acid or monochloramine. Appl Environ Microbiol 69: 5463–5471.
27. Wang H, Edwards M, Falkinham JO 3rd, Pruden A (2012) Molecular survey of the occurrence of Legionella spp., Mycobacterium spp., Pseudomonas aeruginosa, and amoeba hosts in two chloraminated drinking water distribution systems. Appl Environ Microbiol 78: 6285–6294.
28. Regan JM, Harrington GW, Noguera DR (2002) Ammonia- and nitrite-oxidizing bacterial communities in a pilot-scale chloraminated drinking water distribution system. Appl Environ Microbiol 68: 73–81.
29. LeChevallier MW, Cawthon CD, Lee RG (1988) Inactivation of biofilm bacteria. Appl Environ Microbiol 54: 2492–2499.
30. Hoefel D, Monis PT, Grooby WL, Andrews S, Saint CP (2005) Culture-independent techniques for rapid detection of bacteria associated with loss of chloramine residual in a drinking water system. Appl Environ Microbiol 71: 6479–6488.
31. Zhang Y, Edwards M (2009) Accelerated chloramine decay and microbial growth by nitrification in premise plumbing. Journal American Water Works Association 101: 51.
32. Nguyen C, Elfland C, Edwards M (2012) Impact of advanced water conservation features and new copper pipe on rapid chloramine decay and microbial regrowth. Water Res 46: 611–621.
33. Lin YE, Stout JE, Yu VL (2011) Controlling Legionella in hospital drinking water: an evidence-based review of disinfection methods. Infect Control Hosp Epidemiol 32: 166–173.

34. Stout JE, Goetz AM, Yu VL (2011) Hospital Epidemiology and Infection Control; Mayhall CG, editor: Lippincott Williams, & Wilkins.
35. Flannery B, Gelling LB, Vugia DJ, Weintraub JM, Salerno JJ, et al. (2006) Reducing Legionella colonization in water systems with monochloramine. Emerg Infect Dis 12: 588–596.
36. Pryor M, Springthorpe S, Riffard S, Brooks T, Huo Y, et al. (2004) Investigation of opportunistic pathogens in municipal drinking water under different supply and treatment regimes. Water Sci Technol 50: 83–90.
37. Marchesi I, Cencetti S, Marchegiano P, Frezza G, Borella P, et al. (2012) Control of Legionella contamination in a hospital water distribution system by monochloramine. American Journal of Infection Control 40: 279–281.
38. Caporaso JG, Lauber CL, Walters WA, Berg-Lyons D, Huntley J, et al. (2012) Ultra-high-throughput microbial community analysis on the Illumina HiSeq and MiSeq platforms. ISME J 6: 1621–1624.
39. Walters WA, Caporaso JG, Lauber CL, Berg-Lyons D, Fierer N, et al. (2011) PrimerProspector: de novo design and taxonomic analysis of barcoded polymerase chain reaction primers. Bioinformatics 27: 1159–1161.
40. Caporaso JG, Kuczynski J, Stombaugh J, Bittinger K, Bushman FD, et al. (2010) QIIME allows analysis of high-throughput community sequencing data. Nat Methods 7: 335–336.
41. Edgar RC (2010) Search and clustering orders of magnitude faster than BLAST. Bioinformatics.
42. Lozupone C, Knight R (2005) UniFrac: a new phylogenetic method for comparing microbial communities. Appl Environ Microbiol 71: 8228–8235.
43. Morisita M (1959) Measuring of the dispersion of individuals and analysis of the distributional patterns. Memoirs of the Faculty of Science, Kyushu University, Series E (Biology) 2.
44. Horn HS (1966) Measurement of "overlap" in comparative ecological studies. The American Naturalist 100: 419–424.
45. Stout JE, Duda S, Kandiah S, Hannigan J, Yassin M, et al. (2012) Evaluation of a new monochloramine generation system for controlling *Legionella* in building hot water systems. Association of Water Technologies Annual Convention and Exposition.
46. Kandiah S, Yassin MH, Hariri R, Ferrelli J, Fabrizio M, et al. (2012) Control of *Legionella* contamination with monochloramine disinfection in a large urban hospital hot water system. Association for Professionals in Infection Control and Epidemiology Annual Conference.
47. Duda S, Kandiah S, Stout JE, Baron JL, Yassin MH, et al. (2013) Monochloramine disinfection of a hospital water system for preventing hospital-acquired Legionnaires' disease: lessons learned from a 1.5 year study. The 8th International Conference on Legionella.
48. Sydnor ER, Bova G, Gimburg A, Cosgrove SE, Perl TM, et al. (2012) Electronic-eye faucets: Legionella species contamination in healthcare settings. Infect Control Hosp Epidemiol 33: 235–240.
49. Yapicioglu H, Gokmen TG, Yildizdas D, Koksal F, Ozlu F, et al. (2012) Pseudomonas aeruginosa infections due to electronic faucets in a neonatal intensive care unit. J Paediatr Child Health 48: 430–434.
50. Revetta RP, Gomez-Alvarez V, Gerke TL, Curioso C, Santo Domingo JW, et al. (2013) Establishment and early succession of bacterial communities in monochloramine-treated drinking water biofilms. FEMS Microbiol Ecol.
51. Chiao TH, Clancy TM, Pinto A, Xi C, Raskin L (2014) Differential resistance of drinking water bacterial populations to monochloramine disinfection. Environ Sci Technol 48: 4038–4047.
52. Duda S, Kandiah S, Stout JE, Baron JL, Yassin MH, et al. (2014) Evaluation of a new monochloramine generation system for controlling *Legionella* in building hot water systems. Submitted for publication.
53. Jensen N, Whitfield FB (2003) Role of Alicyclobacillus acidoterrestris in the development of a disinfectant taint in shelf-stable fruit juice. Letters in Applied Microbiology 36: 9–14.
54. Revetta RP, Pemberton A, Lamendella R, Iker B, Santo Domingo JW (2010) Identification of bacterial populations in drinking water using 16S rRNA-based sequence analyses. Water Research 44: 1353–1360.
55. Centers for Disease Control and Prevention (2011) Legionellosis–United States, 2000–2009. Morbidity and Mortality Weekly Report: 1083–1086.
56. BSR/ASHRAE (2011) Proposed New Standard 188P, Prevention of Legionellosis Associated with Building Water Systems. Atlanta, GA: American Society of Heating, Refrigerating, and Air-Conditioning Engineers, Inc.

5

A Western Diet Ecological Module Identified from the 'Humanized' Mouse Microbiota Predicts Diet in Adults and Formula Feeding in Children

Jay Siddharth[1¤], Nicholas Holway[2], Scott J. Parkinson[1*]

1 Host Commensal Hub, Developmental and Molecular Pathways, Novartis Institutes for Biomedical Research, Basel, Switzerland, 2 Scientific Computing, NIBR IT, Novartis Institutes Biomedical Research, Basel, Switzerland

Abstract

The interplay between diet and the microbiota has been implicated in the growing frequency of chronic diseases associated with the Western lifestyle. However, the complexity and variability of microbial ecology in humans and preclinical models has hampered identification of the molecular mechanisms underlying the association of the microbiota in this context. We sought to address two key questions. Can the microbial ecology of preclinical models predict human populations? And can we identify underlying principles that surpass the plasticity of microbial ecology in humans? To do this, we focused our study on diet; perhaps the most influential factor determining the composition of the gut microbiota. Beginning with a study in 'humanized' mice we identified an interactive module of 9 genera allied with Western diet intake. This module was applied to a controlled dietary study in humans. The abundance of the Western ecological module correctly predicted the dietary intake of 19/21 top and 21/21 of the bottom quartile samples inclusive of all 5 Western and 'low-fat' diet subjects, respectively. In 98 volunteers the abundance of the Western module correlated appropriately with dietary intake of saturated fatty acids, fat-soluble vitamins and fiber. Furthermore, it correlated with the geographical location and dietary habits of healthy adults from the Western, developing and third world. The module was also coupled to dietary intake in children (and piglets) correlating with formula (vs breast) feeding and associated with a precipitous development of the ecological module in young children. Our study provides a conceptual platform to translate microbial ecology from preclinical models to humans and identifies an ecological network module underlying the association of the gut microbiota with Western dietary habits.

Editor: Dionysios A. Antonopoulos, Argonne National Laboratory, United States of America

Funding: The authors have no support or funding to report.

Competing Interests: All authors are employees of Novartis Institutes for Biomedical Research.

* E-mail: scott.parkinson@novartis.com

¤ Current address: Nestle Institute of Health Sciences, Quartier de l'Innovation, EPFL, Lausanne, Switzerland

Introduction

As part of the acceleration of economic globalization in the last quarter of the 20th century came a realization that the 'Western Lifestyle' is primarily responsible for the forecasted epidemic in chronic diseases. A combination of inactivity and rapid changes in dietary habits are now recognized as major contributing factors to the pathogenesis of cancer, obesity, diabetes, cardiovascular diseases as well as other chronic inflammatory diseases in the developing world. Recent shifts evident in the developing world towards an 'energy dense' diet comprised of animal fat and processed foods alongside reduced complex carbohydrate and dietary fiber intake parallel the predicted dominance of chronic over infectious disease death rates across most of the world [1]. Metabolism of our food is coordinated by the gut microbiota capable of extracting nutrients consumed in the diet [2]. Therefore, there has been growing interest in understanding the relationships between diet, microbes responsible for metabolism and their associated links with chronic diseases. However, comprehension of the molecular mechanisms that link microbial ecology in the context of human health has not kept pace with the publication of technology-driven catalogs of the flora associated with health status [3–5].

Two major challenges associated with the translation of exploratory research findings concerning the microbiota can be summarized with two questions. Firstly, can the microbial ecology of preclinical models be used to predict biological traits in human populations? And secondly, can the plasticity of microbial ecology in humans be harnessed to predict specific populations? With these questions in mind, we decided to take an alternative look at two recent papers investigating the impact of dietary intake. One of these studies investigated a humanized mouse model, where gnotobiotic mice were inoculated with human donor feces and subsequently challenged with a controlled diet. Using principle component analysis, the authors concluded that individual variation in microbiota composition is more significant than microbial changes following dietary changes, consistent with the concept of microbial 'enterotypes' previously put forward [4,5]. The 'enterotype' concept introduced the perception that the individual composition of the microbiota is relatively stable and resistant to environmental influence. Similar conclusions were put forward by Wu et al. following their analysis of the composition of the microbiota from a human dietary intervention study conducted in a controlled environment [3]. The investigators recruited 10 volunteers who were assigned to either a Western or low fat/high fiber dietary group. In the same report, they also looked at the dietary intake of human volunteers in context with the composition of their microbiota. The authors concluded that while dietary change influenced the microbiota of the individual, the composition of the microbiota was dominated by the individual 'enterotype.'

The conclusions of these studies pose some practical challenges with respect to the translation of microbiota research to applications of consequence to human health. Of concern for pharmaceutical applications, is if the source dominates the microbial profile, how can we use preclinical models to support drug-discovery efforts or disease monitoring? In addition, if individual 'enterotypes' dominate microbial composition, how can the microbiota be used to predict response, monitor outcomes or identify specific drug-resistant or sensitive populations?

We took an alternative view to the 'enterotype' approach to specifically address these questions. We considered the composition of the microbiota as a cooperative group of individual bacteria that interact with neighbors to establish an ecological network module (or 'module' for short) whereby they can optimally use nutritional intake to their advantage. Our hypothesis was that if these communities could be identified, they should correlate with specific intervention (in this case dietary intake) regardless of the original composition (or 'enterotype') of the microbiota. We applied this approach to the outlined published studies and identified an interactive module of 9 genera allied with Western diet in the 'humanized' mice. We then applied this module to the controlled dietary study in humans predicting dietary intake in patients. In the same study, the module also correlated with individual dietary intake of 98 volunteers demonstrating the translation of microbial composition from preclinical models to humans as well as harnessing the plasticity in the microbiota across human diversity. Furthermore, in association with dietary intake, the microbial module demonstrated significant correlation with the dietary habits of healthy adults living in diverse geographical locations and children being formula vs breast-fed. Our study provides a conceptual platform to translate microbial ecology from preclinical models to human populations.

Materials and Methods

Public dataset processing: The datasets presented in this study were downloaded from the NCBI SRA database and MG-RAST, the datasets have been summarized in Table S4, including the accession numbers and associated publications. In case of sequences from the NCBI SRA, the datasets were downloaded as sra files and using the protocol described at the SRA site, the sequences were converted to fastq files, using the sra toolkit, deploying the fastq-dump command. In case of MG-RAST, the processed fasta files containing 16S sequences pre-screened for 97% identity to ribosomal genes were retrieved.

The resulting fasta files derived from the two locations were then analyzed identically using the software mother, the primary data cleaning steps were used as described in the 454 SOP. The computational environment was a server running 64 bit Red Hat Enterprise Linux release 5 with a 64 bit mothur ver 1.24 (CentOS precompile version) from the mothur website (http://www.mothur.org/wiki/454_SOP. Accessed 2013 July 1) [6–7]. Specifically the trim.seqs commands with oligos files generated from barcodes present on the NCBI page of the respective sequences. This removed any sequences with ambiguous bases, homopolymers longer than 8 and allowed for barcode difference of one base and a primer difference of two bases. The sequences post quality check and grouping by barcode were assigned taxonomies using the latest RDP template (release 7) from mothur website (http://www.mothur.org/wiki/RDP_reference_files. Accessed 2013 July 1). The command classify.seqs used was with 1000 iterations per sequence.

Data Analysis. Excel files (Tables S5–S8) containing the summarized counts of classified sequences derived above were manually annotated with information from publication or databases, this was done as the deposited data did not have enough metadata on them. This required to cross matching and subsequently merging the data from the associated publication like diet types, breakdown of dietary components etc and deposited metadata with the sequences. The resulting summary files are attached in the Supplementary Materials (Tables S5–S8) for reproducibility. The genus frequency/sample was calculated to normalize for differences in the number of sequences in each sample, which essentially meant deriving the percentage distribution of each genus representation within a sample. The resulting files were imported into TIBCO Spotfire 3.3.2.5 for all data normalizations, calculations, and statistical operations including visualization. Genus to Genus Spearman correlations as seen in figure 1B and Table S1 were calculated with 0 counts removed and cutoffs applied using the Df's calculated within the correlations as indicated.

Results

Identifying A Western Diet Ecological Module In Mice

We considered the concept that the microbial composition of a given ecosystem is the product of multiple dynamic and interactive bacterial populations (or modules) [8–10]. We sought to identify 'modules' of bacteria associated with dietary intake by examining statistical association of individual genera in a study examining association of the Western diet with the microbiota in humanized gnotobiotic mice [5]. This study investigated 'humanized' mice by using fecal transplants from human donors in germ free mice, which were later fed on Western and normal diet, and also included a diet crossover as part of the experimental design. The study aimed at reproducing microbial flora in a surrogate host (mice) and look at microbial flora perturbation following diet changes. The raw 16S sequences were classified and 70 genera of were included in subsequent analysis having been present in ≥25% of the 300 fecal samples. We did not consider culturing prior to inoculation as a factor as the original study did not detect a significant impact on the overall composition of the microbiota. We processed the annotated 16S rRNA sequences derived from two human donors grafted into germ-free mice including 70 genera for further analysis [11].

Examining the composition of individual samples at the Genera level (Figure 1A) samples derived from mice receiving the Western diet (in green) could be distinguished from mice receiving normal chow consistent with the original observations of the authors [5]. In addition, donor-dependent clustering dominated the grouping of the samples supporting the concept of an innate microbial 'enterotype' [3,4]. As anticipated with a Western dietary switch, a trend towards time-dependent segregation of the samples was also observed. The bacterial composition at Western diet day 7 and 14 generally associated independently from the day 1 and 3 samples. The samples collected following reversion back to mouse chow were generally intermingled with those collected prior to the

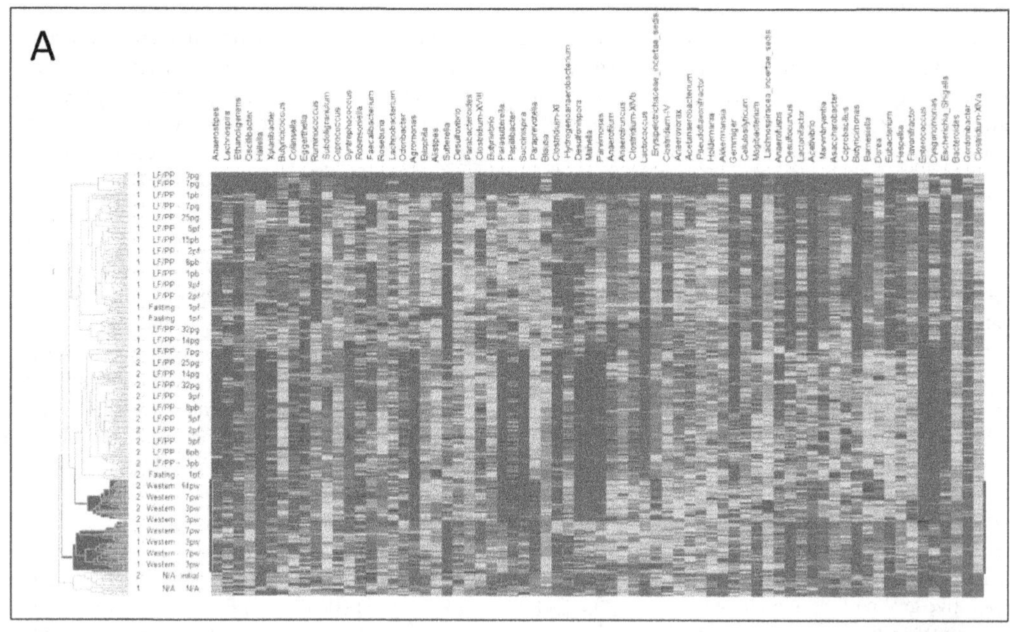
A

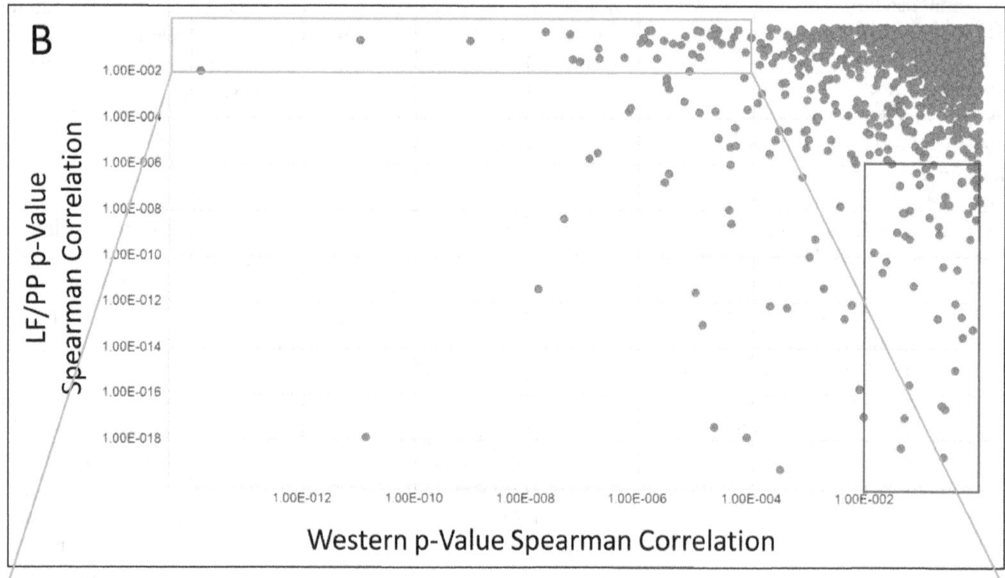
B
1.00E-002
1.00E-004
1.00E-006
1.00E-008
1.00E-010
1.00E-012
1.00E-014
1.00E-016
1.00E-018
LF/PP p-Value
Spearman Correlation
1.00E-012
1.00E-010
1.00E-008
1.00E-006
1.00E-004
1.00E-002
Western p-Value Spearman Correlation

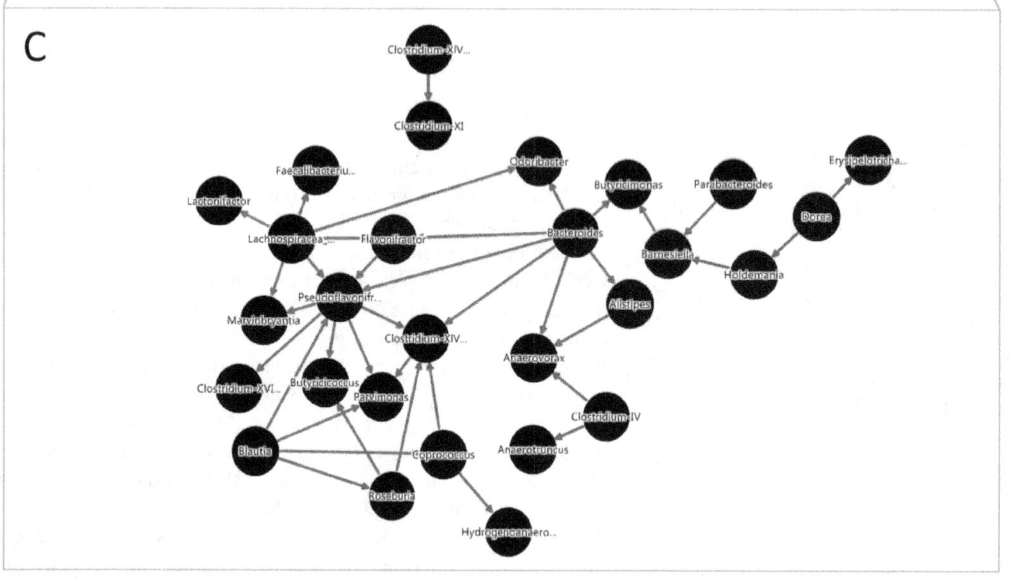
C
Clostridium-XIV...
Clostridium-XI
Faecalibacteriu...
Odoribacter
Butyricimonas
Parabacteroides
Erysipelotricha...
Lactonifactor
Lachnospiracea_...
Flavonifractor
Bacteroides
Dorea
Barnesiella
Holdemania
Pseudoflavonifr...
Marvinbryantia
Alistipes
Clostridium-XIV...
Anaerovorax
Clostridium-XVI...
Butyricicoccus
Parvimonas
Clostridium-IV
Blautia
Coprococcus
Anaerotruncus
Roseburia
Hydrogenoanaero...

Figure 1. Detection of Diet-associated Ecological Communities. A. Hierarchical clustering of fecal 16S rRNA sequences in samples derived from humanized gnotobiotic mice. On the y-axis the human donor (1 or 2), diet (LF/PP; low-fat plant polysaccharide-rich mouse chow, Western; Harlan-Teklad TD96132, Fasting; no food, N/A; human sample prior to inoculation) and time point (indicated by day following dietary change and condition; N/A; human sample cultured prior to inoculation, initial; human fecal sample prior to inoculation, pw; post-Western diet change, pf; post fasting, pg; post-gavage, pb; post-return to mouse chow). On the x-axis the composite bacterial genera are indicated. Heat map; red (high abundance), blue (low abundance), grey (none detected). Hierarchical clustering for both rows and columns was performed using UPGMA clustering method with Euclidean distance measure, ordering weight by average value, normalization on a scale between 0 and 1 with empty value replacement by constant value given as 0. **B**. Identification of diet-related genera/genera interactive pairs. Pairwise genera p-values were calculated by co-occurrence analysis using Spearman correlation. Each point on the graph represents the relationship between two bacterial genera plotted across the p-value observed in the Western diet and the LF/PP (low fat, plant polysaccharide rich)-associated diets. The resulting Western diet-specific (green box) and LF/PP diet-specific (red box) pairs were selected for further analysis. **C**. Network map of Western diet-specific bacterial interactions. The Western diet-specific pairwise associations identified in B (green box) were assembled to visualize the diet-associated ecologic module. Purple edges indicate positive Spearman correlations, blue indicate negative correlations.

switch to Western diet, reflecting a plastic but predictable dynamic of the microbial composition in response to their environment.

Having confirmed the presence of diet-associated dynamic changes in the microbiota, we examined the pair-wise Spearman correlations of all 70 genera to identify co-segregating bacteria, hypothesizing that they would form the building blocks of ecologic modules associated with diet. Generic pairs significantly associated independent of diet, exclusively with Western diet and excluded from Western diet were all identified (Figure 1B, Table S1). For example, *Marvinbryantia* demonstrated a significant correlation with *Lachnospiracae insertae sedis* in the Western diet vs mouse chow ($p = 1.45e^{-14}$ vs $p = 1.14\ e^{-2}$) and similarly *Alistipes* abundance correlated with *Bacteroides* by Spearman correlation ($p = 9.79e^{-12}$ vs $2.69e^{-1}$). Furthermore, *Clostridium XVIII* and *Coprobacillus* had a Spearman correlation of $1.78e^{-19}$ in the mouse chow diet (LF/PP) vs $2.36e^{-1}$ in Western diet samples. These data demonstrate that diet influences the coordinated association of bacteria with each other. This follows conventional wisdom regarding bacterial interaction that interactions between individual bacteria can be synergistic or antagonistic in response to competition for food or the interdependence of groups of microbes to metabolize and/or utilize particular nutrients.

We next investigated the integration of the Western diet-specific pairs into potential ecological modules via building a network map (Figure 1C). The Western diet-associated pairs demonstrated a marked degree of integration. One module of 9 positively correlating genera was observed centered on *Bacteroides* – the focus of one of 3 proposed 'enterotypes' [3,4]. This potential module negatively-associated with a second module of 13 genera surrounding *Pseudoflavonifractor*. Two additional modules (3 genera negatively-associated with the *Bacteroides* module and a fourth, single interaction between *Clostridium XIVa* and *Clostridium XI*) were also observed, however, we focused our attention on those surrounding *Bacteroides* and *Pseudoflavonifractor*.

The next logical step was to investigate whether the identified potential modules reflected dietary intake. We normalized abundance levels of the composite genera of the two potential modules using a Z score calculation to consider each genus with equal weighting and plotted these across the individual samples (Figure 2A, 2B). The abundance of the individual genera comprising the *Bacteroides* module was elevated in the latter Western diet samples (7 and 14 day; 7pw, 14pw) and later the 'fasting' samples when mouse chow was withdrawn for 24 hours (Figure 2A). In both cases genera abundance returned to prior levels when the LF/PP diet was re-initiated suggesting the overall abundance of the module was inhibited by the LF/PP diet. The component genera of the *Pseudoflavonifractor* module demonstrated the opposite temporal pattern being elevated at early time points (day 1 and 3) following the shift to Western diet then decreasing thereafter (Figure 2B). The negative correlation between the two modules is derived from the differential abundance of the modules over time following the switch to Western diet. While this observation could indicate that the *Bacteroides* module represents a 'mature' Western dietary module, both modules demonstrated partitioning with the Western diet. Therefore, we continued to investigate both mature *Bacteroides module* and the *Pseudoflavonifractor module* in our analysis.

As an overall reflection of abundance of the modules, we next calculated the average Z score/sample and assembled the samples in order for both the *Bacteroides* and *Pseudoflavonifractor* modules (Figure 2C). Annotated Western diet samples (yellow) were enriched towards the left of the respective module graphs reflecting an association of the abundance of both modules with the Western diet.

We also generated a network from the pairwise genera associations that were specific to the samples derived from the mouse chow diet (Figure 1B; red box, Figure S1). A central module of 12 genera around a *Barnesiella* hub was observed surrounded by 4 negatively correlating minor modules. *Blautia*, *Syntrophococcus* and *Clostridium XIVb* formed 7, 4 and 4 negative connections, respectively with peripheral members of the *Barnesiella* module. The *Barnesiella* module itself showed no association with diet but did distinguish samples derived from donor 2 (Figure S2). The phenotype of this module, therefore, support the concept of an innate individual enterotype more than an association with diet (Figure 1B, Table S1, Figure S2).

The Bacteroides Module Stratifies Human Samples According To Western Diet

Since the original study was derived from a mouse model inoculated with a human microbiota, we hypothesized that if the modules were physiologically relevant to diet, human subjects receiving a similar diet should be distinguished by their relative average Z score for the module (Figure 3). To do this, we turned our attention to another published study examining the dynamic properties of a controlled dietary change on healthy volunteers [3]. In this particular study healthy human subjects were fed in clinical settings using a high fat/low fiber or a low fat/high fiber diet, which represented a Western and traditional diet, respectively. This was followed up by examination of the feces for microbial perturbation as a result of the diet. The average Z score for the *Bacteroides* and *Pseudoflavonifractor* modules was calculated for the annotated human samples where the *Bacteroides* module demonstrated an enriched distribution in the first quartile correlating with 19/21 samples and a complete absence in the last quartile. In addition, samples from all 5 patients on the high fat/low fiber diet were identified in the top quartile and samples from all 5 patients on the low fat/high fiber diet were identified in the bottom quartile. The *Pseudoflavonifractor* module that associated with donor 2 rather than dietary intake in mice, likewise did not stratify the

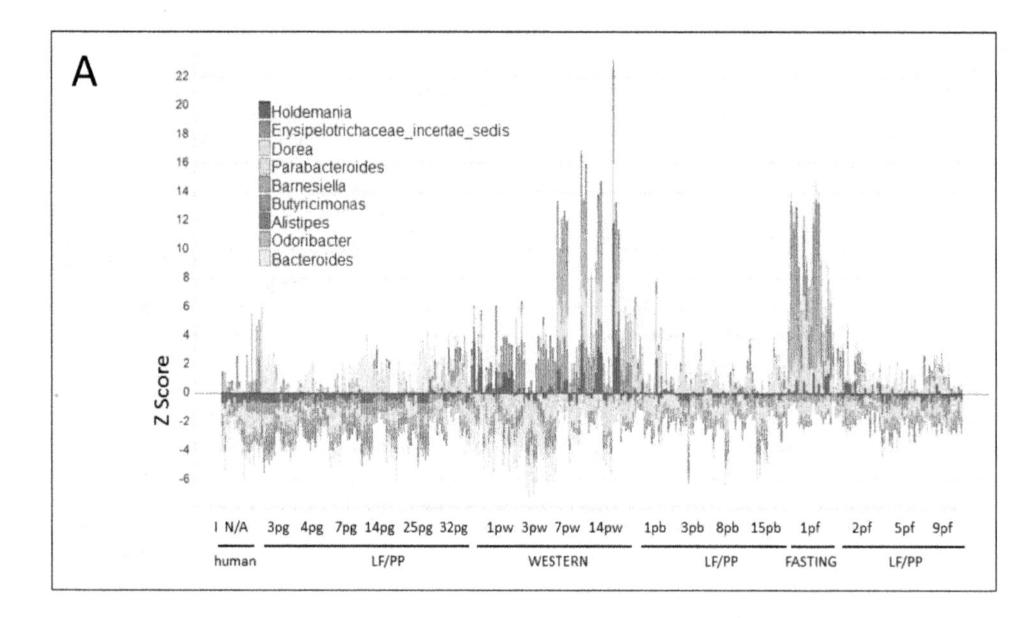
A
Holdemania
Erysipelotrichaceae_incertae_sedis
Dorea
Parabacteroides
Barnesiella
Butyricimonas
Alistipes
Odoribacter
Bacteroides
Z Score
I N/A 3pg 4pg 7pg 14pg 25pg 32pg 1pw 3pw 7pw 14pw 1pb 3pb 8pb 15pb 1pf 2pf 5pf 9pf
human LF/PP WESTERN LF/PP FASTING LF/PP

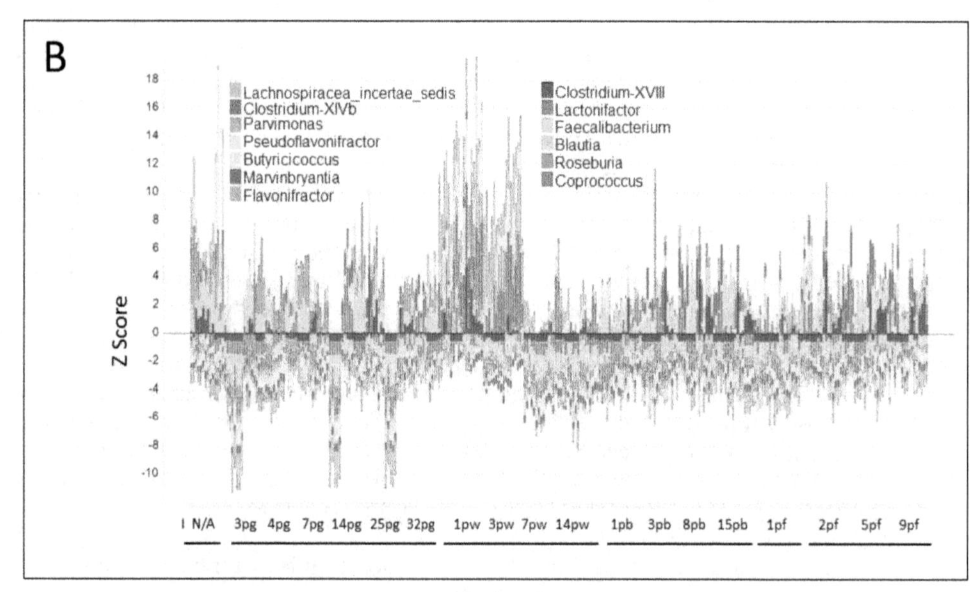
B
Lachnospiracea_incertae_sedis
Clostridium-XIVb
Parvimonas
Pseudoflavonifractor
Butyricicoccus
Marvinbryantia
Flavonifractor
Clostridium-XVIII
Lactonifactor
Faecalibacterium
Blautia
Roseburia
Coprococcus
Z Score
I N/A 3pg 4pg 7pg 14pg 25pg 32pg 1pw 3pw 7pw 14pw 1pb 3pb 8pb 15pb 1pf 2pf 5pf 9pf

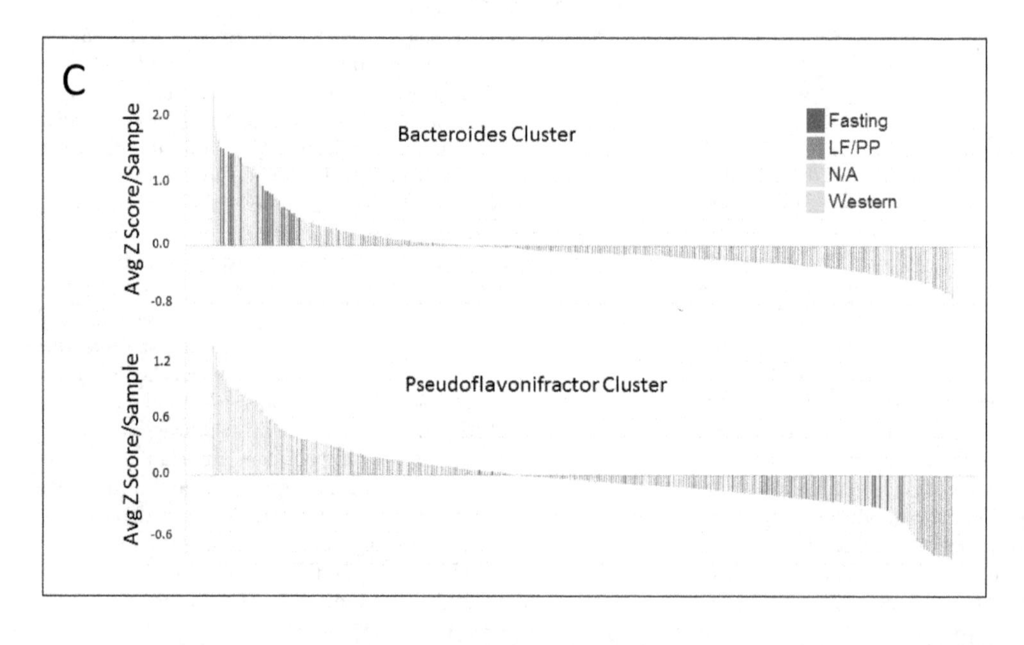
C
Bacteroides Cluster
Fasting
LF/PP
N/A
Western
Avg Z Score/Sample
Pseudoflavonifractor Cluster
Avg Z Score/Sample

Figure 2. Temporal and diet-regulated dynamics of ecological bacterial modules. A. Bacteroides module. Each stacked bar represents the Z score normalized abundance of the component genera indicated in the legend. The diet and time points are indicated on the x-axis and can generally be followed in a temporal manner from left to right. Diets shown are human (undefined starting material), LF/PP (low fat/plant polysaccharide rich chow), WESTERN (Harlan-Teklad TD96132), FASTING (no food). I; initial fecal donor sample, N/A; initial sample prior to inoculation, pg; day following gavage into mouse, pw; day following shift to Western diet, pb: day following return to LF/PP diet, 1pf: day 1 fasting, pf: day after fasting. **B**. Temporal and diet-regulated dynamics of the *Pseudoflavonifractor* module. Each stacked bar represents the Z score normalized abundance of the component genera indicated in the legend. X-axis as for A. **C**. The abundance of the *Bacteroides* and *Pseuodoflavonifractor* modules associate with Western-diet. The average Z score for each fecal sample was calculated for both the *Bacteroides* and *Pseudoflavonifractor*-module component genera and arranged in order of abundance from left to right. Each bar is coloured by the sample diet.

patient population according to fat intake (Figure 3). These data demonstrate that microbial ecology of preclinical models can predict human populations independently of a variance 'enterotype' approach.

We tested whether the association of module abundance with the Western-diet samples simply reflected the individual genus abundance in each sample. In the *Bacteroides* module, the abundance of 6 out of 9 of the members were significantly associated with Western diet in humans – of these only *Holdemania* was significantly lower in the high fat diet samples. In the *Pseudoflavonifractor* module, 6 out of 13 genera were significantly associated with diet. Three of these (*Blautia*, *Flavonifractor* and *Butyricoccus*) were significantly higher in the patients on a low fat diet. One possibility for the association between the gnotobiotic mouse and human study was simply a result of identifying generally high abundance genera in the humanized mice. To examine this, we identified the top 9 genera most significantly associated with Western diet in the humanized mouse study (Table S2). Unsurprisingly, the gnotobiotic mouse samples were partitioned according to diet by the average Z score of this 'Western Abundance module' (Figure 4A). However, when the average Z score of this module of bacteria was applied to the human dietary study it did not associated with diet (Figure 4B). Although the top 10 most abundant samples were all high fat diet-associated, 7 out of 10 were derived from 1 patient and the top quartile contained only 12/21 high fat-diet samples. In addition, the bottom quartile was populated by 13/22 high fat diet samples mostly from two patients, clearly indicating that this 'abundance' module could not distinguish diet-related changes from the variability inherent between individuals. Only 3 of the 'abundance' genera were also significantly different in the human study (*Hydrogenoanaerobacterium*, *Anaerotruncus*, *Clostridium IV*) while *Flavonifractor* was significantly lower in the human high fat vs low fat diet samples. Therefore, abundance of bacterial genera in the absence of ecological context is a poor predictor of the diet-related dynamics of microbial ecology.

The Bacteroides Module Correlates With Dietary Intake In The General Population

The data shown above demonstrate that the *Bacteroides* module can distinguish samples derived from a Western diet under control conditions in both mouse and human. Thus it stands to reason that if the *Bacteroides* module was a biomarker of dietary intake the abundance of the module should correlate with dietary intake in the general population. To test this, we examined another dataset in a human study examining the composition of the microbiota with two dietary questionnaires; one recording patients' recent diet (Recall) and another querying habitual long term diet (FFQ). The study investigated the effect of long term diet and its effect of microbial composition using questionnaires of the habitual long term diet. It was suggested that enterotypes are a result of long

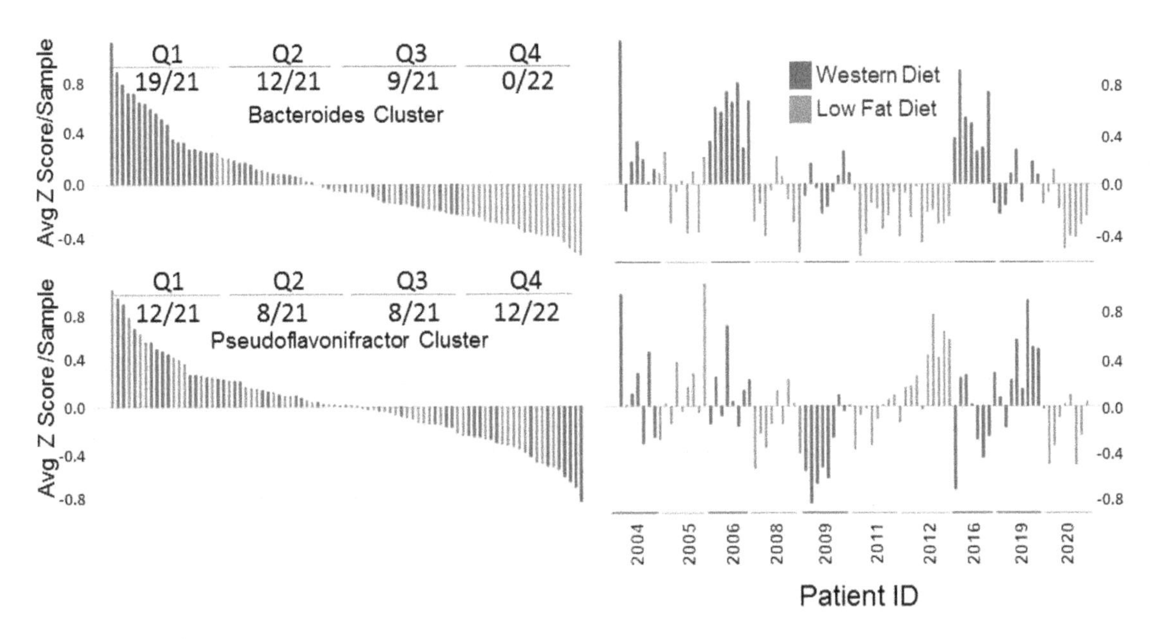

Figure 3. The *Bacteroides* module is associated with Western diet in human patients. On the left is plotted the average Z score for each sample and placed in order of decreasing abundance from left to right. The quartile distribution of high fat diet samples/total is shown for each module. On the right the same average Z scores are plotted with annotation from each patient. The temporal order for each patient is from left to right. Only samples on a controlled Western or low-fat diet (as indicated) were included in the analysis – starting samples prior to commencing the controlled diet were excluded.

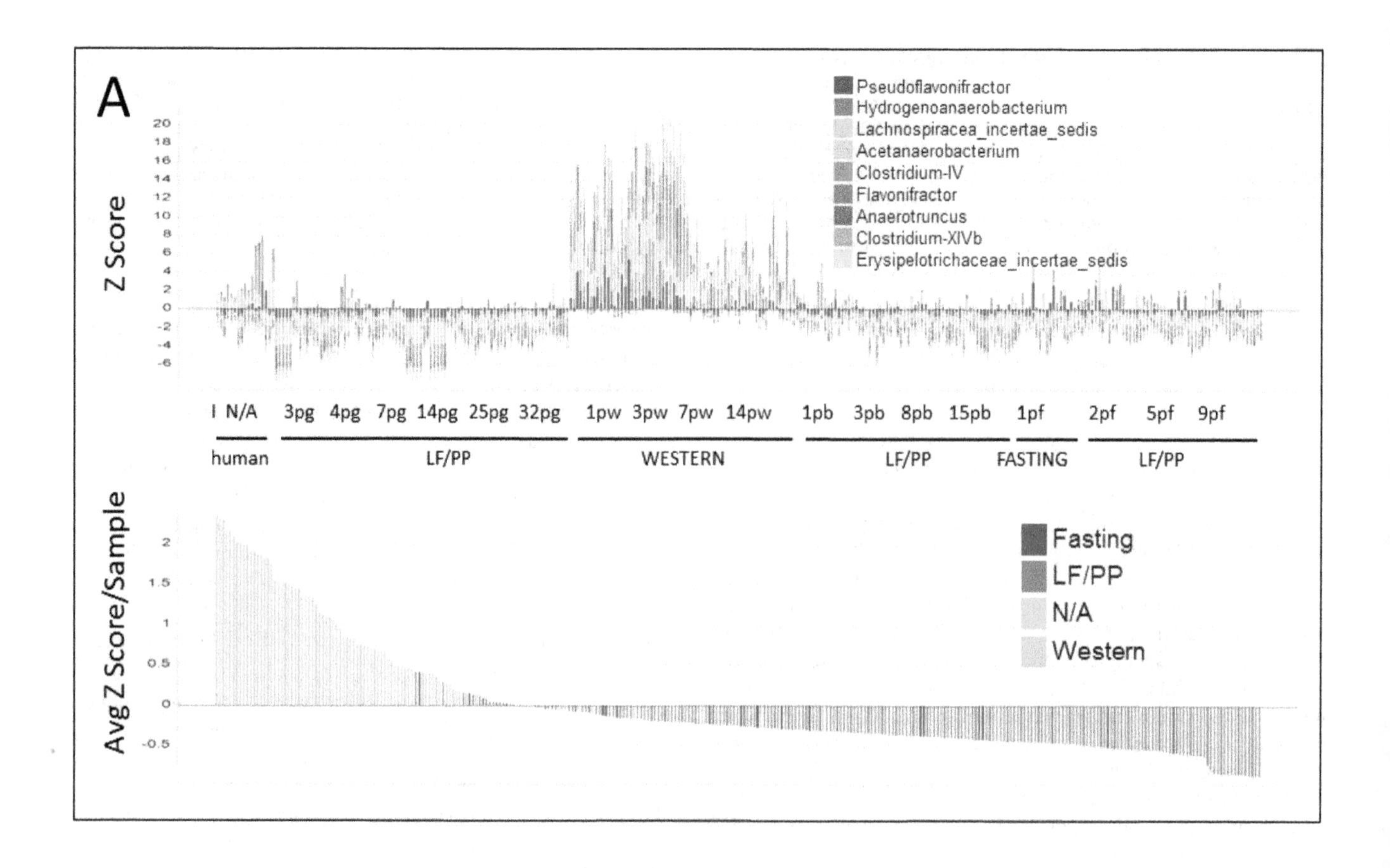

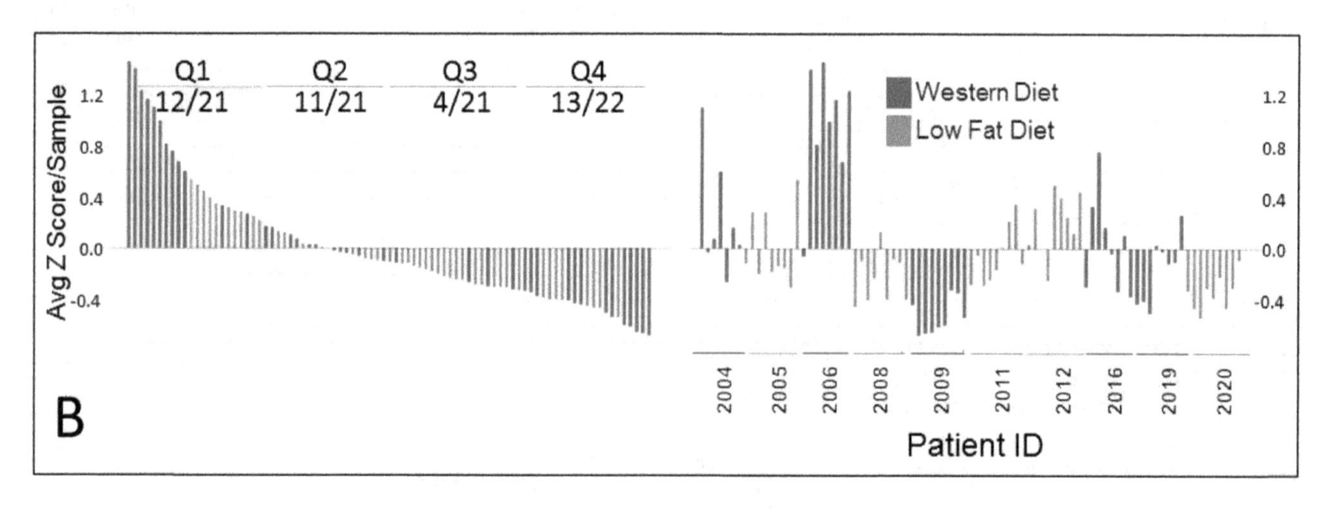

Figure 4. Bacterial abundance in humanized mice is a poor predictor of dietary intake in humans. A. Dynamic association of control 'abundance module' profile with dietary intake in the mouse. The top panel displays the Z score abundance of each of the genera indicated across time and dietary intake in the humanized mouse. The bottom panel shows the average Z scores for each fecal sample placed in order and coloured according to dietary intake of the mouse. **B**. The control 'abundance module' derived from humanized mice does not associate with diet in a controlled human study. Annotation is the same as Figure 3A.

term dietary habits and when it is perturbed for a short period using a diet intervention the core enterotype remains the same. Using the *Pseudoflavonifractor* and 'Western Abundance' modules as controls, the association of the *Bacteroides* module with dietary intake was determined. A total of 24 individual dietary factors from the two questionnaires correlated with the *Bacteroides* module average Z score with a p value <0.01 (Table 1). This compared with no significant correlations found in either the *Pseudoflavonifractor* or the 'Western Abundance' modules. From the 'Recall' questionnaire, positive correlation of the *Bacteroides* module was observed with 6 individual saturated fatty acids (SFA), total saturated fatty acids and percent calories from saturated fatty acids. Negative correlation was seen with the 'Polyunsaturated to Saturated fat ratio', as well as forms of the vegetable-enriched fat-soluble Vitamin E consistent with the emphasis of the Western diet on animal vs vegetable fats [12]. Again, related to the Western shift away from fruit and vegetable intake, negative correlations with the complex carbohydrate pectin, Vitamin C and dietary fiber were also observed consistent with the conclusion that the *Bacteroides*-associated bacterial module is a Western diet-associated dietary ecological module. These data demonstrate that an 'ecological' view of the microbiota can overcome the challenges associated with microbial plasticity and variation in individual populations. This suggests that a similar approach has potential to translate microbial ecology into the clinical setting.

The Bacteroides Module Correlates With Dietary Intake Independent Of Geographical Location

Having established the connection of the *Bacteroides* module with the Western diet in mice, a controlled Western diet study in humans and with recent dietary intake in a Western population, we sought to establish whether it also reflected the dietary habits in diverse geographical locations. To answer this we applied the concept to a recently published study that investigated the fecal microbiota composition across geography by comparing three populations; a Western population from the USA, a population of Guahibo Amerindians residing in Venezuela, and a third population from rural communities in Malawi [13]. The study was comprised of 531 individuals including healthy adults and children, geographical location and their dietary habits. The findings reported similar functional and microbial components in early life, maturing over the first three years of early life depending on local dietary habits. This study demonstrated the common theme of the dietary influence on a stable microbial 'enterotype' that distinguished USA volunteers from those of the developing and underdeveloped world.

The abundance of the *Bacteroides* module was lowest in the Malawian population correlating with a low fat/high fiber diet. However, no significant difference in the abundance of the *Bacteroides* module was found between the Venezuelan and USA populations (Figure 5A). The divergent segregation of the Venezuelan and Malawian samples based on the *Bacteroides* module was supported by the responses to the diet questionnaires contained in the supplementary data of the report [13]. The Venezuelan, but not the Malawian, family food intake included frequent consumption of 'Soda,' 'Milk,' 'Butter,' canned meat and fish, 'Cheese,' and other foods that are perhaps more commonly associated with a Western diet [13]. For example, one Malawi family (of 28 families in total) reported 'thobwa' (a local millet-based drink) and 'soda' use twice daily Indeed, blog references

Table 1. Diet Intake Correlation with Ecological Module: RECALL and FFQ Questionnaires [3].

	BACTEROIDES		*HIGH ABUNDANCE*		*PSEUDOFLAV-ONIFRACTOR*		
Dietary Component	***p-Value***	***r***	***p-Value***	***r***	***p-Value***	***r***	***Quest.***
Vitamin.E.IU.	3.93E-04	−0.3511	8.29E-02	−0.1761	8.44E-01	0.0201	RECALL
Vitamin.E.mg.	4.57E-04	−0.3474	5.91E-02	−0.1913	9.78E-01	0.0029	RECALL
Total.Alpha.Tocopherol.Equivalents	4.67E-04	−0.3469	1.12E-01	−0.1617	7.48E-01	0.0329	RECALL
Phytic.Acid	6.38E-04	−0.3390	1.27E-01	−0.1550	8.00E-01	−0.0260	RECALL
Polyunsaturated.to.Saturated.Fat.Ratio	6.59E-04	−0.3382	7.24E-01	−0.0362	1.82E-02	0.2382	RECALL
Pectins	1.05E-03	−0.3260	2.77E-01	−0.1110	4.62E-01	0.0751	RECALL
Vitamin.C	1.14E-03	−0.3239	1.17E-02	−0.2536	7.12E-01	0.0378	RECALL
Total.Saturated.Fatty.Acids.SFA	1.18E-03	0.3231	3.42E-01	0.0969	4.03E-01	−0.0854	RECALL
Percent.Calories.from.SFA	1.26E-03	0.3212	6.49E-01	0.0466	1.15E-01	−0.1603	RECALL
SFA.myristic.acid	1.73E-03	0.3126	1.23E-01	0.1569	3.91E-01	−0.0877	RECALL
SFA.butyric.acid	2.42E-03	0.3030	2.67E-01	0.1131	6.68E-01	−0.0439	RECALL
glu (glucose)	2.60E-03	−0.3010	3.87E-01	0.0883	4.96E-01	−0.0696	FFQ
Insoluble.Dietary.Fiber	2.79E-03	−0.2989	2.00E-01	−0.1307	9.53E-01	0.0060	RECALL
SFA.caproic.acid	3.15E-03	0.2954	3.00E-01	0.1057	6.38E-01	−0.0482	RECALL
SFA.stearic.acid	3.99E-03	0.2883	6.97E-01	0.0398	2.51E-01	−0.1171	RECALL
Synthetic.Alpha.Tocopherol	3.99E-03	−0.2883	4.15E-02	−0.2063	5.78E-01	−0.0569	RECALL
Total.Dietary.Fiber	4.34E-03	−0.2858	1.59E-01	−0.1432	9.35E-01	0.0084	RECALL
SFA.palmitic.acid	5.03E-03	0.2812	7.01E-01	0.0392	6.31E-01	−0.0492	RECALL
f140 (Myristic fatty acid)	6.15E-03	0.2749	6.14E-01	0.0516	2.46E-01	−0.1183	FFQ
Natural.Alpha.Tocopherol	7.65E-03	−0.2679	9.91E-02	−0.1675	8.55E-01	−0.0187	RECALL
fruct (fructose)	7.87E-03	−0.2670	2.97E-01	0.1064	3.60E-01	−0.0935	FFQ
SFA.capric.acid	8.06E-03	0.2662	2.11E-01	0.1275	3.63E-01	−0.0929	RECALL
t161 (Palmitoleic fatty acid)	8.11E-03	0.2660	7.83E-01	0.0282	4.80E-01	−0.0722	FFQ
Total.Conjugated.Linoleic.Acid.CLA.18.2	8.16E-03	0.2658	9.58E-01	−0.0054	1.83E-01	−0.1357	RECALL

Significant ($p<0.01$) Spearman correlations of '*Bacteroides*' module abundance (Average Z Score) with dietary intake as reported in COMBO. No significant correlations were found for either 'High Abundance' or '*Pseudoflavonifractor*' modules. See Table S3 for complete dataset.

were found of 'thobwa' being referred to locally as 'soda' (http://canteriointernational.org/blog4/2010/06/15/togba-malawi-local-soft-drink. Accessed 2013 July 1). This compared with 19/56 Venezuelan individuals reporting 'Soda' consumption at least once daily including 4/14 children less than 3 years old. Therefore, the epidemiology of the *Bacteroides* module correlated with Western dietary intake independently of geographical location.

The Bacteroides Module Correlates With Dietary Intake (Formula Vs Breast Feeding) In Children

Using the annotation provided with the geographical study, we investigated the association of the *Bacteroides* module with another dietary factor, namely the Western influence of formula vs breast-feeding in children. The development over time of the *Bacteroides* module was firstly investigated using the annotation provided for the USA vs Venezuela vs Malawi study [13]. Consistent with the conclusions of the original study, there was a significant association of the module with age independent of the country of origin likely resulting from the maturation of the gut flora in young children less than 3 years old (Figure S3). The use of formula or breast feeding was also annotated in the original study. Therefore, we examined the association of the *Bacteroides* module with breast feeding in the USA population during this critical developmental period [13]. Children who were formula fed vs breast fed had significantly elevated levels of the *Bacteroides* module (Figure 5B). The association remained significant when age-matched samples (all samples ≤0.8 years old) were included in the analysis from the breast-fed Malawi and Venezuelan children.

In the published annotation, all of the Malawian and Venezuelan children less than 3 years old were breastfed. Malawi has relatively high reported breastfeeding rates contrasting with Venezuela with one of the lowest in the world [14]. Therefore, we critically questioned our observation associating the *Bacteroides* module with formula feeding. In order to ensure that this observation was not an artifact of age, geography or errors in annotation, we therefore decided to extend our analysis to other studies investigating the impact of breast vs bottle feeding on the composition of the microbiota.

Several candidate human studies were investigated, however, the abundance of the component genera making up the *Bacteroides* module were generally rare or absent (due in part to the young age of the participants and in part a lack of depth in the coverage of the sequencing or variable formats such as DGGE or qPCR) making a comparable assessment of our observations in available human datasets impossible. We decided therefore to turn our attention to a preclinical study examining microbiota changes in response to a formula diet in piglets [10,15]. Here piglets were fed formula milk without access to breast milk and their gut microflora was contrasted with the control suckled animals. The data was processed identically as for the human studies determining an average Z score/sample. The highest abundance samples were all derived from the piglets on a formula diet and this was reflected by the significant difference in Z score between the formula fed vs milk fed piglet groups by ANOVA (Figure 5C). Therefore, in a controlled experimental preclinical setting as well as the human populations, the *Bacteroides* module associates with formula feeding. These data support the association of the module with dietary intake across organisms, age and geography.

Discussion

The conclusions from this study demonstrate several important principles. Firstly, preclinical models can be used to predict human biology responding to an often-leveled criticism of microbiota-based studies [15]. While arguments can be made regarding the best preclinical model to use to translate findings into human populations, the technical expertise, infrastructure, and resources surrounding genetically modified mice means they remain the preferred choice for basic and translational research. The conclusive data presented here provide a paradigm to identify biologically-relevant ecological modules in humans. It is reasonable to assume that the approach taken, in this case to investigate Western diet, could equally be applied to identifying bacterial ecology associated with aspects of host genetics, pharmaceutical treatment, drug metabolism and other applications, simplifying the complexities inherent in the analyses of 'top-down' approaches. However, while specific associations with dietary intake have been demonstrated here in the general population, it remains to be tested what metabolic function(s) the *Bacteroides* module might have in this context.

There has been extensive interest in the impact of globalization and its associated dietary influences with the increasing economic and human burden of chronic diseases (such as obesity and diabetes) in the developing world. Death rates from chronic diseases surpass those from infectious diseases in every continent of the world except for Africa [1]. We observed a similar pattern in the association of the *Bacteroides* module with dietary intake in geographical populations. Although our preliminary analysis did not detect any correlation with BMI or other annotated factors independent of diet, it will be interesting to determine whether the *Bacteroides* module also correlates with chronic disease or outcome. Such a link would provide a path from the epidemiology of diet and chronic diseases, microbial ecology in preclinical models progressing into human populations, and provide a clear strategy for identifying therapeutic interventions. In addition, a similar approach may also provide insight into other interventions outside of diet, for example, recovery of the gut flora following antibiotic or NSAID usage. It will also be interesting to identify the critical members of such self-sufficient networks and the degree of perturbation they can resist before the network collapses.

The impact of breast vs bottle feeding on human health is an important and emotive issue [16,17]. While there is general consensus that exclusive breastfeeding helps to prevent infectious diseases, obtaining a clear association with the risk of chronic diseases has been confounded by a number of factors although many studies have highlighted an association of breastfeeding with a decreased odds-ratio of childhood obesity as well as risk factors for cardiovascular diseases in adulthood [18,19]. We demonstrate that the *Bacteroides* module associated with a Western diet in adults is also associated with formula (vs breast) feeding in young children. In adults, there was a significant positive association of the module abundance with saturated fatty acids including linoleic and capric acid intake as well as the ratio of polyunsaturated fatty acids to saturated fatty acids (Table 1). These observations parallel the fatty acid composition over the course of lactation and provide a link between our observations with the *Bacteroides* module, the development of the microbiota and the impact of formula feeding [20,21]. The results of our study come in context of the continuing debate regarding dietary supplementation to infants, questions regarding the link of Western dietary habits to the rising prevalence of chronic diseases and the recently considered role of the microbiota underlying their association.

The concepts that we have used in this study are recognized standards in microbial ecology in other ecosystems such as fresh water lakes, oceans and soil. In these ecosystems, microbes are viewed not as singular life forms working in isolation and complete independence but as interacting and interdependent groups. We have applied these established principles to identify a dietary

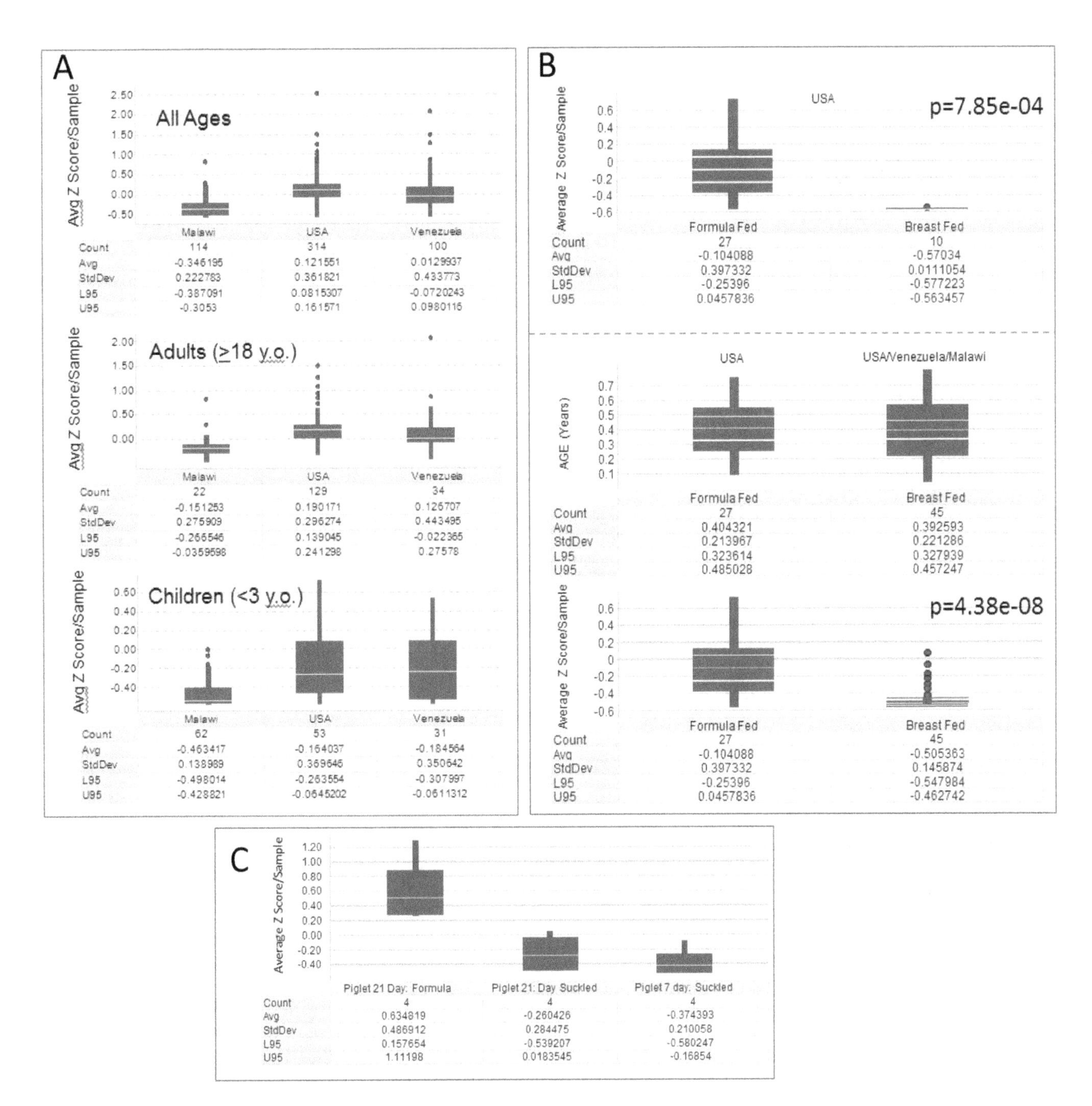

Figure 5. The Bacteroides module distinguishes geographical dietary intake and formula feeding in children and piglets. A. The *Bacteroides* module is endemic to all ages of the developed and developing world. The average Z score per fecal sample was calculated and combined to compare the three populations. The number of samples included with the given criteria (Count), average of the calculated Z scores/ sample (Avg), the standard deviation (StdDev), and the lower 95% and upper 95% confidence limits (L95 and U95, respectively) are displayed in the Tables. The top panel includes all available samples in the study, regardless of age annotation. The middle panel includes only adults with an annotated age 18 years old or more. The bottom panel includes children with an annotated age 3 years old or less. In all cases, the Malawi samples have significantly lower abundance of the *Bacteroides* module than those from the USA and Venezuela. **B.** The Western diet-associated *Bacteroides* module also associates with formula feeding in children. Average Z scores per sample were calculated as in A. The p-values for *Bacteroides* module Z score association with diet by ANOVA are indicated. Top panel: all samples from the USA population that were annotated for formula or breast feeding were included in the analysis. Middle panel: Distribution of ages of annotated breast vs formula fed samples. Age matched samples all ≤0.8 years old (as indicated by top box plot) were analyzed. Only USA samples were annotated as formula fed as indicated. Bottom panel: The association with formula feeding remained significant in age matched samples across populations. **C.** The *Bacteroides* module associates with formula feeding in piglets as well as humans. Piglets suckled for 7 days (right) were switched to formula for 14 days (left), or left to suckle an additional 14 days (middle). The average Z score for the *Bacteroides* module from each sample was plotted.

microbial module that can be translated from preclinical models to humans. Similar microbial correlations with dietary intake may also underlie the association of diet and the chronic disease epidemic and further studies will determine whether this is the case.

Supporting Information

Figure S1 Network map of LF/PP diet-specific bacterial interactions. The LF/PP diet-specific pairwise associations identified in Fig.1B (red box) were assembled to visualize the diet-associated ecologic module. Purple edges indicate positive Spearman correlations, blue indicate negative correlations.

Figure S2 The abundance of the *Barnesiella* cluster associates with human donor rather than diet. An average Z score of the *Barnesiella* cluster component genera was calculated for all fecal samples and arranged in order of abundance from left to right. Top panel: diet related annotation by colour. Bottom panel: donor related annotation as indicated.

Figure S3 The Bacteroides cluster abundance is developmentally regulated. The Spearman correlation of average Z score/sample with age is shown for the three populations studied. Rank ordering of age was performed with all three populations pooled for comparison. The equation of the linear correlation is shown with the r^2 correlation and p-value for each population.

Table S1 'Genus to Genus' Spearman correlations: Western diet vs LFPP. Summary table of calculated statistics of correlated pairwise genera abundance in either Western or LFPP diet-annotated samples from humanized gnotobiotic mice. For the analysis summarized in Fig.1B only pairwise associations with at least 50 Df in the LFPP diet and 16 Df in the Western Diet were considered.

Table S2 Genus abundance ANOVA Statistics: Western diet vs LFPP. Summary table of ANOVA calculated p-values identifying statistically different abundance of genera vs diet. Genera contained within Bacteroides, Pseudoflavonifractor and 'Western Abundance' clusters are indicated. N.S. Not Significant.

Table S3 Spearman correlations of *Bacteroides* cluster abundance (Avg Z Score) with dietary intake as reported in COMBO. All values recorded. Significant ($p<0.01$) associations shown in table 1. *Spearman correlation significant, however association is artifact of 0's in raw data.

Table S4 Original datasets, links and publications analyzed in this study.

Table S5 Summary file of Genus abundance with annotation from Gnotobiotic mouse study. Values shown are frequency per sample (count Genus/total counts).

Table S6 Summary file of Genus abundance with annotation from Human CAFE Diet Study. Both raw counts/sample (top) and Normalized values (frequency/sample: bottom) are shown.

Table S7 Summary file of Genus abundance with annotation from Human COMBO Dietary Intake Study. Both FFQ and Recall dietary annotation is included.

Table S8 Summary file of Genus abundance with annotation from Human Age and Geography study. Both raw counts/sample (top) and normalized values (frequency/sample: bottom) are shown.

Table S9 Summary file of Genus abundance with annotation from formula-fed piglets. Both raw counts/sample (top) and normalized values (frequency/sample: bottom) are shown.

Acknowledgments

We would like to thank the original authors of the analyzed studies for making the raw data and annotation publicly available and in a coherent format. The reference for each of the studies is noted and the accession numbers and location are recorded in the Supplementary Materials. Thanks to Sarah Westcott and Pat Schloss for their input related to our application of MOTHUR as well as Kyle Bittinger, Gary Wu, Rick Bushman and James Lewis for communication surrounding CAFE and COMBO. Thanks to Novartis colleagues Edward Oakeley, Gabi Hintzen and Guglielmo Roma for valuable discussions and Sebastian Bergling for preliminary guidance with analyses. The metanalysis approach was conceived by SJP and JS. Data processing was performed by JS and NH. Data analysis approach was conceived and performed by SJP. Structural and bioinformatics computing support was provided by NH.

Author Contributions

Conceived and designed the experiments: JS SJP. Analyzed the data: JS NH SJP. Contributed reagents/materials/analysis tools: JS NH. Wrote the paper: SJP.

References

1. World Health Organization (2003) Diet, Nutrition and the Prevention of Chronic Diseases: Report of a Joint WHO/FAO Expert Consultation. Geneva: World Health Organization.
2. Cummings JH, Macfarlane GT (1997) Role of intestinal bacteria in nutrient metabolism. *JPEN J* Parenter. Enteral. Nutr. 21: 357–365.
3. Wu GD, Chen J, Hoffmann C, Bittinger K, Chen YY, et al. (2011) Linking long-term dietary patterns with gut microbial enterotypes. Science 334: 105–108.
4. Arumugam M, Raes J, Pelletier E, Le Paslier D, Yamada T, et al. (2011) Enterotypes of the human gut microbiome. Nature 473: 174–180.
5. Goodman AL, Kallstrom G, Faith JJ, Reyes A, Moore A, et al. (2011) Extensive personal human gut microbiota culture collections characterized and manipulated in gnotobiotic mice. Proc. Natl. Acad. Sci. USA 108: 6252–6257.
6. Schloss PD, Westcott SL, Tyabin T, Hall JR, Hartmann M, et al. (2009) Introducing MOTHUR: open-source, platform-independent, community-supported software for describing and comparing microbial communities. Appl. Environ. Microbiol. 75: 7537–7541.
7. Schloss PD, Westcott SL (2011) Assessing and improving methods used in operational taxonomic unit-based approaches for 16S rRNA gene sequence analysis. Appl. Environ. Microbiol. 77: 3219–3226.
8. Little AEF, Robinson CJ, Peterson SB, Raffa KE, Handelsman J (2008).Rules of engagement: interspecies interactions that regulate microbial communities. Annu Rev Microbiol 62: 375–401.
9. Strom SL (2008) Microbial ecology of ocean biogeochemistry: a community perspective. Science 320: 1043–1045.
10. Poroyko V, Morowitz M, Bell T, Ulanov A, Wang M, et al. (2011) Diet creates metabolic niches in the "immature gut" that shape microbial communities. Nutr. Hosp. 26: 1283–1295.
11. Smith P, Siddharth J, Pearson R, Holway N, Shaxted M, et al. (2012) Host Genetics and Environmental Factors Regulate Ecological Succession of the Mouse Colon Tissue-Associated Microbiota. PLoS ONE 7: e30273.
12. McCance R and Widdowson E (2002) Fats and Oils. In: McCance and Widdowson's the composition of foods. London: Royal Society of Chemistry. pp.131–143.

13. Yatsunenko T, Rey FE, Manary MJ, Trehan I, Dominguez-Bello MG, et al. (2012) Human gut microbiome viewed across age and geography. Nature 486: 222–227.
14. UNICEF (2007) *The state of the world's children 2008: child survival.* (UNICEF Publication 978-92-806-4191-2, 2007; http://www.unicef.org/sowc08/docs/sowc08.pdf. Accessed 2013 July 1).
15. Hvistendahl M (2012) Pigs as stand-ins for microbiome studies. Science 336: 1250.
16. Gordon JI, Dewey KG, Mills DA, Medzhitov RM (2012) The human gut microbiota and undernutrition. Sci Transl Med. 4: 137ps12.
17. Fewtrell M, Wilson DC, Booth I, Lucas A (2011) When to wean? How good is the evidence for six months' exclusive breastfeeding. BMJ 342: c5955.
18. Koletzko B, von Kries R, Monasterolo RC (2009) Infant feeding and later obesity risk in *Early nutrition programming and health outcomes in later life: Obesity and beyond* (Springer, location 646, 2009), B.. Koletzko et al eds., 15–29.
19. Scientific Advisory Committee on Nutrition Subgroup on Maternal and Child Nutrition (SMCN) (2011) The influence of maternal, fetal and child nutrition on the development of chronic disease in later life, 2011 www.sacn.gov.uk. Accessed 2013 July 1.
20. Brenna JT, Varamini B, Jensen RG, Diersen-Schade DA, Boettcher JA, et al. (2007) Docosahexaenoic and arachidonic acid concentrations in human breast milk worldwide. Am J. Clin. Nutr. 85: 1457–1464.
21. Gibson RA, Kneebone GM (1981) Fatty acid composition of human colostrum and mature breast milk. Am.J.Clin.Nutr. 34: 252.

6

Much beyond Mantel: Bringing Procrustes Association Metric to the Plant and Soil Ecologist's Toolbox

Francy Junio Gonçalves Lisboa[1,4]*, Pedro R. Peres-Neto[2], Guilherme Montandon Chaer[3], Ederson da Conceição Jesus[3], Ruth Joy Mitchell[4], Stephen James Chapman[4], Ricardo Luis Louro Berbara[1]

1 Soil Science Department, Agronomy Institute, Federal Rural University of Rio de Janeiro, Seropédica-RJ, Brazil, **2** Canada Research Chair in Spatial Modelling and Biodiversity; Université du Québec à Montréal, Département des sciences biologiques, Québec, Canada, **3** Embrapa Agrobiologia, Seropédica-RJ, Brazil, **4** The James Hutton Institute, Craigiebuckler, Aberdeen, United Kingdom

Abstract

The correlation of multivariate data is a common task in investigations of soil biology and in ecology in general. Procrustes analysis and the Mantel test are two approaches that often meet this objective and are considered analogous in many situations especially when used as a statistical test to assess the statistical significance between multivariate data tables. Here we call the attention of ecologists to the advantages of a less familiar application of the Procrustean framework, namely the Procrustean association metric (a vector of Procrustean residuals). These residuals represent differences in fit between multivariate data tables regarding homologous observations (e.g., sampling sites) that can be used to estimate local levels of association (e.g., some groups of sites are more similar in their association between biotic and environmental features than other groups of sites). Given that in the Mantel framework, multivariate information is translated into a pairwise distance matrix, we lose the ability to contrast homologous data points across dimensions and data matrices after their fit. In this paper, we attempt to familiarize ecologists with the benefits of using these Procrustean residual differences to further gain insights about the processes underlying the association among multivariate data tables using real and hypothetical examples.

Editor: Andrew R. Dalby, University of Westminster, United Kingdom

Funding: This work was carried out with the aid of a grant from Conselho Nacional de Desenvolvimento Científico e Tecnológico (CNPq - 563304/2010-3 and 562955/2010-0), Fundação de Amparo à Pesquisa de Minas Gerais (FAPEMIG CRA - APQ-00001-11), and the Inter-American Institute for Global Change Research (IAI-CRN II-021). Francy Lisboa greatly acknowledges a research scholarship from CAPES/EMBRAPA (Carbioma) and Dr. Beata Madari. The funders had no role in study design, data collection and analysis, decision to publish, or preparation of the manuscript.

Competing Interests: The authors have declared that no competing interests exist.

* Email: agrolisboa@gmail.com

Introduction

In multidimensional data analysis, ecologists often encounter situations where they need to choose between two or more numerical approaches that are able to tackle the same question of interest. The preference between approaches is based, among other factors, on the familiarity of the user with the method, which in turn depends on the time a particular method has been available in statistical packages and the ease in implementing and interpreting its results. Another relevant factor to consider is "literature–induced use" in which renowned research groups involved in the development, improvement and generation of statistical ecological approaches have a strong influence on the types of statistical approaches other ecologists use.

Determining the strength of the relationships between multivariate datasets is a routine analysis when trying to understand the environmental factors driving the composition and structure of ecological communities. Two approaches, the Mantel test [1] and Procrustes analysis [2], though considered analogous by the literature in the questions they can tackle [3], have not been used to the same extent. Despite the advantages of Procrustes analysis over the Mantel test [3] regarding greater statistical power in detecting significant relationships (i.e., lower type II errors) and the possibility of analyzing further the patterns of association between multivariate matrices (visually and by further statistical analyses), the Procrustean approach remains relatively unused in tackling questions regarding the relationships between data matrices involving plant and soil information or between soil matrices (Fig. 1).

The Mantel test and the Procrustes approach can be both used in many similar situations where the aim is to assess how multivariate data matrices are associated (correlated), though for unknown reasons they have been used in quite different ways in the ecological literature. For example, while the Mantel test has often been applied when testing the relationship between above and below ground data matrices [4], [5], [6], [7], [8], [9], [10], [11], [12], Procrustes analysis has predominantly been used to contrast the results of different ecological ordinations on the same data [13], [14], [15], [16], to compare fingerprinting tools for assessing microbial communities [17], [18], [19] and for deciding between methodological choices [20], [21]. Indeed the Procrustean framework has been rarely used to make inferences about plant and soil relationships [22], [23], [24], [25], [26] and other types of ecological associations between data sets. However there are instances in which the Procrustean and Mantel tests cannot be used interchangeably. Unlike Mantel, the Procrustean approach can be used to compare multiple data matrices. However, when ecologists are interested in correlating distance (or similarity)

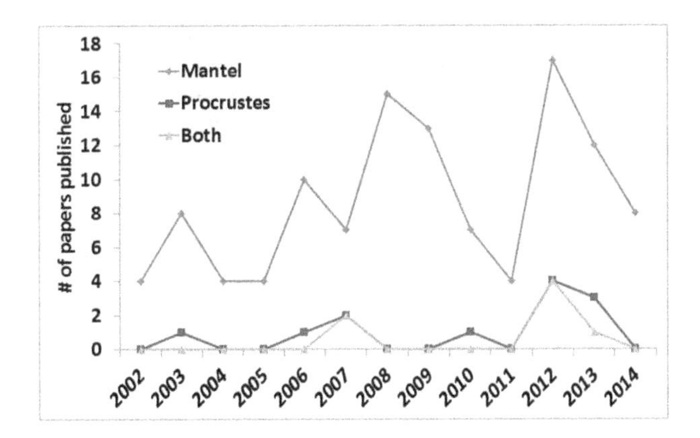

Figure 1. Papers published using Mantel and Procrustes for relating data matrices from soil or plant studies in the ten years since [3] stated the advantages of Procrustes over the Mantel approach. Data obtained using Thompson Reuters database (May, 12, 2014). We searched for papers using uniquely the Mantel approach, uniquely the Procrustes approach and papers using both approaches. The search was based on Procrust* (Procrustean or Procrustes) and PROTEST.

matrices, rather than testing the association among data matrices in their raw form (i.e., not transformed by the property of distance measures), Mantel, rather than Procrustes, is more appropriate. One particular case is the distance-decay of similarity in ecological communities [27] in which one is interested in testing the hypothesis that the similarity in community composition decreases in relation to linear (or log transformed) geographic distance between communities. The differences between raw-based and distance-based approaches have been discussed extensively elsewhere [28], [29].

Despite the relative merits of the Procrustean framework over the Mantel test shown by the relatively well-cited paper by Peres-Neto and Jackson [3], its potential has not yet been tapped. Perhaps the reason for Procrustes analysis not being as popular as the Mantel test among ecologists is the lack of a paper showing that in many situations traditionally investigated by Mantel, the Procrustean analysis can be equally well used. Here, we attempt to familiarize ecologists with the use of Procrustes analysis by using real and hypothetical examples where the Mantel test tends to be preferred. Most importantly, we highlight little explored limits of Procrustes by using its residual vector of association between data tables, hereafter referred as to PAM, in three common statistical approaches: multivariate ordination, variation partitioning and ANOVA.

Procrustes analysis: a foundation for soil and plant ecologists

In ancient Greek mythology there was a character named Procrustes who was a resident of Eleusis Mountain, a known travelers' route. As a "good" host, Procrustes always invited travelers to spend the night at his home; more specifically, he invited them to lie down on his iron bed, which was tailored to fit Procrustes' own body. The guests who did not fit the dimensions of his bed either had their limbs cut off or were stretched until their dimensions approached those of Procrustes's bed. Ironically, none of the guests ever fitted the iron bed because Procrustes secretly had two beds of different sizes [30]. One can easily make a parallel here with ecological data in which data from different sources will almost never easily compare or fit to one another.

Procrustes analysis is based on the search for the best fit between two data tables, hereafter referred to as matrices, where one is kept fixed ("Procrustes' bed" or target matrix), while the other ("Procrustes' guest" or rotated matrix) undergoes a series of transformations (translation, mirror reflection and rotation; [2]) to fit the fixed matrix. Although in this paper we concentrate on fitting two matrices, the extension of Procrustes analysis to multiple matrices is straightforward [3] in which the reference matrix can be either one of the original matrices or their averages (or medians). Hereafter, the target matrix (target) will be referred as to **X**, and the data matrix to be fitted as **Y**. **X** and **Y** are both $n \times p$ matrices, where n is the number of rows and p is the number of columns. The goal of the transformations in **Y** is to minimize the residual sum of squared differences between the corresponding n dimensions between **X** and **Y**; the sum of the squares of these residual differences is termed m^2 (Gower's statistic), representing the optimal fit between the two data matrices, such that the higher the value of m^2, the weaker the relationship between the two data tables is. The significance of m^2 can be estimated through a permutation test (termed PROTEST after [31]; see [3] for further details).

Procrustean association metric (PAM)

The least squares superimposition between the corresponding n observations of **X** and **Y** is one of the main advantages (in addition to the increased statistical power) of the Procrustean framework in contrast to the Mantel test. The Procrustes superimposition generates a ($n \times p$) matrix of residuals that can be further used to contrast the differences between homologous observations (rows) across matrices in the form of a vector (PAM). Given that within the Mantel approach differences between observations across all dimensions are packed down into a single distance, it cannot be used to assess differences across observations across dimensions. Consistent small and large differences across homologous observations across matrices in regard to other factors of interest can further assist in understanding how **X** and **Y** are related. For example, we could use PAM to assess the degree of observation matching between a plant function trait matrix and a composition matrix and assess whether smaller or greater residual values are a function of the time elapsed since some disturbance event.

PAM is simply a vector of residual differences between the corresponding n observations. For example, assuming that an ecologist wants to correlate two matrices of data **X** and **Y**, both of which are formed by four rows (i.e. sites, plots, observational units), Procrustes analysis will generate four residual differences between the **X** and **Y** configurations. The compilation of these residual differences between homologous rows (observations) across dimensions in the form of a vector – PAM – represents a useful way to represent information on the relationship between two matrices and make it available for further statistical analysis, both parametric and non-parametric; this feature is not offered by the Mantel approach.

The use of the residual vector from Procrustes (PAM) has been quite restricted in the plant and soil ecological literature. To our knowledge the first study was by [32] who assessed the plant-pollinator interaction during three consecutive summers in the southeastern portion of California, USA. These authors employed the PAM to identify which pollinating species exhibited the greatest deviation between two consecutive years. Singh et al. [22] used the PAM in a study on soil microbiology to verify the effect of soil pH on the relationship between arbuscular mycorrhizal fungi (AMF) and plant assemblages. These authors employed the following strategy: 1) Procrustes analysis was applied between the matrices representing the AMF community and that representing the plant community; 2) after detecting a significant relationship ($m_{12} = 0.28$; $P < 0.001$), these authors extracted the PAM and used it as a response in a simple regression analysis with the soil pH. No effect of pH on the association between the AMF and plant communities was detected, suggesting that neither the pH nor the identity of the plant species that composed the community affected the AMF community. Other applications can be certainly found (e.g., [24], [25], [26][33]) but its flexibility and general usage remains largely unexplored.

Constructing a practical roadmap for applying PAM

There are few studies in the ecological literature that have used PAM for analyzing relationships between plant and soil datasets. The lack of examples partially explains the low popularity of Procrustes analysis among plant and soil ecologists and ecologists in general as an alternative tool to the more traditional Mantel test. In order to make the possible uses of Procrustean residuals more familiar, we will introduce a number of examples in the form of schematic roadmaps for applying PAM in association with three common statistical approaches: ordination, regression analysis and ANOVA.

Plant and soil ecologists must keep in mind that Procrustes analysis requires that the **X** and **Y** have the same number of rows and columns, though the last dimension is less restricting (see below). Given that the data for both matrices usually originate from the same sites, it is most common in ecology that only the number of columns (descriptors or variables) varies between the two matrices. Therefore, the question arises of how to make the number of columns equal across the two matrices, i.e., how to reduce them to the same dimensionality. Although Procrustes analysis can be performed between matrices having different number of dimensions (i.e., the fit is based on a singular value decomposition (svd) of X^TY, where X and Y are scaled prior to svd and T stands for matrix transpose), traditionally the matrix with the fewer number of columns ("missing columns") is made equal in dimension to the larger matrix by adding columns of zeros in order to keep (Fig. 2a; [2]). Although there are some criticisms related to this practice and alternatives have been suggested [34], the addition of zero columns does not affect the distances between columns among observations and is a convenient device rather than a hurdle [35].

Another convenient way to make **X** and **Y** have the same number of columns is to represent most of the variation in their raw data by matrices formed by the same number of orthogonal axes (Fig. 2b; [3],[35, [36]), i.e., matrices formed by axes derived through ordination methods such as Principal Components Analysis (PCA), Non-metric dimensional scaling, Correspondence Analysis (CA), Principal Coordinate Analysis (PCoA), the choice being dependent on the nature of the data (continuous, presence-absence data, abundance data). Moreover, raw data matrices can be transformed prior to ordination (see [37] for different transformations and their characteristics) or alternatively have pairwise distance matrices calculated from the data matrices that are then orthogonolized via PCoA to extract ordination axes based on the chosen distance measure (e.g., Bray-Curtis, Jaccard, Sorensen, Gower).

Here, for simplicity, we use a PCA in all applications. In cases, where species data (presence/absence or abundance) was used, the data was Hellinger-transformed and PCAs were extracted on species correlation matrix calculated from the transformed data. The Hellinger transformation alleviates the issue of double-zeros in species data matrix transformed into correlation or Euclidean-distance pairwise matrices prior to PCA in which sites sharing no species in common can be found to be more similar than sites sharing a reduced number of species in common (e.g., the horse shoe effect in ordination plots).

The general strategy is as follows:

1) Subject the raw data matrices to an ordination method (here PCA but see above for other strategies);
2) After ordinating **X** and **Y**, use the same number of ordination axes for both matrices (Fig. 2b).

Given that the higher the number of ordination axes used, the higher is the amount of variation explained in **X** and **Y**, it would be interesting run the Procrustean analysis sequentially using matrices made up of an increasing number of ordination axes. It could help ecologists check the consistency of the relationship between **X** and **Y** based on different numbers of ordination axes, which will give more reliability to the results.

The use of PAM in ecological ordination

The first form of PAM shown here is based on ordination methods. Ordination is the graphical representation of the variation of objects (sites), descriptors (species/environmental parameters) or both, in a reduced space formed by orthogonal axes [38].

To illustrate the use of Procrustes analysis associated with ordination we use data derived from Mitchell et al. [39]. This study aimed to compare the plant communities and soil chemistry in their ability to predict changes in the structure of the soil microbial community in three moorland areas established in Northern Scotland called Craggan, Kerrow and Tulchan. The plant community matrices from each area were based on the percent cover. Three matrices for the soil microbial community were obtained for each site: one based on the fatty acid profile of the soil (PLFA analysis), and the other two on the T-RFLP analysis of the communities of fungi and bacteria, respectively. The matrix representing the soil chemistry was based on the concentrations of Na, K, Ca, Mg, Fe, Al, P, total C, total N in addition to pH, loss on ignition and moisture.

There is some consensus that the variation in vegetation can act as a proxy for changes in the soil microbial community, either directly in the case of symbionts, for example, or indirectly via changes in soil chemistry itself. We use Procrustes analysis associated with ordination techniques to verify potential drivers of the soil microbial community and to determine if plant community and soil chemistry are equally related to the microbiological variation. The sequence of analysis was as follows:

1) Ordination analysis: All data matrices (community plant, soil chemistry and soil microbial communities) containing the three chronosequences were subjected to separate PCAs based on correlation matrix. The community plant was Hellinger-transformed prior to PCA. Then, the first six PCA ordination axes from each matrix were retained in order to assemble four PCA matrices representing the variation summarized in the first 3, 4, 5 and 6 PCA axes. Thus, four PCA matrices were obtained from each dataset: plant community, soil chemistry and soil microbial community (PLFA, bacterial and fungal T-RFLP) (Fig. 3a).
2) Procrustes analysis: The PCA matrices of plant community and soil chemistry were used to run Procrustean analyses with the PCA matrices of soil microbial community based on PLFA, and fungal and bacterial T-RFLP datasets.

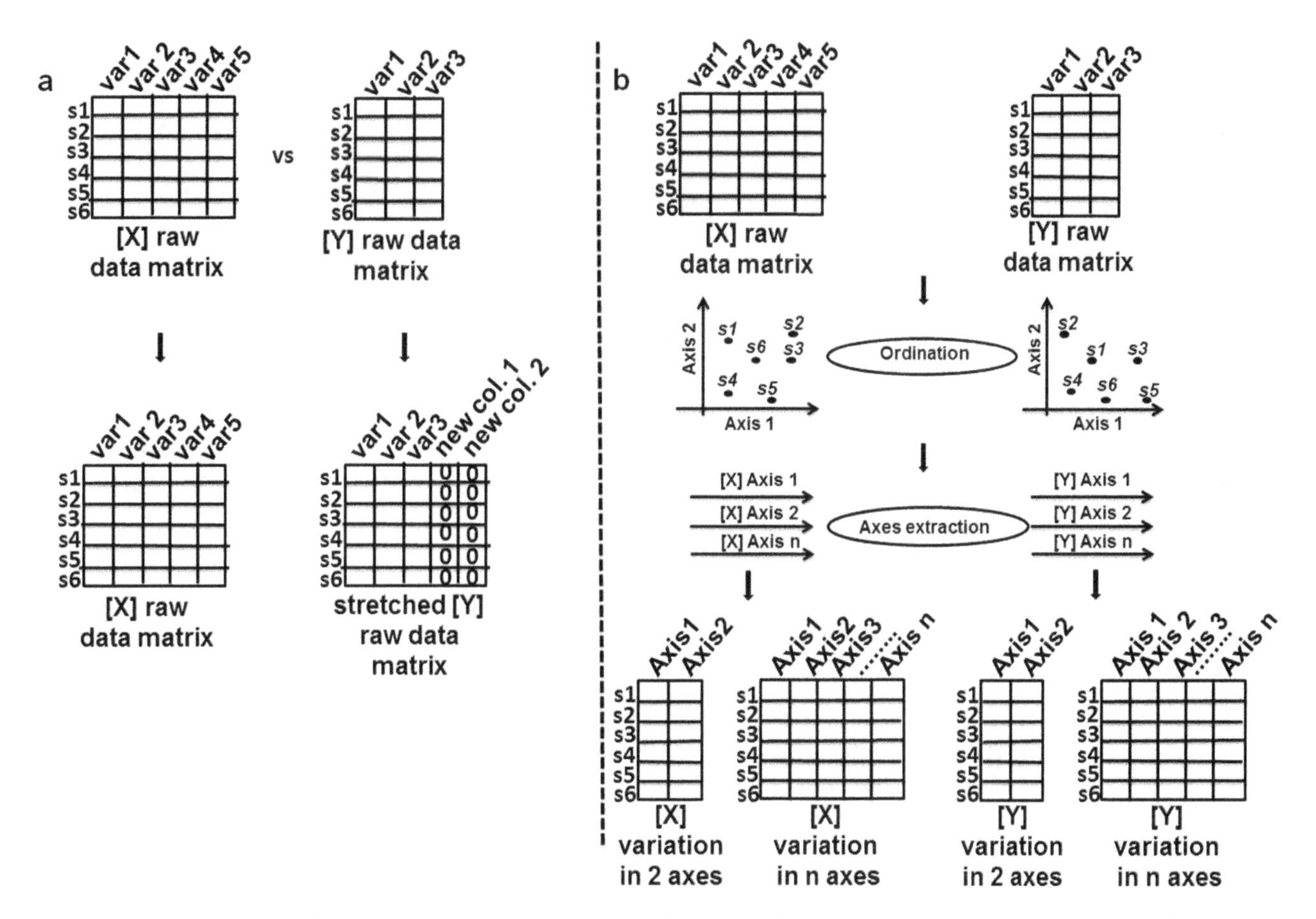

Figure 2. Roadmap for two alternative ways to reach the same dimensionality between matrices, and so relating it by Procrustes analysis. a) Addition of columns containing zeros to the **Y** raw data matrix for matching the **X** raw data matrix dimension; **b)** Application of ordination to raw data matrices to make matrices have equal dimensionality prior to Procrustes analysis.

3) PAM extraction: Since all Procrustean relationships based on PCA matrices with *n* axes were significant, for simplicity, only the PAM obtained from relationships of PCA matrices with 6 axes were used for subsequent analyses. Six PAMs were generated: PAM1 (soil chemistry on PLFA), PAM2 (soil chemistry on bacteria), PAM3 (soil chemistry on fungi), PAM4 (plant on PLFA), PAM5 (plant on bacteria), and PAM6 (plant on fungi) (Fig. 3c).
4) PAM ordination: The PAMs were assembled in a single matrix ("effect matrix") with one PAM per row (Fig. 3c). Therefore, the effect matrix compiled the effects of plant community and soil chemistry on soil microbial community structure derived from the three methods. This effect matrix was submitted to PCA ordination to verify whether the plant community effect on soil microbial community structure differed from the effect of soil chemistry (Fig. 3c).

The results showed that for all chronosequences the plant effect on microbial structure was divergent in relation to the soil chemistry effect, as suggested by the separation along the axis of greatest variation (Fig. 4). Although we cannot apply a proper statistical significance test in one-table based ordination methods (PCA, NMDS, PCoA, etc), visual inferences can be made. For example the Craggan area exhibited a clear distortion between plant community and soil chemistry variation in terms of their effects on the soil microbial community structure depicted by PLFA, bacterial T-RFLP and fungal T-RFLP (Fig. 4a). Also in this area, the response of the microbial community based on PLFA was distant from the response based on molecular data (T-RFLP) (Fig. 4a),

We expected that the effects of soil chemistry on microbial structure were closer to the effect of plant community once the plant community is considered to be a direct and indirect driver for the biotic component of soil [39]. However, these results suggest that plant communities and soil chemistry are acting differently on the soil microbial community structure [24], [40]. They also suggest that the effects of soil chemical properties on the microbial communities may be weakly mediated by above ground alterations [24]. This example shows the usefulness of Procrustes analysis to raise additional evidence in plant and soil ecology studies. (See Text S1 containing the R code used for this example).

The PAM and regression analysis

In regression analysis, 'response' and 'predictor' are common terms. In ecology, predictors can have different natures. Space, time, organic matter and moisture, among other factors, are some examples of predictors. On the other hand the microbial communities are often used as a response variable because they are considered better indicators of a given ecosystem.

Some authors familiar with soil microbial ecology have been using the Mantel test to assess the individual contribution of deterministic and stochastic processes on the soil microbial structure variation [41], [42]. As an example of the utility of the

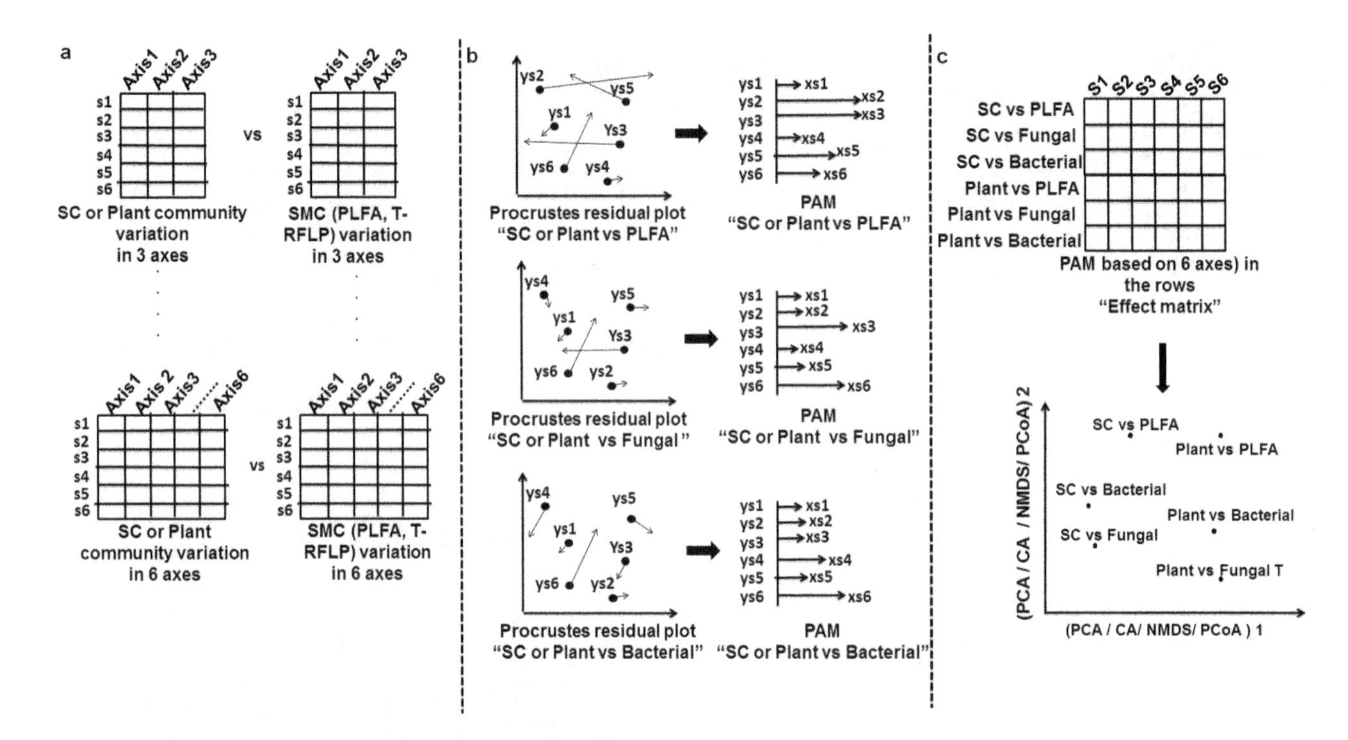

Figure 3. Roadmap for applying the Procrustes association metric (PAM) in the multivariate ordination context using data of [39]. **a)** Assembling matrices with different ordination axes, through Procrustes analysis, soil chemistry (SC) and plant community with soil microbial community (PLFA, and bacterial and fungal T-RFLP); **b)** Extraction of PAM from Procrustean relationships based on matrices with 6 ordination axes; **c)** Assembling of PAM based PCA matrices with 6 axes as rows in a single matrix ("effect matrix"), and using it in an ordination technique (e.g., PCA, PCoA, NMDS) to verify if the different effects diverge.

Procrustes analysis in the context of variation partitioning we can take a hypothetical scenario with four datasets from a given area, corresponding to soil microbial community structure (PLFA), soil microbial functioning (enzyme activities), soil properties and spatial variation. Spatial variation can be represented, for example, by 100 sampling points generated from a 10 m×10 m transects. The matrix of geographical coordinates of the sampling points can be submitted to PCNM (principal coordinates neighbour matrix) analysis generating a matrix of spatial eigenfunctions termed PCNMs [34]. In this scenario, we can assume that the ecologist aims to assess the relative contributions of individual soil properties (deterministic processes) and spatial variation (stochastic event) on the relationship between microbial community structure and soil microbial functioning rather than on these components individually. To use the Procrustean association metric (PAM) in this context, one can use the following steps:

1) Ordinate the two matrices (i.e., the soil microbial community and soil microbial functioning) via PCA (the soil microbial community matrix was Hellinger-transformed) and select a similar number of ordination axes. The multivariate scores of the two matrices across the selected number of axes are subjected to a Procrustes analysis and a PAM was then calculated.
2) Use individual PAMs (based on 2, 3 or more PCA axes) as response variable and soil properties and spatial variation as independent (predictor) variables in a multiple regression framework (Fig. 5b).

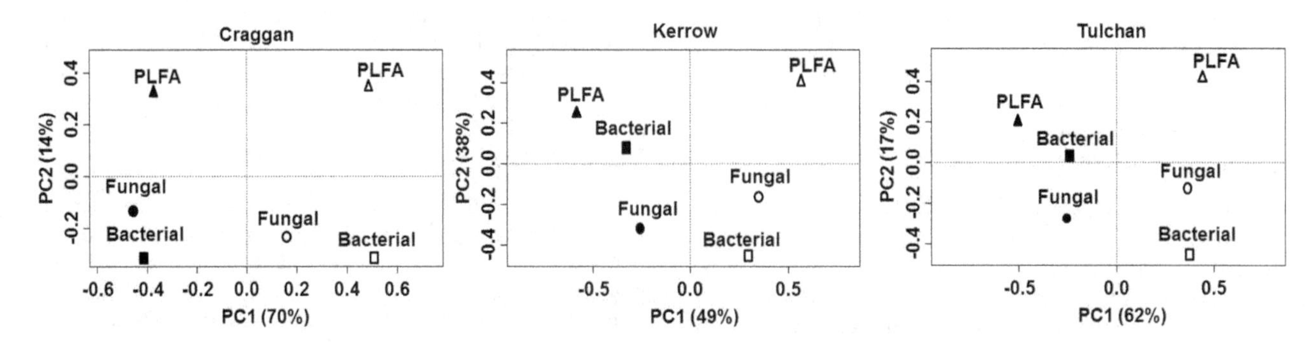

Figure 4. Results from PCA ordination of the Procrustes association metric matrix ("effect matrix") gathering the interactions of soil chemistry and plant community with soil microbial matrices (PLFA, and bacterial and fungal T-RFLP). The filled symbols are the Procrustes relationships between soil chemistry and soil microbial matrices, and the open symbols between plant community and soil microbial matrices. Data from three chronosequences (Craggan, Kerrow and Tulchan) obtained by [39].

3) Finally, the independent contributions of soil properties (independent of space) and unmeasured spatial process and/or factors (spatial variation independent of soil properties) to the microbial structure can be estimated via variation partitioning [43] and represented by a Venn diagram (Fig. 5c). (See Text S2 containing the R code for this example).

The PAM and Analysis of Variance

Although regression and analysis of variance are ultimately the same analysis in which the response is either continuous (regression) or ascribed to factors (ANOVA), we provide examples for each of them in different sections given that often they are seen as distinct forms of analyses. Evaluation of the effects of land use on soil microbial communities has been a common case-study issue in soil ecology. Some of these studies have been carried out using the Mantel approach to assess how land use type effects soil microbial structure and functioning [44], [45]. However, Mantel does not yield a vector of structure – functioning relationship, that is, a continuous variable, able to be partitioned by categorical variables like land use types. In the following example we show how to use PAM to evaluate the effect of land use type on the relationship between microbial community structure and microbial function in the form of PAM.

In a hypothetical scenario, a researcher is interested in studying whether four different land use types within the Amazon biome are affecting the relationship between microbial structure and microbial functioning. In each of the land uses (original forest fragment, silvipastoral system, improved pasture and unimproved pasture) six plots (10 m×10 m) were established and one composite soil sample (0–10 cm) collected per plot (Fig. 6a). The X dataset (soil microbial structure) was represented by PLFA data, and the Y dataset (microbial functioning) by the abundance of genes associated with microorganisms involved in greenhouse gas emission processes, such as nitrifiers, denitrifiers and methanotrophic organisms. The researcher's hypothesis is that in the original forest (non-altered environment) there is a better matching between microbial structure and microbial function. Thus, in anthropogenically disturbed environments (silvipastoral system, improved pasture, and unimproved pasture) the change in microbial structure relative to the original (forest) is not followed by a change in the microbial functioning to the same magnitude. This hypothesis can be tested using an integration of Procrustes analysis and ANOVA through the following steps:

1) Reduce the datasets **X** (soil microbial structure) and **Y** (soil microbial functioning) to similar dimensions using PCA. Then, run the Procrustean analysis between the PCA matrix of the soil microbial community structure and the PCA matrix of soil microbial functioning and extract the PAM (Fig. 6a).
2) Run an ANOVA with land use type as fixed factor and the PAM as the response variable (Fig. 6b).
3) If the F value of ANOVA is significant, a means test can be performed to compare the mean PAMs of the land use types (Fig. 6c). (See Text S3 containing the R code for this example).

Discussion

In this paper we have attempted to show the advantages of the Procrustean analysis over the Mantel test, in which the former can be used for gaining further information on underlying drivers of data table associations. In particular we have shown how the Procrustean association metric (PAM) constructed of the residuals

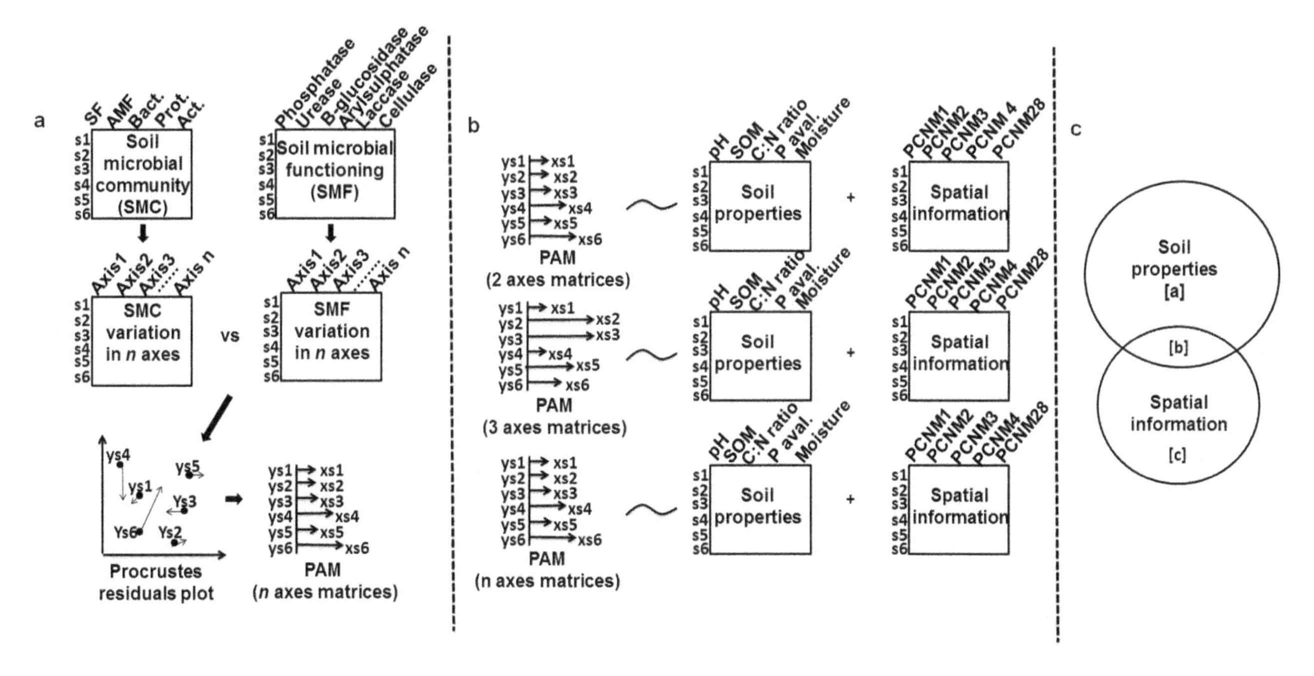

Figure 5. Roadmap for using Procrustes Association Metric (PAM) in a multiple regression analysis framework (variation partitioning). a) Soil microbial community (SMC) and soil microbial functioning (SMF) matrices are submitted to an ordination to reach the same dimensionality, and SMC and SMF matrices formed by 2, 3 and *n* axes related through Procrustes analysis in order to generate PAMs; **b)** PAMs generated were used as response variables in a variation partitioning to verify the individual contribution of soil properties and spatial information (PCNM eigenfunctions) on the SMC-SMF relationship; **c)** Venn diagram depicting the relative contribution of soil properties (niche processes **[a]**) and unmeasured spatial factors (neutral processes **[c]**).

of the vectors after the Procrustes analysis. We concentrated on showing how patterns of concordance between data matrices can be displayed and individual observations contrasted separately using the Procrustean framework, allowing further examination of the common and different association patterns among multiple data matrices. Given that in the Mantel framework, multivariate information is translated into a pairwise distance matrix, we lose the ability to contrast homologous data points across dimensions and data matrices. It is important to notice that it was not our goal to show the statistical advantages of Procrustes over Mantel as done by previous work [3]. Instead, we concentrated on generating different analytical schemes, especially for plant and soil ecologists, to incorporate Procrustes into their statistical toolbox.

What is unique about Procrustean framework? There are at least four characteristics of the approach not shared by others. First, because the approach is correlative rather than regressive, the number of observations (e.g., sites) in the matrices does not have to be greater than the number of columns as in common regression approaches such as RDA and CCA. Second, we can fit as many matrices as we have available; this latter issue is particularly restrictive under a regression approach given the limitation of number of rows versus number of columns. Moreover, all matrices are treated in equal footing as no matrix is treated as response or predictor. Third, the relationships within (only across) matrix columns do not affect the analysis. Fourth, residual values across observations and dimensions can be calculated and explored as shown here. These characteristics should not be necessarily seen as advantages per se over other methods but rather features that are unique and may be useful in many situations. There are certainly other tools that can be used to look at the associations between data sets. RDA and CCA are well-established tools in ecology and are based on regression (asymmetric) methods. Traditionally these approaches may have been thought to be more appropriate for analysis of the examples given in this paper, since they establish relations of cause and effect. However, because these analyses include a regression step, they are limited to situations where the number of rows (sites) in the environmental matrix **X** is higher than the number of columns (variables) [36], [46]. This is not a limitation in Procrustes analysis and moreover, it is not clear how residual variation among homologous observations across dimensions should be explored in the case of RDA and CCA.

At least two other symmetric approaches are similar to the Procrustean approach, namely Co-inertia analysis [47] and symmetric Co-correspondence analysis [48], a form of Co-inertia analysis in which a correspondence analysis is applied to two species matrices prior to the analysis. The main difference resides in the fact that fit is influenced by all variables pairs in Co-inertia analysis (within and between matrices), whereas fit is influenced only by variation between matrices in Procrustean. Co-inertia is always based on ordination within data matrices, whereas in Procrustes either the raw data or their ordination axes can be used. Co-inertia can also take into account row (e.g., sites) and column (e.g., species) weights in the analysis, though the standardization and fit processes in Procrustean analysis could also take these into account [36]. Co-inertia and Procrustean analysis are certainly related in the sense that they both treat matrices as symmetrical during the fitting process, though more studies are necessary to assess in which conditions (e.g., correlation within and across matrices, differences in dimensionality between matrices, outliers within and across matrices) they differ. Finally Dray et al. [36] showed the advantages of merging Co-inertia and Procrustean analysis, where the latter is used as a precursor of the former. In reality, future studies are required to contrast Co-inertia and Procrustean analysis, but in either form of analyses we can produce residual vectors (PAM) that can be further analyzed.

a

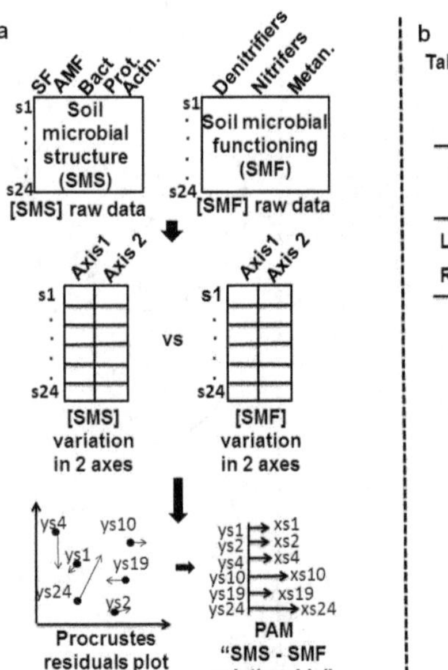

b

Table. One-way ANOVA of the effect of different land use types on the Procrustean relationship between soil microbial structure (SMS) and soil microbial functioning (SMF).

Land use type (fixed factor)	PAM (SMS-SMF relationship)	
	F	*P*
Land use	7.047	0.003**
Residuals	1.345	>0.05

c

Table. Mean effect of different land use types on the Procrustean relationship between soil microbial structure (SMS) and soil microbial functioning (SMF).

Land use types	SMS-SMF relation (PAM in 2 axes)	
	Mean	SE
Original forest	0.021 a	0.003
Silvopastoral system	0.052 b	0.001
Improved pasture	0.074 c	0.003
Unimproved pasture	0.082 c	0.002

Figure 6. Roadmap for using Procrustes association metric (PAM) in an ANOVA context. a) PCA ordination of each SMC and SMF raw data matrices, and then Procrustes correlation from 2 axes-based PCA matrices in order to generate the PAM depicting the SMC-SMF relationship. **b)** Table showing results of a one-way ANOVA for using PAM as response and land use type as fixed factor. **c)** Multiple comparisons test (Tukey, 95%) for means of the Procrustean relationship between soil microbial structure and functioning (PAM in 2 axes) across land use types.

Procrustes can be perhaps best justified when the number of predictors is greater than the number of observations or when **X** and **Y** matrices are equally applicable as explanatory and response variables. In plant-soil ecology, for example, above- and below-ground data matrices can be interchanged as explanatory and response variables. Plant community variation has been shown to be related to variation in below-ground compartments [24]. In addition, soil components such as fertility and the microbial community have been proven to influence aspects of vegetation [49]. Thus, with the literature showing that both types of datasets under analysis can structure each other, the use of Procrustes analysis, as a symmetric canonical analysis method, should be encouraged among plant and soil ecologists and ecologists in general. We hope that this paper has provided enough examples of the potential for using the Procrustes framework as a precursor to further explore ecological data.

Supporting Information

Text S1 R code showing how to use PAM associated to ordination methods (Fig. 4 in the main text). For this example we used data from Mitchell et al. [39].

Text S2 R code showing how to use the PAM in a Regression framework (Fig. 5 in the main text).

Text S3 R code showing how to use the PAM in an ANOVA framework (Fig. 6 in the main text).

Author Contributions

Conceived and designed the experiments: FJGL. Analyzed the data: FJGL. Wrote the paper: FJGL PRPN GMC ECJ SJC RJM RLLB. Contributed with real datasets: SJC RJM.

References

1. Mantel N (1967) The detection of disease clustering and a generalized regression approach. Cancer Res 27: 209–220.
2. Gower JC (1971) Statistical methods of comparing different multivariate analyses on the same data. In: Hodson FR, Kendall DG, Tautu P, editors. Mathematics in the archeological and historical sciences. Edinburgh University Press, Edinburgh. 138–149.
3. Peres-Neto PR, Jackson DA (2001) How well do multivariate data sets match? The advantages of a Procrustean superimposition approach over the Mantel test. Oecologia 129: 169–178.
4. Tuomisto H, Poulsen AD, Ruokolainen K, Moran RC, Quintana C, et al. (2003a) Linking floristic patterns with soil heterogeneity and satellite imagery in Ecuadorian Amazonia. Ecol App 2: 352–371.
5. Tuomisto H, Ruokolainen K, Yli-Halla M (2003b) Dispersal, environment, and the floristic variation of western Amazonian forests. Science 299: 241–244.
6. Tuomisto H, Ruokolainen K, Aguilar M, Sarmiento A (2003c) Floristic patterns along 43-km long transect in an Amazonian rain forest. J Ecol 91: 743–756.
7. Kang S, Mills AL (2004) Soil bacterial community structure changes following disturbance of the overlying plant community. Soil Sci 169: 55–65.
8. Poulsen AD, Tuomisto H, Balsev H (2006) Edaphic and florist variation within a 1-ha plot of lowland Amazonian rain forest. Biotropica 38: 468–478.
9. Fitzsimons MS, Miller RM, Jastrow JD (2008) Scale-dependent niche axes of arbuscular mycorrhizal fungi. Oecologia 158: 117–127.
10. Powers JS, Becknell JM, Irving J, Pères-Aviles D (2009) Diversity and structure of regenerating tropical dry forests in Costa Rica: Geographic patterns and environmental drivers. Forest Ecol Manag 258: 959–970.
11. Castilho-Monroy AP, Bowker MA, Maestre FT, Rodriguez-Echeverria S, Martinez I, et al. (2011) Relationships between biological soil crusts, bacterial diversity and abundance, and ecosystem functioning: insights from a semi - arid Mediterranean environment. J. Veg. Sci 22: 165–174.
12. Pomara LY, Ruokolainen K, Tuomisto H, Young K (2012) Avian composition co-varies with floristic composition and soil nutrient concentration in Amazonian upland forests. Biotropica 44: 545–553.
13. Artz RRE, Chapman SJ, Campbell CD (2006) Substrate utilization profiles of microbial communities in peat are depth dependent and correlate with whole soil FTIR profiles. Soil Biol Biochem, 38: 2958–2962.
14. Trivedi MR, Marecroft MD, Berry PM, Dowson TP (2008) Potential effects of climate change on plant communities in three montane nature reserves in Scotland, UK. Bio Conserv 141: 1665–1675.
15. Jesus EC, Marsh TL, Tiedje JM, Moreira FMS (2009) Changes in land use alter the structure of bacterial communities in Western Amazon soils. ISME J 3: 1004–1011.
16. Merilä P, Lämsa- MM, Stark S, Spetz P, Vierikko K, et al. (2010) Soil organic matter quality as a link between microbial community structure and vegetation composition along a successional gradient in a boreal forest. App Soil Ecol 46: 259–267.
17. Grayston SJ, Campbell CD, Bardgett RD, Mawdsley JL, Clegg CD, et al. (2004) Assessing shifts in microbial community structure across a range of grasslands of differing management intensity using CLPP, PLFA, and community DNA techniques. Appl Soil Ecol 25: 63–84.
18. Singh BK, Munro S, Reid E, Ord B, Potts M, et al. (2006) Investigating microbial community structure in soils by physiological, biochemical and molecular fingerprinting methods. Eur J Soil Sci 57: 72–82.
19. Vinten AJA, Artz RRE, Thomas N, Potts JM, Avery L, et al. (2011) Comparison of microbial community assays for the assessment of stream biofilm ecology. J Microbiol Methods 85: 190–198.
20. Hirst CN, Jackson DA (2007) Reconstructing community relationships: the impact of sampling error, ordination approach, and gradient length. Divers. Distrib 13: 361–371.
21. Poos MS, Jackson DA (2012) Addressing the removal rare species in multivariate bioassessments: The impact of methodological choices. Ecol Indic 18: 82–90.
22. Singh BK, Nunan N, Ridgway KP, Mcnicol J, Young JPW, et al. (2008) Relationship between assemblages of mycorrhizal fungi and bacteria on grass roots. Environ Microbiol 10: 534–542.
23. Burke L, Irwin R (2009) The importance of interannual variation and bottom up nitrogen enrichment for plant – pollinator networks. Oikos 118: 1816–1829.
24. Lisboa FJG, Chaer GM, Jesus EC, Gonçalves FS, Santos FM, et al. (2012) The influence of litter quality on the relationship between vegetation and below-ground compartments: a Procrustean approach. Plant Soil 367 551–562.
25. Landeiro VL, Bini LM, Costa FRC, Franklin E, Nogueira A, et al. (2012) How far can we go in simplifying biomonitoring assessments? An integrated analysis of taxonomy surrogacy, taxonomic sufficiency and numerical resolution in a mega diverse region. Ecol Indic 23: 366–373.
26. Siqueira T, Bini LM, Roque FO, Cottiene K (2012) A metacommunity framework for enhancing the effectiveness of biological monitoring strategies. PLoS One 7: e43626.
27. Nekola JC, White PS (1999) The distance decay of similarity in biogeography and ecology. J Bio 26: 867–878.
28. Legendre P, Borcard D, Peres-Neto PR (2005) Analysing beta diversity: partitioning the spatial variation of community composition data. Ecol Mon 75: 435–450.
29. Tuomisto H, Ruokolainen K (2006) Analysing or explaining beta diversity? Understanding the targets of different methods of analysis. Ecology 87: 2697–2708.
30. Kuehnelt-Leddihn ER (2007) The menace of the herd or Procrustes at large. The Bruce Publishing Company Milwaukee, Auburn. 385 p.
31. Jackson DA (1995) PROTEST: a Procrustean randomization test of community environment concordance. Ecoscience 2: 297–303.
32. Alárcon R, Waser NM, Orlleton J (2008) Year-to-year variation in the topology of a plant - pollinator interaction network. Oikos 117: 1796–1807.
33. Burke L, Irwin R (2009) The importance of interannual variation and bottom up nitrogen enrichment for plant – pollinator networks. Oikos 118: 1816–1829.
34. ten Berge JMF, Kiers HAL, Commandeur JJF (1993) Orthogonal Procrustes rotation for matrices with missing values. B J Math Stat Psyc 46: 119–134.
35. Dijksterhuis GB, Gower JC (1992) The interpretation of Generalized Procrustes Analysis and allied methods. Food Qual Prefer 3: 67–87.
36. Dray S, Chessel D, Thioulouse J (2003) Procrustean Co-inertia analysis for the linking of multivariate datasets. Ecoscience 10: 110–119.
37. Legendre P, Gallagher ED (2001) Ecologically meaningful transformations for ordination of species data. Oecologia 129: 271–280.
38. Legendre P, Legendre L (2012) *Numerical Ecology*, 3rd English edn. Elsevier Science BV, 516 Amsterdam.
39. Mitchell RJ, Hester AJ, Campbell CD, Chapman SJ, Cameron CM (2010) Is vegetation composition or soil chemistry the best predictor of soil microbial community? Plant Soil 333: 417–430.
40. Mitchell RJ, Hester AJ, Campbell CD, Chapman SJ, Cameron CM, et al. (2012) Explaining the variation in the soil microbial community: do vegetation composition and soil chemistry explain the same or different parts of the microbial variation? Plant Soil 351: 355–362.
41. Dumbrell AJ, Nelson M, Helgason T, Dytham C, Fitter AH (2009) Relative roles of niche and neutral processes in structuring a soil microbial community. ISME J doi:10.1038/ismej.2009.122.

42. Zheng YM, Cao P, Fu B, Hughes JM, He JZ (2013) Ecological Drivers of Biogeographic Patterns of Soil Archaeal Community. PloS One 8: e63375.
43. Peres-Neto PR, Legendre P, Dray S, Borcard D (2006) Variation partitioning of species data matrices: estimation and comparison of fractions. Ecology 87: 2603–2613.
44. Chaer GM, Fernandes MF, Myrold DD, Bottomley PJ (2009) Shifts in microbial community composition and physiological profiles across a gradient of induced soil degradation. Soil Sci Soc Am J 73: 1327–1334.
45. Peixoto RS, Chaer GM, Franco N, Reis Junior FB, Mendes IC, et al. (2010) A decade of land use contributes to changes in the chemistry, biochemistry and bacterial community structures of soils in the Cerrado. A Van Leeuw 98: 403–413.
46. Thioulouse J, Simier M, Chessel D (2004) Simultaneous analysis of a sequence of paired ecological tables. Ecology 85: 272–283.
47. Dolédec S, Chessel D (1994) Co-inertia analysis: an alternative method for studying species–environment relationships. Freshwater Biol 31: 277–294.
48. Ter Braak, CJF, Schaffers AP (2004) Co-correspondence analysis: a new ordination method to relate two community compositions. Ecology 85: 834–846.
49. van der Heijden MGA, Klironomos J, Ursic M, Moutoglis P, Streitwolf-Engel R, et al. (1998) Mycorrhizal fungi determines plant biodiversity, ecosystem variability and productivity. Nature 396: 69–72.

7

Discovering the Recondite Secondary Metabolome Spectrum of *Salinispora* Species: A Study of Inter-Species Diversity

Utpal Bose[1], Amitha K. Hewavitharana[1], Miranda E. Vidgen[2], Yi Kai Ng[2], P. Nicholas Shaw[1], John A. Fuerst[2], Mark P. Hodson[3]*

1 School of Pharmacy, The University of Queensland, Brisbane, Queensland, Australia, **2** School of Chemistry and Molecular Biosciences, The University of Queensland, Brisbane, Queensland, Australia, **3** Metabolomics Australia, Australian Institute for Bioengineering and Nanotechnology, The University of Queensland, Brisbane, Queensland, Australia

Abstract

Patterns of inter-species secondary metabolite production by bacteria can provide valuable information relating to species ecology and evolution. The complex nature of this chemical diversity has previously been probed via directed analyses of a small number of compounds, identified through targeted assays rather than more comprehensive biochemical profiling approaches such as metabolomics. Insights into ecological and evolutionary relationships within bacterial genera can be derived through comparative analysis of broader secondary metabolite patterns, and this can also eventually assist biodiscovery search strategies for new natural products. Here, we investigated the species-level chemical diversity of the two marine actinobacterial species *Salinispora arenicola* and *Salinispora pacifica*, isolated from sponges distributed across the Great Barrier Reef (GBR), via their secondary metabolite profiles using LC-MS-based metabolomics. The chemical profiles of these two species were obtained by UHPLC-QToF-MS based metabolic profiling. The resultant data were interrogated using multivariate data analysis methods to compare their (bio)chemical profiles. We found a high level of inter-species diversity in strains from these two bacterial species. We also found rifamycins and saliniketals were produced exclusively by *S. arenicola* species, as the main secondary metabolites differentiating the two species. Furthermore, the discovery of 57 candidate compounds greatly increases the small number of secondary metabolites previously known to be produced by these species. In addition, we report the production of rifamycin O and W, a key group of ansamycin compounds, in *S. arenicola* for the first time. Species of the marine actinobacteria harbour a much wider spectrum of secondary metabolites than suspected, and this knowledge may prove a rich field for biodiscovery as well as a database for understanding relationships between speciation, evolution and chemical ecology.

Editor: Vishnu Chaturvedi, California Department of Public Health, United States of America

Funding: This paper is an output from the Great Barrier Reef Seabed Biodiversity Project, a collaboration between the Australian Institute of Marine Science (AIMS), the Commonwealth Scientific and Industrial Research Organisation (CSIRO), Queensland Department of Primary Industries & Fisheries (QDPIF), and the Queensland Museum (QM); funded by the CRC Reef Research Centre, the Fisheries Research and Development Corporation, and the National Oceans Office; and led by R. Pitcher (Principal Investigator, CSIRO), P. Doherty (AIMS), J. Hooper (QM) and N. Gribble (QDPIF). Utpal Bose was supported by a University of Queensland International Scholarship. Yi Kai Ng was supported by the University of Queensland Research Scholarship. Miranda Vidgen was supported by an Australian Research Council Linkage Award. The funders had no role in study design, data collection and analysis, decision to publish, or preparation of the manuscript.

Competing Interests: The authors have declared that no competing interests exist.

* E-mail: m.hodson1@uq.edu.au

Introduction

Marine bacteria from different phylogenetic groups produce secondary metabolites that play multiple ecological functions within their marine chemical environment [1]. Such secondary metabolites are produced under the pressure of natural selection and may act as agents of interaction (e.g. antagonism or competition) with other microorganisms in a community or as signals for communication within populations of the same species [2], [3]. The production of antibiotics by microbes is a typical example of such a response and is primarily thought to be a defence against microbial competitors; however, these molecules have also been found to have roles as quorum-sensing signals, or other functions that help stabilize microbial communities [3]. Thus, fine scale comparative analysis of secondary metabolite production at the species level provides a means by which to explore and understand species biology, ecology and evolution. In addition, it is one of the keys to understanding the chemical diversity underlying biodiscovery of novel previously unknown pharmaceuticals and other chemical products. Nonetheless, to date limited attention has been paid to such fine-scale comparative analysis of natural product diversity in bacterial species, especially concerning marine bacteria.

The genus *Salinispora* was the first obligate marine actinobacterial genus to be described and members are widely distributed in tropical and sub-tropical marine sediment and marine sponges to depths of 2000 m [4], [5]. The genus comprises three closely related species *S. tropica*, *S. arenicola* and *S. pacifica* [6], [7]. Like their actinobacterial terrestrial counterparts, *Salinispora* produce numerous secondary metabolites with diverse possible pharmaceutical applications. The compound salinosporamide A, isolated from *S. tropica* is currently in Phase 1 clinical trials in patients with multiple myeloma, lymphomas, leukaemia and solid tumours [8]. This genus has also been found to produce secondary metabolites with diverse activities, for instance arenimycin, rifamycins, staurosporine, saliniketal A and B, cyclomarazines and cyclomarins, as well as hydroxamic acid siderophores [4], [8], [9], [10].

Bacterial species classified based on their 16S rRNA gene sequences can vary greatly at the genomic level [11]. These genomic differences are mainly found in isolated islands – regions of the chromosome which are known to contain genes associated with ecological adaptation [11]. To date, most of these reported *Salinispora* species-derived compounds are polyketides or non-ribosomal peptides, or hybrids thereof, and their biosynthesis is accomplished by large multi-enzyme complexes, the polyketide synthases (PKS) or the non-ribosomal peptide synthases (NRPS) [12], [13]. Analyses of the *S. tropica* and *S. arenicola* genomes have revealed the presence of putative natural product biosynthesis genes, comprised of PKS and NRPS, with a large percentage of the genome (8.8% and 10.9% respectively) devoted to secondary metabolite biosynthesis, which is greater than the percentage of such genes in the *Streptomyces* secondary metabolite genome sequence [12]. In 2007, the *S. arenicola* CNS-205 genome sequencing project revealed a 5.8 Mbp genome (CP00850) with at least 30 distinct metabolite gene clusters [12]. These results also suggest that the production of secondary metabolites may be linked to ecological niche adaptation within this group of bacteria, and that the acquisition of natural product biosynthetic genes represents a previously unrecognized influence driving bacterial diversification [14]. Taken together, these findings suggest that species of the genus *Salinispora* possess the capacity to produce a large number of secondary metabolites. Only a limited number of the potentially wide spectrum of such metabolites have actually been detected to date (e.g. represented by the salinosporamides, saliniketals, sporolides, arenimycins, cyclomarins, etc. listed above).

In addition to the production of diverse secondary metabolites, this genus has attracted major interest for the novel phenomenon of species-specific production of such secondary metabolites [15]. Jensen and co-workers (2007) have previously shown that *Salinispora* was the first bacterial genus to be identified as having species-specific secondary metabolite production correlated to their phylogenetic diversity at the species level. Core compounds have been produced by a specific species, for example the compounds salinosporamide A-J, sporolide A and B, and an antiprotealide were only found to be produced by *S. tropica* [14]. However, recent studies have shown that staurosporine, which was previously isolated from *S. arenicola* [16] is also produced by *S. pacifica* [14]. The vertical gene transfer of the *sta*D gene sequences between two sister taxa *S. arenicola* and *S. pacifica*, are responsible for the production of staurosporine in *S. pacifica* [14].

Horizontal gene transfer (HGT), the exchange and stable integration of genetic material from different strains and species, is a major evolutionary force [17]. The process of genetic exchange allows bacterial species to acquire traits from distantly related organisms and as a consequence aids adaptation to the changing environment [18]. Recent research has highlighted that HGT may have occurred throughout a major part of the bacterial genome [12], [15]. Genes acquired by HGT play multiple roles: virulence, metabolism, resistance to antibiotics and the long-term maintenance of organelles [19]. Although it is clear that a large part of the *Salinispora* genome acquired gene clusters for secondary metabolite production via HGT, the ecological and evolutionary significance of these mechanisms remain unclear.

Evidence of HGT in *Salinispora* species comes from a phylogenetic study of the PKS genes associated with the rifamycin biosynthetic gene cluster (*rif*) in *S. arenicola* and *Amycolatopsis mediterranei*, the original source of this compound [20], [21]. Rifamycins are naphthalenic ansamycin antibiotics produced by a number of soil- and marine-derived actinomycetes, for example *Amycolatopsis mediterranei* [22]. These compounds elicit their antibacterial activity through the specific inhibition of RNA synthesis via binding to the beta sub-unit of RNA polymerase [23]. Semi-synthetic rifamycin derivatives, for instance rifampicin, have been used as antibiotic therapies against *Mycobacterium tuberculosis* and *Mycobacterium leprae*, the causative agents for tuberculosis and leprosy, respectively. The naturally occurring variant rifamycin B is the parent molecule for other biologically active rifamycin compounds and rifamycin B is further processed either by natural enzymatic modification or by semi-synthetic mechanisms to produce the biologically active rifamycin analogues O, SV and S [24].

Numerous methods have been used to screen bacterial secondary metabolites for useful biopharmaceuticals [25]. One such method is bioactivity-guided screening, which allows detection of compounds with specific biological activity. For example, salinosporamide A was primarily isolated through bioactivity guided assays [8]. However, this method is clearly biased to specific targets in the activity assays, as well as the repeated "discovery" of known compounds and is limited by the availability of suitable assay methods. An alternative method to identify secondary metabolites is via chemical screening, which can identify the presence of diverse compound classes in a complex set of samples. Traditional screening methods involve separation and isolation of compounds followed by their identification but such methods may require time-consuming optimization of separation conditions for each compound and elaborate methods for identification. Chemical screening is also constrained by marked differences in the chromatographic and spectroscopic properties of the natural compounds. For example, UV/visible light detection by diode array detector-coupled liquid-chromatography is widely used but has limitations for the detection of certain compound classes, may lack specificity and spectra are sparsely represented (and therefore difficult to search) in structural databases [26]. Some other methods offer novel and innovative approaches for screening and/or identifying secondary metabolites, such as phylogenetic analysis at the genomic level [27], genome screening approaches (e.g. PCR-based identification of coding sequences [28] and functional genome analysis [29]. However, these approaches are often limited by the need for conservation of sequence with known genes (e.g. for design of PCR primers) and still require chemical analysis and identification of downstream products arising from the gene expression. Therefore, new approaches are needed to identify chemical signals produced by these microorganisms for selection of potentially useful compound libraries and to better understand microbial chemical ecology and evolution. For example, as it is now clear that variation in metabolite production exists between the two *Salinispora* species [15], more comprehensive analytical and data-mining techniques are needed to explore the secondary metabolite profiles of this genus.

The genome speaks of what compounds could potentially be produced as it codes for the machinery to make production possible. However, it is the phenotypic secondary metabolome that tells the story of which metabolites are actually available to the organism in its chemical arsenal for interacting with its external microbial community and habitat. Metabolomics is the comprehensive analysis of the biochemical content of cells, tissues or biofluids, usually from analysis of extracts [30]. Typically metabolomics experiments have utilised NMR- and/or MS-based analytical techniques to explore the metabolite content of experimental samples. Liquid Chromatography-Quadrupole Time of Flight-Mass Spectrometry (LC-QToF-MS) has received much attention in recent years for microbial metabolic fingerprinting studies as well as in many other fields of biology [25], [31], [32]. In this study, high resolution UHPLC-QToF-MS combined with multivariate/chemometric approaches were applied to investigate the secondary metabolome of *S. arenicola* and *S. pacifica* in order to elucidate the differences between their secondary metabolite profiles and to distinguish these two taxonomically identical species. Here we have highlighted a number of secondary metabolites that are responsible for differentiating two species at the metabolic and taxonomic level. Furthermore, from these data we confirm the first evidence of rifamycin O and W production in *S. arenicola*.

Figure 1. Distribution of sampling sites from which the 46 *S. arenicola* (blue) and *S. pacifica* (red) strains were collected. Bacterial species were spread over both Northern and Southern regions of the Great Barrier Reef (GBR) (over ~2500 km).

Materials and Methods

Sample preparation

This investigation involved the growth and analysis of 46 strains of *Salinispora* from two different species to obtain the secondary metabolite profiles. *Salinispora* isolates were collected at various locations along the Great Barrier Reef (GBR), Queensland, Australia, an area frequently studied for its tropical marine ecosystem (Figure 1). The isolation and taxonomic identification of *Salinispora* isolates has been previously reported [9], [33]. Strains used for rifamycin W identification were reported by Ng and co-workers [34]. In the present study the isolates were cultivated on Difco Marine Agar 2216 for 56 days at 28°C [33], until black pigmentation of the colonies was established in all strains. Between two and five biological replicates were grown for each of the 46 strains. The mycelial cell mass was harvested by scraping it off the growth medium using a scalpel blade and pooled in a pre-weighed 1.5 mL centrifuge tube. The net mass for each sample was recorded and subsequently used to normalise the data obtained from the extracts, based on the precise biomass extracted per sample. The same procedure was repeated using a blank agar plate to obtain a blank extract. To extract the secondary metabolites from the cell mass 1 mL of ethyl acetate was added to the sample and tube, and shaken for 90 minutes at room temperature. The tube was positioned vertically and the layers of ethyl acetate and *Salinispora* extract were allowed to separate before the ethyl acetate layer from each tube was transferred by pipette into three clean 1.5 mL centrifuge tubes. The ethyl acetate was then removed (by evaporation) from the extract using a vacuum centrifuge (Savant Instruments, Hicksville, NY) for 1 hour. The dried extract was resuspended at 15% of the original volume by adding 30 µL methanol and then 120 µL of MilliQ (Millipore, Bedford, MA, USA) water to produce a 20:80 methanol:water solution. The extract solution was stored at −80°C until use. Before HPLC-MS analysis the samples were thawed and filtered using a sterile, 4 mm diameter, 0.2 µm PTFE membrane syringe filter (Phenomenex, Sydney, Australia). After filtration, the samples were kept at ~4°C prior to injection. The injection volume was 20 µL.

UHPLC-QToF-MS analysis. The chromatographic separation of compounds in *Salinispora* extracts was performed using Ultra High Performance Liquid Chromatography (UHPLC) on an Agilent 1290 series system (Agilent Technologies, USA). The UHPLC was coupled to an Agilent 6520 high resolution accurate mass (HRAM) QToF mass spectrometer equipped with a multimode source (Agilent) and controlled using MassHunter acquisition software, (B. 02.01 SP3 - Agilent). Separation was achieved using a 2.1×50 mm, 1.8 µm ZORBAX SB C18 column (Agilent). The chromatographic analysis was performed using 5 mM aqueous ammonium acetate (mobile phase A) and acetonitrile (mobile phase B) at a flow rate of 0.15 mL/min. The column was pre-equilibrated for 15 minutes with 80% (v/v) A and 20% (v/v) B. After injection, the composition of mobile phase was changed from 20% (v/v) B to 100% (v/v) B over a period of 50 min, composition held at 100% (v/v) B for 5 min, and then returned to the starting composition of 20% (v/v) B over next 5 min. The column was re-equilibrated using 20% (v/v) B for 15 min prior to the next injection. The total chromatographic run time was 75 min.

A dual nebulizer electrospray source was used for continuous introduction of reference ions. In MS mode the instrument scanned from *m/z* 100 to 1000 for all samples at a scan rate of 0.8 cycles/second. This mass range enabled the inclusion of two reference compounds, a lock mass solution including purine ($C_5H_4N_4$ at *m/z* 121.050873, 10 µmol L^{-1}) and hexakis (1H, 1H, 3H-tetrafluropentoxy)-phosphazene ($C_{18}H_{18}O_6N_3P_3F_{24}$ at *m/z* 922.009798, 2 µmol L^{-1}). Multimode (i.e. simultaneous Electrospray Ionisation [ESI] and Atmospheric Pressure Chemical Ionization [APCI]) with Fast Polarity Switching (FPS) was employed to ionise compounds after chromatographic separation. The QToF-MS conditions used for these experiments are given in Table S1. Methods for the identification of both rifamycin O and W are detailed in Text S1.

Data analysis and Molecular Formula Generation

Data analysis was performed using Agilent MassHunter Qualitative software (Version B.05.00). The Molecular Feature Extractor (MFE) algorithm within MassHunter Qualitative analysis software was used to extract chemically qualified molecular features from the LC-QToF-MS data files. For empirical formula generation, the Molecular Formula Generator (MFG) algorithm was used. This algorithm uses a wide range of

MS information, for instance accurate mass measurements, adduct formation, multimer formation and isotope patterns to generate a list of candidate compounds. The maximum elemental composition $C_{60}H_{120}O_{30}N_{30}S_{5}Cl_{3}Br_{3}$ was used to generate formulae. MFG can automatically eliminate unlikely candidate compounds and rank the putative molecular formulae according to their mass deviation, isotopic pattern accuracy and elemental composition.

Chemometric analyses. The LC-MS molecular feature-extracted datasets were further processed using Agilent Mass Profiler Professional (MPP) software (Agilent) to extract and align peaks/features from the chromatograms of all extracts of the two *Salinispora* species (observations), resulting in a total of 3341 putative metabolite features (variables). To aid the data-mining process, the LC-QToF-MS data file of the blank sample was also analysed to extract features and to use as a background reference. These reference ions were removed from all samples in the matrix.

The resulting data matrix (144 observations; 3341 variables) was then exported as .csv and imported to SIMCA-P (version 13.0.3.0; MKS Umetrics AB, Umea, Sweden). All data were log transformed, mean centred and scaled to unit variance prior to multivariate analysis. Unsupervised analysis using Principal Component Analysis (PCA) was initially performed to reveal any outliers resulting from technical/instrumental/processing procedures and to assess any groupings or trends in the dataset. Thereafter supervised analysis was performed, where appropriate, whereby predetermined groupings were used to classify the data, using Orthogonal Projection to Latent Structures-Discriminant Analysis (OPLS-DA). The scores and loadings plots from these analyses, which describe the multivariate relationships of the observations and variables respectively, along with other metrics such as S-plots and VIP lists, were then used to determine the features (metabolites) that contribute to the differences between the experimental groups (*Salinispora* species). The selected features were used to generate putative molecular formulae in order to search for and identify compounds via database query, MS/MS fragmentation patterns and comparison to authentic standard(s).

Results

Global metabolic profiling with High Resolution Accurate Mass-Mass Spectrometry (HRAM-MS)

In order to extract the maximum number of secondary metabolites, a non-targeted extraction method was used. Ethyl acetate is a medium polarity solvent that is widely used and is capable of extracting a large percentage of compounds other than those that are extremely polar or non-polar. As an organic solvent, ethyl acetate has the additional benefit of denaturing and thus causing precipitation of macromolecules such as proteins, as well as causing the partition of highly polar analytes to the aqueous layer. Although untargeted, the choice of solvent should dictate that resultant extracts contain much of the secondary metabolome of interest.

A neutral mobile phase pH was used to be applicable to both positive and/or negative ionisation, with the use of fast polarity switching to detect sequentially both positive and negative ions. In addition, a multimode ion source was employed to maximise the ionisation of compounds that preferentially ionise in one source only. Although these strategies undoubtedly reduce the sensitivity of detection for specific ions, the methodology was employed as an initial "one shot" approach to enable the rapid screening of extracts.

Inter-species chemical diversity

The LC-MS chromatograms obtained from two species were compared visually by using Total Ion Current chromatograms (TICs) in the mass range of 100 to 1000 m/z at a retention time window 2–50 minutes in order to detect differences between secondary metabolite production profiles. At the population level, the average number of compounds varied from ~300 in *S. arenicola* to ~150 in *S. pacifica* samples. Figure 2 (A, B) shows typical LC-MS profiles for the two species, *S. arenicola* and *S. pacifica*, and reveals differences between these two species. Figure 2C shows the metabolic profile of three biological replicates obtained from each sample obtained from a single strain. The consistency of these results for all the samples highlights the reproducibility of the biological replicates.

Initially, unsupervised analysis by PCA was used to identify any outliers and assess any groupings or trends in the data set. Results obtained from the PCA scores plot (illustrating the relative similarities or differences of the sample extracts of the two species) in Figure 2D shows that no clearly defined separation exists between the two bacterial species based upon the major sources of variation within their (bio)chemical profiles. The scores plot also confirms that no technical outliers are present but that some biological "outliers" are apparent, such as the sub-cluster of samples from *S. arenicola.* The variables responsible for any groupings or clusters in the data can be determined from the loadings plot Figure 2E; as an example, the presence of a compound with 754.3092 m/z at RT 12 min (Figure 2G) distinguishes the samples in the lower half of Figure 2D from the other samples. This finding is confirmed by extracted ion chromatograms (EICs) created for m/z 754.3092 (Figure 2F) and the box-and-whisker plot (Figure 2H). It is important to note that the multivariate analysis is used to identify important/discriminatory compounds/features within the dataset and the confirmation of their importance should always be achieved by extracting the representative data to ascertain the behaviour of these compounds across the sample set. The 754 ion was identified as rifamycin B (data not shown), a molecule reported in previous work by our group [35]. Identified compounds are listed in Table 1.

The accurate mass m/z values from high resolution measurements highlighted by PCA are used to generate molecular formulae in order to propose putative compounds. As molecular weight increases so does the number of possible molecular formulae [36]. Compound proposals retained after both statistical and visual/manual curation were compiled to a list of accurate mass values, corresponding putative molecular formulae, RTs and IDs (Table S2), and this list was used for future targeted analysis of the 46 strains.

Supervised multivariate analysis methods such as OPLS-DA were used in order to retrieve the variables explaining the differences between the two species only, delineating this information from any interfering or confounding sources of variance. The scores plot Figure 3A from such an analysis shows a complete separation in the predictive (horizontal) component ($R2X(cum) = 21.2\%$, $R2Y(cum) = 92.0\%$, $Q2(cum) = 45.6\%$) which simplifies the interpretation of which variables contribute to a distinct species-specific difference. By this analysis it can be seen that there can indeed be a distinction made between species based on their chemotypes. Variation observed in the orthogonal (vertical) component is unrelated to strain differences but may warrant further investigation given that there are different strains of each species.

The S-plot visualizes the correlation and model influence of the metabolites (variables). As shown in Figure 3B, the variables to the extreme of the lower left quadrant are influential, with good

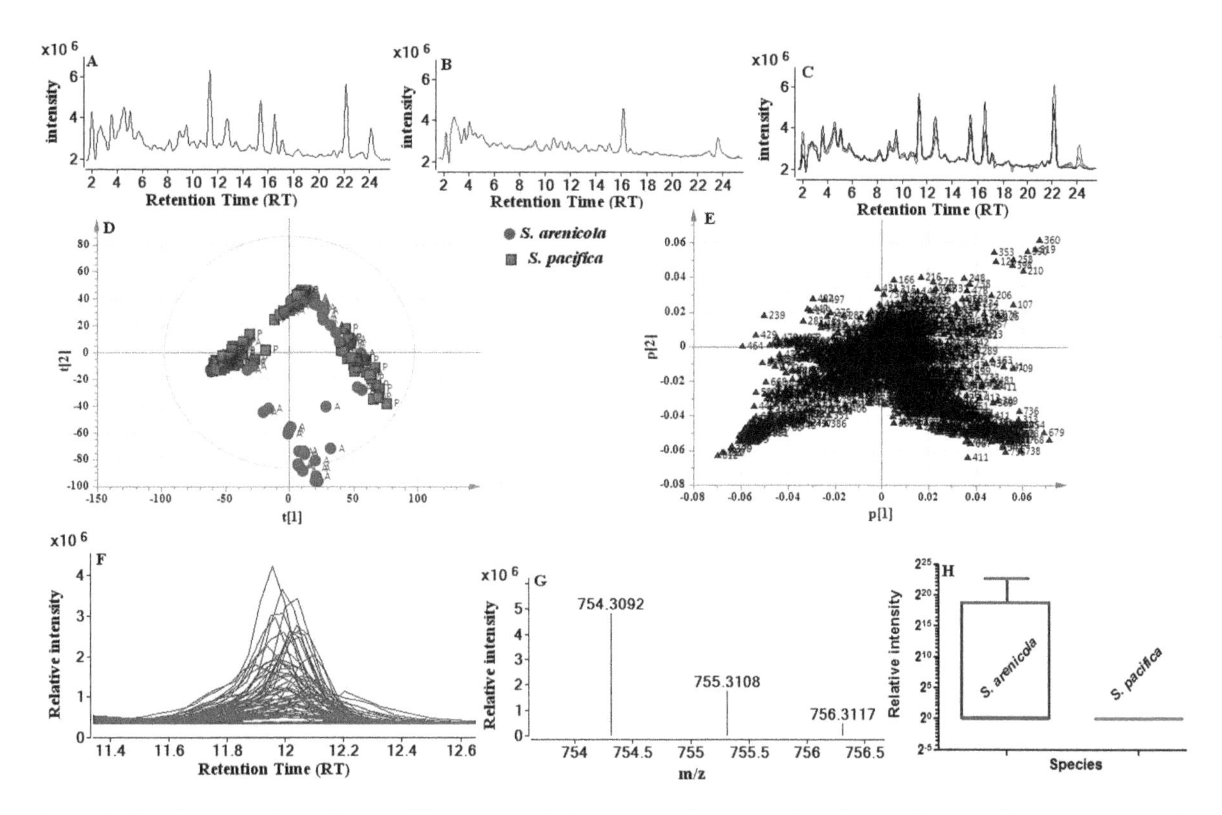

Figure 2. Identification of species-specific compounds in extracts of *S. arenicola* and *S. pacifica* by UHPLC-QToF-MS and PCA. (A) Total ion current (TIC) chromatograms of *S. arenicola* and (B) *S. pacifica* (C) TICs from the same (*S. arenicola*) subset of samples (three replicates) to highlight the reproducibility of acquisition. For ease of interpretation, chromatograms of 2 to 24 minutes are shown to illustrate the period of chromatography during which most compounds elute. Full chromatograms for both species are available in the Supplementary information (Figure S6). (D) Principal component analysis (PCA) scores plot, PC1 (t[1]) versus PC2 (t[2]) showing the variation in the profiles of secondary metabolites from two species: *S. arenicola* (A, blue) and *S. pacifica* (P, red). Each symbol represents one bacterial strain described by all detected metabolites. (E) Inspection of the 2-D loadings plot for PC1 vs. PC2 reveals the variables responsible for the spatial arrangement of samples. (F) Extracted ion chromatogram (EIC) of *m/z* 754.3092 from both species, showing clear species differences in the abundance of this metabolite *S. arenicola* (blue) and *S. pacifica* (red). (G) MS spectrum for EIC peak. (H) Box-and-whisker plot of the abundance of the 754 ion in the two species ($P < .0001$).

reliability (correlation) and the highest (absolute) loadings scores. In the comparison of two groups, *S. arenicola* versus *S. pacifica*, the *S. pacifica* metabolites are comparatively more abundant in the upper right quadrant (with >0 correlation and model influence i.e. loadings), whereas in the lower left quadrant (with <0 correlation and model influence) these metabolites are more abundant in *S. arenicola*. Figures 3C–3H show the EICs, mass spectra and box-and-whisker plots used in the identification of rifamycin S and saliniketal B. From the interrogation of the OPLS-DA loadings plot, S-plot and Variable Importance in Projection (VIP) results, 57 metabolites were listed according to their highest VIP scores Table S2. To further investigate the significance of changes in these metabolites, a combination of approaches was used: t-test, VIP values and EICs, and finally results were combined with PCA results (Table S2). Analysis by the most complex methods used in this section clearly shows a distinction between *Salinispora* species on the basis of large scale secondary metabolite production.

Identification of rifamycin O and W

In order to identify a previously unknown source of rifamycin O and W, an integrated approach was adopted consisting of the following steps: (1) HPLC-QToF-MS analysis with multimode ionisation (ESI and APCI) and fast polarity switching; (2) database searching of monoisotopic masses (tolerance 5 ppm); (3) matching of monoisotopic masses, retention times, mass spectra of molecular and fragment ions of the postulated compound and the authentic reference standards. Rifamycin W was not commercially available and was therefore tentatively identified based on structure using the database matching of the monoisotopic mass in conjunction with the mass spectrum of the molecular and fragment ion.

As shown in Figure S1, LC-UV-Vis (at 430 nm) chromatograms of *S. arenicola* strains MV0318 and MV0472, as well as M413 (ACM 5232) showed two peaks that eluted at times that were different to those found in our previous studies of rifamycins B, S and SV [35]. The bacterial strains previously studied included M403, SW15, M102, M413, SW10, SW17, M414, M412, SW 02 and M101. These two peaks were absent in the chromatograms of blank extracts produced using the sterile culture medium, and in chromatograms of rifamycin standards B and SV run using the same gradients as in this study. Therefore these peaks appear to be specific to the strains analysed in this study and as such are previously undetected metabolites.

The unknown peak eluting at 24.5 min was identified on the basis of its accurate mass, through an in-house database search, as rifamycin O; the monoisotopic mass of the unknown was 752.2985 $[M-H]^-$ and that of rifamycin O in the database was 753.2996 [M]. The negative ionisation mode LC-MS chromatograms and the mass spectra of the peaks of interest for the extracts and rifamycin O standard are shown in Figure 4. Matching of the retention times (24.5 for standard and 24.55 for unknown) and the mass spectra further strengthened the identification through the database search. As described in the experimental section,

Table 1. Identification of six compounds from *S. arenicola* and *S. pacifica.*

Molecular Formula	Overall Score*	Mass**	Polarity	RT (min)	Difference (MFG, ppm)	Compound identification***	Example species
$C_{37}H_{45}NO_{12}$	99.35	695.2937	Negative	21.36	0.68	Rifamycin S	*S. arenicola*
$C_{22}H_{37}NO_6$	94.91	411.2628	Positive	8.12	3.13	Saliniketal B	*S. arenicola*
$C_{22}H_{37}NO_5$	93.72	395.2661	Positive	10.61	2.72	Saliniketal A	*S. arenicola*
$C_{39}H_{49}NO_{14}$	91.34	754.3092	Negative	12.20	0.52	Rifamycin B	*S. arenicola*
$C_{23}H_{31}N_3O_3$	79.43	397.2365	Positive	9.91	2.12	Cyclomarazine A	*S. arenicola*
$C_{17}H_{24}O_4$	61.87	292.1675	Positive	7.87	3.45	Salinipyrone A	*S. pacifica*

The proposed formula obtained after the PCA and OPLS-DA analysis according to high-resolution LC-QTOF-MS measurements.
*Overall sc°re calculated from the empirical formula match with the database search. **Neutral mass calculated for each compound.
***Compound identification performed through the in-house accurate mass database match, MS/MS fragmentation and reference standards.

chromatograms were obtained for both the extract and the standard rifamycin O, using the detection of peaks by MS/MS fragmentation spectra in negative ionisation mode. Figure S2 shows that the observed spectra and the proposed fragmentation pathways for the unknown peak and that of rifamycin O are the same, confirming the identity of the unknown peak to be rifamycin O. Figure S3 demonstrates the fragmentation of the rifamycin molecule to produce the major fragments observed in Figure S2 (m/z 694, 636, 514, 453 and 272).

In MS/MS mode, 752.2985 $[M\text{-}H]^-$ was selected as the precursor ion. Both unknown and standard peaks produced fragment ions with identical m/z values. The presence of m/z 272 (naphthofuran) provides strong evidence for the identification of rifamycin [35]. It is worth mentioning the unique strength of LC-QToF-MS/MS technology in confirming the identity; we were able to detect the unknown peak with only 3 ppm mass difference from the standard rifamycin O, and in MS/MS mode the differences between standard and unknown spectra for 752, 694 and 272 ions were 0, 0 and 0.1 ppm, respectively. The confirmation of the identity in this case was based on several different observations: absorbance at 430 nm; matching the monoisotopic mass with an in-house database entry for rifamycin O; matching of three properties (retention time, mass spectra and fragmentation spectra) with standard rifamycin O; and the likely presence of a naphthofuran system in the fragmentation spectrum.

The unknown peak eluting at 7 min was identified through an in-house database search as rifamycin W; the monoisotopic mass of the unknown was 654.2914 $[M\text{-}H]^-$ (Figure 5) and that of rifamycin W in the database was 655.2992 (5 ppm tolerance; m/z 654.2914 $[M\text{-}H]^-$). Figure 5 shows the negative ionisation LC-MS chromatogram and mass spectrum for the above unknown peak for *S. arenicola* extract of strain M413 (ACM5232). As rifamycin W is not commercially available, it was not possible to obtain the same data for an authentic standard. Following a similar protocol to the rifamycin O above, MS-MS fragmentation spectra for the unknown peak were obtained in negative ionisation mode, as shown in Figure S4. Four major fragments 452, 330, 272 and 245 have been identified and the generation of these product ions from the precursor ion is demonstrated in Figure S5. The confirmation of the identification of the unknown peak as rifamycin W in this case was based on several different observations: absorbance at 430 nm; matching of the mass obtained for the molecular ion in mass spectrum to that in database entry for rifamycin W (0.78 ppm difference); matching the product ions obtained in MS-MS spectrum from rifamycin W with fragmentation patterns; and the presence of the naphthofuran system in the fragmentation spectrum.

Discussion

Analysis of the inter-species level of chemical diversity is important, not only in the context of speciation but also for understanding genetic variation during adaptive evolution within species, and to exploit the full diversity of natural products which may be available for biopharmaceutical screening and dereplication [37]. To date, most of the studies based on *Salinispora* species have focused on a limited number of targeted compounds rather than on chemical or metabolic traits with known functional roles. However, knowledge of the broader scale of secondary metabolite production in this obligate marine actinobacterium is of considerable interest in ecology and evolutionary biology as well as pre-filtering of strains during screening programs for the most likely producers of new biopharmaceuticals. 'Omics' approaches have thus far proven useful in providing some insight when attempting

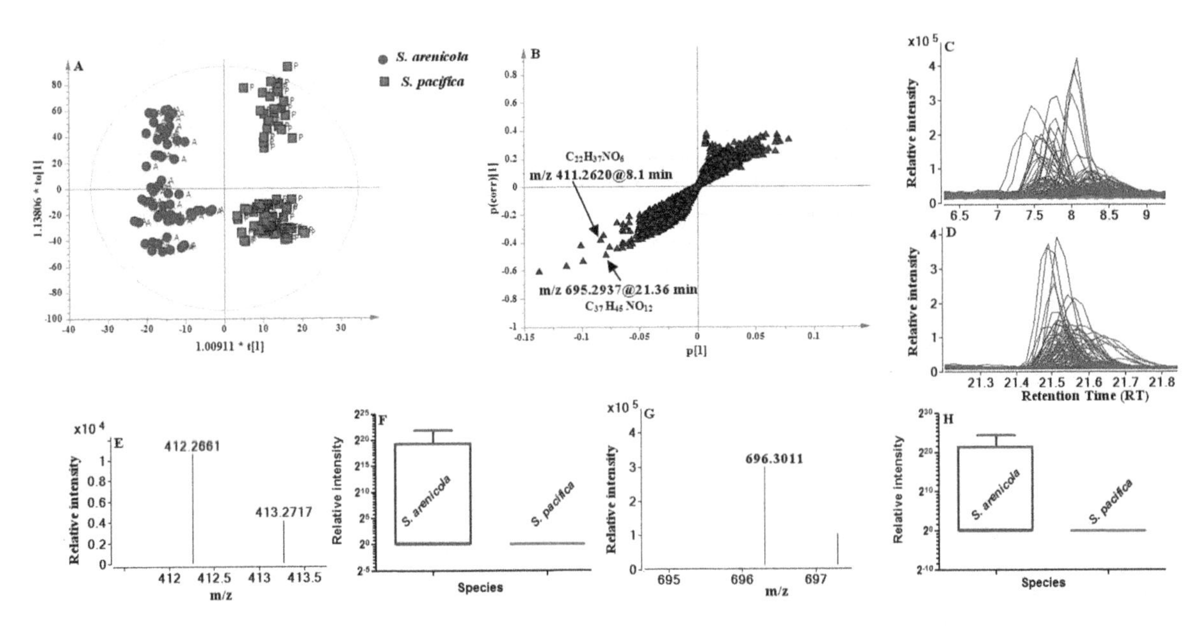

Figure 3. Metabolic profiling distinguishes bacterial samples from two *Salinispora* species *S. arenicola* and *S. pacifica*. (**A**) OPLS-DA scores plot, predictive component t[1] versus orthogonal component (t_o[1]) showing the supervised separation between the two sample classes. The ellipse in A and B represents the Hotelling's T^2 95% confidence interval for the multivariate data. (**B**) Loading S-plots derived from the LC-MS data set of two bacterial species. S-plot shows predictive component p[1] against the correlation p (corr) of variables of the discriminating component of the OPLS-DA model. (**C**) EIC of *m/z* 411 from all samples. (**D**) EIC of *m/z* 695 from two bacterial species. (**E**) Mass spectrum for *m/z* 411. (**F**) Box-and-whisker plot for *m/z* 411 – significantly present in *S. arenicola* but absent from *S. pacifica* (P<.0001) (**G**) Mass spectrum for *m/z* 695. (**H**) Box-and-whisker plot for *m/z* 695 – significantly present in *S. arenicola* but absent from *S. pacifica* (P<.0001).

to answer complex biological questions of a similar nature [38], but more direct metabolomic and phenotypic approaches may also prove productive. Here, we have applied a mass spectrometry-based metabolomics approach to attempt to discover the concealed secondary metabolome in two *Salinispora* species: *S. arenicola* and *S. pacifica*. Our results show that screening a large number of compounds in bacterial species can answer questions relating to which metabolites are produced. This could be a key to answering the question of what their role may be in the adaptation of the organism to its environment and the effects on their immediate community.

With this work we have begun to document the chemical diversity in two *Salinispora* species collected from the Great Barrier Reef off, situated of the north east coast of Australia, an area extending over ~2500 km. There are also similarities in the species-related chemotypes, which indicates the presence of common compounds in these two bacterial species. To remove confounding variation in the dataset and to focus the analysis on the factor of interest, namely species difference, supervised multivariate analysis highlighted that clear chemotypic differences exist between the two bacterial species, as detailed in Table S2. Interestingly, we have found rifamycins and saliniketal A and B to be consistently present of in all of the *S. arenicola* samples. Although we have identified rifamycins and saliniketals in *S. arenicola* species, we were unable to detect any specific compound class as present in *S. pacifica* as a whole. However, the ability of *S. arenicola* to synthesize rifamycins and saliniketals seems to form a definite species-specific character distinguishing *S. arenicola* from *S. pacifica*.

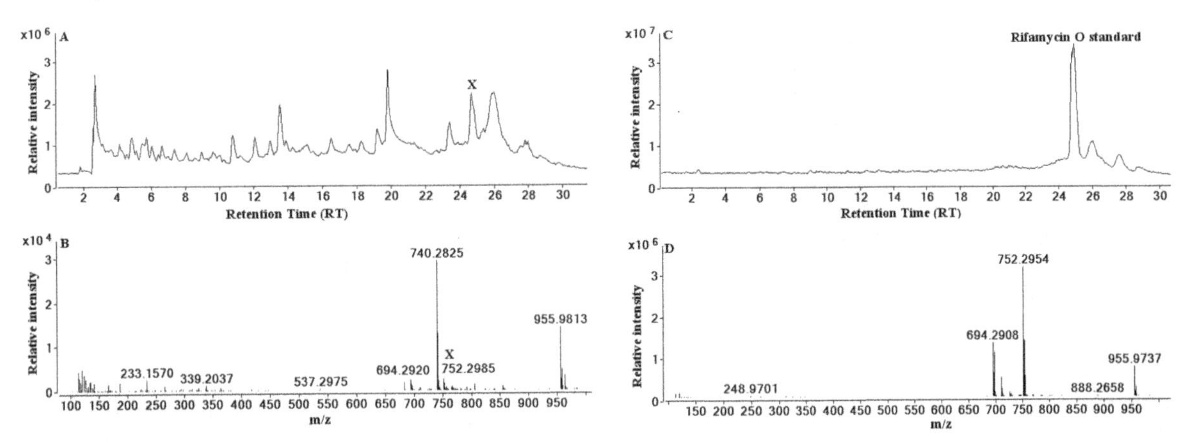

Figure 4. Chromatograms and mass spectra relating to the detection of rifamycin O. (**A**) LC-QToF-MS Total Ion Chromatogram (TIC) for *S. arenicola* strain MV0472 and (**B**) Mass spectrum of peak X and (**C**) Chromatogram for rifamycin O standard and (**D**) Mass spectrum of peak X (lower 2 panels). The retention times of X and rifamycin O standard are 24.5 and 24.55 min, respectively. The *m/z* of molecular ion of X and rifamycin O standards are 752.2985 and 752.2954.

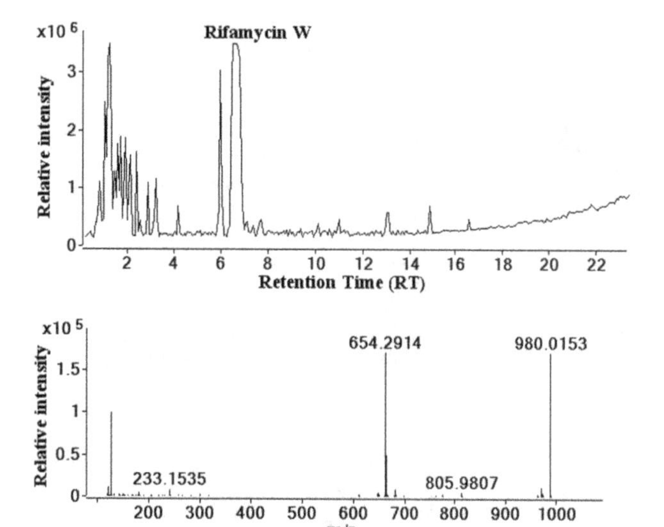

Figure 5. Total Ion Chromatogram for *S. arenicola* strain M413 (ACM5232) and the mass spectrum for peak Y.

The detection of two rifamycin compounds among strains of *S. arenicola* was surprising given that these compounds had previously been observed exclusively in the terrestrial soil actinobacterium *Amycolatopsis mediterranei* [21], [39]. Out of 36 *Salinispora* strains screened in this study, only two strains produced detectable concentrations of rifamycin O, and one produced rifamycin W, as would be expected if a biosynthetic pathway had been inherited from *Amycolatopsis mediterranei* [20]. Nevertheless, most strains showed three peaks at retention times corresponding to previous work relating to other rifamycins (S, SV and B) [35]. Wilson and co-workers have reported that different strains of *Amycolatopsis mediterranei* produce different rifamycins [39]. Moreover, these rifamycins are subject to inter-conversion as a result of their physicochemical properties and vary considerably in their antimicrobial spectrum and the extent of their biological activities [40]; for example, the 16,17,18,19-tetrahydro analogue of rifamycin SV is three-fold less potent than rifamycin SV against *E. coli* polymerase [23]. It has been reported that rifamycin W is the precursor for rifamycin S, SV, B, L and Y [22], [39]. In previous studies we found that rifamycin B production is higher in all the strains and is converted to rifamycin SV and S over time [35]. Similarly, Banerjee and colleagues found that rifamycin B is a precursor for rifamycin Y and other rifamycin analogues [24]. The presence of the enzyme rifamycin oxidase facilitates the biotransformation process from rifamycin B to rifamycin S, a stable analogue with good antimicrobial activity.

There is a high similarity between the *rif* gene sequence found in *Amycolatopsis mediterranei* and *Salinispora* sp. (99% as assessed by neighbour-joining bootstrap values)and the study suggests that the gene has been exchanged horizontally [9]. However, it is not clear at what point during the evolution of the *Salinispora* genus that the pathway may have been acquired. The rifamycin O and W genes were not found to be present in the *S. arenicola* genome from a previous study by Wilson and co-workers [39]. In our study, the presence of rifamycins O and W in the *S. arenicola* secondary metabolome at least in some strains suggests that *rif* O and W genes were acquired by HGT due to the selective advantage these antimicrobial compounds confer in their environment. This evolutionary history is what might be expected if the acquisition of pathways facilitates ecological diversification or a selective sweep [41] resulting from strong selection for the acquired pathway, either of which provide compelling evidence that the production of secondary metabolites has great impact on creating functional traits with important ecological roles. The concept that gene acquisition provides a mechanism for ecological diversification that may finally drive the formation of independent bacterial lineages has been previously proposed [42]. Interestingly the inclusion of secondary metabolism among the functional categories of acquired genes that may have this effect shows the importance of the functional and evolutionary significance of these gene clusters [26]. Inferring the evolutionary histories of the biosynthetic pathways associated with secondary metabolism remains complex but provides a possibility to understand how nature creates structural diversity and the extent to which this diversity is related to phylogenetic grouping.

A previous study of the marine actinobacteria *Salinispora* showed that this genus produces secondary metabolite profiles that are species-specific, for instance Jensen and co-workers have reported core compounds that are identical in each species [15]. To date, traditional screening techniques, for instance bioactivity guided assays [43], and subsequently genome mining techniques [27] have mainly been used to isolate and identify secondary metabolite profiles of *Salinispora*. However, these studies have been constrained to targeted compounds and have not focused on a broader investigation of the wider metabolome. Our results show that the application of UHPLC-QToF-MS and chemometric approaches can be used to successfully differentiate and discriminate between two species and possibly identify new compounds. In this instance the goal of the metabolomic profiling was to discover new biochemical descriptors and was essentially a semi-quantitative analysis. However, these data lend themselves to directed interrogation if specific compounds (e.g. rifamycins) are also of interest, therefore providing a "best-of-both-worlds" outcome. Targetted MS/MS experiments were performed when specific results were required (e.g. rifamycins O and W).

Despite sample quantity and availability being an issue in marine natural product research we found distinct metabolite profiles to be present in each species with some degree of commonality between the two. These findings extend those of Jensen (2007), confirming that species-specific metabolites are produced by *S. arenicola* and *S. pacifica*. The combination of MS-based metabolic profiling and chemometrics enables the discovery and identification of a larger proportion of the secondary metabolome than available with traditional approaches. However, the cross-over between studies is incomplete and this could be due to a number of factors, not least the variability of culture medium conditions [44] as well as geographical differences (even in centimetre scale collections) [26], which undoubtedly play important roles in secondary metabolite production. However, this study provides compelling evidence that a metabolomic profiling approach affords an efficient and effective tool for natural product discovery. Consequently, the distinctions in profiles noted in our study were unrelated to a specific biological activity or aimed at only targeted compounds. This study therefore indicates the benefit of using high resolution accurate mass spectrometry and chemometric analysis in exploring microbial metabolite profiles, as it is rapid and reproducible, and only a small amount of experimental sample is required to obtain valuable information. Most notably, this is the first study to our knowledge to investigate the feasibility of using UHPLC-QToF-MS and chemometrics to explore metabolite profiles in the marine natural product-synthesizing actinobacterium *Salinispora*.

In conclusion, we have shown that the diversity of two *Salinispora* species, based on their metabolomes and thus natural product chemotypes, is significantly greater than suggested by previously identified compounds. Our data reveals that the qualitative variation in the (bio)chemical profiles of the two species provides a major source of differentiation between these species, in addition to previous genetic and targeted chemical classification. We can now appreciate the spectrum of secondary metabolomes in *Salinispora* strains as much wider than already described compounds from these species, providing a rainbow of new natural product 'colours' for biodiscovery from these important marine bacteria.

Supporting Information

Figure S1 LC-UV-Vis (430 nm) chromatograms of *S. arenicola* strains (A) MV0472, (B) MV0318, and (C) M413 (ACM5232) showing 2 new unknown peaks X and Y.

Figure S2 Negative mode MS/MS fragmentation spectra (A) Rifamycin O standard (B) Peak X. Major fragments observed were m/z 694.2872, 636.2470, 514.1703, 392.1159 and 272.0554 for X and m/z 694.2872, 636.2740, 514.1755, 392.1137 and 272.0555 for rifamycin O standard.

Figure S3 Proposed MS/MS fragmentation pattern of rifamycin O in negative ionization mode.

Figure S4 Negative mode MS/MS fragmentation of peak Y. Major fragments are m/z 624.2823, 538.2089, 452.1729, 330.0993, 272.0567 and 245.0333.

Figure S5 Proposed MS/MS negative mode fragmentation pattern of rifamycin W.

Figure S6 Full total ion current (TIC) chromatograms of representative extracts from *S. arenicola* and *S. pacifica*.

Table S1 QToF-MS operational conditions (switching).

Table S2 Empirical formula generated from 46 strains of *S. arenicola* and *S. pacifica* from Great Barrier Reef (GBR) regions. The proposed formula obtained after the PCA and OPLS-DA analysis according to high-resolution LC-QToF-MS measurements.

Acknowledgments

We wish to thank the crew of the FRV Gwendoline May (QDPI&F) and RV Lady Basten (AIMS).

http://www.reef.crc.org.au/resprogram/programC/seabed.

Author Contributions

Conceived and designed the experiments: AKH PNS JAF. Performed the experiments: UB MEV YKN. Analyzed the data: UB MPH. Contributed reagents/materials/analysis tools: JAF AKH PNS MPH. Wrote the paper: UB MPH AKH. Critically reviewed the manuscript: PNS JAF.

References

1. Paul VJ, Puglisi MP, Ritson-Williams R (2006) Marine chemical ecology. Nat Prod Rep 23: 153–180.
2. Phelan VV, Liu W-T, Pogliano K, Dorrestein PC (2011) Microbial metabolic exchange - the chemotype-to-phenotype link. Nat Chem Biol 8: 26–35.
3. Yim G, Wang HH, Davies J (2007) Antibiotics as signalling molecules. Philos Trans R Soc Lond B Biol Sci 362: 1195–1200.
4. Fenical W, Jensen PR (2006) Developing a new resource for drug discovery: marine actinomycete bacteria. Nat Chem Biol 2: 666–673.
5. Mincer TJ, Jensen PR, Kauffman CA, Fenical W (2002) Widespread and persistent populations of a major new marine actinomycete taxon in ocean sediments. Appl Environ Microbiol 68: 5005–5011.
6. Ahmed L, Jensen PR, Freel KC, Brown R, Jones AL, et al. (2013) *Salinispora pacifica* sp. nov., an actinomycete from marine sediments. Antonie van Leeuwenhoek 103: 1069–1078.
7. Maldonado LA, Fenical W, Jensen PR, Kauffman CA, Mincer TJ, et al. (2005) *Salinispora arenicola* gen. nov., sp. nov. and *Salinispora tropica* sp. nov., obligate marine actinomycetes belonging to the family Micromonosporaceae. Int J Syst Evol Microbiol 55: 1759–1766.
8. Fenical W, Jensen PR, Palladino MA, Lam KS, Lloyd GK, et al. (2009) Discovery and development of the anticancer agent salinosporamide A (NPI-0052). Bioorg Med Chem 17: 2175–2180.
9. Kim TK, Hewavitharana AK, Shaw PN, Fuerst JA (2006) Discovery of a new source of rifamycin antibiotics in marine sponge actinobacteria by phylogenetic prediction. Appl Environ Microbiol 72: 2118–2125.
10. Ejje N, Soe CZ, Gu J, Codd R (2013) The variable hydroxamic acid siderophore metabolome of the marine actinomycete Salinispora tropica CNB-440. Metallomics 5(11): 1519–1528
11. Edlund A, Loesgen S, Fenical W, Jensen PR (2011) Geographic distribution of secondary metabolite genes in the marine actinomycete *Salinispora arenicola*. Appl Environ Microbiol 77: 5916–5925.
12. Penn K, Jenkins C, Nett M, Udwary DW, Gontang EA, et al. (2009) Genomic islands link secondary metabolism to functional adaptation in marine actinobacteria. ISME J 3: 1193–1203.
13. Udwary DW, Zeigler L, Asolkar RN, Singan V, Lapidus A, et al. (2007) Genome sequencing reveals complex secondary metabolome in the marine actinomycete *Salinispora tropica*. Proc Natl Acad Sci USA 104: 10376–10381.
14. Freel KC, Nam SJ, Fenical W, Jensen PR (2011) Evolution of secondary metabolite genes in three closely relatedmarine actinomycete species. Appl Environ Microbiol 77: 7261–7270.
15. Jensen PR, Williams PG, Oh D-C, Zeigler L, Fenical W (2007) Species-specific secondary metabolite production in marine actinomycetes of the genus *Salinispora*. Appl Environ Microbiol 73: 1146–1152.
16. Jensen PR, Mincer TJ, Williams PG, Fenical W (2005) Marine actinomycete diversity and natural product discovery. Antonie Van Leeuwenhoek 87: 43–48.
17. Keeling PJ, Palmer JD (2008) Horizontal gene transfer in eukaryotic evolution. Nat Rev Genet 9: 605–618.
18. Doolittle WF (1999) Lateral genomics. Trends Cell Biol 9: M5–M8.
19. Ricard G, McEwan NR, Dutilh BE, Jouany J-P, Macheboeuf D, et al. (2006) Horizontal gene transfer from bacteria to rumen ciliates indicates adaptation to their anaerobic, carbohydrates-rich environment. BMC Genomics 7: 22.
20. Kim TK, Fuerst JA (2006) Diversity of polyketide synthase genes from bacteria associated with the marine sponge *Pseudoceratina clavata*: culture-dependent and culture-independent approaches. Environ Microbiol 8: 1460–1470.
21. Yu T-W, Shen Y, Doi-Katayama Y, Tang L, Park C, et al. (1999) Direct evidence that the rifamycin polyketide synthase assembles polyketide chains processively. Proc Natl Acad Sci USA 96: 9051–9056.
22. Schupp T, Traxler P, Auden J (1981) New rifamycins produced by a recombinant strain of *Nocardia mediterranei*. J Antibiot 34: 965.
23. Aristoff PA, Garcia GA, Kirchhoff PD, Hollis SH (2010) Rifamycins-obstacles and opportunities. Tuberculosis 90: 94–118.
24. Banerjee U, Saxena B, Chisti Y (1992) Biotransformations of rifamycins: process possibilities. Biotechnol Adv 10: 577–595.
25. Krug D, Zurek G, Schneider B, Garcia R, Muller R (2008) Efficient mining of myxobacterial metabolite profiles enabled by liquid chromatography-electrospray ionisation-time-of-flight mass spectrometry and compound-based principal component analysis. Anal Chim Acta 624: 97–106.
26. Krug D, Zurek G, Revermann O, Vos M, Velicer GJ, et al. (2008) Discovering the hidden secondary metabolome of *Myxococcus xanthus*: a study of intraspecific diversity. Appl Environ Microbiol 74: 3058–3068.
27. Eustáquio AS, Nam S-J, Penn K, Lechner A, Wilson MC, et al. (2011) The discovery of Salinosporamide K from the marine bacterium "*Salinispora pacifica*" by genome mining gives insight into pathway evolution. ChemBioChem 12: 61–64.
28. Hornung A, Bertazzo M, Dziarnowski A, Schneider K, Welzel K, et al. (2007) A genomic screening approach to the structure guided identification of drug candidates from natural sources. ChemBioChem 8: 757–766.
29. Zhou J, Kang S, Schadt CW, Garten CT (2008) Spatial scaling of functional gene diversity across various microbial taxa. Proc Natl Acad Sci USA 105: 7768.
30. Oliver SG, Winson MK, Kell DB, Baganz F (1998) Systematic functional analysis of the yeast genome. Trends Biotechnol 16: 373–378.

31. Hodson MP, Dear GJ, Roberts AD, Haylock CL, Ball RJ, et al. (2007) A gender-specific discriminator in Sprague–Dawley rat urine: the deployment of a metabolic profiling strategy for biomarker discovery and identification. Anal Biochem 362: 182–192.
32. Dunn WB, Goodacre R, Neyses L, Mamas M (2011) Integration of metabolomics in heart disease and diabetes research: current achievements and future outlook. Bioanalysis 3: 2205–2222.
33. Vidgen M, Hooper JNA, Fuerst J (2011) Diversity and distribution of the bioactive actinobacterial genus *Salinispora* from sponges along the Great Barrier Reef. Antonie Van Leeuwenhoek 101: 603–618.
34. Ng YK, Hewavitharana AK, Webb R, Shaw PN, Fuerst JA (2012) Developmental cycle and pharmaceutically relevant compounds of *Salinispora* actinobacteria isolated from Great Barrier Reef marine sponges. Appl Microbiol Biotechnol 97: 3097–3108.
35. Hewavitharana AK, Shaw PN, Kim TK, Fuerst JA (2007) Screening of rifamycin producing marine sponge bacteria by LC-MS-MS. J Chromatogr B 852: 362–366.
36. Kind T, Fiehn O (2007) Seven golden rules for heuristic filtering of molecular formulas obtained by accurate mass spectrometry. BMC Bioinformatics 8: 105.
37. Barrett RD, Schluter D (2008) Adaptation from standing genetic variation. Trends Ecol Evol 23: 38–44.
38. Macel M, Van D, Nicole M, Keurentjes JJ (2010) Metabolomics: the chemistry between ecology and genetics. Mol Ecol Resour 10: 583–593.
39. Wilson MC, Gulder TAM, Mahmud T, Moore BS (2010) Shared biosynthesis of the saliniketals and rifamycins in *Salinispora arenicola* is controlled by the sare1259-encoded cytochrome P450. J Am Chem Soc 132: 12757–12765.
40. Zhao W, Zhong Y, Yuan H, Wang J, Zheng H, et al. (2010) Complete genome sequence of the rifamycin SV-producing *Amycolatopsis mediterranei* U32 revealed its genetic characteristics in phylogeny and metabolism. Cell Res 20: 1096–1108.
41. Cohan FM (2002) What are bacterial species? Annu Rev Microbiol 56: 457–487.
42. Ochman H, Worobey M, Kuo C-H, Ndjango J-BN, Peeters M, et al. (2010) Evolutionary relationships of wild hominids recapitulated by gut microbial communities. PLoS Biol 8: e1000546.
43. Feling RH, Buchanan GO, Mincer TJ, Kauffman CA, Jensen PR, et al. (2003) Salinosporamide A: A highly cytotoxic proteasome inhibitor from a novel microbial source, a marine bacterium of the new genus *Salinospora*. Angew Chem Int Ed Engl 42: 355–357.
44. Larsen TO, Smedsgaard J, Nielsen KF, Hansen ME, Frisvad JC (2005) Phenotypic taxonomy and metabolite profiling in microbial drug discovery. Nat Prod Rep 22: 672–695.

8

Intestinal Microbial Ecology and Environmental Factors Affecting Necrotizing Enterocolitis

Roberto Murgas Torrazza[1], Maria Ukhanova[2], Xiaoyu Wang[2], Renu Sharma[3], Mark Lawrence Hudak[3], Josef Neu[1], Volker Mai[2]*

1 Department of Pediatrics, College of Medicine University of Florida, Gainesville, Florida, United States of America, **2** Department of Epidemiology, College of Public Health and Health Professions and College of Medicine and Emerging Pathogens Institute, University of Florida, Gainesville, Florida, United States of America, **3** Department of Pediatrics University of Florida College of Medicine, Jacksonville, Florida, United States of America

Abstract

Necrotizing enterocolitis (NEC) is the most devastating intestinal disease affecting preterm infants. In addition to being associated with short term mortality and morbidity, survivors are left with significant long term sequelae. The cost of caring for these infants is high. Epidemiologic evidence suggests that use of antibiotics and type of feeding may cause an intestinal dysbiosis important in the pathogenesis of NEC, but the contribution of specific infectious agents is poorly understood. Fecal samples from preterm infants ≤32 weeks gestation were analyzed using 16S rRNA based methods at 2, 1, and 0 weeks, prior to diagnosis of NEC in 18 NEC cases and 35 controls. Environmental factors such as antibiotic usage, feeding type (human milk versus formula) and location of neonatal intensive care unit (NICU) were also evaluated. Microbiota composition differed between the three neonatal units where we observed differences in antibiotic usage. In NEC cases we observed a higher proportion of Proteobacteria (61%) two weeks and of Actinobacteria (3%) 1 week before diagnosis of NEC compared to controls (19% and 0.4%, respectively) and lower numbers of Bifidobacteria counts and Bacteroidetes proportions in the weeks before NEC diagnosis. In the first fecal samples obtained during week one of life we detected a novel signature sequence, distinct from but matching closest to *Klebsiella pneumoniae*, that was strongly associated with NEC development later in life. Infants who develop NEC exhibit a different pattern of microbial colonization compared to controls. Antibiotic usage correlated with these differences and combined with type of feeding likely plays a critical role in the development of NEC.

Editor: Dipshikha Chakravortty, Indian Institute of Science, India

Funding: This work was supported by NIH RO1 HD 059143. The funders had no role in study design, data collection and analysis, decision to publish, or preparation of the manuscript.

* E-mail: vmai@ufl.edu

Introduction

Necrotizing Enterocolitis (NEC) is the most devastating intestinal disease in neonates.[1] On the basis of large, multicenter, neonatal network databases from the United States and Canada, the mean prevalence of the disorder is about 7% among infants with a birth weight between 500 and 1500 g but can vary markedly among centers.[2] The most recent data from different Neonatal Intensive Care Units in the National Institutes of Child Health Neonatal Network show a range of 4 to 19 percent in infants less than 28 weeks gestational age[3] and the range from Vermont Oxford Neonatal Intensive Care Units from 2010[4] shows a range from 2.2% to 8.3% (1st and 3rd quartiles) in babies born less than 1500 grams. These inter-unit variances suggest that differences in practice or other variables may contribute to the pathogenesis of this devastating disease.

An integral link between microbial dysbiosis, an exaggerated inflammatory response and NEC was hypothesized over a decade ago.[5] Maintenance of intestinal integrity and promotion of postnatal intestinal growth would accordingly require a delicate balance between intestinal microbiota and the immune system of premature infants, which can be affected by various environmental factors. Aberrant microbial colonization patterns or abnormal responses to normal microbiota might disrupt this balance and contribute to the development of NEC.

Studies using culture-based techniques have demonstrated differences in the intestinal microbiota of patients who subsequently developed NEC versus controls.[6] More recent studies using molecular methods to evaluate fecal microbiota from unaffected preterm infants, as well as some infants in whom NEC developed and from whom samples were obtained before and during NEC, suggest that this disease is associated with unusual intestinal microbes. [7–10] Although various microbes have been cultured from blood and stools in outbreaks of NEC at single institutions,[6,11,12] no single organism has consistently been implicated across centers.[13] The human microbiome project,[14] in conjunction with technologic advances that allow for the molecular identification of a vast array of microbes that are difficult or impossible to culture from the intestine, has given us new tools for generating evidence to test the "abnormal colonization hypothesis".[15] Numerous environmental factors may contribute to intestinal microbial colonization patterns that predispose to NEC. Epidemiologic studies show a correlation between length of prior antibiotic use and the occurrence of NEC.[16,17] The use of human milk versus formula, a factor known to be protective against NEC [18], also is likely to relate to altered microbial colonization.

In our previous study using high throughput sequencing, we demonstrated a bloom of Proteobacteria and several differences in operational taxanomic units (OTUs) prior to the onset of NEC.[9] Furthermore, we detected several operational taxonomic units (OTUs) associated with NEC status that did not have exact matches in Genbank, although they matched closest to *Klebsiella* in the class γ-proteobacteria.

There is little information currently available about variances and dynamics in microbiota associated with neonatal intensive care in different hospitals. Differences in the hospital environment and clinical practice such as routines in antibiotic administration and feeding can contribute to the establishment of distinct microbiota pattern in each unit. This has important implications for our ability to generalize findings regarding specific patterns of microbial dysbiosis in the pathogenesis of NEC in a specific NICU. In this current study, we address these issues by testing the hypotheses that a) the ontogeny of fecal microbiota differs in infants who subsequently develop NEC from those who remain free of NEC and b) environmental factors (e.g., antibiotic exposure, diet) may help explain the variance in NEC prevalence at different NICUs.

Materials and Methods

Ethics Statement

Written informed consent was obtained from the infants' parents and investigations were conducted according to the principles expressed in the Declaration of Helsinki. The study including consent procedure was approved by the UF Health Institutional Review Board 01.

Study Design

Premature infants born at a postmenstrual age ≤32 weeks without major congenital anomalies or malformations were enrolled at three University of Florida affiliated hospitals shortly after birth. Two control infants were selected and matched to each NEC case infant by postmenstrual age, birth weight, hospital of birth, and date of birth (+/− 2 months). We could not match an appropriate second control infant for one of the cases, resulting in a total of 18 case and 35 control infants. NEC cases included only those infants with definite clinical and radiologic signs (pneumatosis intestinalis and/or portal venous gas) or necrotic bowel at surgery.

Weekly stool samples from study infants were collected from diapers beginning with the first available stool (meconium) and continuing until discharge, for immediate storage at −80°C. The samples analyzed from cases included those collected 2 weeks before NEC (15±3; days prior to NEC), 1 week before NEC (8.4±2.6 days prior), and the sample closest to diagnosis of (2.5±2 days prior) NEC. Samples from matched control infants were chosen during the same week of life at which the samples from the cases were obtained. For 12 out of the 18 infants that later developed NEC, and matched controls, one of the samples collected before NEC diagnosis represented the very first stool sample that was obtained during week one of life. These samples were analyzed in a subanalysis.

Written informed consent was obtained from the infants' parents and investigations were conducted according to the principles expressed in the Declaration of Helsinki. The study including consent procedure was approved by the Institutional Review Boards of all three hospitals.

Microbiota Analysis

DNA extraction and quality control by denaturing gradient gel electrophoresis (DGGE). DNA was extracted from 200–300 mg fecal samples using a modified Qiagen stool DNA extraction protocol that included a bead beating step.[19] We used DGGE analysis of the V6–V8 region as described previously for initial quality control.[7]

16S rRNA sequence analysis. DNA was amplified using a primer set based on universal primers 27F (AGAGTTTGATCCTGGCTCA) and 533R (TTACCGCGGCTGCTGGCAC) to which titanium adaptor sequences and barcodes were added. Cleansed PCR products were pooled in equimolar amounts and submitted for sequencing using 454-Titanium chemistry. From the resulting raw data set, low quality sequences or sequences with a length less than 150 nucleotides were removed. We used the ESPRIT-tree algorithm, which maintains the binning accuracy of ESPRIT[20] while improving computational efficiency to bin sequences into Operational Taxonomic Units (OTUs) using similarity levels from 99% (species/strain level) to 80% (phylum level). We used QIIME[21] to calculate Chao rarefaction diversity and UniFrac distances [22] for comparing α and β diversity respectively.

For comparison between the current and the earlier sequence dataset we pooled sequences from both reads and reassigned OTUs using ESPRIT-tree. We then identified OTUs that contained the sequences that were found to differ between cases and controls in either dataset to determine sequence distribution in them.

qPCR based quantification of Bifidobacteria. We used a *Bifidobacteria* specific primer set (F: 5′ TCG CGT C(C/T)G GTG TGA AAG 3′; R: 5′ CCA CAT CCA GC(A/G) TCC AC 3′, annealing temperature 58°C) to quantify the amounts of *Bifidobacteria* genome equivalents in fecal samples. Duplicate vials containing 10 ng of DNA were included in each reaction and DNA purified from *B. longum* was used to generate the standard curve. Samples with less than one genome equivalent/ng of DNA were considered as negative.

Statistics

Paired Student's *t*-test was used for normally distributed data. A chi squared- test and Fisher's exact were used to evaluate demographic data and clinical characteristics as appropriate. Two-tailed *p*-values were calculated and $p<0.05$ was considered to be statistically significant. To test for a difference in the abundance of OTUs a paired chi square test was followed by Fisher exact test. We adjusted for an expected high false discovery rate by increasing the requirement for statistical significance to $p<0.01$.The QIIME package was used to calculate *p*-values for differences in UniFrac distances.

Results

Patient characteristics and clinical outcomes

Baseline characteristics are summarized in table 1. The mean postmenstrual age in case and control infants was 28±2.36 weeks. Mean birth weight was 1187±371 grams for both groups combined, with an almost 20% less mean weight in NEC cases that did not reach statistical significance. The incidence of NEC was 12.4% for Gainesville and 6.8% for Jacksonville during this study period. There were no significant differences in clinical

Table 1. Baseline Characteristics of the Infants. (Mean ± SD).

Characteristic	NEC (N = 18)	Control (N = 35)
Birth weight - g	1073±394	1246±350
Gestational age at birth - wk	27.4±2.6	28.5±2.2
Male sex - no./total no. (%)	12/18 (66.7)	17/35 (48.6)
Type of Milk - no./total no. (%)		
Breast Milk	5/18 (27.8)	20/35 (57.1)
Formula	3/18 (16.7)	3/35 (8.6)
Both	10/18 (55.6)	12/35 (34.2)
Mode of delivery - no./total no. (%)		
Vaginal	9/18 (50)	12/35 (34.3)
C-section	9/18 (50)	23/35 (65.7)
Use of antenatal corticosteroids—no./total no. (%)		
Any	3/18 (16.7)	10/35 (28.6)
Full course	11/18 (61)	20/35 (57)
Prenatal antibiotic exposure	13/18 (72.2)	29/35 (82.9)
Apgar score at 1 min	5±2.8	5.3±2.7
Apgar score at 5 min	7±2.3	7.6±1.7
Positive pressure ventilation (bag and mask)	11/18 (61.1)	16/35 (47)
Continuous positive airway pressure (CPAP)	6/18 (33.3)	13/35 (37.1)
Intubation and mechanical ventilation	10/18 (55.6)	14/35 (41.2)
Day of life of Development of NEC	17.83±12.8	
Total days on antibiotics prior to NEC or sample	6.2±6.9	4.9±3.4

characteristics and major co-morbidities between the two groups (Table 2).

Type of feeding

Of the infants in the control group 57.1% received exclusively maternal milk compared to 27.8% in cases ($p<0.05$).

Antibiotic Exposure

As shown in table 1, there was similar prenatal exposure to antibiotics administered to mothers. Postnatally, the duration of antibiotic exposure prior to NEC did not differ between cases and controls (6.2±6.9 days versus 4.9±3.4 days, mean ± S.D.). We did not detect significant differences in the individual antibiotics prescribed before diagnosis in NEC cases compared with controls (Table 3). While we observed some differences in antibiotic administration between the NICU in Gainesville and the two NICUs in Jacksonville (Figures S1 and S2), with a length of antibiotic therapy of 22 days in Jacksonville and 7 days in Gainesville at the 75th quartile, none of these differences reached statistical significance. Prior to the development of NEC among cases born in Gainesville 73% were exposed to Ampicillin and Gentamicin and 18% to Oxacillin, which was not used in Jacksonville, whereas among NEC cases born in Jacksonville,

Table 2. Major Clinical Outcomes. # cases/# in group (%).

	NEC (N = 18)	Control (N = 35)	P value
Bronchopulmonary dysplasia	5/18 (27.8)	5/35 (14.3)	0.23
Intraventricular hemorrhage	4/18 (22.2)	4/35 (11.4)	0.35
Patent ductus arteriosus	6/18 (33.3)	7/35 (20)	0.26
Periventricular leukomalacia	1/18 (5.6)	3/35 (8.6)	0.12
Retinopathy of prematurity	2/18 (11.1)	8/35 (22.9)	0.27

Table 3. Antibiotic exposure before NEC # prescribed antibiotic/# in group (%).

	NEC	*Control*
Ampicillin	15/18 (83.3)	31/35 (88.6)
Azithromycin	1/18 (5.6)	4/35 (11.4)
Cefazolin	1/18 (5.6)	0/35
Cefotaxime	0/18 (0)	4/35 (11.4)
Ceftazidime	0/18 (0)	1/35 (2.9)
Clindamycin	1/18 (5.6)	0/35 (0)
Fluconazole	0/18 (0)	1/35 (2.9)
Gentamicin	15/18 (83.3)	30/35 (85.7)
Oxacillin	2/18 (11.1)	1/35 (2.9)
Piperacillin/tazobactam	2/18 (11.1)	2/35 (5.7)
Vancomycin	3/18 (16.7)	2/35 (5.7)

100% were exposed to Ampicillin and Gentamicin and to other antibiotics not used in Gainesville such as Vancomycin (40%), Piperacillin/Tazobactam (28%), Azithromycin (14%), Cefazolin (14%), Clindamycin (14%),.

Microbiota analysis

The distribution of dominant OTUs from all three NICUs is shown in Figure 1. The most dominant OTU's detected in both cases and controls matched closest to Enterococcus sp., Staphylococcus sp, Phyllobacterium sp., Bacteroides sp., Escherichia sp., Parabacteroides sp., and Veillonella sp.

Species richness, a measure of alpha (within sample) diversity as determined by a Chao1-based estimate of the total numbers of OTUs present, did not differ between cases and controls at any of the three time-points (Figure 2). This indicates that the total numbers of different bacterial species present in the gut was stable during the observation period.

In contrast, using the UNIFRAC metric of beta (between samples) diversity, we detected a difference in overall microbiota composition between cases and controls two weeks before diagnosis (Figure 3). At this time point, samples from cases clustered together more closely compared to samples from controls ($p<0.05$). This suggests that although the total number of bacteria present did not change over time or differ by case status, the kinds of bacteria present and their proportions did differ. However, no clustering by case status was detected during the following weeks.

The clustering by case status two weeks before diagnosis can be attributed largely to an increased proportion in *Proteobacteria* and a decreased proportion of *Bacteroidetes* in cases, which is most evident when the mean for each group is considered (Figure 4). A matched analysis comparing the distribution of the major phyla showed no difference at week of diagnosis in any of the major phyla. Proteobacteria decreased with advancing age in both cases and controls. At 2 weeks before diagnosis proteobacteria were increased in cases compared to controls, $P<0.001$. At one week before diagnosis Actinobacteria and Proteobacteria was increased in cases compared to controls, $P<0.001$. When we examined the abundances of phyla by center we found no difference in cases but detected in controls a clear difference by center at all time points. During week of diagnosis (in matched cases) we observed in controls from Gainesville a higher proportion of Bacteroidetes, which were almost completely absent in controls from Jacksonville ($P<0.01$), and Proteobacteria at the cost of Firmicutes (Figure 5).

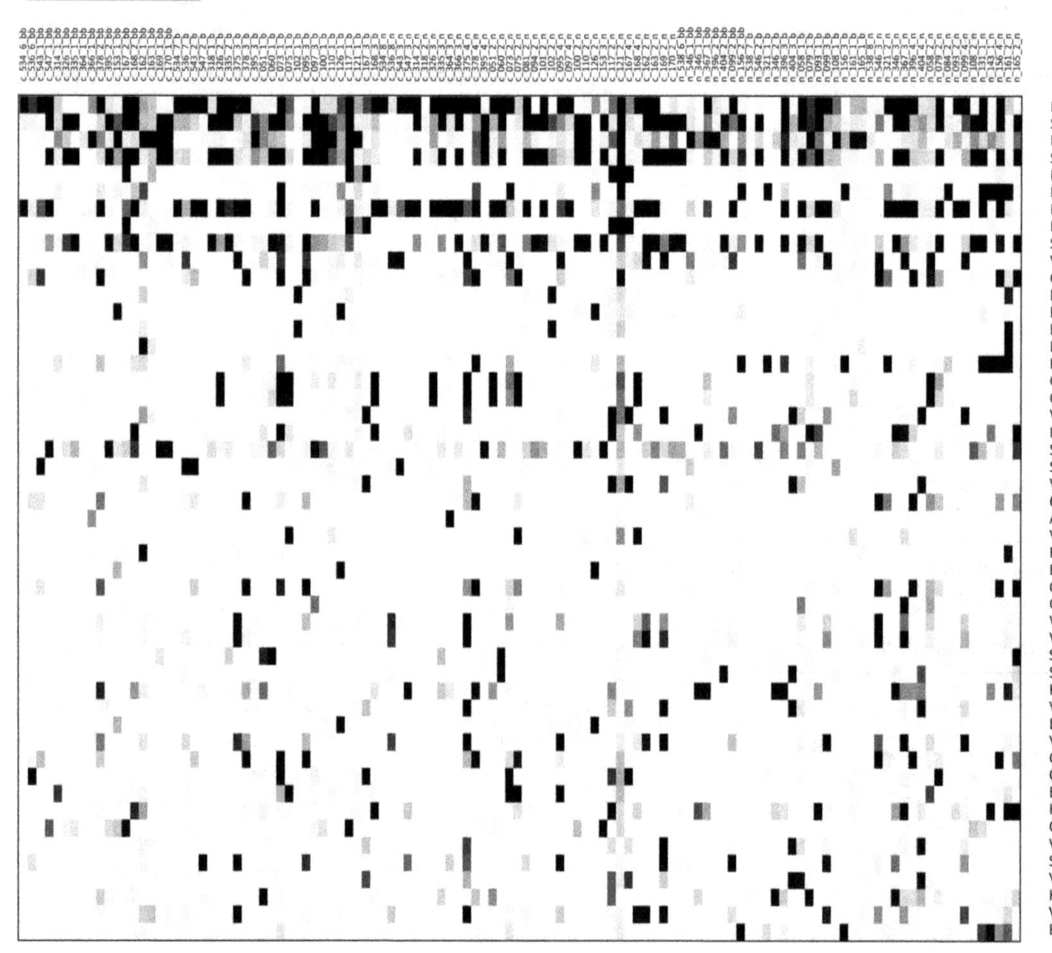

Figure 1. Dominant OTU's. Each row represents a separate OTU with the closest match in the database listed on the right. Each column represents an individual sample from both cases and controls at each of the three time points. OTUs are listed in order of dominance with the most dominant on the top. Darker color indicates higher number of sequences in this OTU in an individual sample (see color code in left corner). Darkest shade indicates 50+ sequence reads obtain for that OTU in that sample.

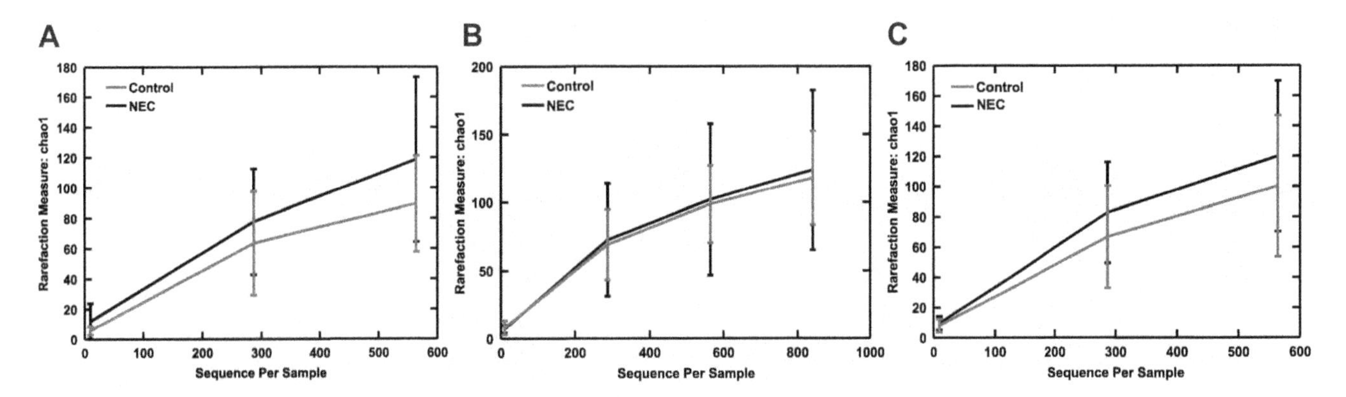

Figure 2. Chao rarefaction diversity. Chao diversity was calculated from the OTU distribution A) 2 weeks before diagnosis of NEC; B) 1 week before diagnosis of NEC; and C) Week of diagnosis of NEC. As a measure of beta (within sample) diversity it is an estimate of the expected total number of OTUs detected in the sample if sequenced to completion.

In the analysis of the distribution of all OTUs between cases and controls at each time point we detected various OTUs, matching to different bacterial phyla/families/classes, that significantly differed in prevalence/frequency by case status. During the week of diagnosis we observed such OTUs in all but one of the cases (Figure 6). These OTUs might represent either novel pathogens contributing to NEC or commensals that can thrive under the conditions in the gut when NEC develops. Multiple OTUs matching closest to the potentially pathogenic *Klebsiella granulomatis*, *Klebsiella pneumoniae* and Clostridium perfringens were detected close to the time of diagnosis more frequently in cases. Other significantly increased OTUs in NEC cases matched closest to *Staphylococcus epidermidis*. Many OTUs frequently observed in controls at early time points were completely missing in cases, suggesting that bacteria normally colonizing the gut of control infants didn't do so in cases. Other OTUs were only detected in cases, some of them in significant numbers especially at the two time points closest to diagnosis.

We used a bifidobacteria targeting qPCR approach to quantify colonization in cases and controls. This was necessary because the 'universal' primer set used in our 16S rRNA sequence analysis is biased against bifidobacteria, a group of bacteria known to frequently colonize the infant gut. While numbers of bifidobacteria appeared lower in cases than in controls during the weeks before diagnosis ($p<0.05$) there was no difference during the week of diagnosis (Figure 7).

We then performed a sub-analysis in all 12 of the NEC cases, and 23 matched controls, for which we had sequence data available for the very first stools collected during week one of life. In these samples we observed a particularly strong association between NEC risk and an OTU distinct from but matching closest to K.p.. This OTU was detected in 11/12 infants that later developed NEC, compared to only 9/23 matched controls ($p<0.01$). Furthermore, in 5/12 NEC cases this OTU represented more than 10% (range 10-63%) of all 16S rRNA sequences that were obtained, compared to only 1/23 controls in which this OTU was >10% (Figure S3).

Discussion

A symbiotic relationship between intestinal microbiota and the host is established soon after birth. Preterm infants have

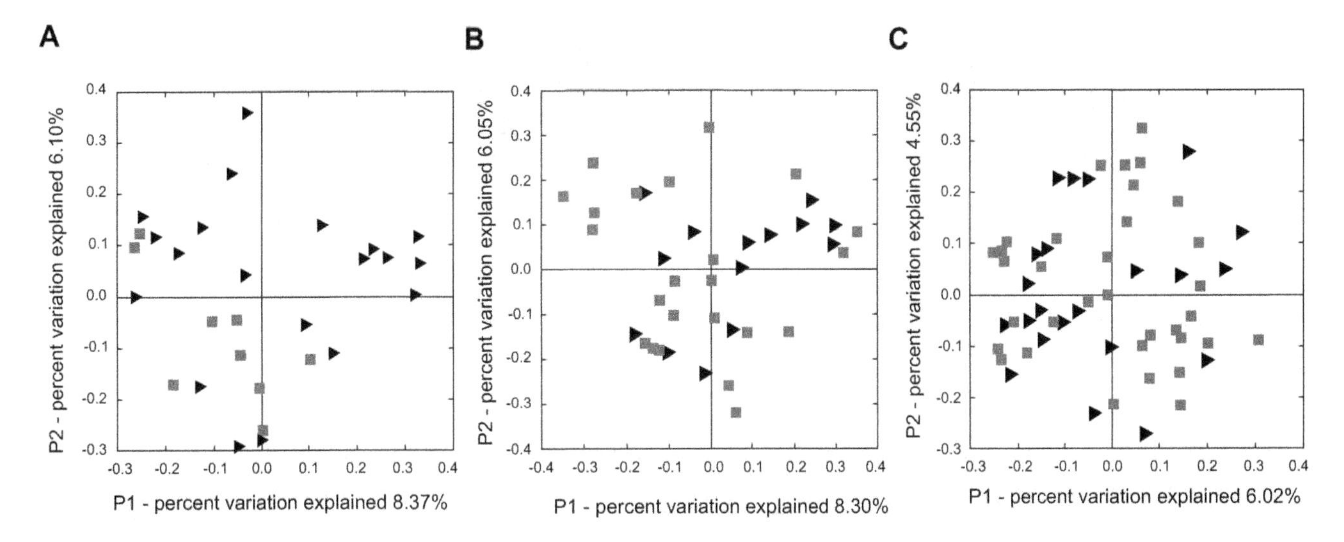

Figure 3. Unifrac diversity measures. Principal component analysis (PCA) of overall diversity based on UniFrac (unweighted) metric A) 2 weeks before; B) 1 week before; and C) week of diagnosis of NEC. Squares represent controls and triangles represent cases. P1 is component 1 and P2 component 2.

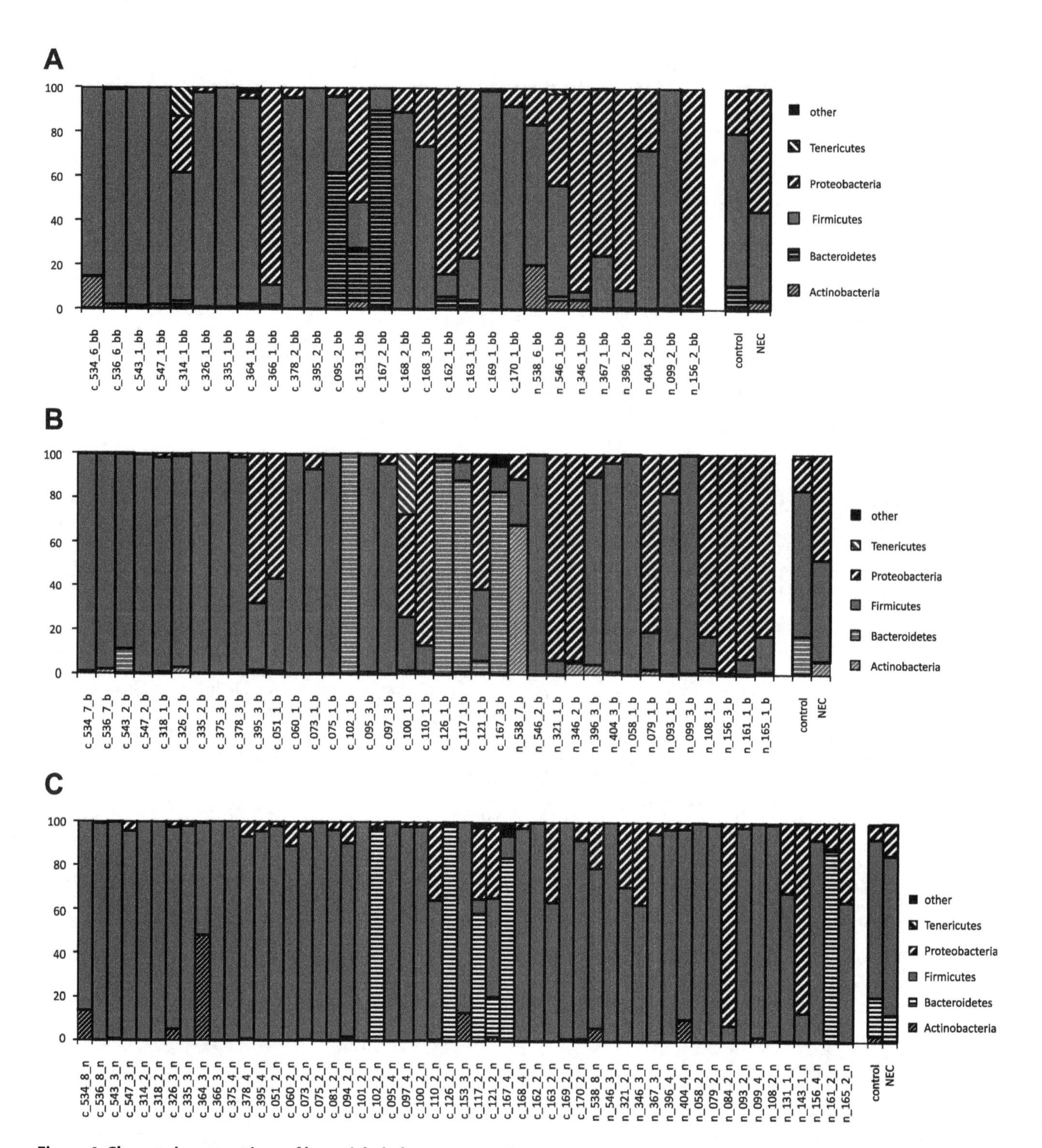

Figure 4. Changes in proportions of bacterial phyla. Proportions of the major bacterial phyla at A) two weeks before, B) one week before, and C) during week of NEC diagnosis for individual samples from controls (c_###) and NEC cases (n_###) and means for samples combined by NEC status (control, NEC).

developmental delays and encounter environmental factors that differ from term infants and challenge the development of a normal symbiosis. Mode of delivery, feeding type (breast versus formula), and antibiotic use may predispose the infant to the development of various diseases including NEC.[23] This prospective study shows that various aspects of intestinal microbiota composition differ in infants who develop NEC compared to controls. Two weeks before diagnosis, there is an increased proportion in *Proteobacteria* and a decreased proportion of *Bacteroidetes* in cases. At one week before diagnosis *Actinobacteria* and *Proteobacteria* were increased in cases compared to controls. *Proteobacteria* decreased with advancing age in both cases and controls. The increased mean proportion of *Actinobacteria* in cases is an observation that differs from our earlier report, but it is mostly

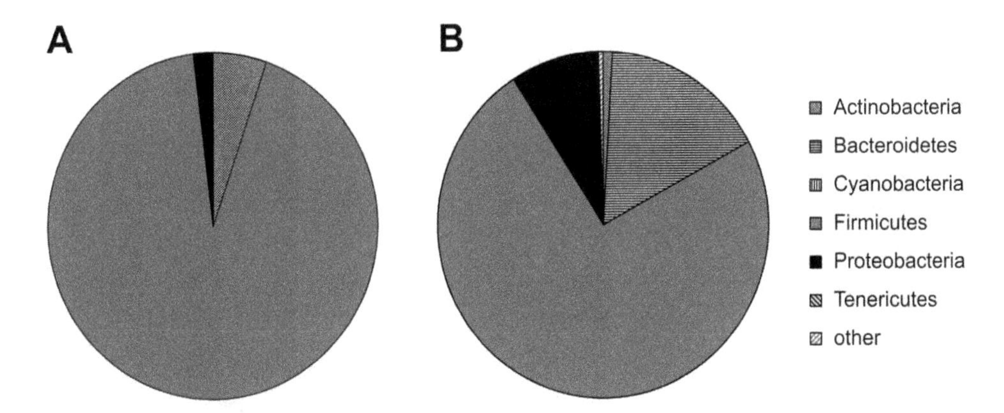

Figure 5. Differences in the proportions of prevalent bacterial phyla, based on 454 16S rRNA sequencing, in controls at week 0 in A) Jacksonville and B) Gainesville.

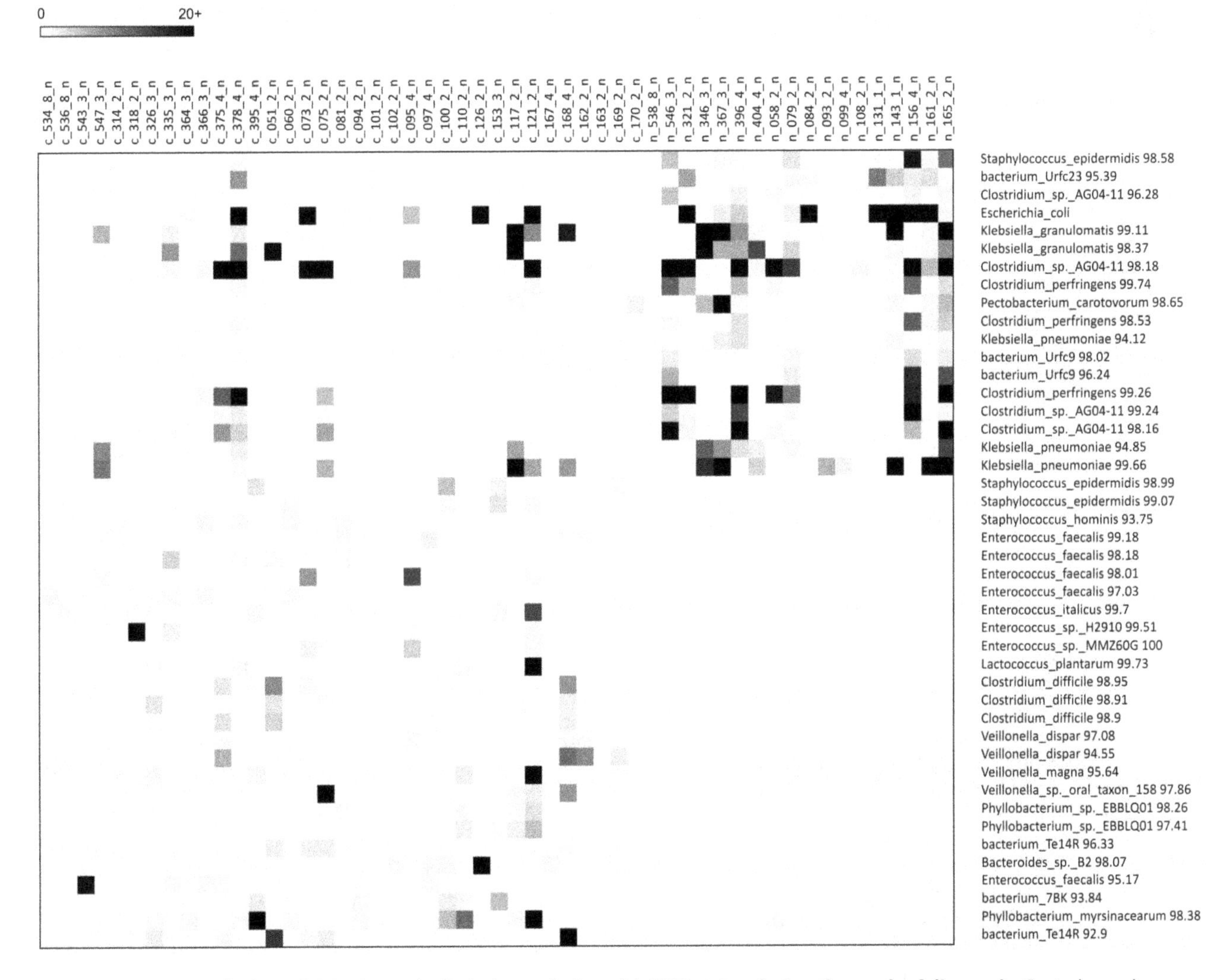

Figure 6. Heat map of selected OTU (98% similarity) correlating with NEC status during the week of diagnosis. Controls are shown on the left (c_###) and NEC cases on the right (n_###). The number of sequences detected per sample for each OTU is indicated by degree of shading, with the darkest shade correlating with the highest number of sequences. OTUs more frequently observed in NEC cases are shown on the top and OTUs less frequently observed are shown on the bottom.

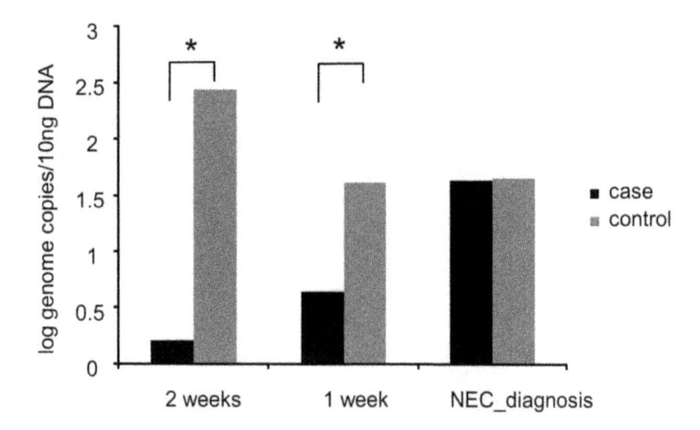

Figure 7. qPCR for fecal counts of Bifidobacteria. * p-value 2 weeks before NEC <0.05; **p-value week of NEC <0.05.

due to a single case in whom *Actinobacteria* contributed the majority of all sequences (Figure 4B). While in cases, the proportion of *Firmicutes* (containing clostridia, lactobacilli and gram positive cocci) consistently increased from 2 wk (34%) to 1 wk (52%) prior to onset of NEC reaching its peak (72%) during the week of NEC there was less of an increase in controls. Many OTUs frequently observed in controls at early time points were lacking in cases, suggesting that bacteria conferring a benefit did not colonize in a timely manner. Other OTUs were significantly increased in cases, suggesting a potentially pathogenic role. Such OTUs include those matching closest to the potentially pathogenic *Klebsiella granulomatis*, *Klebsiella pneumoniae* and *Clostridium perfringens*, but also one that matched *Staphylococcus* epidermidis, a known skin commensal that has previously been associated with NEC risk.[24] Previous studies[11] including our own[9] reported similar observations. The strongest association that we detected was for a K.p.-like OTU that was significantly enriched in the very first stool samples obtained after birth from a subset of infants that later developed NEC. The consistent association in our and other studies with Klebsiella[6,9] and especially OTUs closest to but clearly distinct from K.p. suggests that we might have identified a novel K.p. -like pathogen.

Our data are to some extent consistent with previous studies[6,9] that suggested that the microbiota, particularly amounts of *Proteobacteria*, differ in infants who subsequently develop NEC compared to those who do not.[10] While we did not detect a difference in overall richness (chao-1 alpha diversity) as seen at the time of development of NEC in the Wang study[8], we did detect microbiota clustering, based on UNIFRAC metric, in cases two weeks before diagnosis. Previously, we suggested a *Proteobacteria* bloom associated with NEC onset.[9] In the current study, with a greater number of subjects, such a bloom was not evident, but instead, with increasing age *Proteobacteria* remained high in cases but declined in controls. Differences in gestational age, birth weight and week of NEC onset, in this compared to the earlier report, likely contributed to this discrepancy. Nevertheless, in an earlier report[25] of microbial composition in inflamed intestinal tissue surgically removed from ileum of 24 infants with NEC found that 49% consisted of Proteobacteria, compared to other phyla comprised of: 30.4% were Firmicutes, 17.1% were Actinobacteria and 3.6% Bacteroides. The relative proportion of these phyla is consistent with those seen in the feces of the infants who subsequently developed NEC in both our previous study[9] and the current one. With respect to the longitudinal timeline we believe that the paradox that *Proteobacteria* proportion declined and that the *Firmicutes* increased as the week of NEC approached in the NEC group further impresses the significance that the characteristics of microbiome during the earlier postnatal period drives the susceptibility of premature infants to NEC.

The differences in microbiota compositions observed in controls between the Gainesville and Jacksonville study sites are likely multifactorial, but differences in the administration of individual antibiotics could be a factor contributing to the site differences. However, numbers were too small to make definitive conclusions. Future multi-site studies will need to consider geographic and temporal difference in antibiotic administration. These differences between NICU microbial compositions suggest that we may not be able to extrapolate all data from one NICU to all others.

Many factors can affect a normal pattern of intestinal colonization such as mode of delivery, type of feeding, and antibiotic exposure.[23,26,27] C-section versus vaginal were not related to development of the disease, but consumption of exclusive human milk versus commercial formula was higher in patients who did not develop NEC. Previous studies suggest that infants born via c-section and/or that are predominantly fed infant formula have a similar intestinal pattern of colonization composed predominately of *Proteobacteria* such *Escherichia coli*, and *Firmicutes* including some potential pathogens such as *Clostridia* and *Staphylococcus*.[26,27] In contrast, those infants fed predominantly human milk develop a more desirable "healthy" intestinal flora composed of *mainly Lactobacilli*, *Bacteroides* and *Actinobacteria* (*Bifidobacterium*).[26,28] The current study did not have sufficient number of subjects to compare specifically microbial composition in human milk versus formula fed infants who subsequently developed NEC.

The predominance of *Proteobacteria* one and two weeks before diagnosis of NEC and *Actinobacteria* one week before diagnosis of NEC in cases compared to control could in part suggest that the difference in type of feeding or exposure to antibiotics played a role. Indeed, antibiotic exposure may reduce the diversity of intestinal microbiota, delay the colonization of beneficial bacteria and potentially predispose preterm neonates to NEC.[23,29] Antenatal exposure to antibiotics increased risk of NEC in a retrospective clinical study.[30] In our study antenatal exposure to antibiotics did not differ between the two groups (NEC versus control). We were able to detect a suggestive difference in days of treatment with antibiotics, albeit not statistically significant, when we compared centers. Previous studies [29,31] have shown that duration of antibiotic exposure postnatally is associated with an increased risk of NEC among neonates without prior sepsis. During the study period the incidence of NEC was 12.4% in Gainesville and 6.8% in Jacksonville. As no significant difference in antibiotic use was observed among cases it is possible that the observed differences in antibiotic exposures among controls resulted in the selection for microbes in each NICU unit that correlated with NEC risk.

A recent study[10] found no *Propionibacteria* in 9 NEC infants compared to their presence in 56% of control infants. Our findings from 18 NEC infants appear to differ since the proportion of *Actinobacteria* (includes *Propionibacteria*) was greater in the group that later developed NEC in our study. Whether this was due to differences in timing of sample collection or inter-neonatal intensive care unit differences in microbial ecology remains speculative.

In our study the abundance of *Bifidobacteria* was lower in cases than in controls. This was determined by qPCR analysis as the universal primers used for 16S rRNA sequence analysis contain multiple mismatches with the conserved target region and consequently amplify bifidobacteria poorly. *Bifidobacteria* have been described as beneficial bacteria for intestinal development and function[31] and their higher prevalence may have been related to a greater use of breast milk in control infants.[32].

In summary, it appears that the microbiota of babies who subsequently develop NEC is different from those who do not. The two distinct forms of intestinal dysbiosis prior to the onset of NEC in the two studies from our group suggests that an abnormal pattern of colonization with predominance of *Proteobacteria* early in life or later in the days closer to the development of NEC may be associated with a greater risk for developing NEC. Differences in colonization patterns and NEC were seen depending on neonatal intensive care unit. The use of human milk versus formula also was associated with a lower NEC rate; this finding should be confirmed in other studies. Although the differences in antibiotic usage and human milk feeding could not be correlated to specific changes in microbiota, there were obvious differences prior to the development of NEC. These patterns of altered intestinal microbiota may be modulated by the different modalities of treatment that the infants have undergone. An improved understanding of the factors that establish a healthy intestinal microbiota or that induce dysbiosis offers opportunities for early interventions that reduce the risk of NEC. The consistent finding from now multiple studies that K.p.-like strains appear more frequently observed in NEC cases suggests that an hitherto unknown pathogen might contribute to the etiology in at least a subset of NEC cases.

Acknowledgments

We would like to thank participants and staff at the respective NICU's for their support of this study.

Author Contributions

Conceived and designed the experiments: VM RS MLH JN. Performed the experiments: VM RMT MU XW RS MLH JN. Analyzed the data: VM RMT MU XW RS MLH JN. Contributed reagents/materials/ analysis tools: VM XW. Wrote the paper: VM RMT MU XW RS MLH JN.

References

1. Neu J, Walker WA (2011) Necrotizing enterocolitis. N Engl J Med 364: 255–264.
2. Holman RC, Stoll BJ, Curns AT, Yorita KL, Steiner CA, et al. (2006) Necrotising enterocolitis hospitalisations among neonates in the United States. Paediatr Perinat Epidemiol 20: 498–506.
3. Stoll BJ, Hansen NI, Bell EF, Shankaran S, Laptook AR, et al. (2010) Neonatal outcomes of extremely preterm infants from the NICHD Neonatal Research Network. Pediatrics 126: 443–456.
4. Horbar JD, Soll RF, Edwards WH (2010) The Vermont Oxford Network: a community of practice. Clin Perinatol 37: 29–47.
5. Claud EC, Walker WA (2001) Hypothesis: inappropriate colonization of the premature intestine can cause neonatal necrotizing enterocolitis. FASEB J 15: 1398–1403.
6. Hoy C, Millar MR, MacKay P, Godwin PG, Langdale V, et al. (1990) Quantitative changes in faecal microflora preceding necrotising enterocolitis in premature neonates. Arch Dis Child 65: 1057–1059.
7. Mshvildadze M, Neu J, Schuster J, Theriaque D, Li N, et al. (2010) Intestinal microbial ecology in premature infants assessed with non-culture-based techniques. J Pediatr 156: 20–25.
8. Wang Y, Hoenig JD, Malin KJ, Qamar S, Petrof EO, et al. (2009) 16S rRNA gene-based analysis of fecal microbiota from preterm infants with and without necrotizing enterocolitis. ISME J 3: 944–954.
9. Mai V, Young CM, Ukhanova M, Wang X, Sun Y, et al. (2011) Fecal microbiota in premature infants prior to necrotizing enterocolitis. PLoS One 6: e20647.
10. Morrow AL, Lagomarcino AJ, Schibler KR, Taft DH, Yu Z, et al. (2013) Early microbial and metabolomic signatures predict later onset of necrotizing enterocolitis in preterm infants. Microbiome 1: 13.
11. Boccia D, Stolfi I, Lana S, Moro ML (2001) Nosocomial necrotising enterocolitis outbreaks: epidemiology and control measures. Eur J Pediatr 160: 385–391.
12. Westra-Meijer CM, Degener JE, Dzoljic-Danilovic G, Michel MF, Mettau JW (1983) Quantitative study of the aerobic and anaerobic faecal flora in neonatal necrotising enterocolitis. Arch Dis Child 58: 523–528.
13. Meinzen-Derr J, Morrow AL, Hornung RW, Donovan EF, Dietrich KN, et al. (2009) Epidemiology of necrotizing enterocolitis temporal clustering in two neonatology practices. J Pediatr 154: 656–661.
14. Turnbaugh PJ, Ley RE, Hamady M, Fraser-Liggett CM, Knight R, et al. (2007) The human microbiome project. Nature 449: 804–810.
15. Hattori M, Taylor TD (2009) The human intestinal microbiome: a new frontier of human biology. DNA Res 16: 1–12.
16. Cotten CM, Taylor S, Stoll B, Goldberg RN, Hansen NI, et al. (2009) Prolonged duration of initial empirical antibiotic treatment is associated with increased rates of necrotizing enterocolitis and death for extremely low birth weight infants. Pediatrics 123: 58–66.
17. Alexander VN, Northrup V, Bizzarro MJ (2011) Antibiotic exposure in the newborn intensive care unit and the risk of necrotizing enterocolitis. J Pediatr 159: 392–397.
18. Underwood MA (2013) Human milk for the premature infant. Pediatr Clin North Am 60: 189–207.
19. Mai V, Braden CR, Heckendorf J, Pironis B, Hirshon JM (2006) Monitoring of stool microbiota in subjects with diarrhea indicates distortions in composition. J Clin Microbiol 44: 4550–4552.
20. Sun Y, Cai Y, Liu L, Yu F, Farrell ML, et al. (2009) ESPRIT: estimating species richness using large collections of 16S rRNA pyrosequences. Nucleic Acids Res 37: e76.
21. Caporaso JG, Kuczynski J, Stombaugh J, Bittinger K, Bushman FD, et al. (2010) QIIME allows analysis of high-throughput community sequencing data. Nat Methods 7: 335–336.
22. Lozupone C, Hamady M, Knight R (2006) UniFrac–an online tool for comparing microbial community diversity in a phylogenetic context. BMC Bioinformatics 7: 371.
23. Torrazza RM, Neu J (2013) The altered gut microbiome and necrotizing enterocolitis. Clin Perinatol 40: 93–108.
24. Mollitt DL, Tepas JJ, Talbert JL (1988) The role of coagulase-negative Staphylococcus in neonatal necrotizing enterocolitis. J Pediatr Surg 23: 60–63.
25. Smith B, Bode S, Petersen BL, Jensen TK, Pipper C, et al. (2011) Community analysis of bacteria colonizing intestinal tissue of neonates with necrotizing enterocolitis. BMC Microbiol 11: 73.
26. Harmsen HJ, Wildeboer-Veloo AC, Raangs GC, Wagendorp AA, Klijn N, et al. (2000) Analysis of intestinal flora development in breast-fed and formula-fed infants by using molecular identification and detection methods. J Pediatr Gastroenterol Nutr 30: 61–67.
27. Gronlund MM, Lehtonen OP, Eerola E, Kero P (1999) Fecal microflora in healthy infants born by different methods of delivery: permanent changes in intestinal flora after cesarean delivery. J Pediatr Gastroenterol Nutr 28: 19–25.
28. Penders J, Thijs C, Vink C, Stelma FF, Snijders B, et al. (2006) Factors influencing the composition of the intestinal microbiota in early infancy. Pediatrics 118: 511–521.
29. Jernberg C, Lofmark S, Edlund C, Jansson JK (2007) Long-term ecological impacts of antibiotic administration on the human intestinal microbiota. ISME J 1: 56–66.
30. Weintraub AS, Ferrara L, Deluca L, Moshier E, Green RS, et al. (2011) Antenatal antibiotic exposure in preterm infants with necrotizing enterocolitis. J Perinatol 32: 705–709.
31. Kuppala VS, Meinzen-Derr J, Morrow AL, Schibler KR (2011) Prolonged initial empirical antibiotic treatment is associated with adverse outcomes in premature infants. J Pediatr 159: 720–725.
32. Dai D, Walker WA (1999) Protective nutrients and bacterial colonization in the immature human gut. Adv Pediatr 46: 353–382.

9

Functional Diversity of the Microbial Community in Healthy Subjects and Periodontitis Patients Based on Sole Carbon Source Utilization

Yifei Zhang[1], Yunfei Zheng[2], Jianwei Hu[1], Ning Du[1], Feng Chen[1]*

1 Central Laboratory, School of Stomatology, Peking University, Beijing, P. R. China, **2** Department of Periodontology, School of Stomatology, Peking University, Beijing, P. R. China

Abstract

Chronic periodontitis is one of the most common forms of biofilm-induced diseases. Most of the recent studies were focus on the dental plaque microbial diversity and microbiomes. However, analyzing bacterial diversity at the taxonomic level alone limits deeper comprehension of the ecological relevance of the community. In this study, we compared the metabolic functional diversity of the microbial community in healthy subjects and periodontitis patients in a creative way—to assess the sole carbon source utilization using Biolog assay, which was first applied on oral micro-ecology assessment. Pattern analyses of 95-sole carbon sources catabolism provide a community-level phenotypic profile of the microbial community from different habitats. We found that the microbial community in the periodontitis group had greater metabolic activity compared to the microbial community in the healthy group. Differences in the metabolism of specific carbohydrates (e.g. β-methyl-D-glucoside, stachyose, maltose, D-mannose, β-methyl-D-glucoside and pyruvic acid) were observed between the healthy and periodontitis groups. Subjects from the healthy and periodontitis groups could be well distinguished by cluster and principle component analyses according to the utilization of discriminate carbon sources. Our results indicate significant difference in microbial functional diversity between healthy subjects and periodontitis patients. We also found Biolog technology is effective to further our understanding of community structure as a composite of functional abilities, and it enables the identification of ecologically relevant functional differences among oral microbial communities.

Editor: Michael Glogauer, University of Toronto, Canada

Funding: This work was supported by grants 81300880 and 81200762 from National Natural Science Foundation of China, and by funding from Peking University School of Stomatology (PKUSS20110301). The funders had no role in study design, data collection and analysis, decision to publish, or preparation of the manuscript.

Competing Interests: The authors have declared that no competing interests exist.

* E-mail: moleculecf@gmail.com

These authors contributed equally to this work.

Introduction

Periodontal disease is one of the most common adult diseases, leading to disorders of the supporting structures of the teeth, including the gingivae, periodontal ligaments, and supporting alveolar bone. The accumulation of plaque around the gingival margin triggers gingivitis [1,2]. Once the homeostasis of microbial diversity within the plaque is lost, gingivitis leads to periodontitis [3,4].

Attempts have been made to compare the microbial diversity in patients with periodontitis and healthy subjects [4–7]. Socransky *et al.* (1998) reported that the "orange complex," consisting of Gram-negative, anaerobic species such as *Prevotella intermedia* and *Fusobacterium nucleatum*, was associated with periodontitis, whereas the "red complex," consisting of periodontal pathogens such as *Porphyromonas gingivalis*, *Tannerella forsythia*, and *Treponema denticola*, is detected as the disease worsens [3,8]; Using the Human Oral Microbe Identification Microarray, Colombo *et al.* (2009) detected more species in patients with periodontal disease compared to those without disease.[9]; Using 16S pyrosequencing, Griffen *et al.* (2012) found that community diversity was higher in periodontal disease subjects [10]; Abusleme *et al.* (2013) reported that the shifts in community structure from health to periodontitis are characterized by the emergence of newly dominant taxa without replacement of primary health-associated species [11]. However, these studies merely focused on the taxonomic structure of communities, and thus provided limited insight into the ecological relevance of microbial community structure. Because changes in bacterial types do not necessarily change the function of a community [12], the phenetic status of the microbial community should be considered when the microflora are studied.

Analyses of phenetic characteristics such as microbial metabolism allow for deeper insight into microbial community structure. First, metabolic processes represent a key component in determining the virulence properties of oral pathogens [13,14]. Furthermore, the development of food chains between bacteria and endogenous nutrient metabolism would enhance the diversity of the microflora, which plays a key role in maintaining homeostasis within a microbial community [15]. For example, frequent carbohydrate consumption increases the levels of mutans streptococci and lactobacilli but decreases *Streptococcus sanguinis* levels [16], Aggressive periodontitis appears to be associated with a loss of colonization by *S. sanguinis* [17]. Grenier and Mayrand

(1986) also reported that the nutritional relationships among oral bacteria could explain the mechanisms favoring bacterial succession in periodontal sites [18].

The sole carbon source utilization (SCSU) patterns of microbial samples determined using the Biolog assay (Biolog Inc., Hayward, CA, USA) [19] could be used as a functionally based measure for classifying heterotrophic microbial communities. The Biolog assay for community analysis involves outgrowth of the entire microbiota ecosystem on multiple carbon substrates. Each well contains tetrazolium violet and a minimal amount of proprietary growth media. Color produced from the reduction of tetrazolium violet is used as an indicator of respiration. Commercially available microplates allow for the simultaneous testing of 95 separate carbon sources such that the metabolic response patterns of microbial communities from different habitats can be compared [20–23]. Anderson *et al.* first analyzed the metabolic similarity of experimental dental plaque biofilms [24]. However, the metabolism or activities of the microflora in periodontitis patients and healthy subjects has not been reported.

Here we used Biolog technology to compare the microbial functional diversity between patients with chronic periodontitis and healthy controls to further our knowledge of the ecological basis of periodontal disease.

Materials and Methods

Subjects and sample collection

Plaque samples were collected from 11 patients who had been diagnosed with generalized chronic periodontitis (according to the international Classification of Periodontal Diseases in 1999: more than 30% of sites with a pocket depth >4 mm, inter-proximal attachment loss of >3 mm, bleeding on probing, and radiographic evidence of alveolar bone loss; course >6 weeks) and 12 controls that had been defined as "periodontally healthy" by a licensed periodontist. The following inclusion criteria were used: at least 18 years of age, a minimum of six natural teeth in every quadrant, an absence of other oral diseases such as caries or mucosal disease, and good systemic health. The exclusion criteria were professional periodontal therapy in the 6 months before enrollment, antibiotic use for any purpose within 1 month before entering the study, and smoking. For inclusion in the healthy group, subjects were required to have no clinical signs of inflammation, including redness, swelling, or bleeding on probing, and no pockets with a probing depth >3 mm.

Supragingival and subgingival plaque around the gingival margin was collected 2–3 h after a meal. To avoid salivary contamination, the sample collection sites (≤1 mm above the gingival margin for supragingival plaque collection, and ≤3 mm below the gingival margin for subgingival plaque collection) were isolated with cotton rolls and gently air-dried. Pooled plaque samples were carefully taken from 16 teeth (2 premolars and 2 molars in each quadrant, but not wisdom teeth) of each subject using a scaler and stored in phosphate-buffered saline (PBS; 0.01 M, pH = 7.2–7.4) (Biotop, Huangshang City, China) on ice.

The study protocol was approved by the Institutional Review Board (IRB) of Peking University School and Hospital of Stomatology (Beijing, China) (approval number: PKUSSIRB-2012063). Participants have provided their written informed consent to participate in this study, and the consent procedure was also approved by the IRB of Peking University School and Hospital of Stomatology.

Biolog assays. The collected plaque samples were re-suspended in 11 ml of PBS (0.01 M, pH = 7.2–7.4) and vortexed thoroughly for 60 s. Each plaque suspension was inoculated into Biolog anaerobic-negative (AN) microplates (Biolog Inc.) at 100 μl per well. The Biolog AN plates contained 95 sole carbon sources and a blank well with water only (Table S1 in File S1). The initial optical densities (ODs) of the plaque suspensions were measured before inoculation.

The plates were incubated in a 5% CO_2 incubator at 37°C for up to 4 days. The OD at 590 nm (OD_{590}) in each well was recorded every 24 h using a Biolog microstation and associated software (Biolog OmniLog version 4.1).

Data analyses

Reactions were interpreted as positive or negative using Biolog OmniLog software. Positive wells found in at least 50% of subjects in the healthy and periodontitis groups were defined as core utilized carbon sources.

The overall metabolic activity for a microbial community in the Biolog plates was expressed as average well color development (AWCD) and calculated as follows:

$$AWCD = \sum_{i=1}^{n} \frac{c_i}{n}$$

Where C is the corrected OD value (obtained by subtracting the OD of the control well from that of each experimental well to correct for the background activity) in each well and n is the number of substrates (n = 95). If the result was negative, the OD would be deemed to be zero. An independent *t*-test was performed on the measurement data to compare the mean differences in AWCD between the periodontal disease group and healthy group over 4 days.

Richness and evenness were determined using the Shannon index [25], while dominance was determined using the Simpson index [26]. To avoid negative values with the Simpson index, the control-corrected OD was multiplied by 1000.

To avoid artificial differences (differences caused by varying initial OD values of the plaque suspensions, rather than by the pattern of carbon source consumption) between the communities, we standardized the corrected OD value as proposed previously [19,27] by dividing each corrected OD value by the AWCD of the plate. Differently exploited carbon sources were measured based on standardized OD values using independent *t*-tests and Pearson's correlation [27] (SPSS Statistics version 17.0).

The relationships between the healthy and periodontitis groups based on discriminative carbon sources were determined using cluster analyses (Cluster 3.0, Java TreeView, version 1.1.3) and principle component analyses (PCAs) (Canoco, version 4.5).

Results

Functional diversity in the healthy and periodontitis groups

The demographic and clinical characteristics of the subjects are described in Table 1. The initial OD value of the inoculum was higher in the periodontitis group compared to the healthy group, which reflects the greater amount of accumulated plaque in the periodontitis group. Compared to the healthy group, the periodontitis group yielded greater metabolic responses. Significant differences in the overall rate of color development (AWCD) between the healthy and periodontitis groups were noted during the linear increase stage of inoculation (i.e., the first 24 h in Fig. 1). The curves then presented an asymptotic nature during later incubation periods. No significant difference was found between

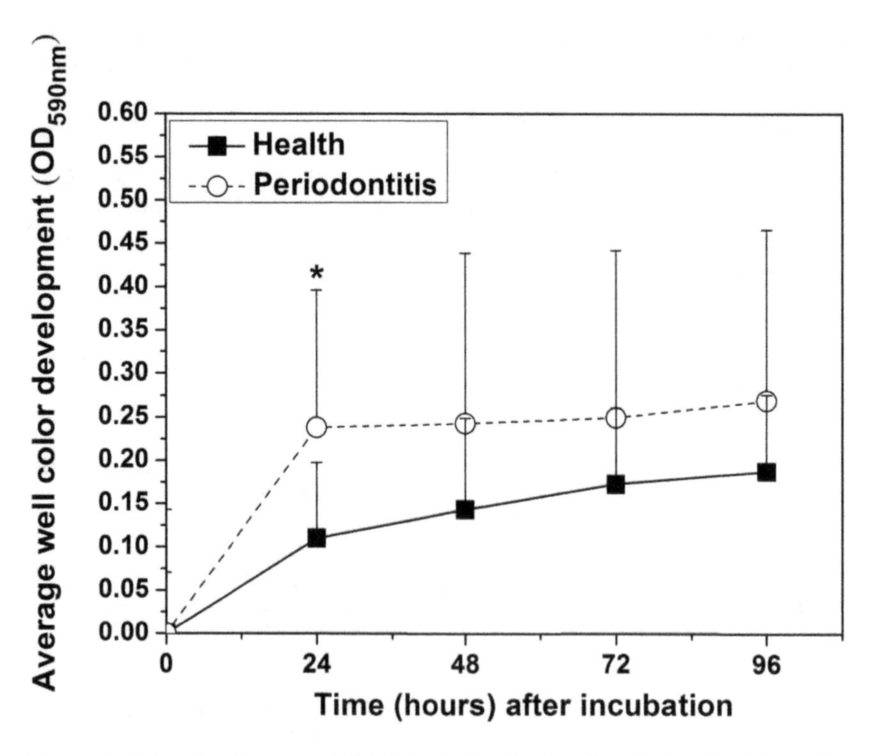

Figure 1. Color development (AWCD) with incubation time in the healthy and periodontitis groups (* indicates a significant difference).

the Shannon indices of diversity between the healthy and periodontitis groups ($p>0.05$, t-test; Table 2). The periodontitis group had a significantly higher Simpson diversity index value than the healthy group at 72 and 96 h after incubation ($p<0.05$, t-test) (Table 2).

Core positive carbon sources utilized by the microbial community in healthy subjects and periodontitis patients

Despite varying intensity levels, a total of 31 carbon sources were positive in most subjects (≥50%) in the healthy or periodontitis group after 96 h incubation, which was defined as core positive carbon sources (Fig. 2). Six of these were found in all subjects: D-fructose, maltose, D-raffinose, maltotriose, D-mannose, and sucrose. Another five sources were present in most healthy subjects, whereas the remaining 17 were preferentially associated with periodontitis (Fig. 2). The periodontitis group harbored more core positive carbon sources than the healthy group (Fig. 2).

Table 1. Demographic and clinical characteristics of the subjects.

Parameter	Healthy subjects	Periodontitis patients	p-value
No. of subjects	12	11	
Age (years)	34.6±7.1	34.9±12.5	NS[a]
Missing teeth	None	None	—
Smoker	None	None	—
Women (%)	50	54.5	NS[b]
Percentage of sites with a PD (%):			
>6 mm	0	27.3	
4–6 mm	0	54.5	
<4 mm	100	18.2	
OD_{630} of plaque suspension	0.04±0.028	0.08±0.03	0.003[a]

[a]Independent t-tests
[b]Chi-square tests
NS: not significant
PD: pocket depth
Quantitative data are presented as the mean±standard deviation (SD); categorical data are presented as percentages.

Table 2. Diversity indices of the periodontally healthy and diseased groups during incubation (mean±SD, $p<0.05$).

Index		Healthy subjects	Periodontitis patients	*p*-value
Shannon	24 h	3.98±0.45	4.08±0.23	0.52
	48 h	3.84±0.37	3.95±0.36	0.44
	72 h	3.77±0.35	3.99±0.40	0.16
	96 h	3.81±0.31	3.95±0.32	0.31
Simpson	24 h	47±22	51±12	0.27
	48 h	37±18	47±18	0.10
	72 h	32±13	50±19	**0.025**
	96 h	33±12	45±15	**0.025**

Correlation between healthy status and carbon source utilization patterns

Analyses of 95-sole carbon source utilization patterns showed 14, 8, 11, and 8 significantly different ($p<0.05$) carbon sources between the two groups at 24, 48, 72, and 96 h, respectively (Table S2 in File S1). Of these, five core positive carbon sources were identified at different time points (Fig. 3): β-methyl-D-glucoside and stachyose were identified at 24 h, maltose and D-mannose were identified at 72 h, and β-methyl-D-glucoside and pyruvic acid at 96 h.

To identify correlations between the pattern of carbon source consumption and healthy status, cluster analyses of 23 subjects from each group were performed based on 14 discriminative carbon sources, which distinguished the healthy group (H) from the diseased group (D) at 24 h (Fig. 4). Our results suggest that the 23 samples could be classified into two groups: 8 of the 12 samples from the healthy group were classified as cluster one, while the 11 samples from the diseased group were classified as cluster two, which coincided with the clinical classification according to the diagnosis. However, certain pairs of samples from different groups were closely associated, such as samples H07 and D04 (Fig. 4). Additionally, the data indicate that some of the samples in the healthy group were similar to those in the diseased group (e.g., H04, H07, H08, and H12) (Fig. 4).

PCAs were performed to identify those carbon sources with key roles in difference between groups at 24 h. We maintained four PCs in the subsequent analysis because they accounted for more than 75% of the total eigenvalue sum (Table 3). Our PCA plots show that PC1 and PC2 accounted for 38.9 and 15% of the total variance, respectively (Fig. 5). Subsequent *t*-tests for the PCs indicated significant differences ($p<0.05$) between the two groups

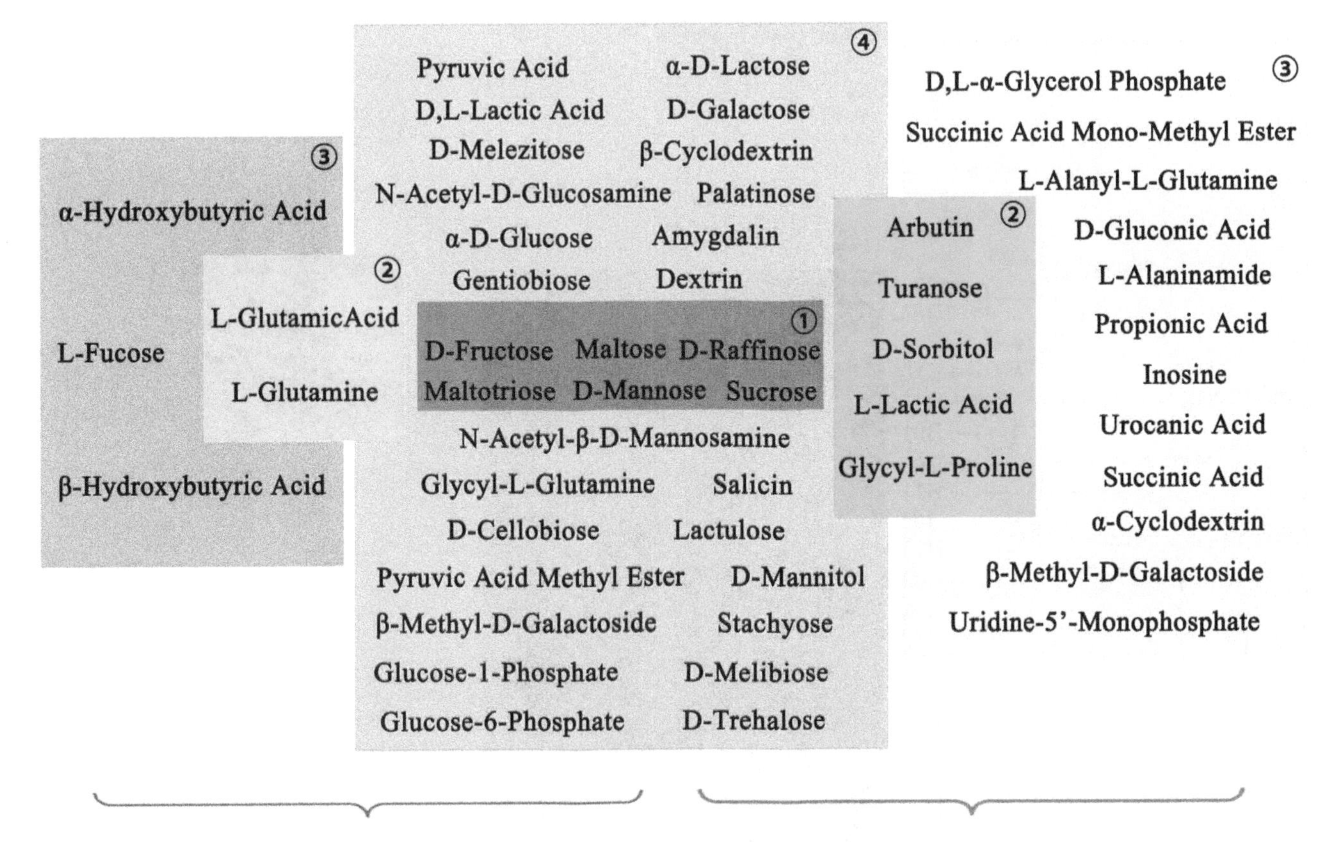

Figure 2. Core positive carbon sources in the healthy subjects and periodontitis patients. Inner box (numbered with 1), positive carbon sources found in all subjects (100%); middle boxes (numbered with 2), present in 71–99% of subjects from each group (H: healthy, P: periodontitis); outer boxes (numbered with 3), present in 50–70% of subjects from the healthy and periodontitis groups; middle box (numbered with 4), positive carbon sources present in at least 50% of subjects in both the healthy and periodontitis groups.

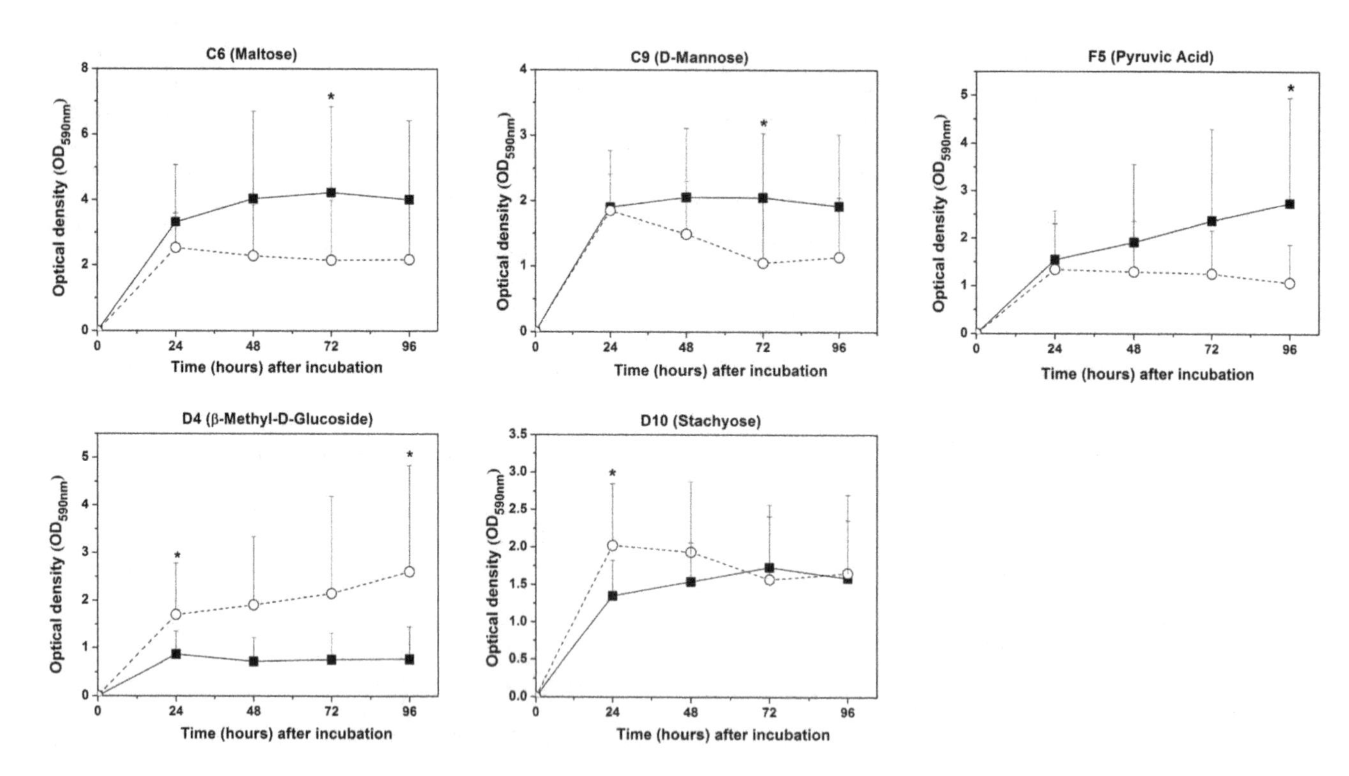

Figure 3. Catabolic kinetics according to inoculation time for discriminative core positive carbon sources (* indicates a significant difference) in the healthy (—■—) and periodontitis (--○--) groups.

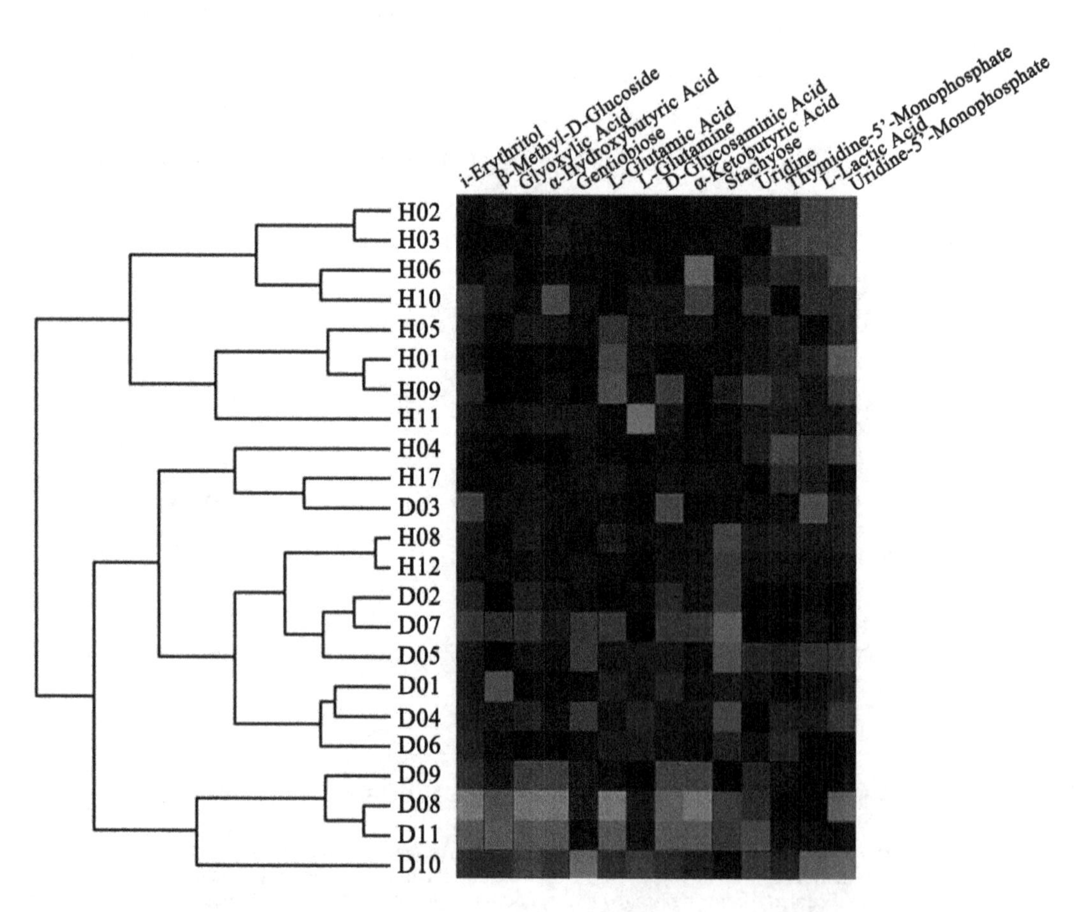

Figure 4. Cluster analyses of 23 clinical plaque samples from healthy (H) and diseased (D) subjects based on standardized OD values of the discriminative carbon sources at 24 h.

on PC1 (Table 3) and identified the carbon sources that best described these differences: E5 (glyoxylic acid), E6 (α-hydroxybutyric acid), E9 (α-ketobutyric acid), and B9 (D-glucosaminic acid) were identified most positively by their PC1 scores, whereas H12 (uridine-5′-monophosphate), H10 (uridine), and H11 (thymidine-5′-monophosphate) were identified most negatively by their PC1 scores (Table 4). The utilization patterns of each microbial community were compared by principal component analysis (PCAs) of the 24 h absorption data. Microbial community samples from different subjects served as objects and the absorbance values of carbon source utilization as variables (Fig.5). It is possible to identify the substrates which contribute to the separation of each sample plot on the ordination biplot (Fig. 5).

Table 3. *t*-tests for PCs extracted from 14 discriminative carbon sources at 24 h ($p<0.05$).

	Healthy subjects	Periodontitis patients	*p*-value
PC1	1.69±1.13	−1.85±1.74	<0.001
PC2	−2.98±1.89	0.33±0.78	0.32
PC3	0.23±0.92	−0.26±1.58	0.38
PC4	0.10±1.03	−0.11±1.43	0.68

Discussion

Plaque around the gingival margin, with its toxic products, stimulates gingival tissue directly and leads to tissue destruction, likely causing periodontal disease [28]. Thus, in this study, we focused on plaque around the gingival margin from patients with periodontitis and orally healthy individuals to determine whether there were functional differences between the two groups. Because many of the anaerobic bacteria isolated from deep subgingival sites are proteolytic and asaccharolytic [28], deeper sampling (>3 mm) would not be optimal because these organisms may not be responsive to the substrates.

Comparisons of overall color development between groups were dependent on both the density and composition of the inoculum: samples with a denser microbial community would produce more effective reactions on the premise that a larger percentage of microorganisms are able to utilize the substrate. However, in this study, we did not achieve equivalent inoculum densities by dilution or concentration of the samples despite performing functional tests on the original community. The microbial community in the periodontitis patients was more metabolically active than that in the orally healthy individuals, indicating that both the amount and composition of plaque around the gingival margin accounted for the functional diversity. Thus, we confirmed the importance of plaque control in the initial therapy of periodontal disease. In addition, we normalized the raw color response data to the AWCD to account for different inoculum densities.

Analyses of sole carbon source consumption levels indicated differences in the metabolism of specific carbohydrates between the healthy and periodontitis groups. Differences in carbon source utilization are related to the color response in a given well, which is related to the number of microorganisms that are able to utilize the substrate within that well. Increased color development in

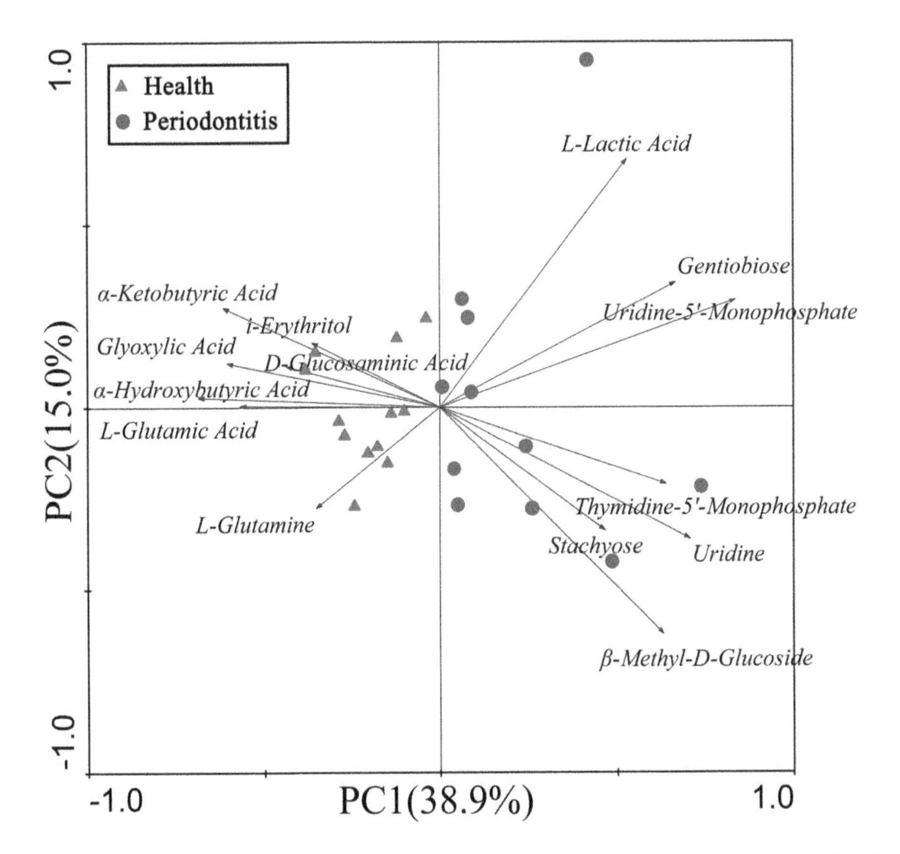

Figure 5. Ordination biplot of principal component analyses of the substrate utilization patterns of the microbial communities in the healthy and periodontitis groups using Biolog AN plate assays. Arrows indicate the directions and relative importance (arrow lengths) of the 14 substrates variables.

Table 4. Substrates with high correlation coefficients (|r|>0.6) for PC1 in a PCA of substrate utilization patterns among microbial communities from periodontally healthy and diseased groups.

	r	*p*-value
E5 (Glyoxylic acid)	0.78	<0.001
E6 (α-Hydroxybutyric acid)	0.76	<0.001
E9 (α-Ketobutyric acid)	0.73	<0.001
H12 (Uridine-5′-monophosphate)	−0.72	<0.001
B9 (D-Glucosaminic acid)	0.66	0.001
H10 (Uridine)	−0.65	0.001
H11 (Thymidine-5′-monophosphate)	−0.64	0.001

certain response wells suggests that the inoculum contained a larger number of microorganisms able to utilize the substrate. The lag phase in color development in a single well (Fig. 2) indicates that a longer period of growth would produce a sufficient density of stained cells; a smaller percentage of microorganisms in the inoculum able to utilize the substrate may be reflected at later incubations times. This could explain the variety of discriminating sole carbon sources at different incubation times (Table S2 in File S1). Moreover, the patterns are a reflection of functional potential rather than *in situ* functional ability because growth plays an important role in this assay.

Cluster analyses were carried out based on the utilization of the discriminative carbon sources at 24 h because the most significant differences in sole carbon source consumption levels were obtained at this phase (linear increase; Fig. 2). In the present study, cluster analyses were used to describe the associations between healthy status and sole carbon source consumption level. Here, we hypothesized that closely associated samples possess similar community-level metabolic functions. This classification further improved the metabolic discrimination of specific carbohydrates between healthy subjects and periodontitis patients, and periodontal health or disease could be distinguished using Biolog AN plates. Furthermore, certain samples in the healthy group were closely associated with those in the disease group. This suggests that the disease-associated metabolic pattern was also found in the healthy subjects. The presence of disease-associated function might be prognostic for future disease in the absence of disease symptoms.

PCA revealed the carbon sources that contributed most to the diversity; it is well established that the biomass and diversity of the microbial community in supragingival and subgingival plaque around the gingival margin are distinguished by the consumption of substrates. In our study, the periodontitis patients resulted in a larger biomass accumulation of the microbial community (indicated by initial OD value) and remarkable changes in its diversity (indicated by AWCD). The close links between the biomass and diversity of the microbial community and the periodontal health status can be reflected by the carbon sources metabolism in plaque samples.

Because oral microbial communities may be dominated by <1000 species-level taxa [10,29], evaluation of the dominant functional characteristics may be more accurate and efficient for classifying and characterizing microbial community structure [9]. Our findings suggest that periodontitis communities result from ecological shifts in community structure [11] as well as shifts in metabolic function. Despite differences in plaque amounts between the healthy subjects and periodontitis patients, most carbon sources utilized in the periodontitis group were also utilized by the healthy group, albeit at varying sole carbon source consumption levels. However, more carbon sources dominated in the periodontitis group compared to the healthy group. The core microbiome has already been defined in healthy individuals and periodontitis patients [6,11]; nevertheless, it may not be adequate to interpret functional shifts. Different bacterial types would have the same function; for example, both *Streptococcus* spp. and *Neisseria* spp. can utilize sucrose. Conversely, different species in the same genera could have different metabolic functions. *Streptococcus mutans* is able to utilize mannitol but *Streptococcus mitis* cannot. Clearly, the concept of core microbial ecology may be better defined with community function measurements rather than taxonomic structure or membership.

The Shannon index is used extensively in the measurement of taxonomic diversity and community structure [30]. However, using the Shannon index to characterize functional microbial diversity may be questionable [31]. In this study, richness was defined as the total number of carbon substrates utilized while evenness was defined as the equability of substrate utilization between all utilized substrates [31]. The Biolog carbon substrate richness and evenness did not vary between the healthy subjects and periodontitis patients. Other variables that can affect microbial diversity such as differences in the dominant species may have contributed to the higher Simpson index in the periodontitis group at later incubation stages, which coincided with the higher number of core positive carbon sources in the periodontitis group. Although carbon source utilization profiles are not a direct representation of bacterial growth, the Biolog substrate catabolic responses for the microbial communities exhibited a lag phase (before 24 h, data not shown), an exponential phase, and a stationary phase, similar to bacterial growth curves. This nonlinearity has important implications for the interpretation of the disparate catabolic kinetics patterns of different carbon sources (as depicted in Fig. 2) and the single fixed-time readings of community Biolog responses.

Microbial community in our oral cavity is not constant, and would be easily affected by environmental factors such as temperature, pH value and so on. Thus analyzing the microbial diversity in a fixed-time point would be a misleading in our comprehension. In our study, the changes in community metabolism according to time had been well demonstrated by Biolog assay (indicated by AWCD and Simpson index), reflecting the changes in community structure.

One limitation of Biolog technology is that it cannot detect microorganisms that do not make use of the carbon sources on the Biolog microplate. Additionally, the response to substrate catabolism requires an effective quantity and activity of the microbial community in the tested samples. Nevertheless, Biolog technology was found to be an effective assay to rapidly visualize community structure as a composite of functional abilities or potential, and it enables the identification of ecologically relevant functional differences among oral microbial communities. Our study showed a distinguishing characteristic of carbonate metabolism, which may provide a potential adjuvant method for the diagnosis of periodontitis. More importantly, certain carbohydrates should be avoided in the diet according to your suggestion, which might prevent susceptible individuals from developing periodontitis.

Supporting Information

File S1 Table S1 & S2. Table S1. Carbon-sources pattern of the Biolog AN microplate. Table S2 Discriminative carbon sources between health (H) group and periodontitis group (P) based on standardized OD values ($p<0.05$).

Author Contributions

Conceived and designed the experiments: Y. Zhang FC. Performed the experiments: Y. Zhang Y. Zheng. Analyzed the data: Y. Zhang JH FC. Contributed reagents/materials/analysis tools: Y. Zheng JH ND. Wrote the paper: Y. Zhang. Obtained biospecimen: Y. Zheng.

References

1. Loe H, Theilade E, Jensen SB (1965) Experimental gingivitis in man. J Periodontol 36:177–187.
2. Theilade E, Wright WH, Jensen SB, Loe H (1966) Experimental gingivitis in man. II. A longitudinal clinical and bacteriological investigation. J. Periodont. Res. 1: 1–13.
3. Socransky SS, Haffajee AD, Cugini MA, Smith C, Kent RL Jr (1998) Microbial complexes in subgingival plaque. J Clin Periodontol 25: 134–144.
4. Moore WE, Moore LV (1994) The bacteria of periodontal diseases. Periodontol 2000 5: 66–77.
5. Aas JA, Paster BJ, Stokes LN, Olsen I, Dewhirst FE (2005) Defining the normal bacterial flora of the oral cavity. J Clin Microbiol 43: 5721–5732.
6. Bik EM, Long CD, Armitage GC, Loomer P, Emerson J, et al. (2010) Bacterial diversity in the oral cavity of 10 healthy individuals. Isme Journal 4: 962–974.
7. Zhang S-M, Tian F, Huang Q-F, Zhao Y-F, Guo X-K, et al. (2011) Bacterial diversity of subgingival plaque in 6 healthy Chinese individuals. Exp Ther Med 2: 1023–1029.
8. Socransky SS, Haffajee AD (2005) Periodontal microbial ecology. Periodontol 2000 38: 135–187.
9. Colombo APV, Boches SK, Cotton SL, Goodson JM, Kent R, et al. (2009) Comparisons of Subgingival Microbial Profiles of Refractory Periodontitis, Severe Periodontitis, and Periodontal Health Using the Human Oral Microbe Identification Microarray. J Periodontol 80: 1421–1432.
10. Griffen AL, Beall CJ, Campbell JH, Firestone ND, Kumar PS, et al. (2012) Distinct and complex bacterial profiles in human periodontitis and health revealed by 16S pyrosequencing. Isme Journal 6: 1176–1185.
11. Abusleme L, Dupuy AK, Dutzan N, Silva N, Burleson JA, et al. (2013) The subgingival microbiome in health and periodontitis and its relationship with community biomass and inflammation. Isme Journal 7: 1016–1025.
12. White DC, Findlay RH (1988) Biochemical markers for measurement of predation effects on the biomass, community structure, nutritional status, and metabolic activity of microbial biofilms. Hydrobiologia 159: 119–132.
13. Shah HN, Seddon SV, Gharbia SE (1989) Studies on the virulence properties and metabolism of pleiotropic mutants of Porphyromonas gingivalis (Bacteroides gingivalis) W50. Oral Microbiol Immunol 4: 19–23.
14. Mazumdar V, Snitkin ES, Amar S, Segre D (2009) Metabolic network model of a human oral pathogen. J Bacteriol 191: 74–90.
15. Marsh PD (1994) Microbial ecology of dental plaque and its significance in health and disease. Adv Dent Res 8: 263–271.
16. Minah GE, Solomon ES, Chu K (1985) The association between dietary sucrose consumption and microbial population shifts at six oral sites in man. Arch Oral Biol 30: 397–401.
17. Stingu CS, Eschrich K, Rodloff AC, Schaumann R, Jentsch H (2008) Periodontitis is associated with a loss of colonization by Streptococcus sanguinis. J Med Microbiol 57: 495–499.
18. Grenier D, Mayrand D (1986) Nutritional relationships between oral bacteria. Infect Immun 53: 616–620.
19. Garland JL, Mills AL (1991) Classification and characterization of heterotrophic microbial communities on the basis of patterns of community-level sole-carbon-source utilization. Appl Environ Microbiol 57: 2351–2359.
20. Haack SK, Garchow H, Klug MJ, Forney LJ (1995) Analysis of factors affecting the accuracy, reproducibility, and interpretation of microbial community carbon source utilization patterns. Appl Environ Microbiol 61: 1458–1468.
21. Schutter M, Dick R (2001) Shifts in substrate utilization potential and structure of soil microbial communities in response to carbon substrates. Soil Biol Biochem 33: 1481–1491.
22. Rusznyak A, Vladar P, Molnar P, Reskone MN, Kiss G, et al. (2008) Cultivable bacterial composition and BIOLOG catabolic diversity of biofilm communities developed on Phragmites australis. Aquatic Botany 88: 211–218.
23. Fisk MC, Ruether KF, Yavitt JB (2003) Microbial activity and functional composition among northern peatland ecosystems. Soil Biol Biochem 35: 591–602.
24. Anderson SA, Sissons CH, Coleman MJ, Wong L (2002) Application of carbon source utilization patterns to measure the metabolic similarity of complex dental plaque biofilm microcosms. Appl. Environ. Microbiol. 68: 5779–5783.
25. Staddon WJ, Duchesne LC, Trevors JT (1997) Microbial diversity and community structure of postdisturbance forest soils as determined by sole-carbon-source utilization patterns. Microbial Ecology 34: 125–130.
26. Kuhn I, Austin B, Austin DA, Blanch AR, Grimont PAD, et al. (1996) Diversity of Vibrio anguillarum isolates from different geographical and biological habitats, determined by the use of a combination of eight different typing methods. Syst Appl Microbiol 19: 442–450.
27. Glimm E, Heuer H, Engelen B, Smalla K, Backhaus H (1997) Statistical comparisons of community catabolic profiles. J Microbiol Methods 30: 71–80.
28. Marsh PD, Martin MV (2009) Oral Microbiology. United Kingdom: Elsevier. 63 p.
29. Dewhirst FE, Chen T, Izard J, Paster BJ, Tanner ACR, et al. (2010) The Human Oral Microbiome. J Bacteriol 192: 5002–5017.
30. Krebs CJ (1994) Ecology: The Experimental Analysis of Distribution and Abundance. New York: Harper Collins College Publishers. 514p.
31. Derry AM, Staddon WJ, Trevors JT (1998) Functional diversity and community structure of microorganisms in uncontaminated and creosote-contaminated soils as determined by sole-carbon-source-utilization. World J Microbiol Biotechnol 14: 571–578.

10

Acanthamoeba polyphaga mimivirus Stability in Environmental and Clinical Substrates: Implications for Virus Detection and Isolation

Fábio P. Dornas[1,9], Lorena C. F. Silva[1,9], Gabriel M. de Almeida[1], Rafael K. Campos[1], Paulo V. M. Boratto[1], Ana P. M. Franco-Luiz[1], Bernard La Scola[2], Paulo C. P. Ferreira[1], Erna G. Kroon[1], Jônatas S. Abrahão[1]*

1 Universidade Federal de Minas Gerais, Instituto de Ciências Biológicas, Laboratório de Vírus, Belo Horizonte, Minas Gerais, Brazil, **2** URMITE CNRS UMR 6236– IRD 3R198, Aix Marseille Universite, Marseille, France

Abstract

Viruses are extremely diverse and abundant and are present in countless environments. Giant viruses of the *Megavirales* order have emerged as a fascinating research topic for virologists around the world. As evidence of their ubiquity and ecological impact, mimiviruses have been found in multiple environmental samples. However, isolation of these viruses from environmental samples is inefficient, mainly due to methodological limitations and lack of information regarding the interactions between viruses and substrates. In this work, we demonstrate the long-lasting stability of mimivirus in environmental (freshwater and saline water) and hospital (ventilator plastic device tube) substrates, showing the detection of infectious particles after more than 9 months. In addition, an enrichment protocol was implemented that remarkably increased mimivirus detection from all tested substrates, including field tests. Moreover, biological, morphological and genetic tests revealed that the enrichment protocol maintained mimivirus particle integrity. In conclusion, our work demonstrated the stability of APMV in samples of environmental and health interest and proposed a reliable and easy protocol to improve giant virus isolation. The data presented here can guide future giant virus detection and isolation studies.

Editor: Richard Joseph Sugrue, Nanyang Technical University, United States of America

Funding: Fundação de Amparo a Pesquisa de Minas Gerais (FAPEMIG)(www.fapemig.br) and the Pró-Reitoria de Pesquisa da Universidade Federal de Minas Gerais (www.ufmg.br/prpq) supported the manuscript language (English) revision. The funders had no role in study design, data collection and analysis, decision to publish, or preparation of the manuscript.

Competing Interests: The authors have declared that no competing interests exist.

* E-mail: jonatas.abrahao@gmail.com

9 These authors contributed equally to this work.

Introduction

Viruses are extremely diverse and abundant and are present in countless environments [1]. In some extreme ecosystems, viruses are the only known microbial predators, and they are powerful agents of gene transfer and microbial evolution [1]. Although viral genomes are ubiquitous in the biosphere, very little is known regarding the ecological roles of viruses in most ecosystems [2,3]. In this context, large nucleocytoplasmic DNA viruses (NCLDVs) emerge as a fascinating research topic for virologists around the world. NCLDVs are frequently found in environmental samples, demonstrating their ubiquity and ecological impact [4]. There is still much to learn about NCLDV host-pathogen relationships and their impact on evolution, ecology and medicine [4].

Acanthamoeba polyphaga mimivirus (APMV), the prototype of the *Mimiviridae* family, was discovered in a hospital water cooling system in Bradford, England, during an outbreak of pneumonia [5]. APMV is an amoeba-associated virus with peculiar features, including a double-stranded DNA, ~1.2 megabase (Mb) genome encoding proteins not previously observed in other viruses, such as aminoacyl-tRNA synthetases and DNA repair chaperones and enzymes, a >700 nm particle diameter and capsid-associated fibers [5,6]. In 2008, a new giant virus named *A. castellanii mamavirus* (ACMV) was isolated from a cooling tower in Paris [7,8]. Other known giant viruses include *Megavirus chiliensis*, isolated from Chilean ocean water [9]; *Lentille virus*, from the contact lens fluid of a patient with keratitis [10]; and *Moumouvirus*, isolated from cooling tower water [11].

Acanthamoeba is believed to be the natural host of *Mimiviridae* [5], though there is evidence of mimivirus replication in vertebrate phagocytes. These amoebae are ubiquitous and have been isolated from aquatic environments, soil, air, hospitals and contact lens fluid. Amoebas are part of vertebrates' normal microbiota, and they are extremely resistant to pH variations, high temperatures and disinfectants [12,13]. Considering the ubiquity of *Acanthamoeba*, giant viruses could hypothetically be found everywhere. Metagenomic analysis demonstrated the presence of mimivirus-like sequences in many aquatic environments [12,14]. There are few data on APMV in animal tissues or its hypothetical role as pneumonia agent, although a recent metagenomic study found mimivirus DNA in bovine serum [15]. Free-living amoebae may potentially propagate pathogens in hospital environments, and hospitalized patients would represent a group of risk for amoeba-associated pneumonia agents, including APMV [5,16–18].

The ubiquity of APMV DNA in the environment but the lack of information about the ecological – and medical – impact of these viruses warrants their isolation and characterization. Research groups trying to "prospect" giant viruses in the laboratory [9,11,19] have difficulty recovering these viruses from environmental samples; there is also no standard protocol for the optimization of isolation techniques [19,20]. Most giant virus prospecting studies rely on direct co-culture of samples with amoebas to propagate viruses. However, the pre-enrichment of environmental samples can be useful for viral isolation, as demonstrated by the discovery of megavirus [9]. In this study, we verified the stability of APMV in hospital and environmental substrates and validated an enrichment protocol for APMV isolation. Our results suggest that the enrichment protocol improves APMV detection from different substrates but does not modify some viral genetic and biological features. Our study may be useful in future giant virus prospecting studies.

Materials and Methods

APMV Preparation

APMV particles were isolated and purified from infected amoebae as previously described [16]. Briefly, *Acanthamoeba castellanii* (ATCC 30234) were grown in 75 cm^2 cell culture flasks (Nunc, USA) in PYG (peptone-yeast extract-glucose) medium supplemented with 7% fetal calf serum (FCS, Cultilab, Brazil), 25 mg/ml Fungizone (amphotericin B, Cristalia, São Paulo, Brazil), 500 U/ml penicillin and 50 mg/ml gentamicin (Schering-Plough, Brazil). After reaching confluence, the amoebas were infected with APMV and incubated at 37°C until the appearance of cytopathic effects. APMV-rich supernatants from the infected amoeba were collected and filtered through a 0.8-micron (Millipore, USA) filter to remove amoeba debris. The viruses were then purified using a Gastrografin gradient (45–36–28%) [5], suspended in PBS and stored at −80°C.

Virus Titration

Samples were serially diluted (1/10) in PYG medium, and 100 µl was inoculated onto 10^5 amoeba seeded in a 96-well Costar® microplate (Corning, NY) on the previous day (8 wells per dilution, 200 µl final volume). Plates were incubated for 2–4 days at 32°C to determine the highest dilution that led to amoebal lysis ($TCID_{50}$/ml) [26].

Virus Recovery from Substrates, Enrichment and Stability Tests

To test the APMV recovery from the assayed substrates, a total of 10^6 $TCID_{50}$ of purified APMV was re-suspended in phosphate buffered saline (PBS) and added to autoclaved salt water (10 ml), fresh water (10 ml) and topsoil (1 g). The fresh water and soil samples were collected from three different Brazilian biomes: Amazon and Mata Atlântica, two very biodiverse rainforests, and Cerrado, a savanna-like biome. The salt water samples were collected from 3 points on the coast of South and Southeast Brazil, for a total of five samples per substrate per biome (Figure 1). The water and land collections were performed with permission of Instituto Chico Mendes (ICM) – protocol numbers: 34293-1 and 33326-2. The field studies did not involve endangered or protected species. Considering the hypothetical role of APMV as pneumonia agent associated with prolonged mechanical ventilation the viral stability in ventilator devices was evaluated. Three different brands of sterile ventilator device (tube) (VD) were used to test APMV recovery (five quadrants of 2×2 cm per brand) [21]. In this case, 10 µl of viral suspension was added to VD quadrants with or without BAV, which were maintained in sterile Petri dishes. After one hour, all the samples were titrated in *A. castellanii* by the $TCID_{50}$ method as described [26]. All the environmental and VD samples were previously tested for APMV DNA and/or infectious particles [5].

APMV stability was analyzed for 12 months in the substrates described above. Samples were maintained in 15 ml tubes at room temperature (Figure S1). Every month, 500 µl aliquots from each substrate were collected, serial diluted in PBS and titrated in amoebae. The viral titer was adjusted to the total assayed volume. The sample enrichment protocol was adapted from Arslan et al. (2011). Briefly, 500 µl of samples were added to 450 ml water-rice medium (40 grains of rice per liter of water) and kept in the dark at room temperature for 5 days. Following this incubation, 5000 pathogen-free amoebas were added to the samples, and 5000 more amoebas were added after twenty days. The samples were titrated in amoebae after thirty days and then every subsequent month for one year. All samples were used in real-time PCRs targeting the conserved hel gene (primers: 5′ACCTGATCCACATCCCA-TAACTAAA3′ and 5′GGCCTCATCAACAAATGGTTTCT-3′). Samples DNA were extracted by Phenol-ChloroformThe real-time PCR was performed with a commercial mix Power SYBr Green (Applied Biosystems, USA), primers (4 mM each) and 1 µl sample in reaction of 10 µl final volume. All reactions were performed in a StepOne thermocycler: 95°C-10 min, 40 cycles-95°C-15 s/60°-15 s, followed by a dissociation step (specific Tm = 73°C).

Transmission Electron Microscopy

A. castellanii were infected at an MOI of 10. Uninfected amoebae were used as controls. At 7 h post-infection, amoebae were washed twice with PBS and fixed with 2.5% glutaraldehyde type 1 (Merck, Germany) for one hour at room temperature. The amoebae culture was dislodged with a cell scraper and centrifuged at 900×g for 5 min. Ultrathin sections were prepared [5] by the Centro de Microscopia, UFMG, Brazil.

One-step Growth Curves

To compare the replication of APMV before and after enrichment, one-step growth curve assays were performed. *A. castellanii* were infected at an MOI of 10. The infectivity was measured by TCID50 after 25 hours (0 to 25 hours) by observing the CPE in amoeba.

Sequencing Analysis

The hel and GlcT genes were amplified by PCR from purified APMV and from samples after enrichment. The hel and GlcT genes were chosen since they have been used as markers in phylogenetic or in evolutionary studies [5,9,11]. The primers were designed based on APMV sequences available in GenBank (Accession NC 014649). The amplicons were directly sequenced in both orientations and in triplicate (Mega-BACE sequencer, GE Healthcare, Buckinghamshire, UK). The sequences were aligned with Genbank references with ClustalW and were manually aligned using MEGA software version 4.1 (Arizona State University, Phoenix, AZ, USA).

Results

APMV is Stable for Long Periods in Different Substrates

To evaluate APMV stability in environmental and hospital substrates, an initial experiment analyzed the recovery of APMV from each substrate. A total of 10^6 $TCID_{50}$ of purified APMV was re-suspended in phosphate buffered saline (PBS) and added to

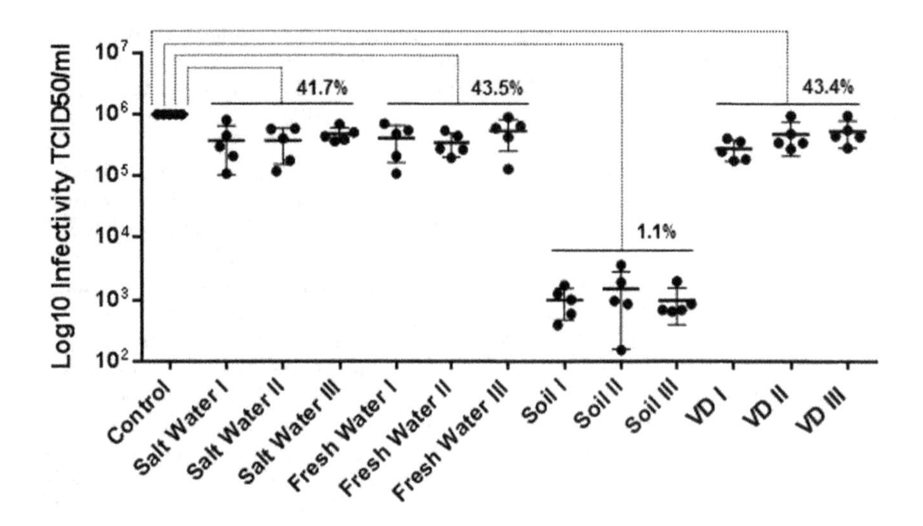

Figure 1. APMV recovery from samples after 1 hour. 10^6 $TCID_{50}$ of purified APMV were added to salt water (10 ml), fresh water (10 ml), topsoil (1 g) and three different brands of VD substrates, and after one hour, virus recovery was evaluated by titration in *A. castellanii*. The fresh water and soil samples were collected from three different Brazilian biomes: Amazon and Mata Atlântica, two highly biodiverse rainforests, and Cerrado, a savanna-like biome. The salt water samples were collected from 3 points on the coast of South and Southeast Brazil, for a total of five samples per substrate per biome. The results are the means+SD of an experiment performed in quintuplicate.

previously autoclaved salt water (10 ml), fresh water (10 ml) and topsoil (1 g). The fresh water and soil samples were collected from three different Brazilian biomes (Amazon, Cerrado and Mata Atlântica), and the salt water samples were collected from 3 points on the coast of South and Southeast Brazil, for a total of five samples per substrate per biome (Figure 1). Three different brands of sterile ventilator device (tube) (VD) were used to test APMV recovery (five quadrants of 2×2 cm per brand) [21]. After one hour, all the samples were titrated in *A. castellanii*. The average values from quintuple replicates showed >40% virus recovery from salt water, fresh water and VD samples, regardless of the biome or brand. In contrast, viral recovery from soil was approximately 1% in all samples (Figure 1).

The stability of APMV was analyzed for 12 months in each substrate. Samples were prepared and maintained in 15 ml tubes at room temperature. Relative humidity and temperature were monitored through the experiment (relative humidity average max-72%; min 56%/temperature average max-25°C; min-19°C) (Figure S1). At monthly intervals, aliquots from each substrate were collected, serially diluted in PBS and titrated in amoebae. Titration assays revealed the long-term stability of APMV in salt and fresh water and VD, as infectious particles were detected in each substrate after 9, 12 and 10 months, respectively (Figure 2A, 2B, 2D). In general, the viral titers decreased less than 1 log until the 6th month, after which time virus titers decreased eventually to undetectable levels. APMV could be isolated from soil samples until the third month, but the viral titer was decreased by >2 log after the first month (Figure 2C).

Enrichment of APMV from Different Substrates

While recovery of APMV depended on the substrate in which the virus was immersed, recovery was never complete, and the viral titer decreased over time in all substrates. To verify that an enrichment protocol following sample collection could improve viral recovery, we prepared substrates as described above containing 10^6 $TCID_{50}$ APMV and then enriched the viruses in these samples [9]. Briefly, 500 µL each sample was added to 450 ml water-rice medium (40 grains of rice per liter of water) and stored in the dark at room temperature for 5 days. Following this incubation, 5000 amoebas were added to the samples, and 5000 more amoebas were added after twenty days. After thirty days, the samples were titrated in amoebae; samples were then titrated again monthly for one year. The enrichment process improved APMV recovery from all the analyzed substrates. APMV was detected after one year in salt and fresh water (Figure 3A and 3B), seven months in soil (Figure 3C) and one year in VD samples (Figure 3D). When compared with the results shown in Figure 2, the overall viral titer and the viral recovery time were increased after the enrichment. In enriched soil samples, virus could be detected up to seven months; this duration was much improved from the unenriched maximum detection time of three months.

APMV Stability in VD Containing BAL Samples

Detection of APMV in VD is clinically relevant and could establish this virus as a human pathogen [16]. One factor that might impact viral detection in VD is the presence of broncho-alveolar lavage (BAL). To verify if BAL impairs APMV recovery from VD samples, 10^6 $TCID_{50}$ of APMV were added to VD containing BAL, which were then enriched. While BAL generally reduced APMV recovery (Figure 4A), the enrichment protocol increased the initial viral titers and allowed the isolation of APMV up to the 9th month of the experiment (Figure 4B).

Real-time PCR of APMV in Environmental and Hospital Substrates

As shown above, the optimal recovery APMV results from pre-enrichment of substrates. Viral isolation from environmental samples is vital for the characterization of new giant viruses, but it is not absolutely required for some ecological, evolutionary or epidemiological studies. Molecular viral detection, e.g., PCR, is faster, cheaper and less laborious than viral isolation. To verify that enrichment improves APMV detection by PCR, the samples described in the previous experiments were used as templates for real-time PCRs designed to detect APMV helicase (hel). APMV DNA was detected in all water and VD samples, with or without enrichment. APMV DNA was detected up to the eleventh month in non-enriched salt water samples and up to the twelfth month in

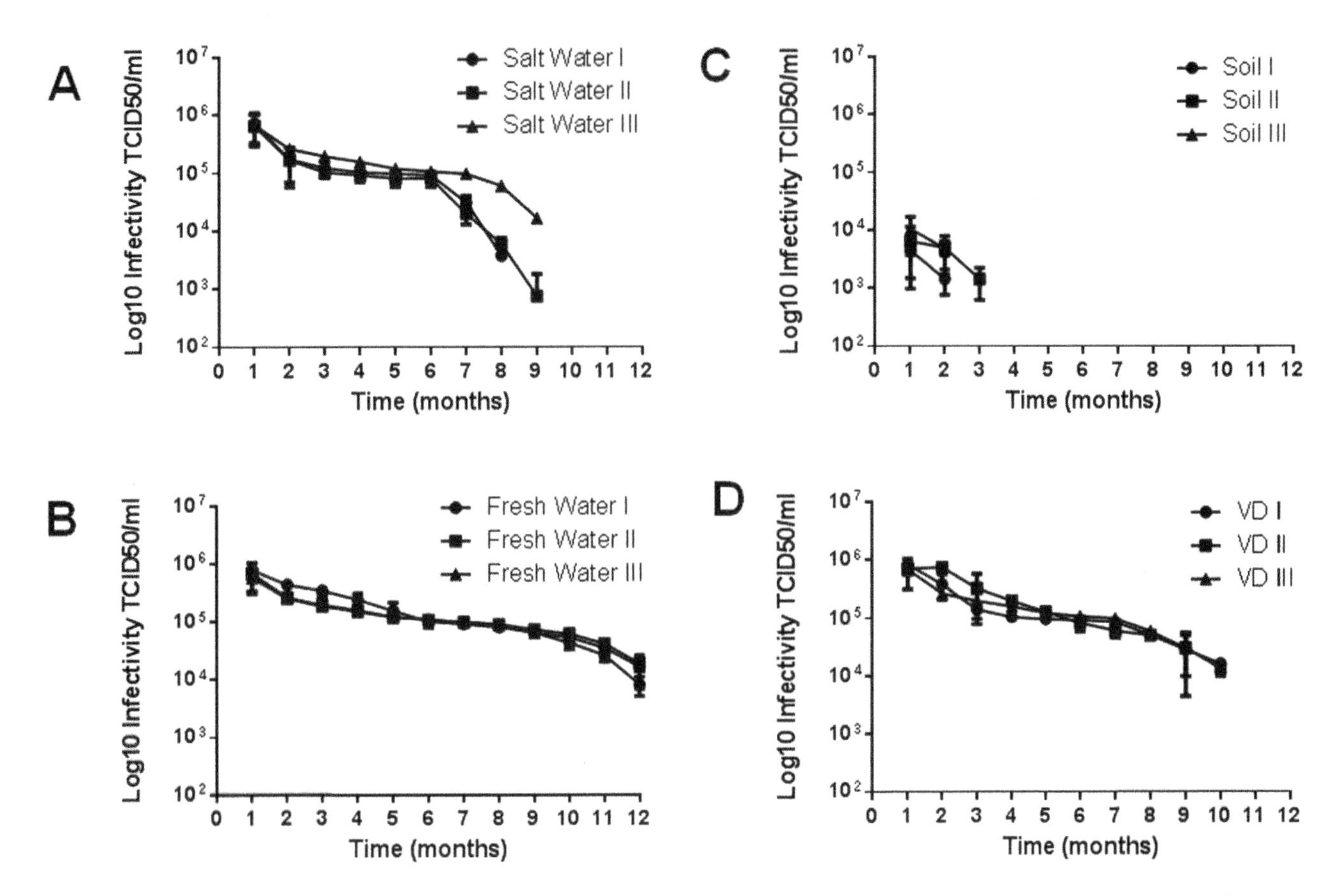

Figure 2. APMV stability in different substrates. To evaluate the long-term stability of APMV in different substrates for one year, 10^6 $TCID_{50}$ of purified APMV were added to salt water, soil, fresh water and VD substrates, which were maintained in 15 ml tubes at room temperature. At monthly intervals, the samples were titrated in amoebae. (A) Salt water; (B) Fresh water; (C) Soil; (D) VD. The results are the means+SD of an experiment performed in quintuplicate. I, II and III represent independent experiments performed with samples collected from distinct locations.

enriched samples. In soil samples, APMV DNA was detected up to five months and nine months in unenriched and enriched samples, respectively. Enrichment also improved APMV detection in VD containing BAL samples, with a detection time shift from the eighth to the eleventh month (Table 1).

APMV Biology, Morphology and DNA Sequences are not Changed by Enrichment

Virus isolation in laboratory conditions could promote genotype and/or phenotype changes by artificial selection. Boyer et al. (2011) showed dramatic changes in APMV genomes, replication and morphology after sub-culturing APMV in a germ-free amoeba host. To evaluate possible changes in APMV caused by the enrichment process, morphological, virological and genetic assays were performed. No differences were observed in one-step growth curves of virus recovered from all assayed samples, suggesting that no biological modifications occurred during enrichment (Figure 5). These samples were also observed by electron microscopy, and no morphological changes were detected when compared to APMV images previously published [5,7]. (Figure S2). In addition, the sequences of the APMV hel and GlcT genes from enriched samples were 100% identical to the APMV reference sequences in Genbank (Figures S3 and S4).

Enrichment Applicability Tests

To test the applicability of the enrichment protocol, water samples were collected from 2 urban lakes and from a river in Southeast and North Brazil. A total of 475 samples of 5 ml each were collected from the water surface. Aliquots of the samples were enriched or directly inoculated onto *A. castellanii* monolayers for virus isolation. In parallel, real-time PCRs to detect APMV were performed.

From the 475 samples, six giant viruses were isolated with the enrichment protocol (Table 2). These viruses induced cytopathic effects in amoebas, including cell rounding and lysis, and were positive in PCR assays. In contrast, only one of the six virus isolates was propagated by direct inoculation of the water samples in amoebas. Sequencing of hel gene confirmed that all virus isolates belonged to the *Megavirales* order (data not shown).

Discussion

Virus isolation has technical limitations, and classical protocols based on filtration have delayed the detection of giant viruses [22]. Another problem for giant virus isolation is their limited known host range, which restricts the cellular systems that can be used for *in vitro* culture [22]. Thus, most information on giant viruses comes from environmental metagenomic studies and not from virus isolation [22]. This approach must be made with extreme caution, as comparing fragmented environmental sequences with those found in databases can result in false assumptions about giant virus complexity and HGT events [14]. Isolation of these viruses is imperative for understanding their roles in the environment and as potential vertebrate pathogens.

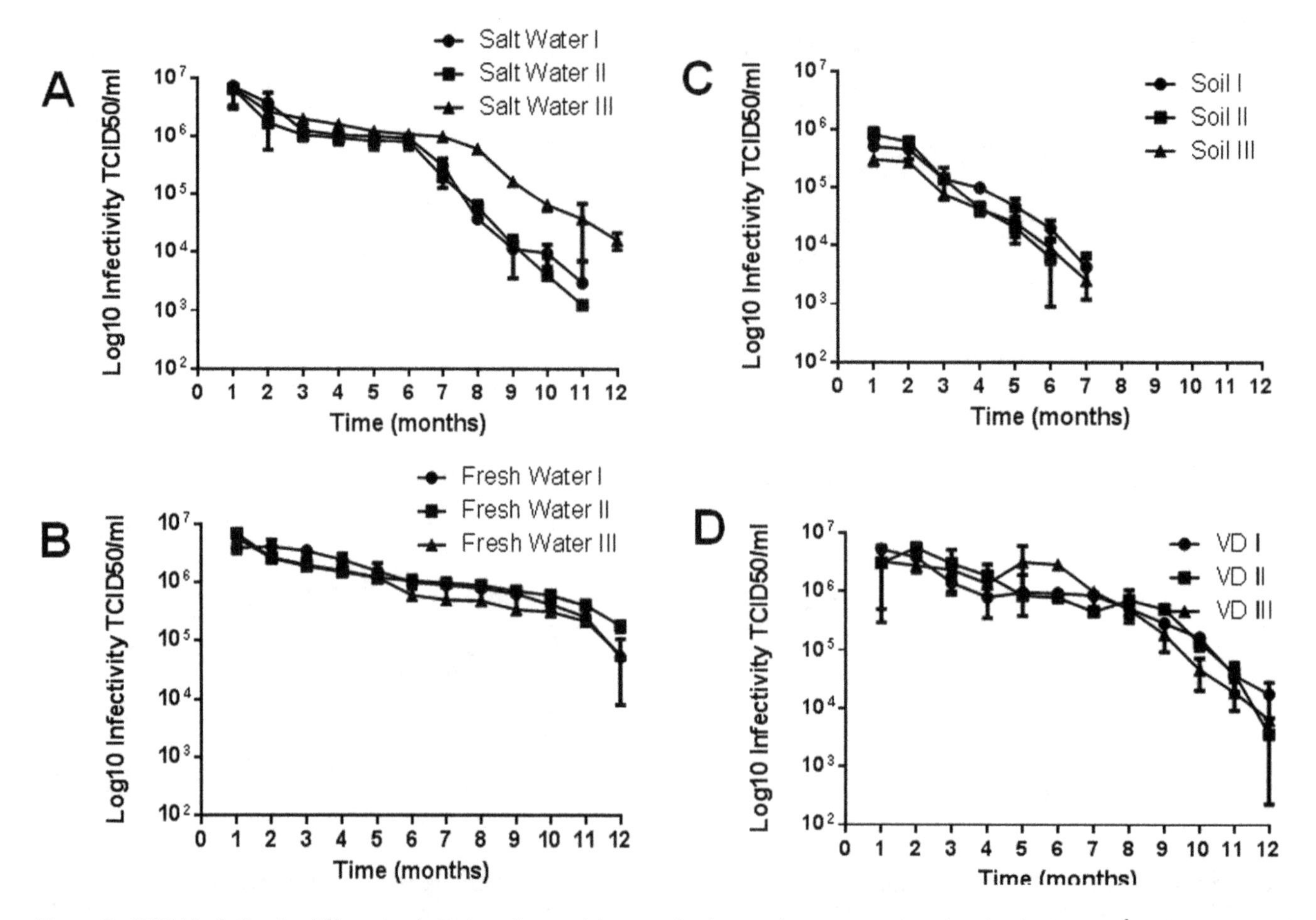

Figure 3. APMV isolation in different substrates after enrichment. As above, substrates were inoculated with APMV (10^6 viral particles) and then enriched. At one-month intervals, the samples were titrated in amoebae. (A) Salt water; (B) Fresh water; (C) Soil; (D) VD. The results are the means+SD of an experiment performed in quintuplicate. I, II and III represent independent experiments performed with samples collected from distinct locations.

The ubiquity of amoebae implies the probable ubiquity of *Mimiviridae*, which has been confirmed by indirect viral detection methods. Despite their remarkable abundance, giant viruses are not easily isolated. Many of these viruses are unknown, and their stability in environmental samples is not determined. Furthermore, there are few described approaches for viral isolation or protocols that favor viral isolation from complex substrates. Our investigation estimates APMV stability in several relevant substrates and

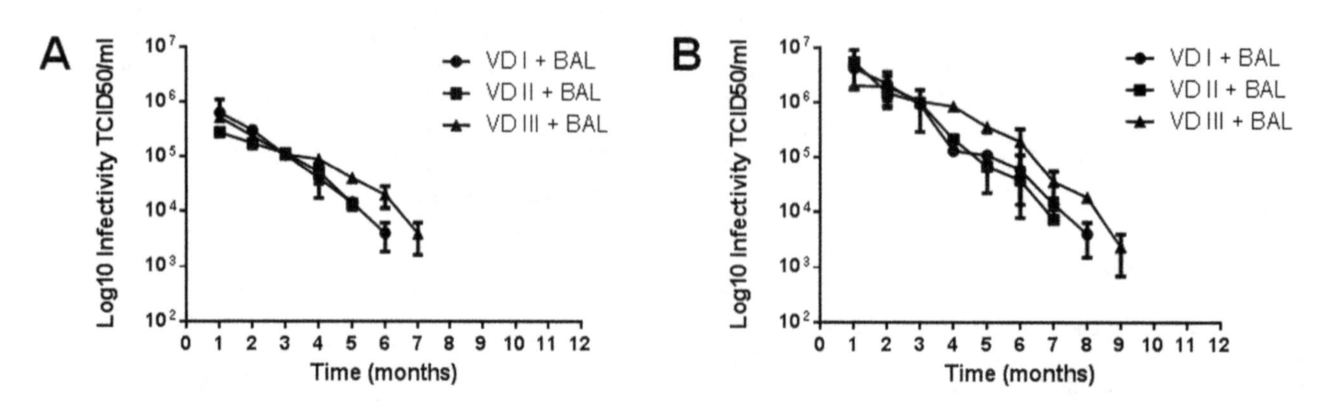

Figure 4. APMV stability in VD enriched after BAL addition. Purified APMV (10^6 viral particles) were added to VD samples, and subsequently, BAL samples were added. The BAL-VD samples were or were not enriched and maintained at room temperature. The samples were titrated in amoebae monthly. (A) Samples without enrichment; (B) Enriched samples. The results are the means+SD of an experiment performed in quintuplicate.

Table 1. APMV detection by real-time PCR in samples with or without enrichment.

	APMV detection by PCR[a,b]	
Sample	**With no Enrichment (months)**	**Post - Enrichment (months)**
Salt Water	1st to 11th	1st to 12th
Fresh Water	1st to 12th	1st to 12th
Soil	1st to 5th	1st to 9th
VD	1st to 12th	1st to 12th
VD+BAL	1st to 8th	1st to 11th

[a]Real-time PCR - Helicase gene;
[b]Results presented considering all replicates.

Table 2. Enrichment applicability tests: detection of giant viruses with no and post-enrichment by viral isolation and real-time PCR.

		Positive samples [total of positive (%)][a]	
Sample	**Total of samples (n)**	**With no Enrichment**	**Post - Enrichment**
Lake 1	325	1 (0.3%)	2 (0.6%)
Lake 2	88	0 (0%)	2 (2.27%)
River	35	0 (0%)	2 (5.71%)
Total	**475**	**1 (0.21%)**	**6 (1.2%)**

[a]Isolation in *A. castellani* monolayer+positivity in hel gene real-Time PCR.

shows that enrichment prior to isolation optimizes viral stability and recovery from any substrate.

In our first analysis, APMV recovery was reduced at one hour after its addition to different substrates. Although viral recovery from fresh water, salt water or VD was approximately 40% of the initial viral load, viral recovery from soil samples was much lower, representing less than 2% of the input viruses (Figure 1). The reasons for these recovery ranges include microbial influence and physical or chemical characteristics of the substrate. These factors interfere with survival through viral particle aggregation and virucidal activity. Particularly for APMV, viral aggregation is important since through their capsid surrounded by fibrils can occurs easily aggregation and influence in the filtration process required for viral isolation [12,23,24]. Following these results, we measured the stability of APMV in the substrates mentioned above for twelve months. Virus was recovered from fresh water at all time points tested, and virus was recovered from salt water, VD and soil until the ninth, tenth and third months of the experiment, respectively (Figure 2). Viral titers dropped over time in all substrates due to the reasons stated above.

In theory, an enrichment protocol could increase the number of giant viruses in a sample (viral replication) and, thus, their likelihood of recovery. This protocol favors the proliferation of heterotrophic organisms that are consumed by amoebas, and these amoebas are used for giant virus replication. After establishing the stability of APMV in different substrates, we added an enrichment protocol prior to viral isolation. The enrichment improved the method sensitivity, as viruses were isolated at later time points compared with samples without enrichment. Furthermore, enrichment increased virus yields (Figure 3).

Among the analyzed substrates, VD is particularly interesting for the hypothetical pathogenic aspect of *Mimiviridae*, as APMV has been detected in pneumonia patients. Devices such as VD are sources of nosocomial infections and could be sources for giant virus infections as well. We showed that APMV had long-term stability in VD and that the pre-enrichment improved viral detection. In a hospital setting, VDs would most likely be filled with clinical specimens that could interfere with APMV stability and recovery. To test this possibility, we added BAL to VDs before adding APMV and then determined viral recovery and stability with or without enrichment. BAL affected APMV stability; APMV was recovered from BAL-containing VDs only up to seven months without enrichment. In this case, enrichment allowed APMV recovery at nine months (Figure 4). BAL decreased APMV stability but did not completely neutralize the virus, so these samples were still potentially infectious. As shown previously, enrichment improved APMV recovery from BAL-containing VDs and was thus useful for isolating the giant virus from these samples.

APMV in the enriched or unenriched samples was also quantified by real-time PCR, which is more sensitive than cell

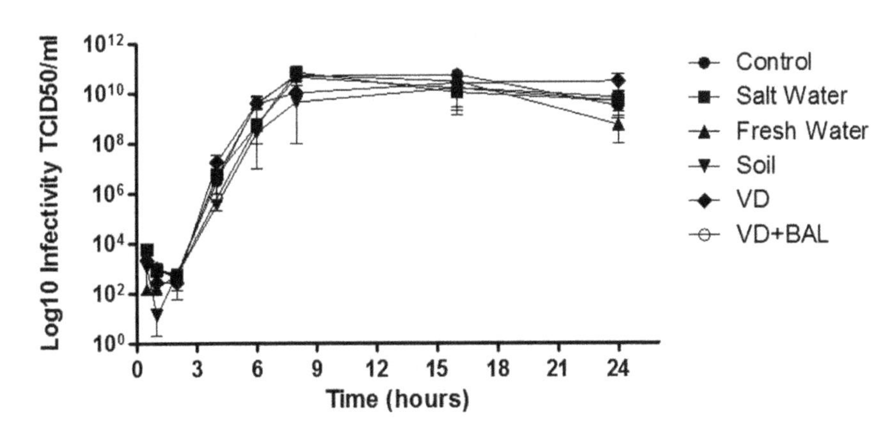

Figure 5. APMV one-step growth curves with or without enrichment. Purified APMV (10^6 viral particles) was added to salt water, soil, fresh water and VD substrates, which were then enriched. After viral isolation from each substrate, *A. castellanii* were infected at an MOI of 10. The infectivity was measured by $TCID_{50}$ after 25 hours (0 to 25 hours) by observing CPE in amoeba. The results are the means of experiments performed in duplicate.

culture isolation and is also improved by enrichment (Table 1). While isolation needs viable viral particles in the sample, PCR detection needs viral DNA from any viral particles, whether viable, inactivated or defective. Thus, genomic studies can also benefit from the enrichment protocol described in this paper.

In allopatric conditions, successive passages of APMV in amoebae drastically reduce the viral genome, resulting in morphological and genetic changes [25]. To verify that enrichment protocol did not modify the original virus, we compared one-step growth curves from viruses obtained after enrichment to the prototype virus. The curves were similar, suggesting that the enrichment did not cause any biological alterations (Figure 5). We also verified the absence of morphological and genetic changes in APMV by electron microscopy and sequencing. All viruses were similar in size and had the characteristic APMV fibers and internal membranes (Figure S2). Analysis of hel and GlcT revealed that enrichment did not alter these genes, as they were identical to reference sequences (Figure S3).

In conclusion, our work determines the stability of APMV in samples of environmental and clinical interest and proposes a reliable and easy protocol to improve giant virus isolation. This protocol can be used for giant virus prospecting studies.

Supporting Information

Figure S1 Temperature and relative humidity averages during the 12 experimental months.

Figure S2 APMV do not exhibit changes in viral morphology after enrichment. Purified APMV (10^6 viral particles) were added to salt water, soil, fresh water and VD substrates, which were then enriched. After viral isolation from each substrate, *A. castellanii* were infected at an MOI of 10 and analyzed by EM at 8 hours post-infection. A, B, C and D: APMV recovery from enriched salt water, fresh water, soil and VD, respectively.

Figure S3 APMV do not show changes in the GlcT gene after enrichment. PCR for the GlcT gene was performed using enriched APMV samples as templates, and the resulting amplicon was sequenced. The DNA sequences were aligned with APMV reference sequences from Genbank using the ClustalW method and were manually aligned using MEGA software version 4.1 (Arizona State University, Phoenix, AZ, USA).

Figure S4 APMV do not show changes in the hel gene after enrichment. PCR for the hel gene was performed using enriched APMV samples as templates, and the resulting amplicon was sequenced. The DNA sequences were aligned with APMV reference sequences from Genbank using the ClustalW method and were manually aligned using MEGA software version 4.1 (Arizona State University, Phoenix, AZ, USA).

Acknowledgments

We thank João Rodrigues dos Santos, Gisele Cirilo dos Santos, and colleagues from Gepvig and the Laboratório de Vírus for their excellent technical support. We also thank Fundação de Amparo a Pesquisa de Minas Gerais (FAPEMIG) and the Pró-Reitoria de Pesquisa da Universidade Federal de Minas Gerais.

Author Contributions

Conceived and designed the experiments: FPD LCFS PVMB JSA. Performed the experiments: FPD LCFS GMA RKC PVMB APMFL. Analyzed the data: FPD LCFS GMA BL EGK PCPF JSA. Contributed reagents/materials/analysis tools: EGK PCPF JSA. Wrote the paper: FPD LCFS GMA BL EGK JSA.

References

1. Rohwer F, Prangishvili D, Lindell D (2009) Roles of viruses in the environment. Environ Microbiol 11: 2771–2774.
2. Breitbart M, Miyake JH, Rohwer F (2004) Global distribution of nearly identical phage-encoded DNA sequences. FEMS Microbiol Lett 236: 249–256.
3. Short CM, Suttle CA (2005) Nearly identical bacteriophage structural gene sequences are widely distributed in both marine and freshwater environments. Appl Environ Microbiol 71: 480–486.
4. Van Etten JL, Lane LC, Dunigan DD (2010) DNA viruses: the really big ones (giruses). Annu Rev Microbiol 64: 83–99.
5. La Scola B, Audic S, Robert C, Jungang L, de Lamballerie X, et al. (2003) A giant virus in amoebae. Science 299: 2033.
6. Moreira D, Brochier-Armanet C (2008) Giant viruses, giant chimeras: the multiple evolutionary histories of Mimivirus genes. BMC Evol Biol 8: 12.
7. La Scola B, Desnues C, Pagnier I, Robert C, Barrassi L, et al. (2008) The virophage as a unique parasite of the giant mimivirus. Nature 455: 100–104.
8. Desnues C, Raoult D (2010) Inside the lifestyle of the virophage. Intervirology 53: 293–303.
9. Arslan D, Legendre M, Seltzer V, Abergel C, Claverie JM (2011) Distant Mimivirus relative with a larger genome highlights the fundamental features of Megaviridae. Proc Natl Acad Sci U S A 108: 17486–17491.
10. Desnues C, La Scola B, Yutin N, Fournous G, Robert C, et al. (2012) Provirophages and transpovirons as the diverse mobilome of giant viruses. Proc Natl Acad Sci U S A 109: 18078–18083.
11. Yoosuf N, Yutin N, Colson P, Shabalina SA, Pagnier I, et al. (2012) Related giant viruses in distant locations and different habitats: Acanthamoeba polyphaga moumouvirus represents a third lineage of the Mimiviridae that is close to the megavirus lineage. Genome Biol Evol 4: 1324–1330.
12. Monier A, Larsen JB, Sandaa RA, Bratbak G, Claverie JM, et al. (2008) Marine mimivirus relatives are probably large algal viruses. Virol J 5: 12.
13. Siddiqui R, Khan NA (2012) Biology and pathogenesis of Acanthamoeba. Parasit Vectors 5: 6.
14. Ghedin E, Claverie JM (2005) Mimivirus relatives in the Sargasso sea. Virol J 2: 62.
15. Hoffmann B, Scheuch M, Höper D, Jungblut R, Holsteg M, et al. (2012) Novel orthobunyavirus in Cattle, Europe, 2011. Emerg Infect Dis 18: 469–472.
16. La Scola B, Marrie TJ, Auffray JP, Raoult D (2005) Mimivirus in pneumonia patients. Emerg Infect Dis 11: 449–452.
17. Raoult D, La Scola B, Birtles R (2007) The discovery and characterization of Mimivirus, the largest known virus and putative pneumonia agent. Clin Infect Dis 45: 95–102.
18. Ghigo E, Kartenbeck J, Lien P, Pelkmans L, Capo C, et al. (2008) Ameobal pathogen mimivirus infects macrophages through phagocytosis. PLoS Pathog 4: e1000087.
19. Boughalmi M, Saadi H, Pagnier I, Colson P, Fournous G, et al. (2012) High-throughput isolation of giant viruses of the Mimiviridae and Marseilleviridae families in the Tunisian environment. Environ Microbiol.
20. La Scola B, Campocasso A, N'Dong R, Fournous G, Barrassi L, et al. (2010) Tentative characterization of new environmental giant viruses by MALDI-TOF mass spectrometry. Intervirology 53: 344–353.
21. Campos RK, Andrade KR, Ferreira PC, Bonjardim CA, La Scola B, et al. (2012) Virucidal activity of chemical biocides against mimivirus, a putative pneumonia agent. J Clin Virol 55: 323–328.
22. Van Etten JL, Lane LC, Dunigan DD (2010) DNA viruses: the really big ones (giruses). Annu Rev Microbiol 64: 83–99.
23. Gerba CP, Schaiberger GE (1975) Effect of particulates on virus survival in seawater. J Water Pollut Control Fed 47: 93–103.
24. Wetz JJ, Lipp EK, Griffin DW, Lukasik J, Wait D, et al. (2004) Presence, infectivity, and stability of enteric viruses in seawater: relationship to marine water quality in the Florida Keys. Mar Pollut Bull 48: 698–704.
25. Boyer M, Azza S, Barrassi L, Klose T, Campocasso A, et al. (2011) Mimivirus shows dramatic genome reduction after intraamoebal culture. Proc Natl Acad Sci U S A 108: 10296–10301.
26. Raoult D, Audic S, Robert C, Abergel C, Renesto P, et al. (2004) The 1.2-megabase genome sequence of Mimivirus. Science 306: 1344–1350.

11

Composition and Similarity of Bovine Rumen Microbiota across Individual Animals

Elie Jami[1,2], Itzhak Mizrahi[1]*

1 Department of Ruminant Science, Institute of Animal Sciences, Agricultural Research Organization, Bet Dagan, Israel, 2 Department of Molecular Microbiology and Biotechnology, The George S. Wise Faculty of Life Sciences, Tel Aviv University, Ramat-Aviv, Israel

Abstract

The bovine rumen houses a complex microbiota which is responsible for cattle's remarkable ability to convert indigestible plant mass into food products. Despite this ecosystem's enormous significance for humans, the composition and similarity of bacterial communities across different animals and the possible presence of some bacterial taxa in all animals' rumens have yet to be determined. We characterized the rumen bacterial populations of 16 individual lactating cows using tag amplicon pyrosequencing. Our data showed 51% similarity in bacterial taxa across samples when abundance and occurrence were analyzed using the Bray-Curtis metric. By adding taxon phylogeny to the analysis using a weighted UniFrac metric, the similarity increased to 82%. We also counted 32 genera that are shared by all samples, exhibiting high variability in abundance across samples. Taken together, our results suggest a core microbiome in the bovine rumen. Furthermore, although the bacterial taxa may vary considerably between cow rumens, they appear to be phylogenetically related. This suggests that the functional requirement imposed by the rumen ecological niche selects taxa that potentially share similar genetic features.

Editor: Purificación López-García, Université Paris Sud, France

Funding: The authors have no support or funding to report.

Competing Interests: The authors have declared that no competing interests exist.

* E-mail:itzhakm@agri.gov.il

Introduction

A significant proportion of domesticated animal species worldwide—the source of most meat and dairy products—are ruminants. Chief among these are dairy cattle. Ruminants are herbivores, and their digestive system allows them to absorb and digest large amounts of plant material. This capacity is of enormous significance to man, as ruminants essentially convert the energy stored in plant mass to digestible food products [1]. The ability to absorb and digest the plant material resides in the ruminants' foregut, the rumen, which is essentially a chambered anaerobic compartment. The rumen is inhabited by a high density of resident microbiota, consisting of bacteria, protozoa, archaea and fungi, which degrade the consumed plant materials [2]. The rumen microorganisms, of which bacteria are the most abundant and diverse (~95% of the total microbiota [3]), ferment and degrade the plant fibers in a coordinated and complex manner which results in the conversion of plant materials into digestible compounds, such as volatile fatty acids and bacterial proteins. These, in turn, define the quality and composition of milk and meat and their production yields [4–6]. Hence, the rumen microbiota is essential to the animals' well being and productivity, and consequently mankind. Therefore an understanding of these complex microbial populations and their interactions is of great importance.

Several cultivation-free methods have been used to study rumen microbial communities in both domesticated and wild ruminants [4,5,6,7,8]. In a recent study, used denaturing gradient gel electrophoresis (DGGE) analysis to investigate the effect of rumen sampling location and timing on ruminal bacterial diversity [9]. That study revealed high similarity between samples taken from different locations and time points for each individual cow, but lower similarity between samples taken from different host animals [9]. Other studies have focused on changes occurring in the microbial community and gene expression following changes in diet [10,11].

In a study examining the changes in ruminal bacterial communities during the feeding cycle, it was implied that cows fed the same diets can exhibit substantial differences in bacterial community composition [4]. Differences in rumen microbial composition were further emphasized in a recent metagenomic study exploring the ruminal fiber-adherent microbial populations of three steers, one of which had a microbiome and metagenome which were remarkably different from the other two [3]. These observations raise important and fundamental questions regarding ruminal bacterial populations, among them: How similar are the ruminal bacterial populations across individual animals fed the same diet in terms of composition, abundance and occurrence? Are there specific populations which are present across all individual rumens? If so, what is the extent and composition of these populations?

We addressed these questions by analyzing the compositions and similarities of bacterial populations from 16 animals' rumens using amplicon pyrosequencing of the V2 and V3 regions of the 16 S rRNA gene with a total of 162,000 reads, ~10,000 reads per sample. We present a study characterizing the similarities in identity and abundance of the rumen bacterial populations across all samples, as well as of specific populations that were present in all rumen samples examined.

Results

Identity of the ruminal bacterial composition

We sampled the ruminal contents of 16 Holstein Friesian lactating cows fed the same diet ad libitum for several months and held under the same experimental conditions for 6 weeks. Samples were taken 1 h after feeding as described by Brulc *et al.* [3]. Microbial cells were separated from the rumen samples and their DNA was extracted using a protocol described by Stevenson and Weimer [6]. We then identified and characterized the overall ruminal bacterial composition as well as the taxa shared by all cows, by using bacterial tag-encoded amplicon pyrosequencing generated from the V2 and V3 regions of the 16 S rRNA gene. In total, 162,000 reads were generated with an average of 9587±2059 reads per sample. We used the QIIME pipeline [12] to filter the reads and for quality control, as well as for some of the data analyses. After filtering, quality control and chimera removal (see materials and methods), the total number of operational taxonomic units (OTUs) detected by the analysis reached 4986, with an average 1800±324 OTUs per rumen sample (an OTU was defined as a read sharing ≥97% nucleotide sequence identity) (Tables S1, S2). We performed a sample-based rarefaction test to assess whether our sampling and sequencing efforts provided efficient OTU coverage. After the tenth sample, the number of OTUs was saturated, as revealed by the asymptotic nature of the sample rarefaction curve (Figure 1A). Taxonomic assignment showed that the dominant ruminal bacterial phyla, summing to 93% of total bacterial reads, were Firmicutes and Bacteroidetes, representing 42% and 51% of total OTUs, respectively; 5.21% of the reads were attributed to the phylum Proteobacteria, 0.87% to Actinobacteria, and 0.68% to Tenericutes. Other phyla were also present but at lower percentages (Figure 2A). Examining each sample composition at the phylum level, we observed noticeable differences between individual cows reflected by changes in the abundance of Firmicutes, Bacteroidetes and Proteobacteria (Figure 2B).

Core bacterial community shared by all cows

We analyzed our data for the distribution of each OTU across all samples. Occurrence of each OTU across the samples was evaluated and grouped into different categories according to prevalence. Figure 3 exhibits the percentage of OTUs shared by each sample occurrence category. This analysis revealed that 35% of the OTUs are present in only 10 to 20% of the samples and 14% are present in 20 to 30% of the samples, resulting in almost 50% of the OTUs being shared by a small proportion of the samples. This analysis also revealed a group with ~4% of OTUs shared by all samples and a group of ~2% shared by 90 to 99% of the samples. We next performed a genus-level analysis of the composition and abundance of the core bacterial community shared by 100% of the samples, and identified 32 genera that were shared by all samples (Figure 4). Some of the shared genera were highly abundant in the overall rumen bacterial community across the samples, such as the genus *Prevotella* which accounted for an average 52% of all rumen bacterial genera, while others, although shared by all of the samples, accounted for an average of 0.1% of the total rumen bacterial genera, such as the genus *Oscillospira* (Figure 4). Most of the shared genera varied in abundance across the samples. We further analyzed this core community at the species level (≥97%). This analysis revealed 157 OTUs shared by all samples with the highest representation from the following taxa: genus *Prevotella* (80 OTUs), family Lachnospiraceae (14 OTUs), genus *Butyrivibrio* (16 OTUs) (Table S3).

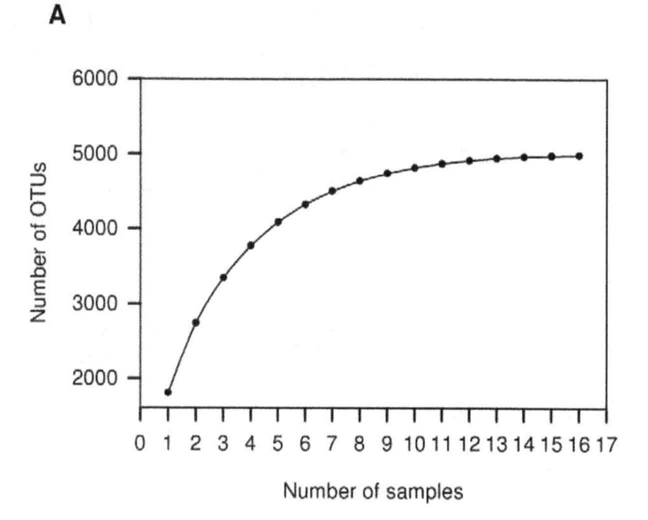

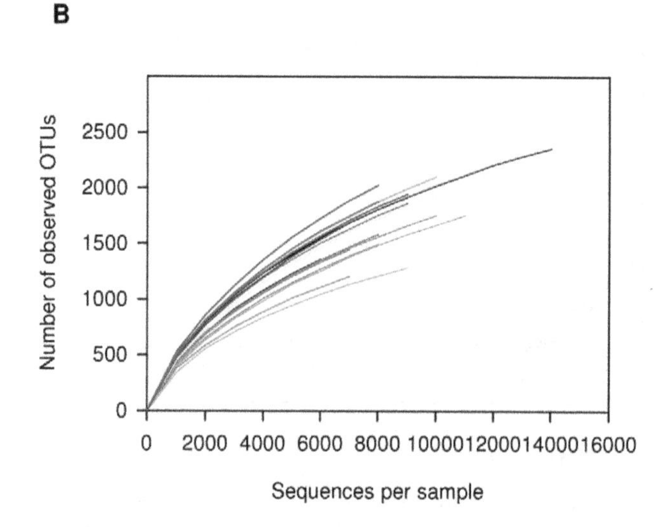

Figure 1. Rarefaction analysis for the assessment of OTU coverage. (**A**) Sample-based rarefaction curve showing the increase in OTU numbers as a function of the number of individuals sampled. Each added sample adds OTUs to the plot which has not yet been seen in previous samples. The curve becomes asymptotic as the OTU number saturates, and each sample adds an increasingly smaller number of new OTUs, indicating adequate coverage for the environment being tested. (**B**) Individual rarefaction curves for each rumen sample taken.

We also employed quantitative real-time PCR to monitor the presence and abundance of species belonging to genera which were not present in all samples but are considered to be important for the rumen ecosystem (Table S4). Most of the species examined were detected by the real-time analysis in all samples except *Ruminobacter amylophilus H18*, which was not found in all samples and exhibited very low abundance, just above detection level, when it was observed (Table S4).

Similarity between cows

To assess the degree of similarity between the samples, we performed a pairwise similarity analysis in which the distances of each sample were paired and then averaged, giving the similarity of a specific sample to all others. To this end, we used the Bray-Curtis metric, as well as the weighted UniFrac metric which also measures the distance between communities based on their phylogenetic lineages (using the pyNAST QIIME

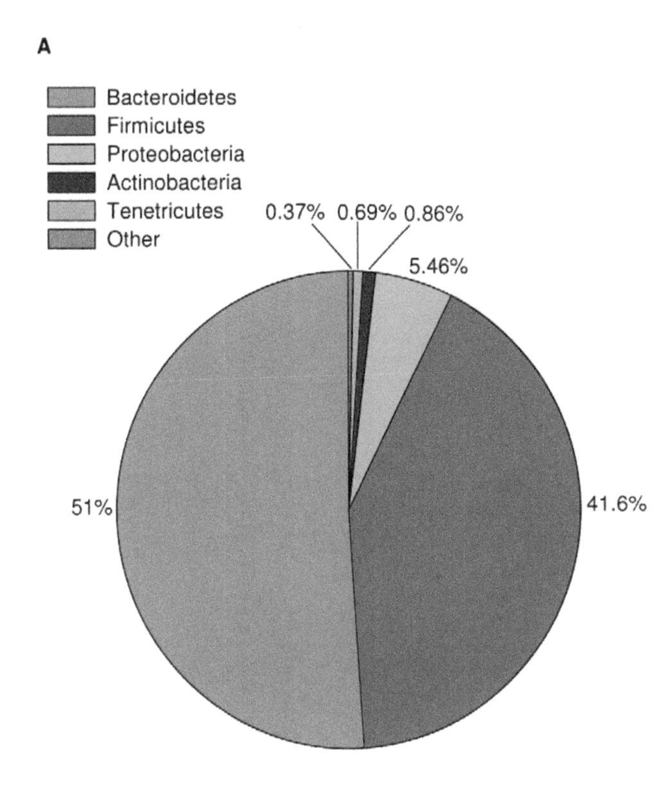

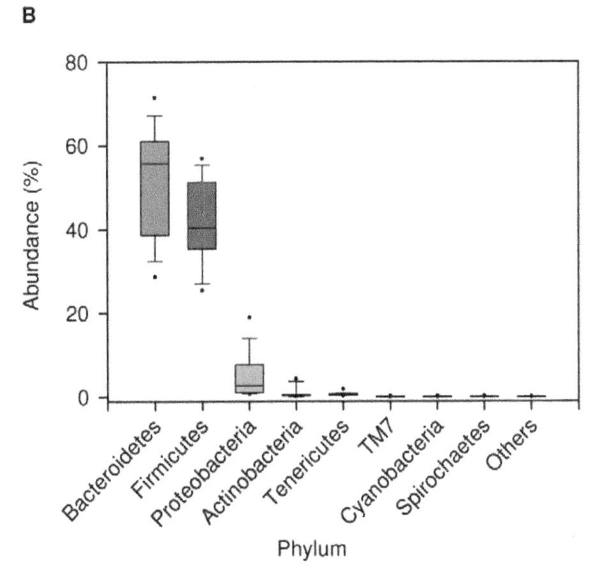

Figure 2. Composition and abundance of bacterial taxa, as determined by pyrosequencing of the 16 S rDNA gene. (**A**) Pie chart showing the average distribution of the phyla across all ruminal samples. (**B**) Box plot showing the relative abundance of each phylum, represented as percentage on the Y-axis. The boxes represent the interquartile range (IQR) between the first and third quartiles (25th and 75th percentiles, respectively) and the vertical line inside the box defines the median. Whiskers represent the lowest and highest values within 1.5 times the IQR from the first and third quartiles, respectively. Samples with a relative abundance of a given phylum exceeding those values are represented as points beside the boxes (color-coded).

implementation for sequence alignment and tree building [13]). Figure 5 shows the pairwise metric values for each individual cow and the average of each of the metrics for all possible cow pairs.

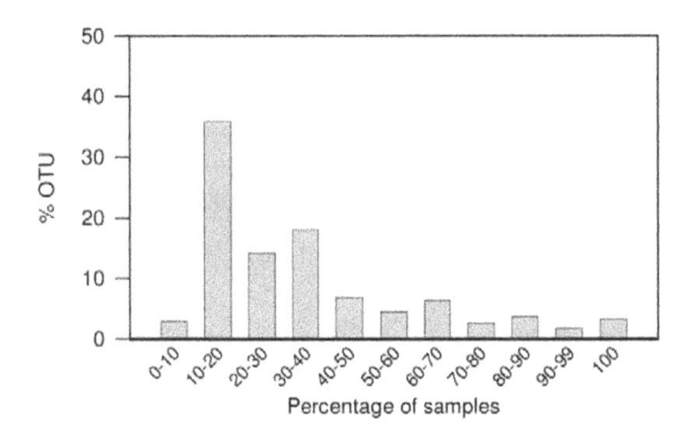

Figure 3. OTU occurrence across samples. Different OTUs were summed into categories according to their frequency of occurrence across different ruminal samples and binned accordingly, from OTUs shared by up to 10% of the samples to those shared by all samples. The X-axis represents the percentage of cows sharing a specific OTU. The Y-axis represents the percentage of OTUs found in each category.

An average 51% similarity between each pair of samples was calculated by the Bray-Curtis metric, 82% using the weighted UniFrac metric. The datasets calculated by the two metrics were significantly different (t-test, $P<0.001$).

Discussion

An understanding of the microbial ecosystem in the rumen is of great importance for the general study of microbial communities and their symbiosis with multicellular organisms, as well as for our everyday lives. A fundamental aspect in understanding any ecosystem is to identify its permanent and temporary residents. Therefore, the objectives of this study were to assess the general ruminal bacterial population in terms of total number of possible bacterial OTUs and taxa present, and to characterize the similarities across different cow rumens in terms of taxa's temporal and universal occurrence. To cover most of the possible bacterial OTUs found in the ruminal ecosystem of our study, we characterized the bacterial communities across a relatively large array of animal rumens and used tag-encoded amplicon pyrosequencing of the V2 and V3 regions of the 16 S rRNA gene at a depth of 10,000 reads per sample.

A sample-based rarefaction test on our data revealed that after the tenth sample (approximately 100,000 reads), most of the possible bacterial OTUs present in our overall rumen samples were covered. Note that this analysis is limited to the OTUs that were covered by the pyrosequencing procedure (PCR amplification, primers etc.). This finding confirms the conclusion of a recent in-silico study in which the collective microbial diversity in the rumen was examined by meta-analysis of all curated 16 S rRNA gene sequences deposited in the RDP database. The authors of that study estimated that 80,000 reads would cover all possible OTUs in the rumen [14]. Furthermore, in that study, the bacterial sequences were assigned to 5271 OTUs at the species level (≥97% similarity), which is also in agreement with our data in which a total of 4896 OTUs were assigned at ≥97% similarity.

An analysis of the prevalence of each OTU across all animals revealed that about 50% of them occur only in 0% to 30% of the animals sampled (Figure 3). Thus, despite the strict maintenance of similar experimental conditions, diet and sampling procedures, a large fraction of the OTUs occurred in only a small number of samples, contributing to the differences between the ruminal bacterial populations.

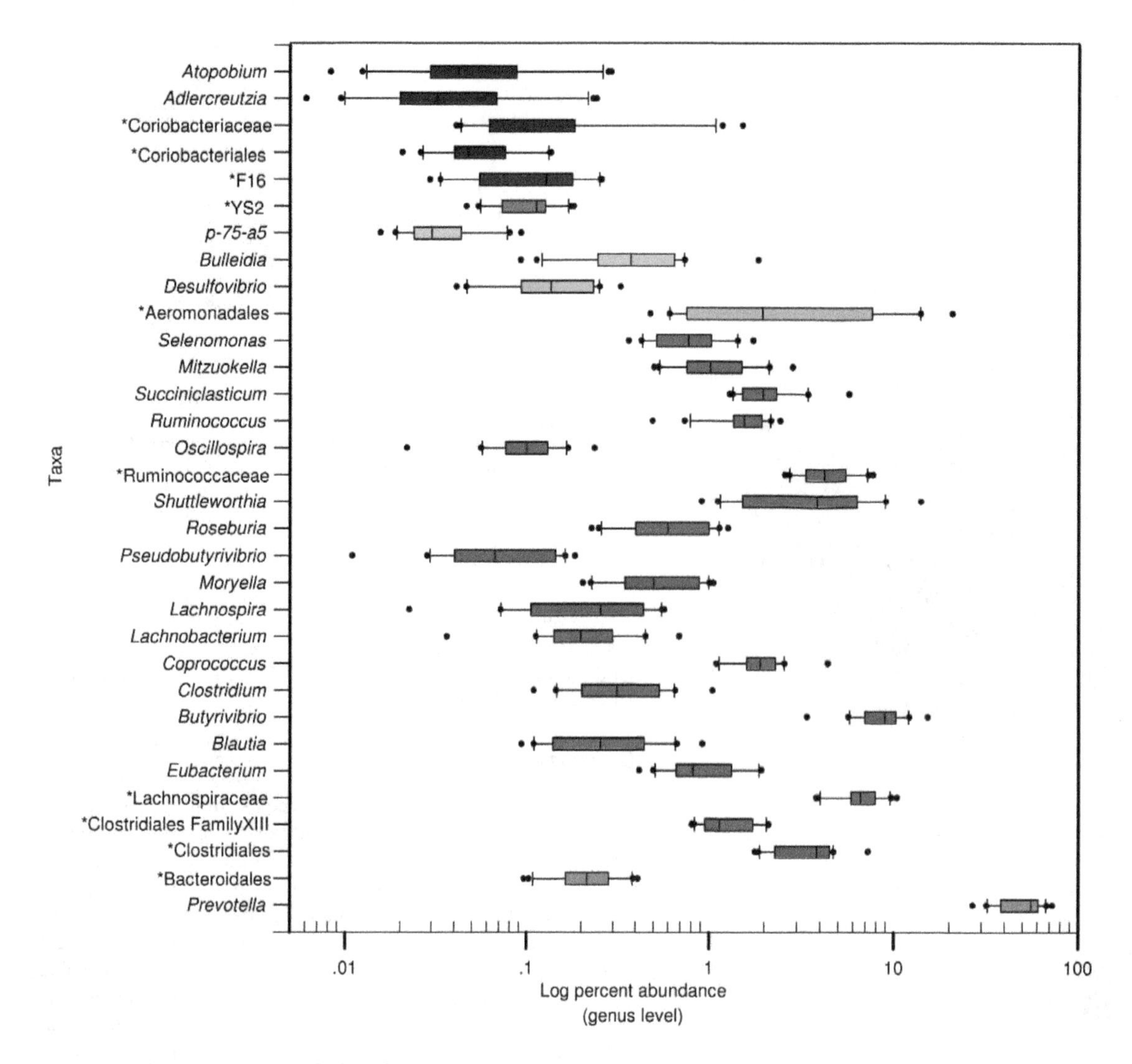

Figure 4. Shared genera and abundance across samples. Box plot showing the relative abundance of the bacterial genera shared by all samples, represented as log percentage on the X-axis. The boxes represent the interquartile range (IQR) between the first and third quartiles (25th and 75th percentiles, respectively) and the vertical line inside the box defines the median. Whiskers represent the lowest and highest values within 1.5 times the IQR from the first and third quartiles, respectively. Samples with a relative abundance of a given taxon exceeding those values are represented as points beside the boxes. The box color denotes the phylum of the genera: Bacteroidetes (blue), Firmicutes (red), Proteobacteria (green), Tenericutes (light blue), Cyanobacteria (orange), TM7 (gray), Actinobacteria (purple). Taxa not indentified at the genus level are identified by an asterisk and their highest taxonomic identification.

This was also seen in pairwise comparisons using the Bray-Curtis metric which indicated an average 51% similarity between the samples (Figure 5). However, when the weighted UniFrac metric was used to calculate the similarity between samples, an average value of approximately 82% was measured. The significant increase in similarity, compared to the Bray-Curtis metric, suggests that although a large number of OTUs differ between samples, they are phylogenetically related. This can be explained by the notion that the phylogenetically related OTUs have similar genetic profiles, enabling them to occupy proximal or similar ecological niches.

The average composition of the rumen bacterial community consisted mainly of the phyla Firmicutes and Bacteroidetes, 43% and 50% of all reads, respectively; the Proteobacteria accounted for 5.455% of the reads, and Actinobacteria and Tenericutes for 0.9% and 0.7%, respectively (Figure 2A). Interestingly, when we examined community composition for each individual sample, the ratios between the different phyla changed considerably among samples (Figure 2B): the Bacteroidetes were highly variable, with an abundance ranging from 26% to 70% of all reads. This observation concurs with a recent study of the human gut microbiome in which the Bacteroidetes also varied greatly among samples [15]. Similarly, the relative abundance of Proteobacteria varied considerably among samples, from 0.5% of all reads to as high as 20% in some samples. It is important to note that studies examining the phylum distribution in Holstein cows' 16 S rDNA clone libraries have observed distributions similar to some of the samples measured here [16].

An important finding of this work was that all sampled animals shared a group of bacterial taxa consisting of 32 genera which varied considerably in abundance across samples (Figure 4). The representation of these shared genera in the overall ruminal bacterial community was highly diverse, as low as 0.01% for some of the genera and up to 50% for others (Figure 4). It is tempting to speculate that even though some genera represent only a small fraction of the total rumen bacteria, the fact that they are shared

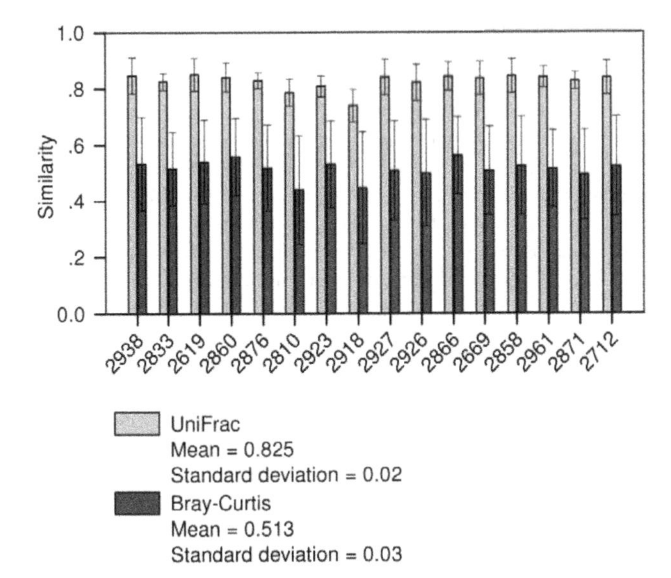

Figure 5. Pairwise similarity calculation. The average of pairwise comparisons of each sample to all others was calculated. The QIIME pipeline was used to compute the Bray-Curtis metric (gray bars) and the weighted UniFrac metric (black bars). The X-axis denotes the serial number of the cow from which the rumen sample was taken, and the Y-axis represents the degree of similarity: the closer the similarity to 1, the more similar the samples. At the bottom, average and standard deviation of each similarity metric calculated across all cow pairs are shown.

by all animal rumens might indicate that they fill an important function in the rumen ecosystem or that they occupy a special ecological niche in the rumen; however, this speculation requires further examination, including a determination of their presence in animals at other growth stages and on different diets. Furthermore, the present study describes samples from the same breed and herd, and there is likely to be greater divergence between breeds, herds and geographical locations.

Interestingly, some bacterial taxa were absent from the core groups identified by pyrosequencing, even though some species of these taxa are considered crucial for fiber degradation in the rumen. Notably the phylum Fibrobacteres, which includes one of the main cellulolytic bacteria—*Fibrobacter succinogenes*—which is thought to be of great importance for rumen function, was found in only half of the samples. Several studies of the rumen microbiome have suggested that the abundance of this phylum, and in particular *F. succinogenes*, varies considerably across cows and diets. This was evident in a recent metagenomic study in which this phylum was completely absent from the fiber-adherent and total overall rumen microbiome [3]. In two recent studies, this phylum was shown to be influenced by fiber content in the diet: in one study examining 16 S clone libraries from the rumen, it was represented in 2 out of 647 clones in a low-fiber diet, increasing to 19 out of 620 clones in a high-fiber diet [17]; this trend was also observed in the other study using real-time PCR analysis [8].

To assess the possibility of dietary influence on the abundance of *Fibrobacter* in the current study, we pyrosequenced metagenomic DNA from a cow fed a higher content of fiber (50% instead of 30%). The higher-fiber diet led to 0.48% representation of the phylum Fibrobacteres, a 24-fold increase over the average 0.02% representation in the group of cows examined in this study (Figure S1). Nevertheless, the apparent absence of this phylum and other important rumen bacteria from some of the samples could be due to low abundance of these taxa, below the detection level of the pyrosequencing method. Therefore, to assess the possible presence of these rumen bacteria across the samples we used the more sensitive real-time PCR method. Indeed, when we measured the abundance of some rumen bacterial species known to be important for rumen function, including *F. succinogenes*, we found that most of these species exist in low numbers in all of the samples (Table S4). Bacterial species showing a relative abundance of less than 0.01% in our samples using real-time PCR were generally not detected by the pyrosequencing. Nonetheless, in most samples there was agreement in abundance between the pyrosequencing and real-time PCR results (Table S4). The genus *Succinovibrio* appeared to be consistently underrepresented when using the pyrosequencing method, with relative abundance values up to tenfold lower than the real-time PCR values, which could indicate either that the primers used for real-time PCR amplify other organisms or that the primers used for pyrosequencing are less adapted to amplification of this genus. This requires further investigation. The genus *Prevotella* was highly represented in our shared microbial community: it was the most abundant bacterial genus with an average 50% of all reads. This finding is consistent with a previous study by Stevenson and Weimer [9] in which several bacterial species were quantified in ruminal samples from two lactating cows using real-time PCR. That study reported the predominance of *Prevotella* members, which comprised 42% to 60% of the bacterial rRNA gene copies in the samples [6]. It is interesting to note that although relative quantification by real-time PCR was performed in that study, our pyrosequencing results were in the same range. The high abundance of this genus is interesting from an ecological point of view: it might be the result of a metabolic niche that is wide enough to be occupied by bacteria that have similar metabolic capabilities, due to genetic relatedness or to high genetic variability that enables members of this genus to occupy different ecological niches within the rumen. Indications from previous studies imply the latter, as members of this genus are considered to exhibit a remarkable degree of genetic diversity [18,19,20]. Nevertheless, these two possibilities need to be further examined and distinguished.

The changes in the overall rumen bacterial population, together with changes in the shared bacterial communities, are an important subject for further research. The nature of these variations and their effects on the animals warrant further careful characterization. For example, the stability of these differences over time in each individual animal should be further established, and their effects on rumen metabolic parameters and animal physiological characteristics should be examined as well.

The work presented here describes the composition of the overall bacterial communities of the rumen ecosystem, and their similarities and differences across individual cows fed the same diet. It also reveals a microbial community that was present in all rumen samples tested, and shows that while a large number of species are not shared by all samples, there is high phylogenetic similarity between the communities. These observations increase our understanding of this important microbial ecosystem, and raise new questions for further study.

Materials and Methods

Animal handling and sampling

The experimental procedures used in this study were approved by the Faculty Animal Policy and Welfare Committee of the Agricultural Research Organization (ARO) approval number IL-168/08, Volcani Research Center, and were in accordance with the guidelines of the Israel Council on Animal Care.

Israeli Holstein Friesian lactating cows (n = 16) were housed at the ARO's experimental dairy farm in Bet Dagan, Israel, in one shaded corral with free access to water. The cows were fed a diet consisting of 30% roughage and 70% concentrate as described in Table S5 ad libitum, provided once a day. The cows were kept on this diet for a few months prior to the experiment as this is the standard diet fed to lactating cows at the experimental farm. The corral holding the animals was specially designed to keep the animals as a group, thus maintaining normal herd behavior, while allowing individual feeding and monitoring: each cow had an electronic chip that opens its individual feeding area. The cows were kept in this facility on the above diet for 6 weeks prior to sample collection. The samples were taken 1 hour after the morning feeding: 500 ml of ruminal contents was collected via the cow's mouth using a stainless-steel stomach tube with a rumen vacuum sampler. The pH was determined immediately and was on average 6.51±0.37 across all samples. Samples were transferred to CO_2-containing centrifuge bottles to maintain anaerobic conditions, and kept on ice. Immediately after collection, the ruminal samples were processed in the laboratory, located 100 m away.

After assessing several protocols for isolation of rumen microbes from the samples, we selected the one described by Stevenson and Weimer [6], as it exhibited the highest number of OTUs in an automated ribosomal intergenic spacer analysis (ARISA) (data not shown), as well as a good ability to detach the fiber-adherent bacteria, as reflected by enrichment of known fiber-adherent species quantified by real-time PCR analysis (data not shown). Isolation of the rumen microbial populations, including detachment of the fiber-adherent microbial populations and planktonic populations from the fibers, was performed with some minor modifications that included mixing the fiber-adherent microbial populations with the planktonic ones. Briefly, following 2 minutes of blender homogenization, the homogenate was centrifuged at 10,000 *g* and the pellet was dissolved in extraction buffer (100 mM Tris-HCl, 10 mM ethylenediaminetetraacetic acid [EDTA], 0.15 M NaCl pH 8.0): 1 g of pellet was dissolved in 4 ml of buffer and incubated at 4°C for 1 hour, as chilling has been shown to maximize the release of particle-associated bacteria from ruminal contents [6]. The suspension was then centrifuged gently at 500 *g* for 15 minutes at 4°C to remove ruptured plant particles while keeping the bacterial cells in suspension [21]. The supernatant was then passed through four layers of cheesecloth and centrifuged (10,000 *g*, 25 minutes, 4°C), and the pellets were kept at −20°C until DNA extraction.

DNA extraction

The DNA extraction was performed as described by Stevenson and Weimer [6]. Briefly, cell lysis was achieved by bead disruption with phenol followed by phenol/chloroform DNA extraction. The final supernatant was precipitated with 0.6 vol isopropanol and resuspended overnight in 50 to 100 µl Tris-EDTA buffer, then stored at 4°C for short-term use, or archived at −80°C.

Real-time PCR

Quantitative real-time PCR analysis was performed to investigate the relative abundance of specific bacterial species through amplification of their copy of the 16 S rRNA gene using the primers shown in Table S6 [6,22]. A standard curve was generated for each bacterial strain selected. By amplifying a serial twofold dilution of gel-extracted PCR products obtained by the amplification of each amplicon, we generated individual standard curves suitable for the quantification of each bacterial strain individually. A standard curve was also generated for the total bacterial 16 S rRNA gene in the samples by amplifying 10-fold dilutions of the gel-purified PCR product of one rumen sample. The standard curves were obtained using four dilution points, and were calculated using Rotorgene 6000 series software (Qiagen, Germany). Subsequent quantifications were calculated with the same program using the standard curve generated in each run (equating to one bacterial species), and at least one known purified product dilution used for the standard curves was added to each quantification reaction in order to assess the reproducibility of the reactions. All obtained standard curves met the required standards of efficiency (R^2>0.99, 90%>E>115%). Real-time PCR was performed in a 10-µl reaction mixture containing 5 µl Absolute Blue SYBR Green Master Mix (Thermo Scientific), 0.5 µl of each primer (10 µM working concentration), 3 µl nuclease-free water and 2 µl of 10 ng DNA templates. Amplification involved one cycle held at 95°C for 15 minutes for initial denaturation and activation of the hot-start polymerase system, and then 40 cycles at 95°C for 10 seconds followed by annealing for 15 seconds at 60°C and extension at 72°C for 20 seconds. To determine the specificity of amplification, a melting curve of PCR products was monitored by slow heating from 60 to 99°C (alternating 1°C increments with holding for 10 seconds), with fluorescence collection at 1°C intervals. Quantification of the selected bacteria was performed by dividing the specific bacterial count obtained for each bacterium, using the appropriate set of primers, by the total bacterial count obtained by amplification with the universal bacterial primers.

454 tag amplicon pyrosequencing and data analyses

454 amplicon pyrosequencing of the ruminal DNA samples was performed by the Research and Testing Laboratory (Lubbock, TX) using primers covering the 103- to 530-bp region of the 16 S rRNA gene sequence which corresponds to the V2 and V3 regions (107 F: 5′-GGCGVACGGGTGAGTAA-3′ and 530 R: 5′-CCGCNGCNGCTGGCAC-3′). The tagging and sequencing protocol was as described by Dowd *et al.* [23].

Data quality control and analyses were mostly performed using the QIIME pipeline [12]: 256,000 raw reads were assigned to their designated rumen sample using the split_library.py script which also performs quality filtering based on length (<200 bp) and quality of the reads, resulting in 162,000 reads of 351 bp on average (all read parameters per sample are listed in Table S2). The next step was to align the obtained sequences to define OTUs, in order to eventually assign taxonomy to them. Different OTU-generation methods have been reported to give different estimates of OTU number [24]. Therefore, we used three different clustering methods for OTU generation: UCLUST [25], ESPRIT-tree [26] and CD_HIT_OTU [27] (Table S1), which have been proven to generate satisfactory and comparable numbers of OTUs [24]. These analyses showed that the CD_HIT_OTU provides a significantly higher number of OTUs compared to the UCLUST method. The ESPRIT-tree method resulted in a slightly lower number of OTUs which was not significantly different from the UCLUST results (P>0.05). Therefore, the UCLUST method was selected to further analyze the data as it was the better well-adapted to the QIIME pipeline. The degree of similarity between sequences was defined as ≥97% to obtain OTU identity at the species level. Next we used the Chimera Slayer algorithm [28] for chimeric sequence removal. OTUs which clustered only one or two reads were manually removed. These processes resulted in 153,000 reads for analysis.

After constructing an OTU table, taxonomy was assigned using the BLAST algorithm and the reference database found at: http://blog.qiime.org designated "most recent Greengenes

OTUs". In parallel, phylogeny was calculated using the pyNAST [29] algorithm to create a tree which would enable the generation of UniFrac similarity measurements between samples. For the similarity measurement between the bacterial communities in the samples, two similarity indices were used: a Bray-Curtis comparison between samples according to both the presence and absence of OTUs and the abundance of OTUs between the samples, and the UniFrac metric, which uses the phylogenetic tree of the created OTUs to compare the phylogenetic closeness of the bacterial community between samples.

Supporting Information

Figure S1 Pie chart showing the phylum distribution of one rumen sample taken from a cow fed a 50% fiber diet.

Table S1 Comparison of different clustering methods and number of OTUs generated by each method.

Table S2 Number of reads and mean length of reads per sample that were used for the analysis (after quality filtering and removal of chimeras, singletons and doubletons).

Table S3 Taxonomic identification of the core OTUs (97% similarity) found in 100% of the samples and the number of OTUs associated with each specific taxon.

Table S4 Comparative analysis of real-time PCR and pyrosequencing. Comparative analysis of real-time PCR results from five specific bacterial species and the abundance observed for their respective genera using pyrosequencing. The mean and standard error of the mean (S.E.M) were also calculated for both methods. Pearson correlation was calculated between the pyrosequencing and real-time PCR results for each taxon.

Table S5 Formulated ingredients (g/kg DM) of the basic total mixed rations.

Table S6 PCR primers used for detection of rumen bacteria in this study by real-time PCR [6,22].

Acknowledgments

We would like to thank to the academic editor and reviewers for their excellent remarks which helped improve this manuscript greatly.

We would also like to thank Uri Gophna for his helpful advices and the critical reading of the manuscript.

Author Contributions

Conceived and designed the experiments: IM EJ. Performed the experiments: EJ. Analyzed the data: EJ IM. Contributed reagents/materials/analysis tools: EJ IM. Wrote the paper: EJ IM.

References

1. Flint HJ, Bayer EA, Rincon MT, Lamed R, White BA (2008) Polysaccharide utilization by gut bacteria: potential for new insights from genomic analysis. Nat Rev Microbiol 6: 121–131.
2. Flint HJ (1997) The rumen microbial ecosystem–some recent developments. Trends Microbiol 5: 483–488.
3. Brulc JM, Antonopoulos DA, Miller ME, Wilson MK, Yannarell AC, et al. (2009) Gene-centric metagenomics of the fiber-adherent bovine rumen microbiome reveals forage specific glycoside hydrolases. Proc Natl Acad Sci U S A 106: 1948–1953.
4. Welkie DG, Stevenson DM, Weimer PJ (2009) ARISA analysis of ruminal bacterial community dynamics in lactating dairy cows during the feeding cycle. Anaerobe 16: 94–100.
5. Sundset MA, Edwards JE, Cheng YF, Senosiain RS, Fraile MN, et al. (2009) Molecular diversity of the rumen microbiome of Norwegian reindeer on natural summer pasture. Microb Ecol 57: 335–348.
6. Stevenson DM, Weimer PJ (2007) Dominance of *Prevotella* and low abundance of classical ruminal bacterial species in the bovine rumen revealed by relative quantification real-time PCR. Appl Microbiol Biotechnol 75: 165–174.
7. Wanapat M, Cherdthong A (2009) Use of real-time PCR technique in studying rumen cellulolytic bacteria population as affected by level of roughage in swamp buffalo. Curr Microbiol 58: 294–299.
8. Tajima K, Aminov RI, Nagamine T, Matsui H, Nakamura M, et al. (2001) Diet-dependent shifts in the bacterial population of the rumen revealed with real-time PCR. Appl Environ Microbiol 67: 2766–2774.
9. Li M, Penner GB, Hernandez-Sanabria E, Oba M, Guan LL (2009) Effects of sampling location and time, and host animal on assessment of bacterial diversity and fermentation parameters in the bovine rumen. J Appl Microbiol 107: 1924–1934.
10. Callaway TR, Dowd SE, Edrington TS, Anderson RC, Krueger N, et al. (2010) Evaluation of bacterial diversity in the rumen and feces of cattle fed different levels of dried distillers grains plus solubles using bacterial tag-encoded FLX amplicon pyrosequencing. J Anim Sci 88: 3977–3983.
11. Taniguchi M, Penner GB, Beauchemin KA, Oba M, Guan LL (2010) Comparative analysis of gene expression profiles in ruminal tissue from Holstein dairy cows fed high or low concentrate diets. Comp Biochem Physiol Part D Genomics Proteomics 5: 274–279.
12. Caporaso JG, Kuczynski J, Stombaugh J, Bittinger K, Bushman FD, et al. (2011) QIIME allows analysis of high-throughput community sequencing data. Nat Methods 7: 335–336.
13. Lozupone C, Knight R (2005) UniFrac: a new phylogenetic method for comparing microbial communities. Appl Environ Microbiol 71: 8228–8235.
14. Kim M, Morrison M, Yu Z (2011) Status of the phylogenetic diversity census of ruminal microbiomes. FEMS Microbiol Ecol 76: 49–63.
15. Arumugam M, Raes J, Pelletier E, Le Paslier D, Yamada T, et al. (2011) Enterotypes of the human gut microbiome. Nature 473: 174–80.
16. Tajima K, Aminov RI, Nagamine T, Ogata K, Nakamura M, et al. (1999) Rumen bacterial diversity as determined by sequence analysis of 16 S rDNA libraries. FEMS Microbiology Ecology 29: 159–169.
17. Fernando SC, Purvis HT, 2nd, Najar FZ, Sukharnikov LO, Krehbiel CR, et al. (2010) Rumen microbial population dynamics during adaptation to a high-grain diet. Appl Environ Microbiol 76: 7482–7490.
18. Ramsak A, Peterka M, Tajima K, Martin JC, Wood J, et al. (2000) Unravelling the genetic diversity of ruminal bacteria belonging to the CFB phylum. FEMS Microbiol Ecol 33: 69–79.
19. Purushe J, Fouts DE, Morrison M, White BA, Mackie RI, et al. (2010) Comparative genome analysis of *Prevotella ruminicola* and *Prevotella bryantii*: insights into their environmental niche. Microb Ecol 60: 721–729.
20. Avgustin G, Wallace RJ, Flint HJ (1997) Phenotypic Diversity among Ruminal Isolates of Prevotella ruminicola: Proposal of *Prevotella brevis* sp. nov., *Prevotella bryantii* sp. nov., and *Prevotella albensis* sp. nov. and Redefinition of Prevotella ruminicola. Int J Syst Bacteriol 47: 284–288.
21. Dehority BA, Grubb JA (1980) Effect of short-term chilling of rumen contents on viable bacterial numbers. Appl Environ Microbiol 39: 376–381.
22. Walter J, Tannock GW, Tilsala-Timisjarvi A, Rodtong S, Loach DM, et al. (2000) Detection and identification of gastrointestinal *Lactobacillus* species by using denaturing gradient gel electrophoresis and species-specific PCR primers. Appl Environ Microbiol 66: 297–303.
23. Dowd SE, Callaway TR, Wolcott RD, Sun Y, McKeehan T, et al. (2008) Evaluation of the bacterial diversity in the feces of cattle using 16 S rDNA bacterial tag-encoded FLX amplicon pyrosequencing (bTEFAP). BMC Microbiol 8: 125.
24. Sun Y, Cai Y, Huse SM, Knight R, Farmerie WG, et al. (2011) A large-scale benchmark study of existing algorithms for taxonomy-independent microbial community analysis. Brief Bioinform 13: 107–21.
25. Edgar RC (2010) Search and clustering orders of magnitude faster than BLAST. Bioinformatics 26: 2460–2461.
26. Cai Y, Sun Y (2011) ESPRIT-Tree: hierarchical clustering analysis of millions of 16 S rRNA pyrosequences in quasilinear computational time. Nucleic Acids Res 39: e95.
27. Li W, Jaroszewski L, Godzik A (2001) Clustering of highly homologous sequences to reduce the size of large protein databases. Bioinformatics 17: 282–283.

28. Haas BJ, Gevers D, Earl AM, Feldgarden M, Ward DV, et al. (2011) Chimeric 16 S rRNA sequence formation and detection in Sanger and 454-pyrosequenced PCR amplicons. Genome Res 21: 494–504.

29. Caporaso JG, Bittinger K, Bushman FD, DeSantis TZ, Andersen GL, et al. (2009) PyNAST: a flexible tool for aligning sequences to a template alignment. Bioinformatics 26: 266–267.

12

Microbial Ecology of Thailand Tsunami and Non-Tsunami Affected Terrestrials

Naraporn Somboonna[1]*, Alisa Wilantho[2], Kruawun Jankaew[3], Anunchai Assawamakin[4], Duangjai Sangsrakru[2], Sithichoke Tangphatsornruang[2], Sissades Tongsima[2]

1 Department of Microbiology, Faculty of Science, Chulalongkorn University, Bangkok, Thailand, **2** Genome Institute, National Center for Genetic Engineering and Biotechnology, Pathumthani, Thailand, **3** Department of Geology, Faculty of Science, Chulalongkorn University, Bangkok, Thailand, **4** Department of Pharmacology, Faculty of Pharmacy, Mahidol University, Bangkok, Thailand

Abstract

The effects of tsunamis on microbial ecologies have been ill-defined, especially in Phang Nga province, Thailand. This ecosystem was catastrophically impacted by the 2004 Indian Ocean tsunami as well as the 600 year-old tsunami in Phra Thong island, Phang Nga province. No study has been conducted to elucidate their effects on microbial ecology. This study represents the first to elucidate their effects on microbial ecology. We utilized metagenomics with 16S and 18S rDNA-barcoded pyrosequencing to obtain prokaryotic and eukaryotic profiles for this terrestrial site, tsunami affected (S_1), as well as a parallel unaffected terrestrial site, non-tsunami affected (S_2). S_1 demonstrated unique microbial community patterns than S_2. The dendrogram constructed using the prokaryotic profiles supported the unique S_1 microbial communities. S_1 contained more proportions of archaea and bacteria domains, specifically species belonging to Bacteroidetes became more frequent, in replacing of the other typical floras like Proteobacteria, Acidobacteria and Basidiomycota. Pathogenic microbes, including *Acinetobacter haemolyticus*, *Flavobacterium* spp. and *Photobacterium* spp., were also found frequently in S_1. Furthermore, different metabolic potentials highlighted this microbial community change could impact the functional ecology of the site. Moreover, the habitat prediction based on percent of species indicators for marine, brackish, freshwater and terrestrial niches pointed the S_1 to largely comprise marine habitat indicating-species.

Editor: Vasu D. Appanna, Laurentian University, Canada

Funding: The research was supported by Research Funds from the Faculty of Science, Chulalongkorn University, Under the A1B1-NS (RES-A1B1-NS-01), and Thai Aviation Refuelling Co., Ltd. The funders had no role in study design, data collection and analysis, decision to publish, or preparation of the manuscript.

Competing Interests: The authors received funding from Thai Aviation Refuelling Co., Ltd.

* E-mail: Naraporn.S@chula.ac.th

Introduction

Phra Thong island, Phang Nga province of southern Thailand (Figure 1), represents a location for comparative studies of tsunami (S_1) and non-tsunami (S_2) affected terrestrial ecosystems. The $S_1$and S_2 shared nearby geographies separated by a hill, whereby S_1 terrain was inundated by the Indian Ocean tsunami on 26 December 2004 and S_2 unaffected; otherwise both were comparable based on geological characteristics [1,2]. The tsunami left an Andaman Sea-facing, S_1, distinguished terrestrial layer that was classified by geologist as a sand layer of 5–20 cm thick (layer A in Figure 1; [1]). Interestingly, geological evidence indicated three historic tsunamis also occurred prior to the 2004 tsunami at S_1, and none to S_2. The youngest recorded historic tsunami predating the 2004 tsunami was approximately 600 years ago (600yo) (layer B in Figure 1; [1]).

Each tsunami occurrence could affect the S_1 terrestrial characteristics due to the massive impact of seawater with marine organisms and garbage [1,3,4]. Studies comparing the 2004 tsunami affected versus non-affected (or pre-affected) terrestrials and terrestrial water reported the greater salinity, acidity, conductivity, turbidity and organic contents following the tsunami occurrence [5–7]. Studies also reported widespread disease-carrying vectors, such as mosquitoes, trematodes and snails, after the 2004 tsunami [3,4]. Several bacterial and fungal infections involved skin and respiratory disorders were documented among repatriated tourists [8] and people working in the tsunami affected area [9]. In addition, the 2004 tsunami sediments consisted of higher concentrations of Mercury and Thallium [10,11]. Together, this chance of terrestrial characteristics could affect the microbial biodiversity and functional ecology.

Nonetheless, the impact of tsunamis on microbial diversity and ecology function remains ill-defined. The present study thereby analyzed the microbial biodiversity and their potential functional composites in the tsunami impacted S_1 terrain, in comparison to the non-affected S_2 site, using 16S and 18S rRNA genes pyrosequencing derived metagenomic DNA approach. For each site, the data included the prokaryotic and eukaryotic diversity profiles categorized into different depth levels corresponding to the terrestrial ages: 2004 tsunami, 1–300yo (pre-dating the 2004), 300–600yo, 600yo tsunami, and >600yo, respectively (starting from the top layer to a deeper layer), and also the amalgamated profiles for each site. Geologists determined the terrestrial age period from its depths below the land surface [1]. The overall results represent for the first time the use of metagenomics in analysing the prokaryotic and eukaryotic microbial biodiversity of the 2004 tsunami and non-tsunami affected terrestrials. Unlike

98° 00' E
98° 30' E
99° 00' E
20 km
PT
9° 00' N
PHANG NGA
ANDAMAN
SEA
8° 00' N
PHUKET
S1
S2
A
B

Figure 1. Index map of Phra Thong island relative to Phuket and terrestrial sites where samples were collected. The lower left photograph shows the pit wall of tsunami affected site (S_1). Light color sheets **A** and **B** represent 2004 tsunami and 600yo tsunami affected terrestrial layers, respectively [1]. The lower right photograph shows the pit wall of non-tsunami affected site (S_2) of the parallel geolography, and samples of equivalent depths to those of S_1 were collected. Time period of the terrestrial is determined via sample depth [1].

traditional biodiversity study that is conducted via cultivation technique and could reveal merely less than 1% of the true microbiota, metagenomics is a culture-independent technique that has been proved worldwide a robust, reliable and comprehensive tool for obtainment of entire microbiota from diverse environmental and clinical samples [12–17].

Materials and Methods

Sample collection

The owners of the lands gave permission to conduct the study on these sites. We confirm that the study did not involve endangered or protected species.

Phra Thong island provides a location for comparative tsunami (S_1: N9.13194 E98.26250) and non-tsunami (S_2: N9.07250 E98.27222) affected terrestrial studies based on geological evidences (Jankaew, personal communication) [1,2]. S_1 and S_2 are 6.73 km apart. S_1 is 0.40 km from the sea, and S_2 is 2.26 km from the sea (Figure 1). Approximately 1 kg samples were collected, each in sterile containers, between 11:00–15:00 hours during 23–24 March 2011. S_1 samples comprised: 2004 tsunami (14.5 cm), 1–300yo (22 cm), 300–600yo (29 cm), 600yo tsunami (38 cm), and >600yo (46 cm); S_2 samples comprised: S21 (14.5 cm), S22 (22 cm), S23 (29 cm), S24 (38 cm), and S25 (46 cm). The number in parenthesis represents the depth level where the sample was collected, and that entailed the approximate age of the sample relative to the year 2004. On-site records for color, texture and pH were done. All samples were transported in ice chest, stored in 4°C and processed for the next steps within 14 days.

Metagenomic DNA extraction and DNA quality examination

Each sample was mixed with a sterile spatula, and 15 g each was used for metagenomic DNA extraction [18]. Two independent metagenomic DNA extractions were performed per sample. The samples were dissolved in an extraction buffer (Epicentre, Wisconsin, USA) with Tween 20, low-speed centrifuged to remove large debris, and poured through four-layered sterile cheesecloth to remove particles and organisms of >30 µm in size. Microorganisms between 0.22 and 30 µm were collected by filtering over a sterile 0.22 µm filter membrane (Merck Millipore, Massachusetts, USA) [15]. Total nucleic acid from each sample was extracted using Meta-G-Nome DNA Isolation Kit (Epicentre) following the manufacturer's protocols. Metagenomic DNA quality was assessed using agarose gel electrophoresis. The DNA concentration and purity was further analysed by A_{260} and the ratio of A_{260}/A_{280} spectrophotometry, respectively.

PCR generation of pyrotagged 16S and 18S rDNA libraries

Table 1 lists forward and reverse pyrotagged 16S and 18S rRNA gene primers. For broad-range 16S and 18S rRNA genes amplification, universal prokaryotic 338F (forward) and 803R (reverse) primers [19–21], and universal eukaryotic 1A (forward) and 516R (reverse) primers [15,22,23] were used. Italics denote the eight nucleotides pyrotag sequences, functioning to specify sample names [24]. A 50-µl PCR reaction comprised 1× EmeraldAmp GT PCR Master Mix (TaKaRa, Shiga, Japan), 0.3 µM of each primer, and 100 ng of the metagenome. PCR conditions were 95°C for 4 min, and 30–35 cycles of 94°C for 45 s, 50°C for 55 s and 72°C for 1 min 30 s, followed by 72°C for 10 min. To generate the pyrotagged 16S or 18S rDNA libraries with minimized stochastic PCR biases, two to three independent PCRs were performed per extracted metagenomes, and two extracted metagenomes per sample, resulting in a minimum of four PCR products to be pooled for pyrosequencing per sample.

Gel purification and pyrosequencing

PCR products (~473 bp for 16S rDNAs; ~577 bp 18S rDNAs) were excised from agarose gels, and purified using PureLink Quick Gel Extraction Kit (Invitrogen, New York, USA). The 454-sequencing adaptors were ligated to all 16S and 18S rDNA fragments, the reactions were purified by MinElute PCR Purification Kit (Qiagen), and the samples were pooled for pyrosequencing on an eight-lane Roche picotiter plate, 454 GS FLX system (Roche, Branford, CT) at the in-house facility of the National Center for Genetic Engineering and Biotechnology, according to the recommendations of the supplier.

Sequence annotation and bioinformatic analyses

After removal of unreliable sequences, including sequences that failed the pyrosequencing quality cut-off and sequences shorter than 50 nucleotides, the sequences were categorized based on the appended pyrotag sequences. Sequences corresponding to the same sample category were inspected for domain and taxon compositions using mg-RAST [25,26] with default parameters. Species were identified by BLASTN [27] with E-value $\leq 10^{-5}$ against 16S rDNA databases including NCBI non-redundant [28], RDP [29] and Greengenes [30], and for 18S rDNAs the databases included NCBI non-redundant [28], EMBL [31,32] and SILVA [33]. Evolutionary distances and phylogenetic tree were computed with default thresholds (E-value $\leq 10^{-8}$, similarity score ≥80%). Species (or phylum) prevalence was determined by dividing the frequency of reads in the species (or phylum) by the total number of the identifiable reads. The differences in community structures were compared using Yue & Clayton theta similarity coefficients (Thetayc) and Morisita-Horn dissimilarity index, in mothur [34–36]. Low Thetayc and Morista-Horn inferred high community similarity. An unweighted pair group method with arithmetic mean (UPGMA) clustering was constructed using Thetayc, in mothur [36]. Furthermore, functional subsystems and functional groups of the prokaryotic profiles were determined using SEED-based assignments in mg-RAST server [25,26,37]. For habitat classification, the data were compared against a World Register of Marine Species (WoRMS) database [38].

Results

Metagenome abundances and compositions at domain and kingdom levels

On-site records for physical characteristics of S_1 and S_2 were as shown in Figure 1. Alternating layers of black soil-like and grey sand-like comprised most of S_1, whereby more homogeneous layers of grey sand-like predominated in S_2. Differences in pH

Table 1. Pyrotagged16S and 18S rRNA genesuniversal primers.

Sample names	Forward primers (5′-3′)	Reverse primers (5′-3′)
16S rRNA gene universal primers		
2004 tsunami	*TAGTAGCG*ACTCCTACGGGAGGCAGCAG	*TAGTAGCG*CTACCAGGGTATCTAATC
1-300yo	*AGACGACG*ACTCCTACGGGAGGCAGCAG	*AGACGACG*CTACCAGGGTATCTAATC
300–600yo	*ACTCGTAG*ACTCCTACGGGAGGCAGCAG	*ACTCGTAG*CTACCAGGGTATCTAATC
600yo tsunami	*ACATCGAG*ACTCCTACGGGAGGCAGCAG	*ACATCGAG*CTACCAGGGTATCTAATC
>600yo	*ACGCTATC*ACTCCTACGGGAGGCAGCAG	*ACGCTATC*CTACCAGGGTATCTAATC
S21	*TACTACGC*ACTCCTACGGGAGGCAGCAG	*TACTACGC*CTACCAGGGTATCTAATC
S22	*AGCAGAGC*ACTCCTACGGGAGGCAGCAG	*AGCAGAGC*CTACCAGGGTATCTAATC
S23	*TCAGCTAC*ACTCCTACGGGAGGCAGCAG	*TCAGCTAC*CTACCAGGGTATCTAATC
S24	*AGAGCGAC*ACTCCTACGGGAGGCAGCAG	*AGAGCGAC*CTACCAGGGTATCTAATC
S25	*ATGCTCAC*ACTCCTACGGGAGGCAGCAG	*ATGCTCAC*CTACCAGGGTATCTAATC
18S rRNA gene universal primers		
2004 tsunami	*AGATAGCG*CTGGTTGATCCTGCCAGT	*AGATAGCG*ACCAGACTTGCCCTCC
1-300yo	*TGTAGACG*CTGGTTGATCCTGCCAGT	*TGTAGACG*ACCAGACTTGCCCTCC
300–600yo	*TGCAGTAG*CTGGTTGATCCTGCCAGT	*TGCAGTAG*ACCAGACTTGCCCTCC
600yo tsunami	*TCTGCGAG*CTGGTTGATCCTGCCAGT	*TCTGCGAG*ACCAGACTTGCCCTCC
>600yo	*ATCAGCAG*CTGGTTGATCCTGCCAGT	*ATCAGCAG*ACCAGACTTGCCCTCC
S21	*ATACAGTC*CTGGTTGATCCTGCCAGT	*ATACAGTC*ACCAGACTTGCCCTCC
S22	*ATCATATC*CTGGTTGATCCTGCCAGT	*ATCATATC*ACCAGACTTGCCCTCC
S23	*TGCGATGC*CTGGTTGATCCTGCCAGT	*TGCGATGC*ACCAGACTTGCCCTCC
S24	*ATCGCAGC*CTGGTTGATCCTGCCAGT	*ATCGCAGC*ACCAGACTTGCCCTCC
S25	*TATACTAC*CTGGTTGATCCTGCCAGT	*TATACTAC*ACCAGACTTGCCCTCC

Italic sequence denotes the 8 nt-pyrotagged sequence.

were also evident between the two ecosystems where S_1 ranged from 6–7 (more acidity), while S_2 ranged from 7–7.5.

Following total nucleic acids extraction of 0.22–30 μm sizes, S_1 and S_2 had average metagenomic concentration of 23.16 ng and 27.02 ng per gram of soil, respectively. Libraries of pyrotagged 16S and 18S rRNA gene fragments were constructed, and pyrosequenced to obtain the culture-independent prokaryotic and eukaryotic profiles of the sites. After removal of unreliable sequences, 21,592reads for S_1 and 33,308 reads for S_2 remained for BLASTN species identification. Significant E-values ($\leq 10^{-5}$) were identified for 20,555 reads for S_1 (95.20%) and 31,946 reads for S_2 (95.91%). Reads with non-significant E-values ($>10^{-5}$) were omitted from the analyses.

The domain compositions of S_1 indicated a lower proportion of eukaryotes and higher proportion of prokaryotes than S_2 (Figure 2). The proportion of eukaryotic species was reduced by almost half in S_1. Among prokaryotes, the greatest divergence between S_1 and S_2 was evident among archaea compared to bacteria where found increased in S_1 (Figure 3: 1.15-fold for bacteria, 4.07-fold archaea). These greater representations somewhat caused the reduced diversity of the other 4 kingdoms of lives.

Diversity of prokaryotic phyla and species

Major prokaryotic phyla for S_1 included Proteobacteria, Bacteroidetes, and Actinobacteria; and S_2 included Proteobacteria, Acidobacteria, and Actinobacteria, in diminishing order (Figure 4A). Consistent with Figure 3, S_1 demonstrated greater prokaryotic biodiversity than S_2. S_1 contained many new species belonging to uncultured species, i.e. OP3, GNO4 and SC3, whereas S_2 still contained high proportion of common environmental phyla, including Proteobacteria and Acidobacteria (Figure 4A). Different sample periods showed slight variation of prokaryotic phyla profiles in S_1, whereas in S_2 more variation among the phyla distributions was evident (Figure 4B). For instances, Proteobacteria comprised 62.23% in the S_1 layer of 2004 tsunami, 58.56% 1–300yo, 59.18% 300–600yo, 60.69% 600yo tsunami, and 62.54% >600yo; while S_2 comprised 46.07% in the S21 layer, 65.72% S22, 77.64% S23, 75.05% S24, and 67.41% S25. Actinobacteria comprised 14.48% in S_1 layer of 2004 tsunami, 13.22% 1–300yo, 13.53% 300–600yo, 13.82% 600yo tsunami, and 13.57% >600yo; while S_2 layers demonstrated 14.32% S21, 7.59% S22, 6.52% S23, 8.92% S24, and 10.47% S25 (Figure 4B). Further distinguished differences were diagnosed upon analysis based on species distributions: no parallel-age pairs of S_1 and S_2 showed similar species distribution pattern with the >600 yo and S25 pair showing the least differences. Examples of predominated species in S_1 were *Acinetobacter haemolyticus*, *Polynucleobacter* sp., *Polynucleobacter necessaries* and *Flavobacterium denitrificans*; and S_2 were *Burkholderia* sp. and *Silvimonas terrae* (Figure 5). Subsequently, the computed Thetayc and Morisita-Horn dissimilarity indices revealed dissimilar community structures between S_1

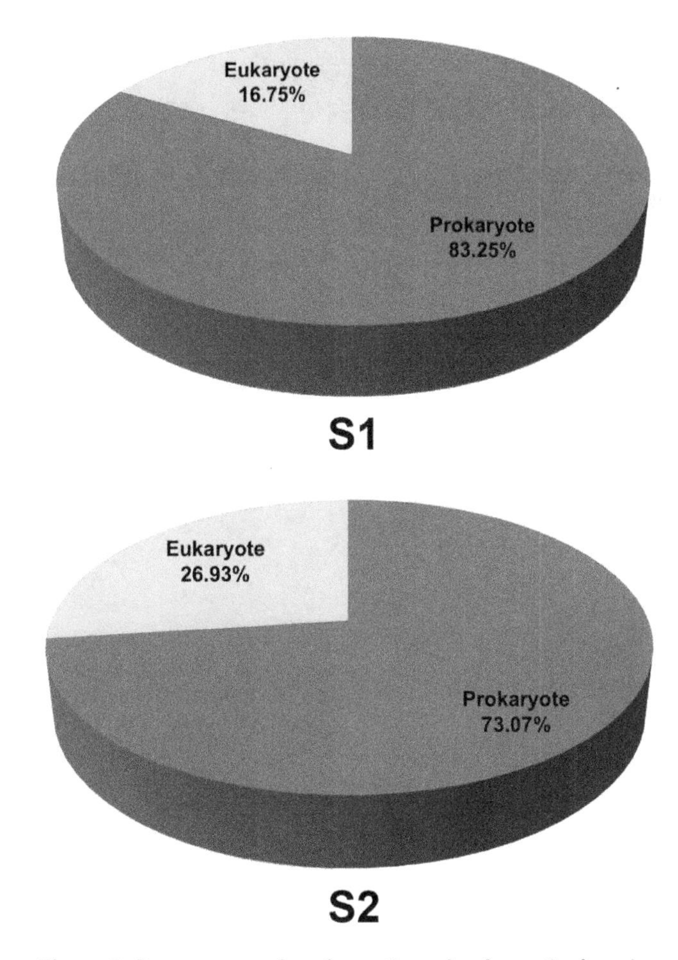

Figure 2. Percentages of prokaryotic and eukaryotic domains in S_1 and S_2.

and S_2. As shown in UPGMA dendrogram in Figure 6A: terrestrial layers corresponding to S_1 site were clustered together separated from the S_2 layers. Although S21 was grouped with the S_1 layers, it was an outer related branch to the S_1 group.

Potential metabolic system analysis of prokaryotic communities

From a total of 28 possible metabolic subsystems by mg-RAST [25,26,37], the prokaryotic communities of S_1 contained 19 subsystems, and S_2 contained 11 subsystems. The predominant subsystems in the S_1 included: regulation and cell signalling (14.35%), cell wall and capsule (10.05%), protein metabolism (8.13%), and sulfur metabolism (3.83%) (Figure 7). Overall, S_1 inhabited prokaryotic communities with high metabolic potentials for cell wall and capsule (i.e. gram-negative cell wall components by TIdE/PmbA protein), protein metabolism (i.e. protein degradation by prolyl endopeptidase), regulation and cell signaling (i.e. regulation of virulence by two-component system response regulator), and carbohydrates (i.e. sugar alcohols and central carbohydrate metabolism by HPr kinase/phosphorylase). S_2 inhabited prokaryotic communities with high metabolic potentials for respiration (i.e. electron donating and accepting reactions by Type cbb3 cytochrome oxidase biogenesis protein), clustering based subsystems (i.e. copper-translocating P-type ATPase), and virulence, disease and defence (resistance to antibiotics and toxic compounds by acriflavin resistance protein).

Diversity of eukaryotic phyla and species

Dominant eukaryotic phyla for S_1 were in kingdom Animalia: Brachiopoda (47.82% in 2004 tsunami, 32.53% 1–300yo, 37.32% 300–600yo, 49.15% 600yo tsunami, 25.00% >600yo), and Mollusca (28.88% 2004 tsunami, 29.39% 1–300yo, 31.34% 300–600yo, 30.51% 600yo tsunami, 7.14% >600yo); and kingdom Protozoa: Dinophyta for particularly the >600yo layer (51.99%) (Figures 8A and 8B). For S_2, although Brachiopoda and Mollusca were dominant, fungal phylum Basidiomycota (0.08% S21, 66.51% S22, 7.47% S23, 12.82% S24, 12.86% S25) and animal phylum Arthropoda (3.05% S21, 3.42% S22, 28.33% S23, 7.70% S24, 2.86% S25) were the most prevalent. When analyzing the data into individual sample periods, similar finding to Figure 4B were found. Different sample periods of S_1 demonstrated less phyla pattern variation than those of S_2 (Figure 8B). Distinguished phyla pattern of S_2 from S_1 were displayed apparently in S22, S23 and S24 layers (Figure 8B), resulting in their divergence from S_1 and the other S_2 communities by the UPGMA dendrogram constructed using Thetayc dissimilarity indices (Figure 6B). Similar to prokaryotes (Figure 6A), the eukaryotic communities corresponding to the S_1 site were relatively clustered together (Figure 6B). Analysis at the species level identified a more diverse fungal and animal species among S_2 layers (Figure 9).

Habitat classification

The S_1 and S_2 prokaryotic and eukaryotic profiles were matched against WoRMS database [38] to further characterize their microbial ecology: how each is related to marine, brackish water, freshwater, and terrestrial species communities. Figure 10 exhibited a substantially higher abundance of marine species habitat with S_1 ($S_1 = 24.11\%$, $S_2 = 13.33\%$), and terrestrial species with S_2 ($S_1 = 0.11\%$, $S_2 = 1.42\%$). Examples of abundant marine prokaryotes in S_1 were: *Lutaonella thermophilus*, *Shewanella aquimarina*, *Erythrobacter ishigakiensis* and *Thalassobacter stenotrophicus*. Abundant marine eukaryotes in S_1 included: *Dinophysis acuminata*, *Ctenodrilidae sp.*, *Remanella sp.*, *Nemertinoides elongatus*, *Skeletonema grethae*, *Crassostrea gigas*, *Hymenocotta mulli*, *Diplodasys ankeli*, *Pinna muricata*, *Arenicola marina*, *Limopsis marionensis* and *Haliplanella lucia*.

Discussion

On-site records indicated the greater turbidity (Figure 1) and acidity of S_1 were in agreement with previous reports [1,2,5–7]. Together with many other tsunami studies, these different soil types suggested terrestrial component changes following tsunami inundation, which could affect the microbial ecology of the site. The terrestrial microbiome representing the 2004 tsunami-affected site has never been studied. Our findings represent the first to utilize metagenomics in gaining databases of these entire terrestrial microbiomes, including prokaryotes and eukaryotes, in tsunami-affected (S_1) and non-tsunami affected (S_2) sites of Phra Thong island, as of March 2011. The data helped characterize the microbial biodiversity and its impact by tsunami occurrence. This knowledge is essential for scientists and engineers involved with land management and environmental bio-improvement.

Diminished total nucleic acids from S_1 suggested a less populated microbial community. Although some fossil DNA and fragments of DNA from live animals (known as extracellular "dirt" DNA) could be included in the extracted metagenomes, and might partly complicate the analysis. Andersen et al. [39] found extracellular "dirt" DNA from the terrestrial surface could reflect an overall taxonomic richness and relative abundance of species of a site at the time of investigation. Hence, some extracellular "dirt"

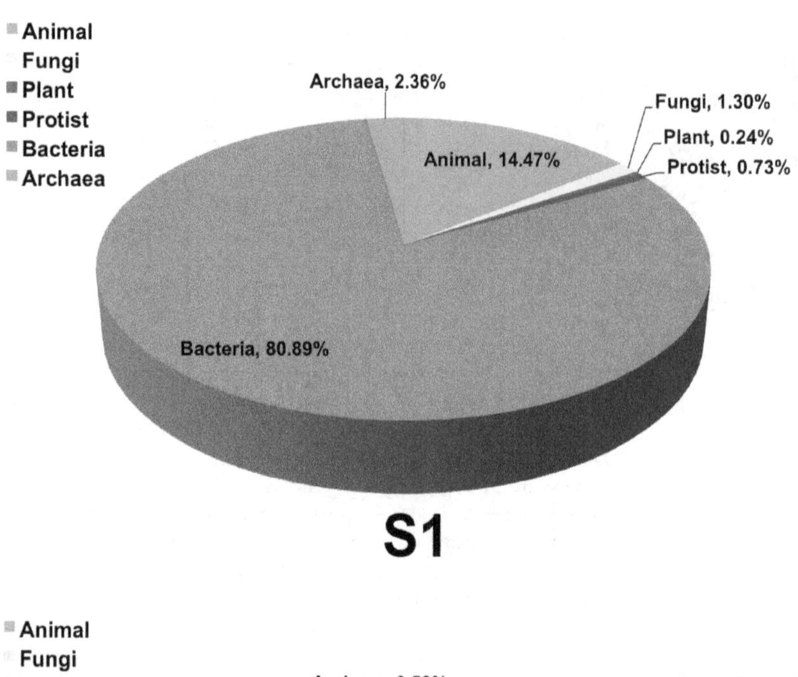

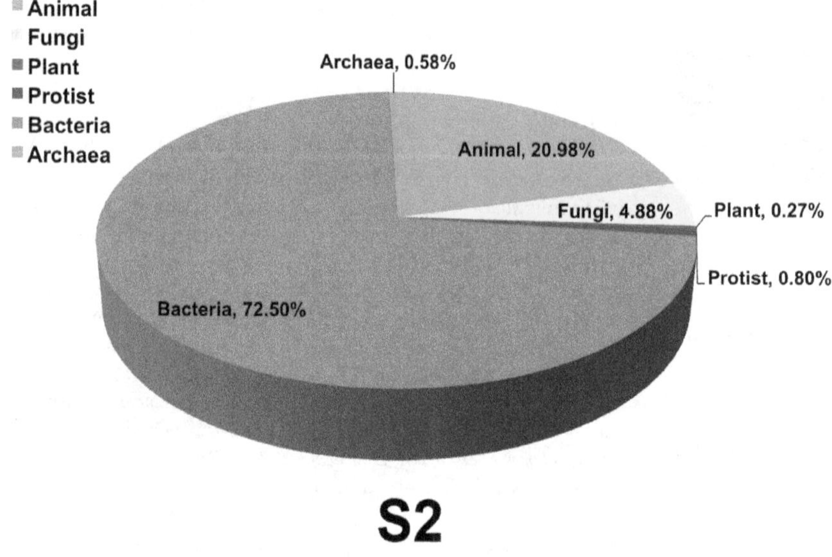

Figure 3. Percentages of 6 kingdoms of lives in S_1 and S_2.

DNA in our extracted metagenome should also reflect an overall biodiversity.

Libraries of pyrotagged 16S and 18S rRNA gene fragments were successfully constructed and pyrosequenced: 21,592 reads for S_1 and 33,308 reads for S_2 were retrieved after removal of unreliable sequences. For BLASTN species identification, greater than 95% of the S_1 and S_2 reads were identified with $\leq 10^{-5}$ E-values. The amount of reads should be sufficient to recapture the relationships among the samples, as Caporaso et al. [40] reported 2,000 reads could recapture the same relationships among samples as did with the full dataset. Additionally, many studies discovered variable regions 3 and 4 of 16S rRNA gene analyses were more effective than random sequence reads analyses in estimating the biodiversity and relationships among the samples [41–43].

In S_1, domain of prokaryotes became highly present (Figure 2). In particular, S_1 had a richer archaeal population (4.07-fold increase), meanwhile fungi, animals, plants, and protists were decreased (Figure 3). This finding was consistent with the fact that tsunami inundation might leave a terrestrial site inhospitable, causing archaea and bacteria to be more common due to their flexible life activities and requirements [44–46]. The greater biodiversity of kingdoms in S_2 supported the more hospitable terrestrial habitat than S_1.

Phyla and species distribution patterns between S_1 and S_2 were different. Changing the prokaryotic pattern of major phyla, precisely S_1 was predominantly comprised of Bacteroidetes with a lower prevalence of Proteobacteria, Actinobaceria and Acidobacteria (Figure 4A), highlighted the modified microbial ecology. Wada et al. [47] reported the similar change of bacterial floras in the sludge brought ashore by the 2011 East Japan earthquake. Bacteroidetes was more evident than Proteobacteria in the affected coastal water area. Additionally, numerous sulfate-reducing bacteria were evident in the sludge, which corresponded with high concentrations of sulfate ions in the sludge and the affected water area. The latter report was consistent with our finding of the higher sulfur metabolism in S_1 (Figure 7). Further, among the Bacteroidetes, flavobacteria predominanted which is generally

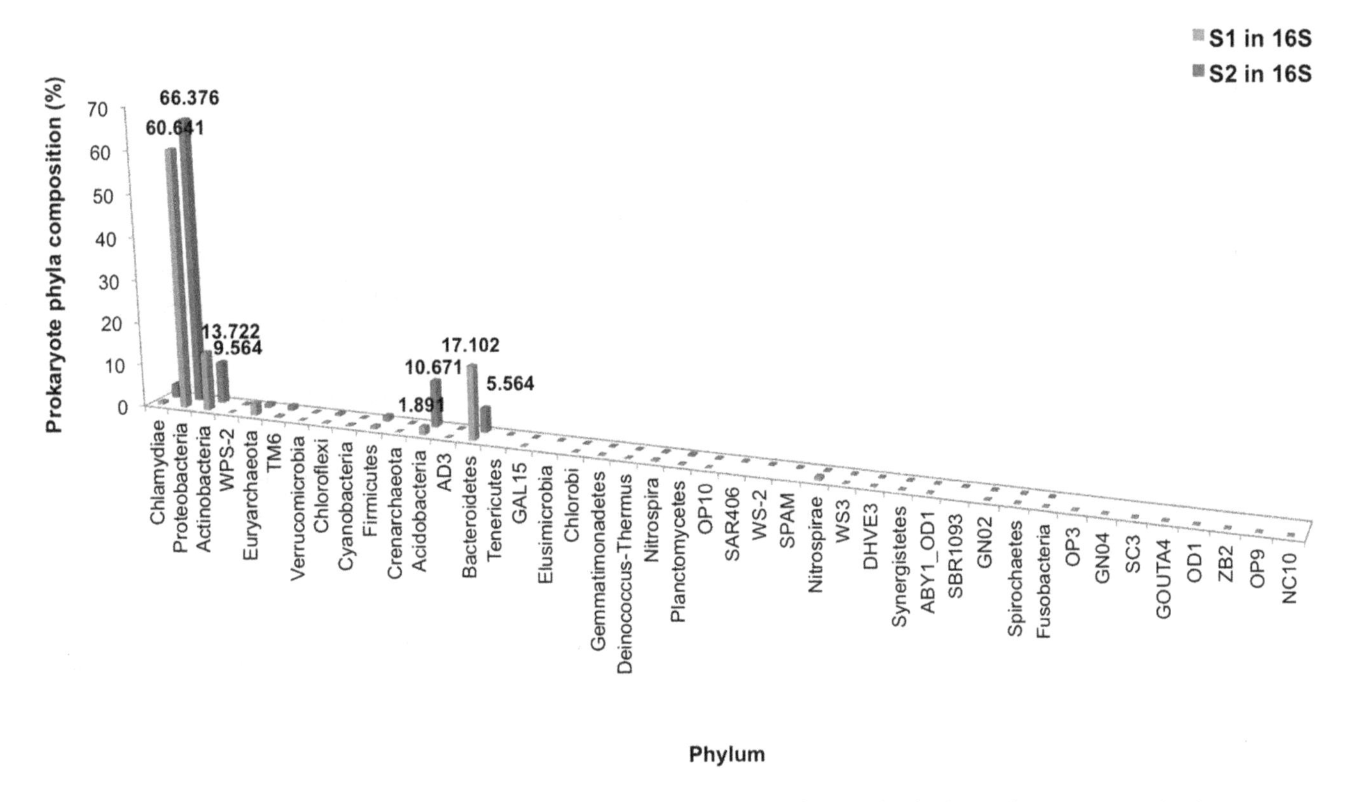

Figure 4. Distribution of prokaryotic phyla in S_1 and S_2, without (A) and with (B) individual sample ages categorization.

classified as an environmental bacterium with both commensal and pathogen species of marine animals and humans. Banning et al. [48] discovered several flavobacteria strains in various marine environments could function as predators on other bacteria. These bacteria have minimal growth requirements, only sea salt and the utilization of the lysed bacteria. The marine Flavobacteria thus could have critical consequences on microbial ecology as they could eradicate certain microbial communities [48]. Additionally, *Polynucleobacter* sp. are bacterioplankton that could survive broad ecological niches due to their ability to obtain energy by consuming organic materials from other organisms through nitrogen fixation, nitrification, remineralisation and methanogenesis. *Photobacterium* sp. is also a genus with metabolic versatility, which can degrade chitin and cellulose for carbohydrates [49]. Consequently, the flavobacteria, bacterioplankton and photobacteria activities could partly support the high metabolic subsystems of carbohydrates and protein metabolism in S_1 (Figure 7), albeit the overall poor living condition. Note the increased regulation and cell signalling, and cell wall and capsule subsystems (Figure 7) could in part symbolize the growth activities of these bacteria, given capsule lies outside the bacterial cell wall and considered a virulent factor. Bacterial capsule protects the bacteria against some hostile environment, such as desiccation, and prevents phagocytosis by host immune cells [50]. For examples, *Acinetobacter haemolyticus* in contaminated seafood produces shiga toxin that causes bloody diarrhea [51], and *Flavobacterium* sp. cause cold water disease in salmon and other fish species [52]. *Photobacterium*, a genus in family *Vibrionaceae*, is primarily marine microorganisms that evolved to become pathogenic to marine animals, causing mortality in crabs and fish, and indirect pathogens of humans through contact or consumption [49,53]. Hence residents and workers in these areas were recommended to minimize direct contact with the affected soil, sludge and water, to prevent their risk of infection, and frequent hand wash [3,8,9,47]. Note the many new uncultured species in S_1 (Figure 4A) further emphasized its environmental change resulting in new identified species. Our 16S rRNA gene analyses supported the high dissimilarity indices between S_1 and S_2 prokaryotic community structures (Figure 6A).

The metabolic potentials in Figure 7 supported the prior results, showing advanced metabolic subsystems of regulation and cell signalling, cell wall and capsule, protein metabolism, sulfur metabolism, and carbohydrates in S_1. In contrast, S_2 microbial communities carried high metabolic potentials for pathways of respiration, photosynthesis, and drug and bioactive compound production. This finding supported the diversified biodiversity in the non-affected terrestrials, and highlighted the more abundant pharmaceutical related microbial producers in the naturally undisturbed environments [15], like S_2.

For eukaryotic phyla and species distribution patterns, while both mollusks and brachiopods predominated in both terrestrials, given both animals were marine animals and were more prominent in tsunami-inundated S_1 site, fungi Basidiomycota and animal Arthropoda were only highly proportionate in the S_2 area (Figure 8A). Like prokaryotic phyla distribution patterns among various terrestrial depths (Figure 4B), the more similar eukaryotic phyla distribution patterns among various terrestrial depths were evident in S_1 (Figure 8B) highlighted the factor that a massive tsunami hit could destroy the biodiversity within microbial ecosystems. Basidiomycota, which were found more evident in S_2, are higher fungi that play important roles as carbon recycler and nutrient decomposer, and posed the chief source of bioactive natural products [54–56].

Since the 2004 tsunami up to our study period, an effect of microbial population mixing and microbial change due to human or animal activity on S_1 and S_2 sites in Phra Thong island should

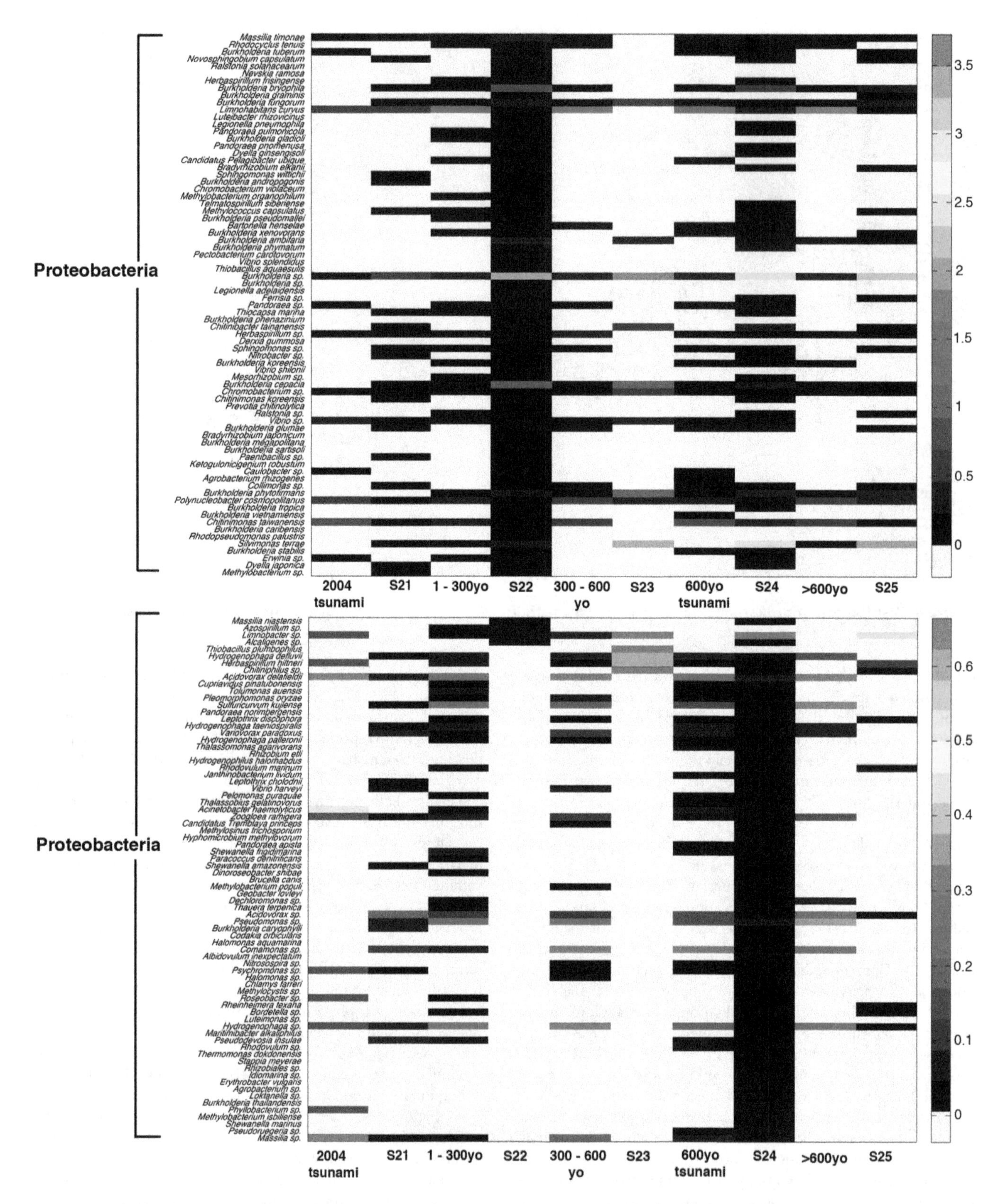

Figure 5. Distribution of prokaryotic species in S_1 and S_2, categorized by individual sample ages. Different color on the diagram represents a different relative abundance, based on the percent frequency chart on the right.

be minimal. The reasons are because, after the 2004 tsunami, Phra Thong island remains almost no human inhabited and no human activity. The place becomes part of the wildlife sanctuary. Microbial population mixing through time could only be by rainfall and plant root penetration (mostly grass at both sites); hence it should be minimal and in vertical direction only.

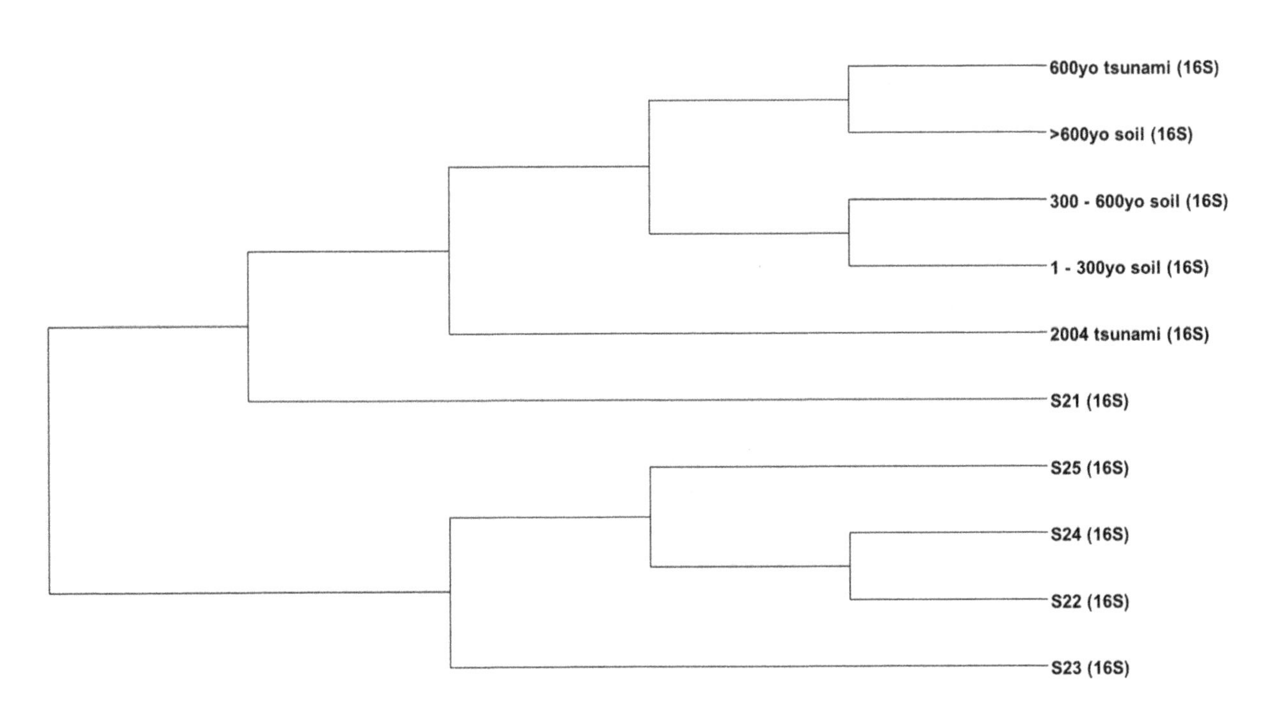

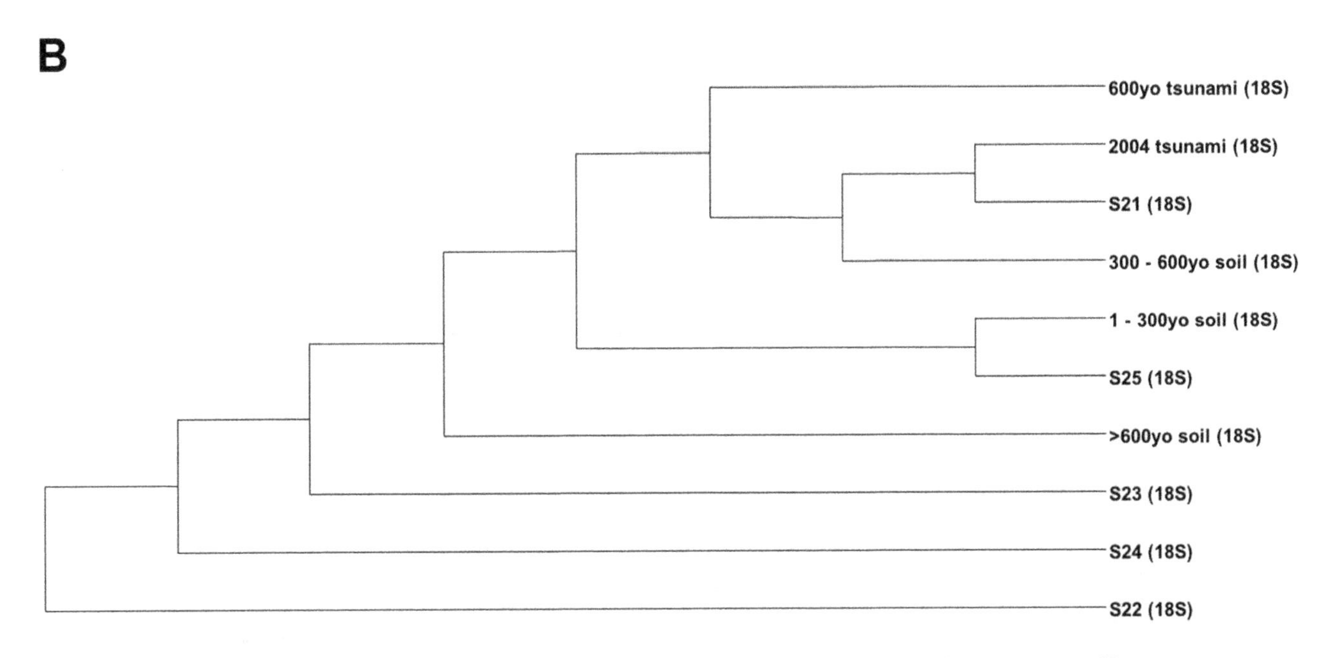

Figure 6. UPGMA clustering comparing relatedness among S_1 and S_2 prokaryotic (A) and eukaryotic (B) profiles.

Together, the 16S and 18S rRNA gene profiles indicated both terrestrials marine habitat, corresponding to the fact that the two sites were located on a small island in Andaman Sea of Thailand. Nevertheless, twice the greater prediction for marine habitat for S_1 (Figure 10) highlighted its much higher number of prokaryotic and eukaryotic species that represent marine habitat species-indicators, and tsunami inundation.

Conclusion

During the past decades, tsunamis have occurred more frequently though the correlation between tsunami disturbance and change of terrestrial microbial ecology remain poorly defined. The present study provided a culture-independent prokaryotic and eukaryotic analyses representing 0.22–30 μm metagenomes belonging to Thailand tsunami and non-tsunami affected terrestrials.

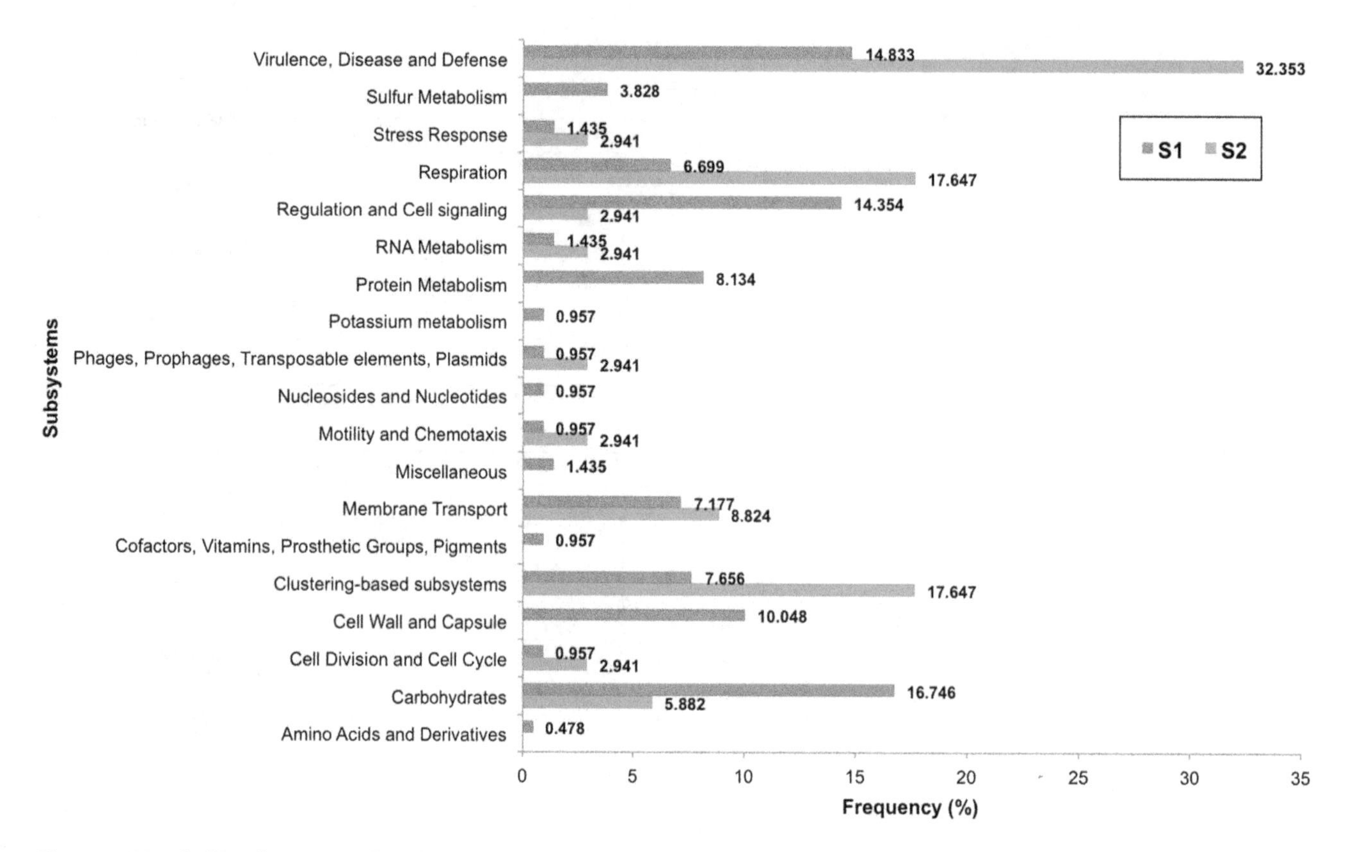

Figure 7. Metabolic subsystems of prokaryotic communities in S_1 and S_2.

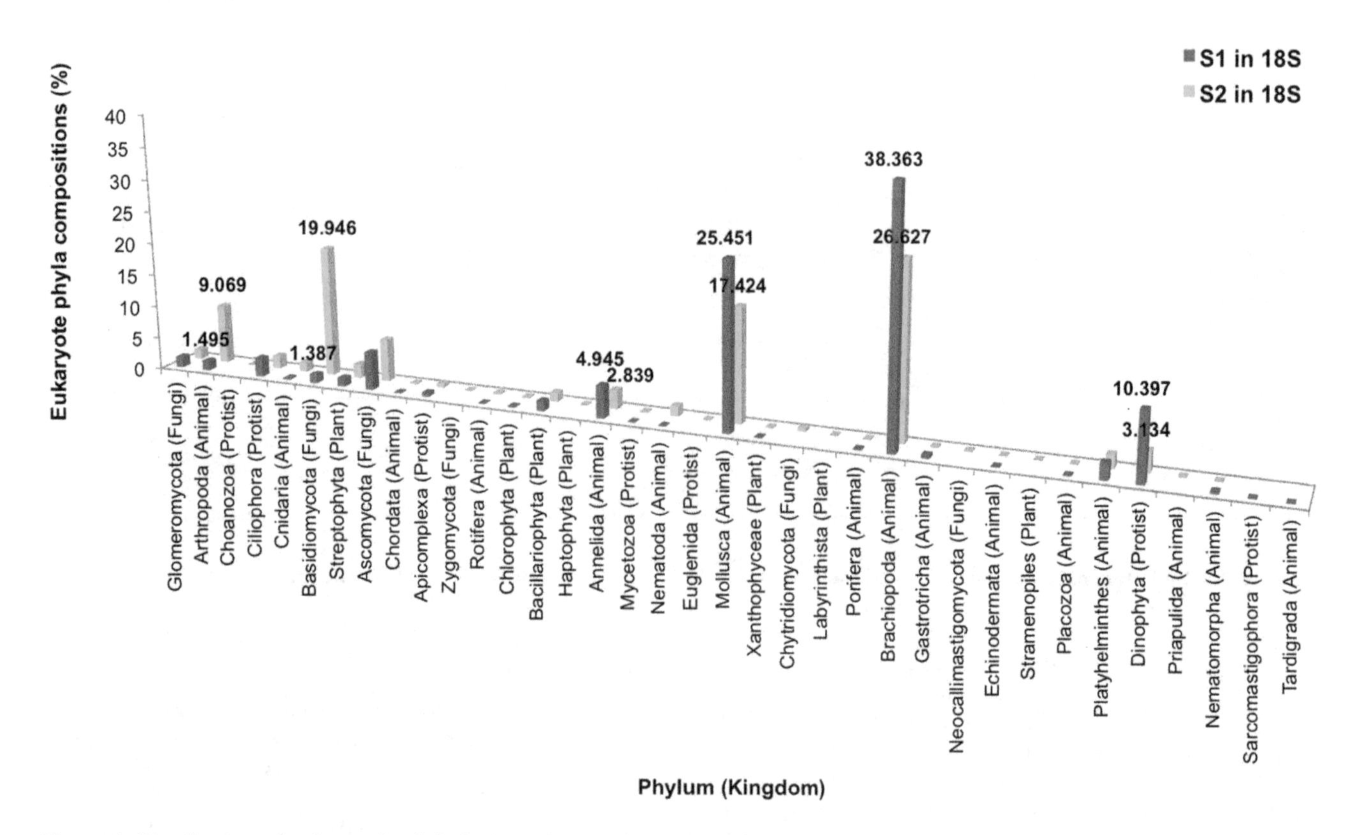

Figure 8. Distribution of eukaryotic phyla in S_1 and S_2, without (A) and with (B) individual sample ages categorization.

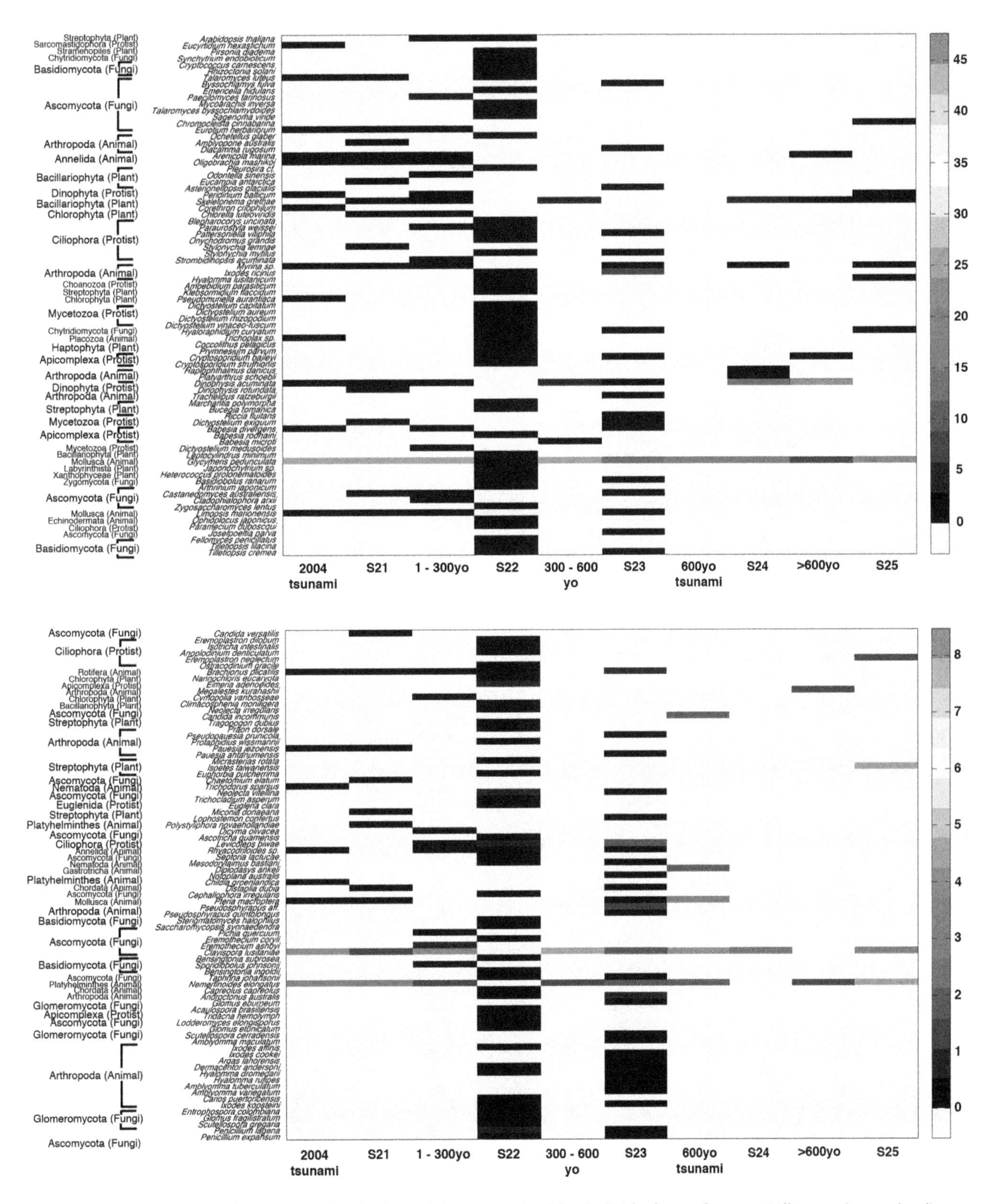

Figure 9. Distribution of eukaryotic species in S_1 and S_2, categorized by individual sample ages. Different color on the diagram represents a different relative abundance, based on the percent frequency chart on the right.

The different prokaryotic and eukaryotic profiles highlighted the differences due to tsunami, and helped fulfill our knowledge of diverse terrestrial microbial ecologies. The biodiverse species of S_1 distinguished its microbial communities and metabolic potentials. For instances, the finding of predator and prey bacterial relationship, and cell wall and capsule subsystem were common for S_1, whereas bioactive compound producers were more common in S_2.

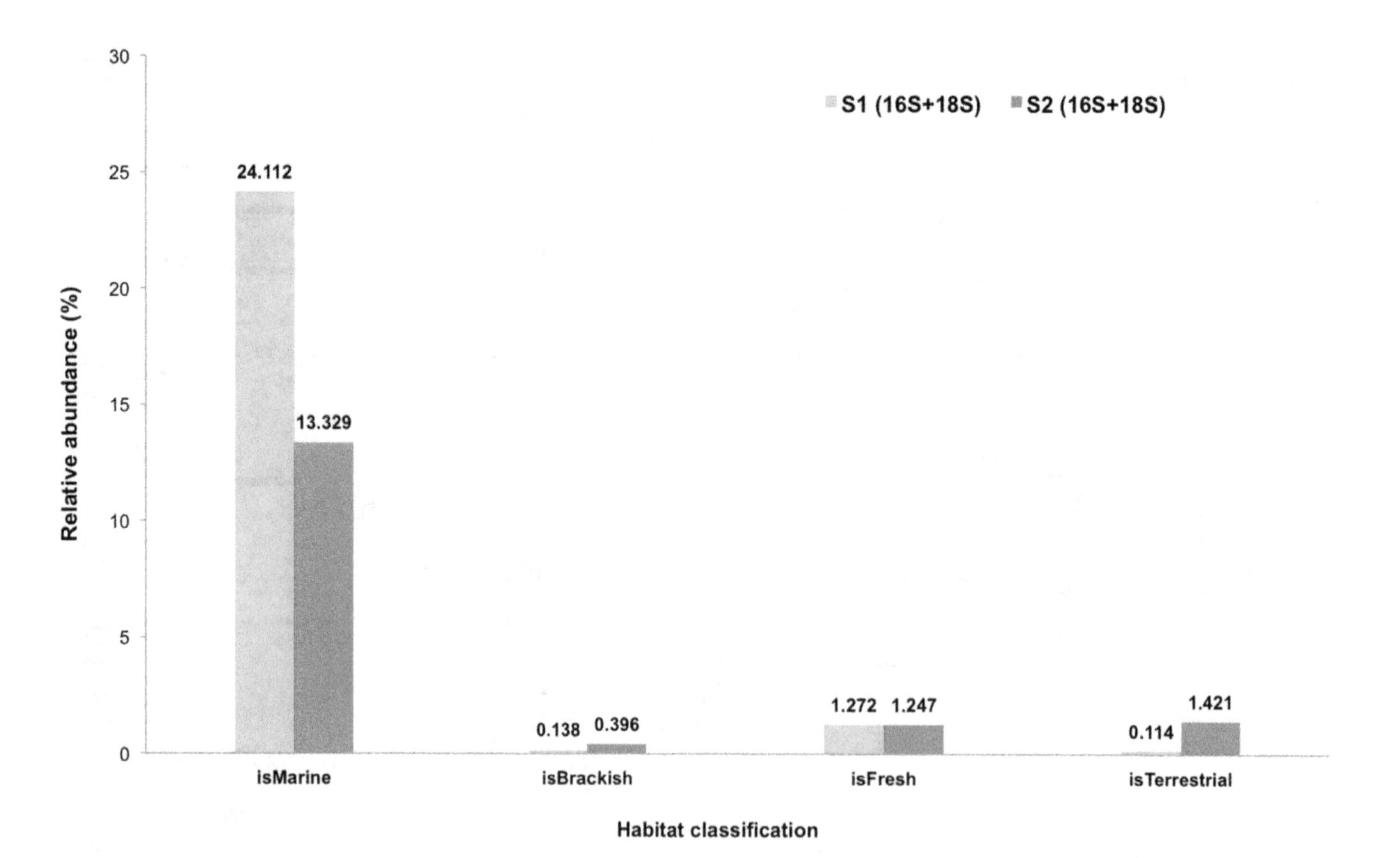

Figure 10. Habitat classification for S_1 and S_2. Prokaryotic and eukaryotic species profiles of S1 and S2 were matched against WoRMS database for habitat classification.

Further, the marine habitat analysis demonstrating the greater percent of marine prokaryotic and eukaryotic species in S_1 could perhaps help serve as another biomarker for the geological history of a terrestrial site. Nonetheless, more researches are required to utilize these as biomarkers for estimating times and presence of any geological incidences. Presently, identification of historic tsunami was restricted to examination of marine and brackish diatoms while silica shell of diatom could be dissoluted through time, especially in hot weather of a tropical country [1,2,57,58].

Acknowledgments

The authors thank Sarah Hof and Khunaluck Kidmoa for general help, Dominik Brill for helping with field work, and Troy Skwor for manuscript editing.

Author Contributions

Conceived and designed the experiments: NS AA. Performed the experiments: NS DS S. Tangphatsornruang. Analyzed the data: NS AW. Contributed reagents/materials/analysis tools: NS KJ S. Tongsima. Wrote the paper: NS. Revised the manuscript: NS KJ S. Tongsima.

References

1. Jankaew K, Atwater BF, Sawai Y, Choowong M, Charoentitirat T, et al. (2008) Medieval forewarning of the 2004 Indian ocean tsunami in Thailand. Nature 455:1228–1231.
2. Sawai Y, Jankaew K, Martin ME, Prendergast A, Choowong M, et al. (2009) Diatom assemblages in tsunami deposits associated with the 2004 Indian Ocean tsunami at Phra Thong Island, Thailand. Mar Micropaleontol 73:70–79.
3. Sri-Aroon P, Chusongsang P, Chusongsang Y, Pornpimol S, Butraporn P, et al. (2010) Snails and trematode infection after Indian ocean tsunami in Phang-Nga province, southern Thailand. Southeast Asian J Trop Med Public Health 41: 48–60.
4. Prummongkol S, Panasoponkul C, Apiwathnasorn C, Lek-Uthai U (2012) Biology of *Culex sitiens*, a predominant mosquito in Phang Nga, Thailand after a tsunami. J Insect Sci 12: 11. doi:10.1673/031.021.1101
5. Tharnpoophasiam P, Suthisarnsuntorn U, Worakhunpiset S, Charoenjai P, Tunyong W, et al. (2006) Preliminary post-tsunami water quality survey in Phang-Nga province, southern Thailand. Southeast Asian J Trop Med Public Health 37: 216–220.
6. Collivignarelli C, Tharnpoophasiam P, Vaccari M, De Felice V, Di Bella V, et al. (2008) Evaluation of drinking water treatment and quality in Takua Pa, Thailand. Environ Monit Assess 142: 345–358.
7. Collivignarelli C, Tharnpoophasiam P, Vaccari M, De Felice V, Di Bella V, et al. (2008) Water monitoring and treatment for drinking purposes in 2004 tsunami affected area-Ban Nam Khem, Phang Nga, Thailand. Environ Monit Assess 147: 191–198.
8. Chastel C (2007) Assessing epidemiological consequences two years after the tsunami of 26 December 2004. Bull Soc Pathol Exot 100: 139–142.
9. Huusom AJ, Agner T, Backer V, Ebbehøj N, Jacobsen P (2012) Skin and respiratory disorders following the identification of disaster victims in Thailand. Forensic Sci Med Pathol 8: 114–117.
10. Boszke L, Astel A (2007) Fractionation of mercury in sediments from coastal zone inundated by tsunami and in freshwater sediments from the rivers. J Environ Sci Health A Tox Hazard Subst Environ Eng 42: 847–858.
11. Lukaszewski Z, Karbowska B, Zembrzuski W, Siepak M (2012) Thallium in fractions of sediments formed during the 2004 tsunami in Thailand. Ecotoxicol Environ Saf 80: 184–190.
12. Rusch DB, Halpern AL, Sutton G, Heidelberg KB, Williamson S, et al. (2007) The Sorcerer II Global Ocean Sampling expedition: northwest Atlantic through eastern tropical Pacific. PLoS Biol 5:e77.
13. Yooseph S, Nealson KH, Rusch DB, McCrow JP, Dupont CL, et al. (2010) Genomic and functional adaptation in surface ocean planktonic prokaryotes. Nature 468:60–66.
14. Zinger L, Ammaral-Zettler LA, Fuhman JA, Horner-Devine MC, Huse SM, et al. (2011) Global patterns of bacterial beta-diversity in seafloor and seawater ecosystems. PLoS ONE 6: e24570. doi:10.1371/journal.pone.0024570

15. Somboonna N, Assawamakin A, Wilantho A, Tangphatsornruang S, Tongsima S (2012) Metagenomic profiles of free-living archaea, bacteria and small eukaryotes in coastal areas of Sichang island, Thailand. BMC Genomics 13:S29. doi:10.1186/1471-2164-13-S7-S29
16. Auld RR, Myre M, Mykytczuk NCS, Leduc LG, Merritt TJS (2013) Characterization of the microbial acid mine drainage microbial community using culturing and direct sequencing techniques. J Microbiol Methods 93: 108–115.
17. Redel H, Gao Z, Li H, Alekseyenko AV, Zhou Y, et al. (2013) Quantitation and composition of cutaneous microbiota in diabetic and nondiabetic men. JID 207: 1105–1114.
18. Taberlet P, Prud'Homme SM, Campione E, Roy J, Miquel C, et al. (2012) Soil sampling and isolation of extracellular DNA from large amount of starting material suitable for metabarcoding studies. Mol Ecol 21:1816–1820.
19. Baker GC, Smith JJ, Cowan DA (2003)Review and re-analysis of domain-specific 16S primers. J Microbiol Methods 55:541–555.
20. Humblot C, Guyoet J-P (2009) Pyrosequencing of tagged 16S rRNA gene amplicons for rapid diciphering of the microbiomes of fermented foods such as pearl millet slurries. Appl Environ Microbiol 75: 4354–4361.
21. Nossa CW, Oberdorf WE, Yang L, Aas JA, Paster BJ, et al. (2010) Design of 16S rRNA gene primers for 454 pyrosequencing of the human foregut microbiome. WJG 16:4135–4144.
22. Grant S, Grant WD, Cowan DA, Jones BE, Ma Y, et al. (2006) Identification of eukaryotic open reading frames in metagenomic cDNA libraries made from environmental samples. Appl Environ Microbiol 72:135–143.
23. Bailly J, Fraissinet-Tachet L, Verner M-C, Debaud J-C, Lemaire M, et al. (2007) Soil eukaryotic functional diversity, a metatranscriptomic approach. ISME J 1:632–642.
24. Meyer M, Stenzel U, Hofreiter M (2008) Parallel tagged sequencing on the 454 platform. Nat Prot 3:267–278.
25. Overbeek R, Begley T, Butler RM, Choudhuri JV, Chuang HY, et al. (2005) The subsystems approach to genome annotation and its use in the project to annotate 1000 genomes. Nucleic Acids Res 33: 5691–5702.
26. Meyer F, Paarmann D, D'Souza M, Olson R, Glass EM, et al. (2008) The metagenomics RAST server - a public resource for the automatic phylogenetic and functional analysis of metagenomes. BMC Bioinformatics 9: 386. doi:10.1186/1471-2105-9-386
27. Altschul SF, Madden TL, Schäffer AA, Zhang J, Zhang Z, et al. (1997) Gapped BLAST and PSI-BLAST: a new generation of protein database search programs. Nucleic Acids Res 25: 3389–3402.
28. Sayers EW, Barrett T, Benson DA, Bolton E, Bryant SH, et al. (2010) Database resources of the National Center for Biotechnology Information. Nucleic Acids Res 38: D5–D16.
29. Maidak BL, Cole JR, Liburn TG, Parker CT Jr, Saxman PR, et al. (2001) The RDP-II (ribosomal database project). Nucleic Acids Res 29: 173–174.
30. McDonald D, Price MN, Goodrich J, Nawrocki EP, DeSantis TZ, et al. (2012)An improved Greengenes taxonomy with explicit ranks for ecological and evolutionary analyses of bacteria and archaea. ISME J 6:610–618.
31. Brunak S, Danchin A, Hattori M, Nakamura H, Shinozaki K, et al. (2002) Nucleotide sequence database policies. Science 298: 1333.
32. Leinonen R, Akhtar R, Birney E, Bower L, Cerdeno-Tárraga A, et al. (2011) The European nucleotide archive. Nucleic Acids Res 39: D28–D31.
33. Pruesse E, Quast C, Knittel K, Fuchs BM, Ludwig W, et al. (2007) SILVA: a comprehensive online resource for quality checked and aligned ribosomal RNA sequence data compatible with ARB. Nucleic Acids Res 35: 7188–7196.
34. Yue JC, Clayton MK (2005) A similarity measure based on species proportions. Commun Stat Theor Methods 34: 2123–2131.
35. Chao A, Chazdon RL, Colwell RK, Shen T-J (2006) Abundance-based similarity indices and their estimation when there are unseen species in samples. Biometrics 62: 361–371.
36. Schloss PD, Westcott SL, Ryabin T, Hall JR, Hartmann M, et al. (2009) Introducing mothur: open-source, platform-independent, community-supported software for describing and comparing microbial communities. Appl Environ Microbiol 75: 7537–7541.
37. Aziz RK, Bartels D, Best AA, Dejongh M, Disz T, et al. (2008) The RAST server: rapid annotations using subsystems technology. BMC Genomics 9: 75. doi:10.1186/1471-2164-9-75
38. Appeltans W, Bouchet P, Boxshall GA, De Broyer C, de Voogd NJ, et al. (eds.) (2012) World Register of Marine Species [http://www.marinespecies.org]
39. Andersen K, Bird KL, Rasmussen M, Haile J, Breuning-Madsen H, et al. (2012) Meta-barcoding of 'dirt' DNA from soil reflects vertebrate biodiversity. Mol Ecol 21:1966–1979.
40. Caporaso JG, Lauber CL, Walters WA, Berg-Lyons D, Lozupone CA, et al. (2010) Global patterns of 16S rRNA diversity at a depth of millions of sequences per sample. *PNAS*: doi:10.1073/pnas.1000080107
41. Manichanh C, Chapple CE, Frangeul L, Gloux K, Guigo R, et al. (2008) A comparison of random sequence reads versus 16S rDNA sequences for estimating the biodiversity of a metagenomic library. *Nucleic Acids Res* 36: 5180–5188.
42. Maurice CF, Haiser HJ, Turnbaugh PJ (2012) Xenobiotics shape the physiology and gene expression of the active human gut microbiome. *Cell* 152: 39–50.
43. Ridaura VK, Faith JJ, Rey FE, Cheng J, Duncan AE, et al. (2013) Gut microbiota from twins discordant for obesity modulate metabolism in mice. *Science* 341. doi:10.1126/science.1241214
44. Henry EA, Devereux R, Maki JS, Gilmour CC, Woese CR, et al. (1994) Characterization of a new thermophilic sulfate-reducing bacterium *Thermodesulfovibrio yellowstonii*, gen.nov. and sp.nov.: its phylogenetic relationship to *Thermodesulfobacterium commune* and their origins deep within the bacterial domain. *Arch Microbiol* 161: 62–69.
45. Huang LN, Zhou H, Chen YQ, Luo S, Lan CY, et al. (2002) Diversity and structure of the archaeal community in the leachate of a full-scale recirculating landfill as examined by direct 16S rRNA gene sequence retrieval. *FEMS Microbiol Lett* 214:235–240.
46. Kendall MM, Liu Y, Sieprawska-Lupa M, Stetter KO, Whitman WB, et al. (2006) *Methanococcus aeolicus* sp. Nov., a mesophilic, methanogenic aechaeon from shallow and deep marine sediments. *Intl J Syst Evol Microbiol* 56: 1525–1529.
47. Wada K, Fukuda K, Yoshikawa T, Hirose T, Ikeno T, et al. (2012) Bacterial hazards of sludge brought ashore by the tsunami after the great East Japan earthquake of 2011. *J Occup Health* 54: 255–262.
48. Banning EC, Casciotti KL, Kujawinski EB (2010) Novel strains isolated from a coastal aquifer suggest a predatory role for flavobacteria. *FEMS Microbiol Ecol* 73: 254–270.
49. Vezzi A, Campanaro S, D'Angelo M, Simonato F, Vitulo N, et al. (2005) Life at depth: *Photobacterium profundum* genome sequence and expression analysis. *Science* 307: 1459–1461.
50. Yoshida K, Matsumoto T, Tateda K, Uchida K, Tsujimoto S, et al. (2000) Role of bacterial capsule in local and systemic inflammatory responses of mice during pulmonary infection with *Klebsiella pneumoniae*. *J Med Microbiol* 49: 1003–1010.
51. Grotiuz G, Sirok A, Gadea P, Varela G, Schelotto F (2006) Shiga toxin 2-producing *Acinetobacter haemolyticus* associated with a case of bloody diarrhea. *J Clin Microbiol* 44: 3838–3841.
52. Starliper CE (2011) Bacterial coldwater disease of fishes caused by *Flavobacterium psychrophilum*. *J Adv Res* 2: 97–108.
53. Osorio CR, Toranzo AE, Romalde JL, Barja JL (2000) Multiplex PCR assay for *ureC* and 16S rRNA genes clearly discriminates between both subspecies of *Photobacterium damselae*. *Dis Aquat Org* 40: 177–183.
54. Rosa LH, Machado KM, Rabello AL, Souza-Fagundes EM, Correa-Oliverira R, et al. (2009) Cytotoxic, immunosuppressive, trypanocidal and antileishmanial activities of Basidiomycota fungi present in Atlantic rainforest in Brazil. *Antonie Van Leeuwenhoek* 95: 227–237.
55. Qadri M, Johri S, Shah BA, Khajuria A, Sidiq T, et al. (2013) Identification and bioactive potential of endophytic fungi isolated from selected plants of the Western Himalayas. *SpringerPlus* 2: 8.
56. Jaszek M, Osińska-Jaroszuk M, Janusz G, Matuszewska A, Stefaniuk D, et al. (2013) New bioactive fungal molecules with high antioxidant and antimicrobial capacity isolated from *Cerrena unicolor* idiophasic cultures. *BioMed Res Intl* 2013. doi:10.1155/2013/497492
57. Hemphill-Haley E (1995) Diatoms evidence for earthquake-induced subsidence and tsunami 300 yr ago in southern coastal Washington.*GSA Bulletin* 107: 367–378.
58. Hemphill-Haley E (1996) Diatoms as an aid in identifying late-Holocene tsunami deposits. Holocene6: 439–448.

13

Bacterial Community Dynamics and Taxa-Time Relationships within Two Activated Sludge Bioreactors

Reti Hai, Yulin Wang, Xiaohui Wang*, Yuan Li, Zhize Du

Department of Environmental Science and Engineering, Beijing University of Chemical Technology, Beijing, China

Abstract

Background: Biological activated sludge process must be functionally stable to continuously remove contaminants while relying upon the activity of complex microbial communities. However the dynamics of these communities are as yet poorly understood. A macroecology metric used to quantify community dynamic is the taxa-time relationship (TTR). Although the TTR of animal and plant species has been well documented, knowledge is still lacking in regard to TTR of microbial communities in activated sludge bioreactors.

Aims: 1) To characterize the temporal dynamics of bacterial taxa in activated sludge from two bioreactors of different scale and investigate factors affecting such dynamics; 2) to evaluate the TTRs of activated sludge microbial communities in two bioreactors of different scale.

Methods: Temporal variation of bacterial taxa in activated sludge collected from a full- and lab-scale activated sludge bioreactor was monitored over a one-year period using pyrosequencing of 16S rRNA genes. TTR was employed to quantify the bacterial taxa shifts based on the power law equation $S = cT^w$.

Results: The power law exponent w for the full-scale bioreactor was 0.43 ($R^2 = 0.970$), which is lower than that of the lab-scale bioreactor ($w = 0.55$, $R^2 = 0.971$). The exponents for the dominant phyla were generally higher than that of the rare phyla. Canonical correspondence analysis (CCA) result showed that the bacterial community variance was significantly associated with water temperature, influent (biochemical oxygen demand) BOD, bioreactor scale and dissolved oxygen (DO). Variance partitioning analyses suggested that wastewater characteristics had the greatest contribution to the bacterial community variance, explaining 20.3% of the variance of bacterial communities independently, followed by operational parameters (19.9%) and bioreactor scale (3.6%).

Conclusions: Results of this study suggest bacterial community dynamics were likely driven partly by wastewater and operational parameters and provide evidence that the TTR may be a fundamental ecological pattern in macro- and microbial systems.

Editor: Julio Vera, University of Erlangen-Nuremberg, Germany

Funding: This study was supported by the Fundamental Research Funds for the Central Universities (ZY1306) and special fund of State Key Joint Laboratory of Environment Simulation and Pollution Control (3K06ESPCT). The funders had no role in study design, data collection and analysis, decision to publish, or preparation of the manuscript.

Competing Interests: The authors have declared that no competing interests exist.

* E-mail: wangxiaohui2008@gmail.com

Introduction

Biological activated sludge process is the most widely used biological process to treat municipal and industrial wastewater. The efficient and stable operation of biological wastewater treatment plant (WWTP) relies upon the relative abundance or activity of these microbial populations within it [1]. Because variations in microbial community composition are often associated with changes in the functional capabilities of those communities, understanding microbial temporal patterns of activated sludge can be critical for understanding ecosystem processes, and then to enhance the treatment performance and stability [2,3].

In recent years, a growing collection of studies in the microbial ecology of activated sludge suggests that microbial communities of activated sludge exhibit a wide range of discernible temporal patterns, particularly within specific microbial subpopulations such as nitrifiers [4,5], denitrifiers [6], phosphorus-accumulating organisms [7], and methanogens [8]. Such temporal variations in microbial communities are thought to be influenced by the deterministic environmental and operational variables and/or stochastic factors [9–11]. By contrast, in a full-scale WWTP, Kim *et al.* [12] found that some microbial communities were composed of core members that exhibit minimal temporal variability and rarer taxa that exhibit more pronounced fluctuations in abundance over time. Further, some microbial communities have the capacity to recover quickly after disturbance events, either to the pre-disturbance state or to an alternative stable state [13,14]. Overall, these and other time series highlight that microbial communities are dynamic and exhibit temporal patterns that can reflect underlying biotic and abiotic processes.

Many previous studies have been performed in lab-scale bioreactors which treated synthetic wastewater, where selective pressures likely differ dramatically from those in full-scale plants [10]. An equilibrium model based on island biogeography also predicts that the scale of bioreactors affect microbial communities within them [3]. Indeed, a very limited number of studies have characterized microbial population dynamics in activated sludge from full-scale WWTPs. To data, there is no quantitative comparison of the change rate of microbial population within activated sludge from different scale bioreactors. Furthermore, nearly all of the previous studies employed clone library or fingerprinting methods to characterize the temporal dynamics of microbial communities in bioreactors. These techniques are lack of ability to detect rare species within a given habitat. For example, denaturing gradient gel electrophoresis (DGGE) exhibits method-dependent detection threshold in absolute abundance and it has been estimated that populations for $<10^3$ individuals per sample within a given community will be below the detection threshold [11,15]. Fortunately, the ongoing development of high-throughput sequencing technologies has made it feasible to describe the temporal dynamics of microbial communities in higher resolutions that were previously unattainable [16].

Activated sludge bioreactors are excellent test beds for fundamental questions in microbial ecology [5]. One of the fundamental objectives of ecology is to understand how biodiversity is generated and maintained across spatial and temporal scales [11,16]. Patterns of species diversity provide important insights into the underlying mechanisms that regulate biodiversity, and are central to the development of ecological models and theories. One such pattern is the species-area relationship (SAR), which describes the tendency that species richness increases with area. SAR plays a central role in biodiversity research, and recent work has increased awareness of its temporal analogue, the species-time relationship (herein referred to as the taxa-time relationship [TTR]) [17]. The TTR describes how the species richness of a community increases with the time span over which the community is observed [18]. It is similar in form to the SAR: $S=cT^w$, where S is the taxa richness, T is the time of observation, c is an empirically derived constant, and w is a scaling exponent that reflects species turnover [10]. Although the TTR of animal and plant species has been well documented [18], it is only recently that TTRs have been tested in a limited number of microbial systems [10,16,19]. To data, very few studies have applied the power-law TTR to activated sludge bioreactors, and they found that the accumulation of observed taxa within the bioreactor followed a power law relationship [10,11]. However, Comparisons of the temporal scaling exponent (w) among various microbial populations and between bioreactors at different scale have not yet been explored.

The primary objectives of this study were thus two-fold: 1) to characterize the temporal dynamics of bacterial taxa in activated sludge from two bioreactors of different scale and investigate factors affecting such dynamics; and 2) to evaluate the TTRs of activated sludge microbial communities in two bioreactors of different scale, and compare the temporal scaling exponent w among different microbial populations within the two bioreactors.

Materials and Methods

Sample collection and site description

In this study, activated sludge samples were collected from two bioreactors of different scale: one is a full-scale wastewater treatment system and the other is a lab-scale bioreactor which was used to simulate the full-scale bioreactor. The two reactors received identical wastewater and were operated with same process: anaerobic/anoxic/aerobic (A^2O). Both of them were located in the same WWTP and were operated by the Beijing Drainage Group Co. Ltd.

Activated sludge samples were monthly collected from the aerobic zone of the two bioreactors from May 2010 to April 2011. For archiving, each 1.5 ml sample was dispensed into a 2 ml sterile Eppendorf tube and centrifuged at $14{,}000\times g$ for 10 min. The supernatant was decanted, and the pellets were stored at $-20°C$ prior to analysis. No specific permits were required for the described field studies. We confirm that: i) the locations were not privately-owned or protected in any way; and ii) the field studies did not involve endangered or protected species.

DNA extraction

The pellets of the activated sludge samples were washed three times by centrifugation using sterile high-purify water for 5 min at $15{,}000\times g$. DNA was extracted using a FastDNA SPIN Kit for Soil (MP Biotechnology, USA) according to the manufacturer's protocol.

PCR amplification and purification

The extracted DNA samples were amplified with a set of primers targeting the hypervariable V4 region of the 16S rRNA gene. The forward primer is 5′-AYTGGGYDTAAAGNG-3′ and the reverse primers are an equal portion mixture of four primers, i.e. 5′-TACCRGGGTHTCTAATCC-3′, 5′-TACCAGAGTATCTAATTC-3′, 5′-CTACDSRGGTMTCTAATC-3′, and 5′-TACNVGGGTATCTAATCC-3′ [20,21]. Barcodes that allow sample multiplexing during pyrosequencing were incorporated between the 454 adapter and the forward primer.

Pyrosequencing and sequence analysis

The composition of the PCR products of V4 region of 16S rRNA genes was determined by pyrosequencing using the Roche 454 FLX Titanium sequencer (Roche, Nutley, NJ, USA). Samples in this study were individually barcoded to enable multiplex sequencing. The raw reads have been deposited into the NCBI Sequence Read Archive (Accession Number: SRR952788 and SRR954283). After pyrosequencing, Python scripts were written to remove sequences containing more than one ambiguous base ('N') and the sequences shorter than 150 bps. The RDP Classifier (Version 10.31) was used to assign all effective sequences to taxonomic ranks with a set confidence threshold of 50%.

Data analysis

Temporal variation of bacterial taxa was assessed graphically by the ordination method of nonmetric multidimensional scaling (NMDS) using a statistical software PC-ORD version 6 (MJM Software Design, Gleneden Beach, OR, USA). In the present analysis, a matrix of the OTUs for each sample was used as input data. NMDS ordination was generated based on the Sorenson similarity matrix which was constructed for all pairs of samples.

Moving-window analysis was used to characterize the change rate of bacterial community in this study [22]. Firstly, a similarity matrix for each activated sludge sample was calculated based on Pearson product-moment correlation coefficient using SPSS version 17.0 software (IBM Corporation, Chicago, IL, USA). Then, each similarity percentage value was subtracted from the 100% similarity value to get the change values. Finally, moving-window analysis was performed by plotting the change values between month x and month x−1 [23]. The average change value

Table 1. Number of taxa classified by different taxonomic levels from the full-and lab-scale bioreactor.

Taxonomic level	bioreactor	The number of taxa at each level												Range	Average	Standard deviation
		1	2	3	4	5	6	7	8	9	10	11	12			
Phylum	Full[a]	12	13	13	16	16	14	16	15	14	14	15	14	12–16	14.3	1.2
	Lab[b]	12	13	14	15	16	14	16	15	16	14	15	14	12–16	14.5	1.2
Class	Full	23	23	26	29	31	28	27	28	23	26	26	27	23–29	26.4	2.4
	Lab	21	24	27	23	32	29	29	22	23	24	21	23	21–32	24.8	3.4
Order	Full	51	53	52	64	65	61	57	57	56	50	51	51	50–65	55.7	5.1
	Lab	49	60	62	58	66	64	64	53	57	53	47	51	49–66	57.0	6.1
Family	Full	65	78	67	84	92	76	62	73	64	63	68	65	63–92	71.4	9.0
	Lab	63	71	73	79	87	70	69	62	78	65	59	64	59–87	70.0	7.9
Genus	Full	219	274	231	278	281	256	249	237	215	201	229	205	201–281	239.6	26.8
	Lab	208	242	251	267	278	245	231	205	261	237	197	223	197–278	237.1	24.4
OTUs	Full	3611	3043	2787	3878	3592	3317	2706	2847	3452	3027	2779	2513	2513–3878	3055.8	408.0
	Lab	2454	2983	3150	3275	3742	2773	2870	2697	3130	2993	2137	2452	2137–3275	2874.8	426.2

[a]Full-scale bioreactor.
[b]Lab-scale bioreactor.

Δt (one month) was calculated as the average and standard deviation for the respective change values [4].

Canonical Correspondence Analysis (CCA) was used to reveal relationships between bacterial community dynamics and operational and environmental parameters. Statistically important explanatory variables were identified by the forward selection method using a Monte Carlo permutation test (499 permutations under the full model). Operational variables that failed to contribute significant improvement ($P<0.05$) to a model's explanatory power were excluded from final CCA analyses. The contributions of wastewater characteristics (W), operational parameters (O) and scale of bioreactor (S) to the variances of bacterial communities were assessed with variance partitioning analysis (VPA) using CCA. All wastewater characteristics, operational parameters, and bioreactor scale data were $\log_2$ (x+1) transformed for standardization. CCA and VPA were performed using the VEGAN package in R (v.2.15.1; http://www.r-project.org/)

Results

Bioreactor performance and operational conditions

For the one year period (May 2010 to April 2011), 13 operational and environmental parameters were monitored from the two activated sludge bioreactors. Variations in these parameters are summarized in Table S1 and S2. Performance of the full-scale bioreactor was relatively stable across the sampling period. Although biological oxygen demand (BOD) in the influents varying from 150 to 288 mg/L, the BOD removal efficiency was always excellent (>93%) over the duration of study. The average BOD concentration in the effluent was below 9 mg/L. Total nitrogen (TN) in the influent ranged from 35.9 to 69.3 mg/L, while the ammonium concentration was between 32.8 and 54.2 mg/L. The average removal efficiencies of TN and ammonia were 62% and 97%, giving final concentrations in effluent of less than 20.9 mg/L and 1.5 mg/L, respectively. The temperature of the bioreactor showed a seasonal pattern from 16.3°C (winter) to 25.1°C (summer). The bioreactor was maintained with relatively stable pH (7.1±0.3), hydraulic residence time (HRT) (8.2±1.3 h), DO (3.5±0.5 mg/L) and mixed liquor suspended solids (MLSS) (3,775±640 mg/L), while solid retention time (SRT) (8.7±2.1 days) exhibited a larger variation comparatively.

The lab-scale bioreactor was also relatively functional stable during the study period. The influent characteristics were the same as the full-scale bioreactor. The average BOD, TN and ammonia removal capacities were 94%, 61%, and 97%, giving final average concentrations in effluent of less than 7 mg/L, 22 mg/L and 2 mg/L, respectively. Other environmental and operational parameters (pH, DO, MLSS, HRT and SRT) were kept relatively stable (Table S2).

Bacterial community composition

By using 454 pyrosequencing, 13422–31151 effective sequence tags were yielded for the 24 samples. The library size of each sample was normalized to 13422 sequences by randomly removing the extra sequences, which was the smallest number of sequencing reads among the 24 samples, to conduct the downstream analyses for different samples at the same sequencing depth.

RDP Classifier was used to assign these sequence tags into different OTU with 3% of nucleotide cutoff. A total of 10223 OTUs were recovered from the full-scale bioreactor and 14791 OTUs were from the lab-scale bioreactor. Individual samples contained much smaller number of OTUs from 2513 to 3878 in the full-scale bioreactor and 2137 to 3275 within the lab-scale bioreactor (Table 1).

RDP Classifier was also used to assign these sequence tags into different phylogenetic bacterial taxa. The threshold for the bootstrap cutoff was set to 50% to ensure the classification reliability [12]. Table 1 summarizes the numbers of taxa at different levels of classification within the two bioreactors. The bacterial community composition of the two bioreactors showed typical activated sludge communities [21,24,25]. At the phylum level, *Proteobacteria* was the predominant phylum in the full-scale bioreactor, constituting between 19 and 51% of all detected OTUs. *Bacteroidetes*, *Acidobacteria*, and *Chloroflexi* were the subdominant groups, each containing 13–52%, 1–22% and 1–14% of the detections respectively. These four phyla represented approximately 73–95% of bacteria detected within the full-scale bioreac-

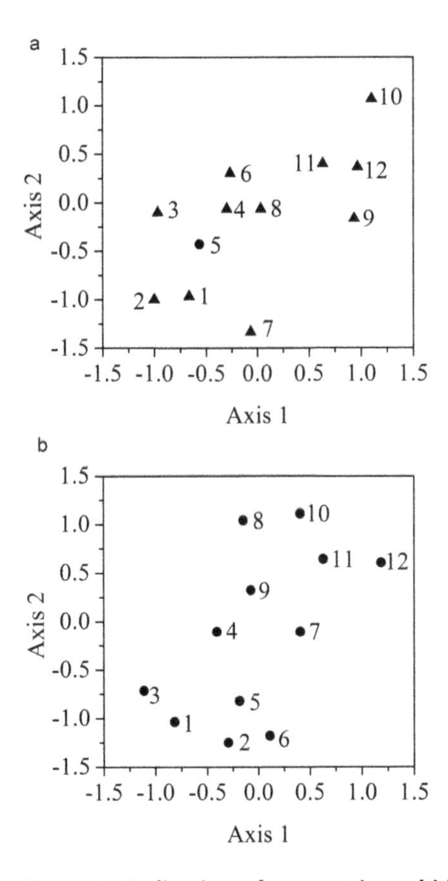

Figure 1. Ordination of nonmetric multidimentional scaling based on OTUs (3% cutoff) identified from a full- (a) and lab-scale (b) bioreactors. Sample numbers: 1- May 2010, 2- June 2010, 3-July 2010, 4- August 2010, 5-September 2010, 6- October 2010, 7-November 2010, 8-December 2010, 9-January 2011, 10- February 2011, 11-March 2011, 12-April 2011.

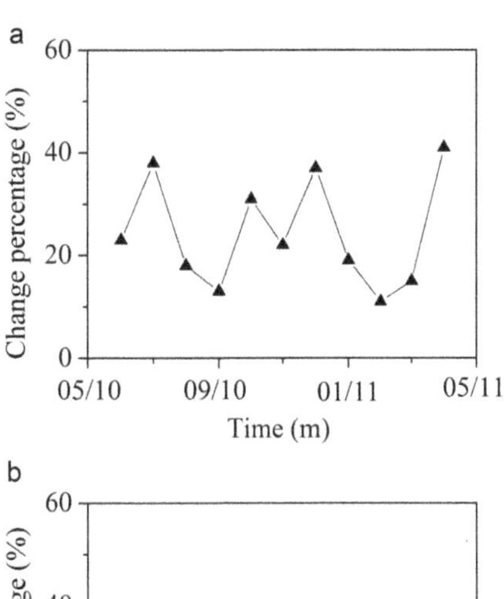

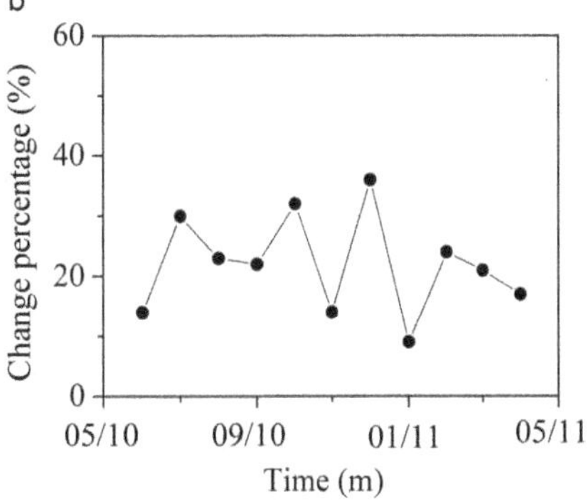

Figure 2. Moving-window analysis based on pyrosequencing data for a full-scale (a) and lab-scale (b) bioreactors. Each data point in the graph is a comparison between two consecutive dates, as it represents the correlation between the samples of month x and month x−1.

tor. Within *Proteobacteria* the β-subdivision was the predominant group (26–54%), followed by α-*Proteobacteria* (6–43%), γ-*Proteobacteria* (7–37%) and δ-*Proteobacteria* (1–19%). Within the β-*Proteobacteria*, six taxa were identified. *Rhodocyclales* is the dominant group within a range of 25–68% of all 12 samples, followed by *Burkholderiales* and *Burkholderiales*, representing 14–52% and 8–32% of each population respectively. The other three detected groups (*Burkholderiales*, *Rhodocyclales*, and *Burkholderiales*) had fewer detections in all samples and constituted less than 10% of β-*Proteobacteria*.

Similar to the full-scale bioreactor, in the lab-scale bioreactor *Proteobacteria* was also the predominant phylum (22–49%), following *Bacteroidetes* (17–43%), *Acidobacteria* (2–23%), *Chloroflexi* (1–19%). Within *Proteobacteria* β-subdivision was the predominant group (17–49%), followed by α-*Proteobacteria* (7–46%), γ-*Proteobacteria* (5–39%) and δ-*Proteobacteria* (3–25%).

Tracking bacterial community dynamics

The temporal dynamics of the bacterial compositions within the two bioreactors was evaluated using an indirect gradient ordination technique nonmetric multidimensional scaling (NMDS) ordination (Fig. 1) which was constructed based on the observed OTUs at a 3% cutoff. At a two-dimensional solution, the stress values of the ordination for the full- and lab-scale reactors were 16.7 and 14.8, respectively, demonstrating that the reduced NMDS ordinates preserve patterns in activated sludge bacterial community dynamics. In general, samples collected in similar time periods were clustered closely together in the ordination. This ordination pattern suggests a gradual succession within the overall bacterial community over time.

Moving window analysis was used to evaluate the change rate of bacterial community [26]. The correlation coefficients of two consecutive dates (one month) from the full-scale bioreactor were between 64% and 91%. Thus, changes were from 9% to 36% (Fig. 2), and the $\Delta_{t\ (\text{one month})}$ was 21.4%±9.5%. While the change rate for the lab-scale bioreactor were 13%–41%, with the $\Delta t_{(\text{one month})}$ 25.1%±11.7%, which is higher than the full-scale bioreactor.

Correlations between bacterial community dynamics and operational and environmental variables

CCA was performed to discern the possible relationship between bacterial community structure and operational and environmental variables (Fig. 3). Based on variance inflation factors (VIF) with 999 Monte Carlo permutations, four significant environmental variables: temperature, influent BOD, bioreactor scale and DO were selected in the CCA biplot. The length of an environmental parameter arrow in the ordination plot indicates the strength of the relationship of that parameter to community composition. As such, temperature, influent BOD, DO and Bioreactor scale appears to be the most important environmental parameters.

VPA was further performed to assess the contributions of wastewater characteristics (BOD, TN, TP, pH), operational parameters (temperature, SRT, HRT, MLSS), and bioreactor

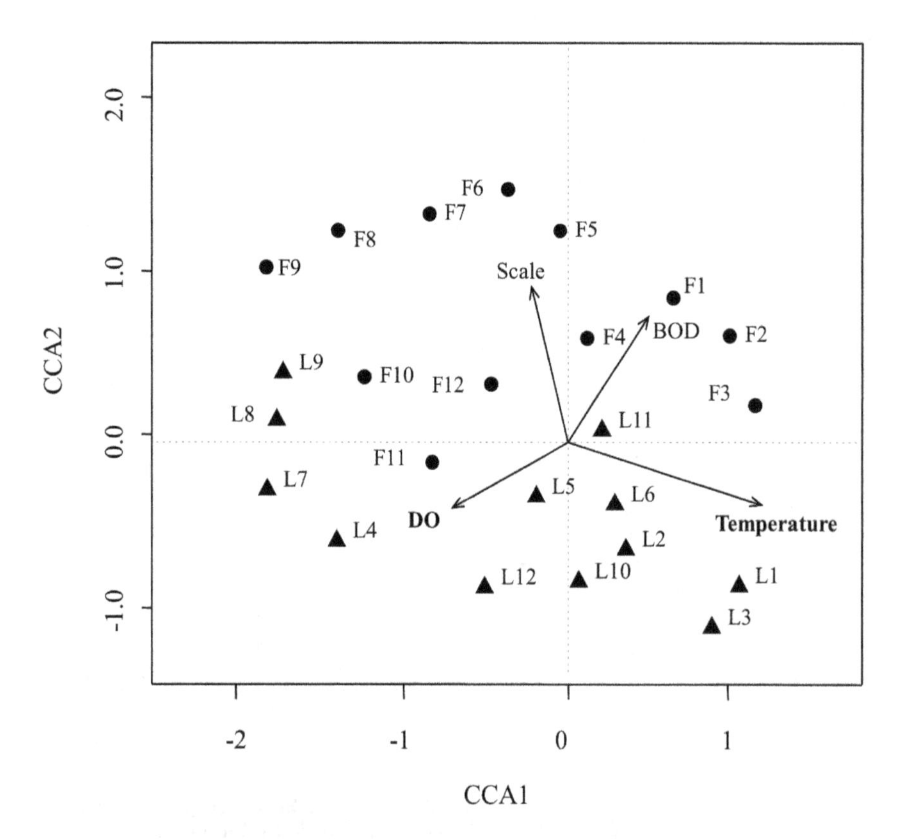

Figure 3. Canonical correspondence analysis (CCA) of pyrosequencing data and measurable variables in a full- and lab-scale bioreactors. Arrows indicate the direction and magnitude of measurable variables associated with bacterial community structures. Circles and triangles represent different bacterial community structures from the full- and lab-scale bioreactor, respectively. Samples are named with "F" (Full-scale bioreactor) or "L" (Lab-scale bioreactor) and numbers. Sample numbers: 1- May 2010, 2- June 2010, 3-July 2010, 4- August 2010, 5-September 2010, 6- October 2010, 7-November 2010, 8-December 2010, 9-January 2011, 10- February 2011, 11-March 2011, 12-April 2011.

scale to the bacterial community variance. Fig. 4 indicated that 49% of the variance could be explained by these three components. Wastewater characteristics, operational parameters and bioreactor scale could independently explain 20.3%, 19.9%, and 3.6% of the variation of bacterial communities, respectively. Interactions among the three major components seemed to have

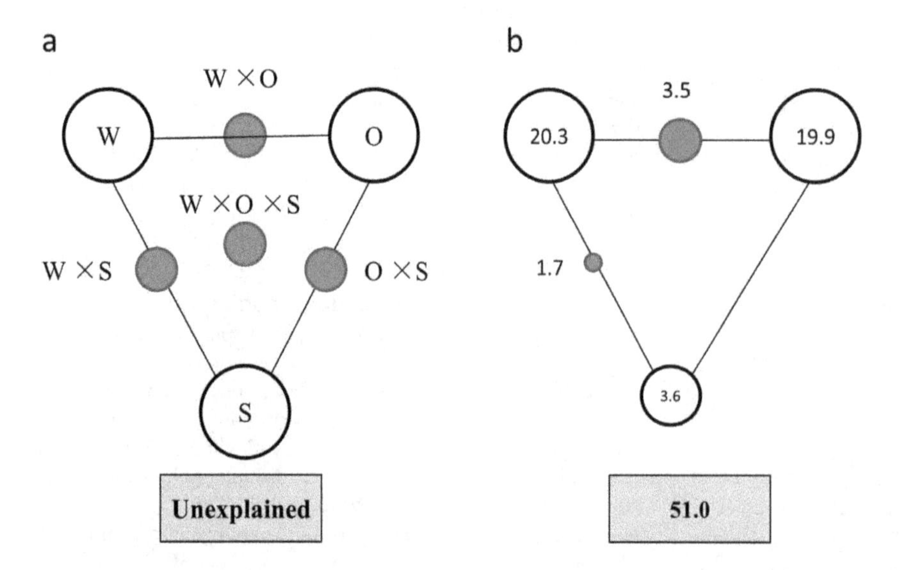

Figure 4. Variation partitioning analysis of microbial community explained by wastewater characteristics (W), operational parameters (O), and Scale of bioreactor (S). (a) General outline; (b) bacterial communities. Each diagram represents the biological variation partitioned into the relative effects of each factor or a combination of factors, in which geometric areas were proportional to the respective percentages of explained variation. The edges of the triangle represent the variation explained by each factor alone. The sides of the triangles represent interactions of any two factors, and the middle of the triangles represent interactions of all three factors.

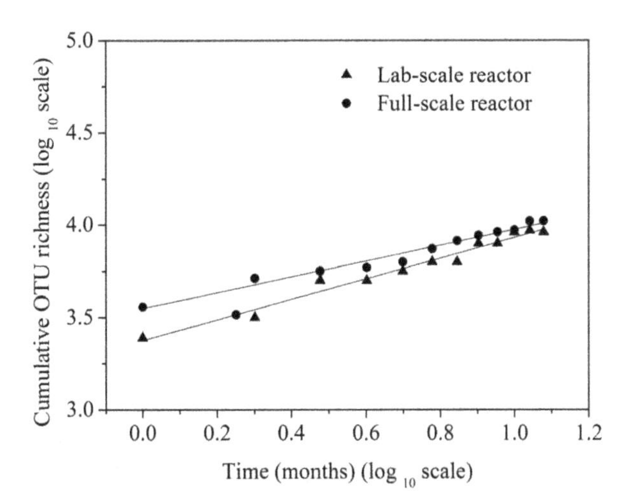

Figure 5. Taxa-time relationships for a full-scale bioreactor (circles) and lab-scale bioreactor (triangles). The lines are fitted to a power law equation $S=cT^w$, where S is the number of observed taxa, c is the constant, T is the time, and w is the taxa-time relationship exponent.

less influence than did individual components, and were only observed between wastewater characteristics and operational parameters (3.5%) and between wastewater characteristics and bioreactor scale (1.7%).

Taxa-time relationship (TTR)

The bacterial TTRs for the two bioreactors were visualized using the power law equation $S=cT^w$, as described above, and plotted in log-log space in Fig. 5. The TTR relationship exponent (w) is the slope of the linear regression line fitted to the log–log plots, which can be considered a measure of temporal turnover of microbial taxa. The exponents (i.e., temporal turnover rate) for the full-scale bioreactor was 0.43 ($R^2=0.970$), which is lower that of the lab-scale bioreactor ($w=0.55$, $R^2=0.971$).

The TTRs for phyla level was further analyzed. The results showed that the power law exponents of 13 phyla ranged from 0.32 to 0.57 in the two bioreactors (Table 2), suggesting that bacterial phyla have different temporal turnover rates. Interestingly, the exponents of dominant phyla were general lower than that of the rare phyla (Table 2). For example, the average exponents of five most dominant phyla (*Proteobacteria*, *Bacteroidetes*, *Acidobacteria*, *Chloroflexi* and *Verrucomicrobia*) within the full- and lab-scale bioreactor were 0.35 and 0.36, while that of the five rarest phyla (*Synergistetes*, *Fusobacteria*, *TM7*, *Euryarchaeota* and *OP10*) were 0.50 and 0.55.

Discussion

Understanding the factors that shape microbial community structure in WWTPs could potentially enhance treatment performance and control. In this study, CCA ordination analysis indicated that temperature was an important variable influencing microbial community structures. This agrees with the findings of Wells et al. [5], who suggested, based on the survey of bacteria community variance via terminal restriction fragment length polymorphism (T-RFLP) in a full-scale WWTP, that water temperature is one of the most influential variables on bacterial community variance. Similar results have also been obtained in an expanded granular sludge bed (EGSB) bioreactor by Siggins *et al.* [27]. As with temperature, influent BOD was also significantly linked to bacterial community structures. BOD provides carbon and energy sources to heterotrophic bacteria and influences the growth rate of the bacteria [12]. Previous studies have reported that BOD (or organic loading) was an important factor to mediate the bacterial community structures [28]. In addition to temperature and influent BOD, DO was also strongly and significantly linked to bacterial community variance in CCA analyses. DO is well recognized as a critical process parameter in biological wastewater treatment processes, due to its impact on bacterial

Table 2. The power law exponent of taxa-time relationship for each phylum within two activated sludge bioreactors.

Phyla	Full-scale bioreactor			Lab-scale bioreactor		
	Relative abundance (%)[a]	*w*	r^2	*Relative abundance(%)*	*w*	r^2
Proteobacteria	25.8	0.33	0.951	43.1	0.34	0.954
Bacteroidetes	41.0	0.32	0.965	27.1	0.35	0.955
Acidobacteria	11.7	0.37	0.952	4.6	0.36	0.944
Chloroflexi	1.9	0.38	0.949	9.0	0.37	0.983
Verrucomicrobia	5.6	0.37	0.968	8.5	0.38	0.957
Planctomycetes	1.8	0.42	0.967	1.1	0.48	0.962
Firmicutes	2.5	0.42	0.975	3.7	0.46	0.967
Actinobacteria	0.3	0.41	0.971	0.6	0.47	0.971
Chlamydiae	0.6	0.46	0.954	0.3	0.52	0.966
Spirochaetes	0.6	0.46	0.969	0.4	0.54	0.958
Synergistetes	0.1	0.49	0.972	0.1	0.54	0.973
Fusobacteria	0.1	0.47	0.955	0.1	0.55	0.982
TM7	0.1	0.52	0.967	0.1	0.57	0.981
Euryarchaeota	0.1	0.53	0.981	0	0.55	0.979
OP10	0.2	0.51	0.977	0.1	0.56	0.978

[a]It is an average of the relative abundance of each phylum within 12 samples.

activity and the high operational costs of aeration, but little is known about the specific selection of distinct bacterial lineages by DO concentration [29]. The results of this study showed that DO had a significant effect in shaping the bacterial community structure in wastewater treatment systems. In two lab-scale bioreactors with high and low DO concentrations, Park *et al.* [30] have also demonstrated that DO concentration was an important structuring factor based on the T-RFLP analysis of bacterial community structures. Bioreactor scale is also one of the important factors mediating the bacterial community. To data, there are rare studies have examined the effect of bioreactor scale on the microbial community structure and diversity. In seven membrane bioreactors (MBR) of increasing size, van der Gast *et al.* [31] observed a significant linear relationship between bacterial taxa richness and reactor size, and they also found a gradient of greater evenness in community structure as MBR volume increased. Additional research is warranted to establish a firm understanding of how bioreactor sizes affect microbial communities in activated sludge systems.

The VPA results showed that 49% of the community variances were explained by these three components. Thus 51% of the community variance could not be explained by these three components. It is reasonable to expect that some unmonitored wastewater and operational variables may play an influential role in mediating bacterial community structures in WWTPs. In addition to the deterministic factors (wastewater and operational variables), the neutral factors (random immigration and births/deaths) may also affect the structure of the bacterial community. Ofiteru *et al.* [9] demonstrated that, in a full-scale WWTP, the variation of bacterial community was consistent with neutral community assembly, where chance and random immigration played an important and predictable role in shaping the communities. In four activated sludge bioreactors, Ayarza and Erijman [32] illustrated that both neutral and deterministic effects operated simultaneously in the assembly of bacterial floc. Similar results were also observed in other studies [33,34] [35]. The relative influence of deterministic environmental and stochastic factors in structuring microbial communities within bioreactors warrants future investigation.

We demonstrated that accumulation of observed bacterial taxa in activated sludge over time followed a power-law relationship, with a power-law exponent of $w = 0.43$ and 0.55 for a full- and lab-scale bioreactors, respectively. The values fell within the typical values determined previously. Shade *et al.* [16] conducted a meta-analysis of temporal dynamics in microbial communities, including 76 sites representing air, aquatic, soil, brewery wastewater treatment, human- and plant-associated microbial biomes. They found that there was a very consistent TTR across microbial communities, and the power law exponents ranged between 0.24 and 0.61. Our observed scaling exponent is also well in line with the few previous studies that have examined TTRs of bacterial taxa in activated sludge systems. van der Gast et al. [11] employed lab-scale reactors to examine the impact of different percentages of municipal and industrial wastewater on TTR exponent, and found that the power law exponent decreased pronounced from 0.512 to 0.162 as selective pressure (industrial wastewater concentration) increased. Wells *et al.* [10] detailed a power law exponent of $w = 0.209$ for bacterial communities in a full-scale WWTP, based on T-RFLP of bacterial 16S rRNA gene.

In our study, the power law exponents of dominant phyla were general lower than that of the rare phyla. Similar results were also obtained by Kim *et al.* [12], who reported that in a full-scale activated sludge bioreactor, the exponents for the general and rare bacterial taxa were 0.23 and 0.55, respectively.

The power law exponent of the full-scale bioreactor is lower than the value for the lab-scale bioreactor. It is tempting to speculate that this discrepancy is associated with the more than five order of magnitude variation in volume between these two reactors. Indeed, White *et al.* [18] has demonstrated that the exponent of TTR in macroecology was negatively correlated the spatial scale of observation, and van der *et al.* [11] suggested that this may be the case in microecolgy as well. Moreover, recent work has demonstrated the interdependence of spatial and temporal accumulation of species in the species-time-area relationship (STAR) in various systems [36]. Additional research is warranted to study the relationship between spatial and temporal turnover in microbial systems.

In conclusion, our results showed that accumulation of observed taxa in activated sludge over time followed a power-law relationship. The power-law exponent w for the full-scale bioreactor was 0.43, which is lower that of the lab-scale bioreactor ($w = 0.55$). The power law exponents of dominant phyla were general lower than that of the rare phyla within the two bioreactors. Overall, our results suggest that bacterial communities of activated sludge exhibited TTRs similar to those observed previously for plant and animal communities. These results highlight that a continued integration of microbial ecology into the broader field of ecology.

Author Contributions

Conceived and designed the experiments: XW RH. Performed the experiments: YW ZD. Analyzed the data: XW XL. Wrote the paper: XW.

References

1. Gentile ME, Jessup CM, Nyman JL, Criddle CS (2007) Correlation of functional instability and community dynamics in denitrifying dispersed-growth reactors. Appl Environ Microb 73: 680–690.
2. Rittmann BE, Hausner M, Loffler F, Love NG, Muyzer G, et al. (2006) A vista for microbial ecology and environmental biotechnology. Environ Sci Technol 40: 1096–1103.
3. Briones A, Raskin L (2003) Diversity and dynamics of microbial communities in engineered environments and their implications for process stability. Curr Opin Biotech 14: 270–276.
4. Wang X, Wen X, Xia Y, Hu M, Zhao F, et al. (2012) Ammonia oxidizing bacteria community dynamics in a pilot-scale wastewater treatment plant. Plos One 7: e36272.
5. Wells GF, Park HD, Yeung CH, Eggleston B, Francis CA, et al. (2009) Ammonia-oxidizing communities in a highly aerated full-scale activated sludge bioreactor: betaproteobacterial dynamics and low relative abundance of Crenarchaea. Environ Microbiol 11: 2310–2328.
6. Gentile ME, Nyman JL, Criddle CS (2007) Correlation of patterns of denitrification instability in replicated bioreactor communities with shifts in the relative abundance and the denitrification patterns of specific populations. ISME J 1: 714–728.
7. Slater FR, Johnson CR, Blackall LL, Beiko RG, Bond PL (2010) Monitoring associations between clade-level variation, overall community structure and ecosystem function in enhanced biological phosphorus removal (EBPR) systems using terminal-restriction fragment length polymorphism (T-RFLP). Water Res 44: 4908–4923.
8. Fernandez AS, Hashsham SA, Dollhopf SL, Raskin L, Glagoleva O, et al. (2000) Flexible community structure correlates with stable community function in

methanogenic bioreactor communities perturbed by glucose. Appl Environ Microb 66: 4058–4067.
9. Ofiteru ID, Lunn M, Curtis TP, Wells GF, Criddle CS, et al. (2010) Combined niche and neutral effects in a microbial wastewater treatment community. Proc Natl Acad Sci USA 107: 15345–15350.
10. Wells GF, Park HD, Eggleston B, Francis CA, Criddle CS (2011) Fine-scale bacterial community dynamics and the taxa-time relationship within a full-scale activated sludge bioreactor. Water Res 45: 5476–5488.
11. van der Gast CJ, Ager D, Lilley AK (2008) Temporal scaling of bacterial taxa is influenced by both stochastic and deterministic ecological factors. Environ Microbiol 10: 1411–1418.
12. Kim T-S, Jeong J-Y, Wells GF, Park H-D (2013) General and rare bacterial taxa demonstrating different temporal dynamic patterns in an activated sludge bioreactor. Appl Microbiol Biot 97: 1755–1765.
13. Shade A, Peter H, Allison SD, Baho DL, Berga M, et al. (2012) Fundamentals of microbial community resistance and resilience. Front Microbiol 3: 417.
14. Werner JJ, Knights D, Garcia ML, Scalfone NB, Smith S, et al. (2011) Bacterial community structures are unique and resilient in full-scale bioenergy systems. Proc Natl Acad Sci USA 108: 4158–4163.
15. Cocolin L, Bisson LF, Mills DA (2000) Direct profiling of the yeast dynamics in wine fermentations. Fems Microbiol Lett 189: 81–87.
16. Shade A, Gregory Caporaso J, Handelsman J, Knight R, Fierer N (2013) A meta-analysis of changes in bacterial and archaeal communities with time. ISME J 7: 1493–1506.
17. Adler PB, White EP, Lauenroth WK, Kaufman DM, Rassweiler A, et al. (2005) Evidence for a general species-time-area relationship. Ecology 86: 2032–2039.
18. White EP, Adler PB, Lauenroth WK, Gill RA, Greenberg D, et al. (2006) A comparison of the species-time relationship across ecosystems and taxonomic groups. Oikos 112: 185–195.
19. Matthews B, Pomati F (2012) Reversal in the relationship between species richness and turnover in a phytoplankton community. Ecology 93: 2435–2447.
20. Ye L, Shao MF, Zhang T, Tong AHY, Lok S (2011) Analysis of the bacterial community in a laboratory-scale nitrification reactor and a wastewater treatment plant by 454-pyrosequencing. Water Res 45: 4390–4398.
21. Wang X, Hu M, Xia Y, Wen X, Ding K (2012) Pyrosequencing analysis of bacterial diversity in 14 wastewater treatment systems in china. Appl Environ Microb 78: 7042–7047.
22. Marzorati M, Wittebolle L, Boon N, Daffonchio D, Verstraete W (2008) How to get more out of molecular fingerprints: practical tools for microbial ecology. Environ Microbiol 10: 1571–1581.
23. Wang XH, Wen XH, Yan HJ, Ding K, Zhao F, et al. (2011) Bacterial community dynamics in a functionally stable pilot-scale wastewater treatment plant. Bioresource Technol 102: 2352–2357.
24. Zhang T, Shao MF, Ye L (2012) 454 Pyrosequencing reveals bacterial diversity of activated sludge from 14 sewage treatment plants. ISME J 6: 1137–1147.
25. Hu M, Wang X, Wen X, Xia Y (2012) Microbial community structures in different wastewater treatment plants as revealed by 454-pyrosequencing analysis. Bioresource Technol 117: 72–79.
26. Wittebolle L, Vervaeren H, Verstraete W, Boon N (2008) Quantifying community dynamics of nitrifiers in functionally stable reactors. Appl Environ Microb 74: 286–293.
27. Siggins A, Enright AM, O'Flaherty V (2011) Temperature dependent (37-15 degrees C) anaerobic digestion of a trichloroethylene-contaminated wastewater. Bioresource Technol 102: 7645–7656.
28. Pholchan MK, Baptista JD, Davenport RJ, Curtis TP (2010) Systematic study of the effect of operating variables on reactor performance and microbial diversity in laboratory-scale activated sludge reactors. Water Res 44: 1341–1352.
29. Liu G, Wang J (2013) Long-Term Low DO Enriches and Shifts Nitrifier Community in Activated Sludge. Environ Sci Technol 47: 5109–5117.
30. Park HD, Noguera DR (2004) Evaluating the effect of dissolved oxygen on ammonia-oxidizing bacterial communities in activated sludge. Water Res 38: 3275–3286.
31. van der Gast CJ, Jefferson B, Reid E, Robinson T, Bailey MJ, et al. (2006) Bacterial diversity is determined by volume in membrane bioreactors. Environ Microbiol 8: 1048–1055.
32. Ayarza JM, Erijman L (2011) Balance of Neutral and Deterministic Components in the Dynamics of Activated Sludge Floc Assembly. Microbial Ecol 61: 486–495.
33. Curtis TP, Sloan WT (2006) Towards the design of diversity: stochastic models for community assembly in wastewater treatment plants. Water Sci Technol 54: 227–236.
34. Sloan WT, Lunn M, Woodcock S, Head IM, Nee S, et al. (2006) Quantifying the roles of immigration and chance in shaping prokaryote community structure. Environ Microbiol 8: 732–740.
35. Graham DW, Knapp CW, Van Vleck ES, Bloor K, Lane TB, et al. (2007) Experimental demonstration of chaotic instability in biological nitrification. ISME J 1: 385–393.
36. McGlinn DJ, Palmer MW (2009) Modeling the sampling effect in the species-time-area relationship. Ecology 90: 836–846.

14

Carbohydrate-Free Peach (*Prunus persica*) and Plum (*Prunus domestica*) Juice Affects Fecal Microbial Ecology in an Obese Animal Model

Giuliana D. Noratto[1☯¤], Jose F. Garcia-Mazcorro[2☯], Melissa Markel[3], Hercia S. Martino[1], Yasushi Minamoto[3], Jörg M. Steiner[3], David Byrne[4], Jan S. Suchodolski[3], Susanne U. Mertens-Talcott[1,5]*

1 Department of Nutrition and Food Science, Texas A&M University, College Station, Texas, United States of America, **2** Facultad de Medicina Veterinaria y Zootecnia, Universidad Autónoma de Nuevo León, General Escobedo, Nuevo León, México, **3** Gastrointestinal Laboratory, Texas A&M University, College Station, Texas, United States of America, **4** Department of Horticultural Sciences, Texas A&M University, College Station, Texas, United States of America, **5** Veterinary Physiology and Pharmacology, Texas A&M University, College Station, Texas, United States of America

Abstract

Background: Growing evidence shows the potential of nutritional interventions to treat obesity but most investigations have utilized non-digestible carbohydrates only. Peach and plum contain high amounts of polyphenols, compounds with demonstrated anti-obesity effects. The underlying process of successfully treating obesity using polyphenols may involve an alteration of the intestinal microbiota. However, this phenomenon is not well understood.

Methodology/Principal Findings: Obese Zucker rats were assigned to three groups (peach, plum, and control, n = 10 each), wild-type group was named lean (n = 10). Carbohydrates in the fruit juices were eliminated using enzymatic hydrolysis. Fecal samples were obtained after 11 weeks of fruit or control juice administration. Real-time PCR and 454-pyrosequencing were used to evaluate changes in fecal microbiota. Over 1,500 different Operational Taxonomic Units at 97% similarity were detected in all rats. Several bacterial groups (e.g. *Lactobacillus* and members of Ruminococcacea) were found to be more abundant in the peach but especially in the plum group (plum juice contained 3 times more total polyphenolics compared to peach juice). Principal coordinate analysis based on Unifrac-based unweighted distance matrices revealed a distinct separation between the microbiota of control and treatment groups. These changes in fecal microbiota occurred simultaneously with differences in fecal short-chain acids concentrations between the control and treatment groups as well as a significant decrease in body weight in the plum group.

Conclusions: This study suggests that consumption of carbohydrate-free peach and plum juice has the potential to modify fecal microbial ecology in an obese animal model. The separate contribution of polyphenols and non-polyphenols compounds (vitamins and minerals) to the observed changes is unknown.

Editor: Heidar-Ali Tajmir-Riahi, University of Quebect at Trois-Rivieres, Canada

Funding: This research was supported by the Vegetable and Fruit Improvement Center (http://vfic.tamu.edu/) at Texas A&M University. Hercia S. Martino thanks the National Counsel of Technological and Scientific Development (CNPq-Brazil) for the scholarship (reference number: 200382/2011-0). Jose F Garcia-Mazcorro is financially supported by CONACYT (Mexico) through the National System of Researchers (SNI, for initials in Spanish) program (http://www.conacyt.mx/index.php/el-conacyt/sistema-nacional-de-investigadores). The funders had no role in study design, data collection and analysis, decision to publish, or preparation of the manuscript.

* Email: SMTalcott@tamu.edu

☯ These authors contributed equally to this study.

¤ Current address: School of Food Science, Washington State University, Pullman, Washington, United States of America

Introduction

Obesity is a critical health issue worldwide affecting both industrialized and developing nations. Several factors have been associated with the increasing prevalence of obesity, including diminished physical exercise and an increased consumption of saturated fats and refined carbohydrates. Obesity is associated with multiple clinical complications and diseases including insulin resistance, hypertension, inflammation, oxidative stress, and dyslipidemia [1–4].

Polyphenols are a diverse group of compounds that are ubiquitous in the plant kingdom [5]. Over the last few years, the beneficial effects associated with the consumption of polyphenols have been widely studied [6–9]. Several *in vitro* and *in vivo* studies have demonstrated the anti-oxidant and anti-inflammatory activities of polyphenolics [10–12], some of which have also been shown to possess anti-lipidemic and anti-obesity effects, including suppression of adipogenesis and adipocyte proliferation, inhibition of fat absorption, as well as modulation of energy metabolism and inflammation [6,13]. Interestingly, a growing number of investigations suggest that dietary polyphenols can modulate the composition and metabolic activity of intestinal microorganisms [14–21], which may be, at least in part, involved in the underlying mechanisms for the associated health benefits. This hypothesis is supported by the close association between energy harvest, obesity, and the complex assembly of microorganisms residing in the intestinal tract [22,23].

Obesity has been linked to the composition of the gut microbiota but this relationship is not completely understood. Moreover, dietary interventions aiming to treat obesity have mostly focused on non-digestible carbohydrates [23]. Although the effect of polyphenols on the intestinal microbiota has been studied using culture and molecular techniques [15,16,24,25], research is needed to determine whether these widely available compounds are capable of modulating the gut microbiota in obese individuals. Additionally, the gut microbiota consists of hundreds of microbial taxa, an ecosystem that can only be fully approached using high-throughput sequencing systems. Unfortunately, very few papers are available that have made use of these technologies to obtain a better insight on the effect of polyphenolics-rich fruits on the intestinal microbiota [26].

The use of animal models is common to study the gut microbiota because mammals (humans included) share the most predominant gut phylotypes and therefore the obtained results may help guide future interventions, either dietary or therapeutic, in human populations. Zucker rats possess a mutation in the leptin receptor and develop metabolic syndrome symptoms, including insulin resistance and dyslipidemia, at 4–5 weeks of age. This animal model has been very well characterized as a model of obesity and therefore makes it attractive for studies of the gut microbiota [27]. The present study aimed to investigate the effect of carbohydrate-free peach and plum juice on fecal microbial ecology using obese Zucker rats as the animal model. Animals were assigned to three groups (peach, plum and control obese), the lean wild-type was used as control lean. Quantitative real-time PCR revealed a significantly higher abundance of the phylum Bacteroidetes, the family Ruminococcacea, and the genera *Faecalibacterium*, *Lactobacillus*, and *Turicibacter* in the plum group (3 times more polyphenolics than peach) when compared to the control and the lean groups. These changes were accompanied by a significant difference between control and treatment groups in principal coordinate analysis (based on Unifrac-based unweighted distance matrices), differences in fecal fatty acids among the animal groups as well as by a significantly lower body weight in the plum group.

Material and Methods

Ethics statement

Experiments were approved by the Institutional Animal Care and Use Committee at Texas A&M University (AUP#2010-138). This research complies with the 'Animal Research: Reporting of *In Vivo* Experiments' (ARRIVE) guidelines (Checklist S1) [28].

Study design

Male Zucker-Lepr^{fa}/Lepr^{+} heterozygotes rats were used to evaluate the effects of peach and plum juice on the obese fecal microbiota. The lean Zucker-Lepr+ (Wild Type) rats were used as negative controls. Animals were purchased from Harlan Laboratories (Houston, TX) at 5–6 weeks age and maintained in a ventilated rack system with food and water provided *ad libitum*. All obese Zucker rats were the same age and arrived at the same time in our laboratory. After an acclimation period of seven days, the obese Zucker rats were allocated to three groups (n = 10 each) namely control, peach, and plum. The wild type Zucker rat group (n = 10) was named lean. The control and lean groups received a control beverage containing water with glucose in the same concentration as the average concentration of reducing sugars in peach and plum juices (2.4%±0.1). Additionally, pH was adjusted to match the pH of juices using citric acid. Animals were housed in pairs (2 rats per cage) at 22–25°C under a 12 hours light cycle. All rats were visually inspected every day and body weight was recorded from all animals once a week.

Preparation of peach and plum juices

The commercial varieties "Angeleno" plum and "Crimson Lady" peach were collected at a mature, firm stage of development from commercial packing houses near Fresno, CA and shipped next day to the Department of Horticultural Sciences, Texas A&M University, College Station, TX. Fruits were stored at 4°C on the day of arrival whereby the stone was removed and the edible flesh stored at −80 °C until juice preparation. Peach and plum juices were prepared by enzymatic hydrolysis of pureed pulp obtained with a food processor. In brief, fruit puree was heated up to 90°C to inactivate polyphenoloxidase enzymes, cooled down to 50–55°C and subjected to enzymatic hydrolysis for 2 h with a mixture of food-grade enzymes multicellulase complex and hemicellulases (ValidaseTRL), pectin esterase, depolymerase, cellulases, hemicellulases, and arabinase (Crystalzyme 200XL) kindly supplied by Valley Research (South Bend, IN). After enzymatic hydrolysis, clarified peach and plum juices were obtained by centrifugation at 5000 rpm for 5 min.

Reducing sugars and total polyphenols

Reducing sugars were determined using dinitrosalicilic acid as a reagent against a standard curve of glucose [29]. Peach and plum juices contained 2.3±0.3% and 2.5±0.4% of reducing sugars respectively. Total polyphenols were quantified with Folin-Ciocalteu reagent (Fisher Scientific, Pittsburgh, PA) against a standard curve of gallic acid and expressed as mg gallic acid equivalents (GAE)/L [30]. Peach and plum juices contained 430±6.3 and 1,270±12.6 mg GAE/mL respectively.

Fecal collection and DNA extraction

Fresh fecal samples were obtained from all rats at the end of the study (11 weeks of consumption of sugary water or peach or plum juices) and stored at −80 C until analysis. Total DNA was extracted and purified from 100 mg of fecal sample using a bead-beating phenol-chloroform method as previously described [31].

Quantitative real-time PCR (qPCR)

The primary experimental outcome was the abundance of fecal microbiota, as determined by qPCR and pyrosequencing. qPCR analyses were performed to first investigate changes in specific bacterial groups among the animal groups. Briefly, PCR reaction mixtures (total of 10 µL) contained 5 µL of SsoFast EvaGreen supermix (Biorad Laboratories), 2.6 µL of water, 0.4 µL of each primer (final concentration: 400 nM), and 2 µL of adjusted (5 ng/µL) DNA. PCR conditions were 95°C for 2 min and 40 cycles at 95°C for 5 s and 10 s at the optimized annealing temperature (Table 1). A melt curve analysis was performed to verify the specificity of the primers using the following conditions: 1 min at 95°C, 1 min at 55°C, and 80 cycles of 0.5°C increments for 10 s each. Raw PCR data was normalized to the qPCR data for the total bacteria (universal primers F341 and R518) and all samples were run in duplicate as performed elsewhere [33].

Table 1. Oligonucleotides used in this study for qPCR analysis.

qPCR primers	Sequence (5′-3′)	Target	Annealing (°C)	Reference
UniF	CCTACGGGAGGCAGCAG	All bacteria	59	[32]
UniR	ATTACCGCGGCTGCTGG			
RumiF	ACTGAGAGGTTGAACGGCCA	Family Ruminococcaceae	59	[33]
RumiR	CCTTTACACCCAGTAAWTCCGGA			
FaecaliF	GAAGGCGGCCTACTGGGCAC	*Faecalibacterium*	60	[33]
FaecaliR	GTGCAGGCGAGTTGCAGCCT			
LacF	AGCAGTAGGGAATCTTCCA	*Lactobacillus*	58	[34]
LacR	CACCGCTACACATGGAG			
TuriciF	CAGACGGGGACAACGATTGGA	*Turicibacter*	63	[35]
TuriciR	TACGCATCGTCGCCTTGGTA			
CFB555f	CCGGAWTYATTGGGTTTAAAGGG	Bacteroidetes	60	[36]
CFB968r	GGTAAGGTTCCTCGCGTA			
BifF	TCGCGTCYGGTGTGAAAG	*Bifidobacterium*	60	[34]
BifR	CCACATCCAGCRTCCAC			

454-pyrosequencing

Bacterial tag-encoded FLX-titanium amplicon pyrosequencing (bTEFAP) was performed using the primers 28F (GAGTTTGATCNTGGCTCAG, forward) and 519R (GTNTTACNGCGGCKGCTG, reverse) targeting a semi-conserved region of the 16S rRNA gene at the Research and Testing Laboratory (Lubbock, TX). The Quantitative Insights in Microbial Ecology (QIIME) software platform (version 1.5.0) was used for processing and analysis of the sequences [37]. The process included chimera removal and denoising using UCHIME [38] and USEARCH [39], respectively, as well as removal of sequences that had low quality tags, primers, or ends, and failed to be at least 250 bp in length. The operational taxonomic units (OTUs) were defined as sequences with at least 97% similarity using the RDP classifier [40] in QIIME. Alpha and beta diversity measures were calculated using an equal number of sequences (2489, lowest number of sequences in a sample after removal of chimeric sequences) also using QIIME. Collection and sequence information has been submitted to the Sequence Read Archive (SRP029310).

Fecal fatty acids analysis

Short-chain fatty acids (SCFA) and branched-chain fatty acids (BCFA) were measured in fecal samples in order to obtain a better understanding of the effect of peach and plum juice on the metabolic activity of the intestinal microbiota. Concentrations of SCFA (acetate, propionate, butyrate), and BCFA (isobutyrate, isovalerate, valerate) in feces were measured using a stable isotope dilution gas chromatography-mass spectrometry (GC-MS) assay as previously described [41], with some modifications. Briefly, the fecal samples were weighed and diluted 1:5 in extraction solution (2N hydrochloric acid). After homogenization for 30 min at room temperature, fecal suspensions were centrifuged for 20 min at 2,100 g at 4°C. Supernatants were then collected using serum filters (Fisher Scientific Inc., Pittsburgh, Pa). Of each sample, 500 µl of supernatant were mixed with 10 µl of internal standard (200 mM heptadeuterated butyric acid) and extracted using a C18 solid phase extraction column (Sep-Pak C18 1 cc Vac Cartridge, Waters Corporation, Milford, MA). Samples were derivatized using *N*-tert-Butyldimethylsilyl-*N*-methyltrifluoroacetamide (MTBSTFA) at room temperature for 60 minutes. A gas chromatograph (Agilent 6890N, Agilent Technologies Inc, Santa Clara, CA) coupled with a mass spectrometer (Agilent 5975C, Agilent Technologies Inc, Santa Clara, CA) was used for chromatographic separation and quantification of the derivatized samples. Separation was achieved using a DB-1ms capillary column (Agilent Technologies Inc., Santa Clara, CA). The GC temperature program was as follows: 40°C held for 0.1 min, increased to 70°C at 5°C/min, 70°C held for 3.5 min, increased to 160°C at 20°C/min and finally increased to 280°C for 3 min at 35°C/min. The total run time was 20.5 min. The mass spectrometer was operated in electron impact positive-ion mode with selective ion monitoring at mass-to-charge ratios (M/Z) of 117 (acetate), 131 (propionate), 145 (butyrate and isobutyrate), 152 (deuterated butyrate; internal standard), and 159 (valerate and isovalerate). Quantification was based on the ratio of the area under the curve of the internal standard and each fatty acid. Results are reported as micromoles (µmol) per gram of wet feces.

Statistical analysis

The experimental unit in this study was individual rats. Pyrosequencing data was used to determine any significant differences to the control using an analysis of similarities (ANOSIM) on the unweighted Unifrac distance matrix in PAST [42]. An unweighted Pair Group Method with Arithmetic Mean (UPGMA) hierarchical clustering was generated using QIIME to visualize clustering of samples. Differences in relative proportions of sequences (including the Firmicutes/Bacteroidetes ratio), alpha diversity indices, fecal fatty acids, body weight, and qPCR data were analyzed using an analysis of variance (ANOVA) or its non-parametric counterpart Kruskal-Wallis using JMP 9.0.0 (SAS Institute Inc.), depending on sample size, type of data, and/or normality of the residuals from the ANOVA. Multiple comparisons were adjusted by the Tukey-Kramer or the Dunn's method. A $p<0.05$ was considered for statistical significance. QIIME, JMP and R (version 2.15.2) were used to generate graphs.

Results

Throughout the study (11 weeks) control and lean groups consumed an average of 50.6±8.7 and 46.0±7.9 mL water/animal-day, respectively, peach and plum groups consumed an average of 47.5±9.0 and 45.2±11.8 mL juice/animal-day respectively (Table S1). All rats remained clinically healthy during the study.

qPCR analyses

qPCR analyses were performed on 6 samples from the lean group, 8 samples from the obese control group, 7 samples from the peach group, and 9 samples from the plum group. The reason for using a subset of samples obeyed availability of fecal DNA for all analysis. The abundance of Bacteroidetes (phylum) and the genera *Faecalibacterium*, *Lactobacillus*, and *Turicibacter* were found to be significantly higher in the plum group when compared to all other groups ($p<0.05$; Figure 1). The abundance of the family Ruminococcaceae was found to be significantly higher in the plum group when compared to both the control and the lean groups. Additionally, Ruminococcaceae was also significantly higher in the peach group when compared to the control group (Figure 1).

bTEFAP

Pyrosequencing was performed in an effort to investigate differences in the overall phylogenetic composition of the fecal microbiota among the animal groups. For this analysis, we analyzed 4 fecal DNA samples from the obese control group, 4 samples from the peach group and 4 samples from the plum group. Additionally, we also included one fecal DNA sample from a lean subject but the results from this separate analysis of all samples (control obese, peach, plum and the lean subject) are only provided as supporting information (Figures S1–S3, Table S2). A total of 60,798 non-chimeric good-quality 16S rRNA gene sequences were analyzed (average: 5,067 ±1,666 sequences per sample). The fecal microbiota of all rats was composed by 1,549 OTUs (97% similarity) from 12 distinctive bacterial phyla. Despite the high bacterial diversity, only four phyla (Firmicutes, Bacteroidetes, Verrucomicrobia, and Proteobacteria) accounted for more than 90% of all the obtained sequences (Figure 2). The Firmicutes/Bacteroidetes ratio was not significantly different among the control obese and treatment groups ($p=0.209$, Figure 2).

A heat map of the most abundant OTUs (≥ 500 total in all samples analyzed) suggested differences in the relative abundance of various bacterial groups among the different animal groups (Figure 3) that confirmed the qPCR results (see above). Specifically, the relative abundance of OTUs from Turicibacteraceae was found to be high only in samples from the plum group. Moreover, most animals in the plum and the peach group had a high abundance of one unclassified Ruminococcaceae, and OTUs from several Bacteroidetes were also high only in the treatment groups (Figure 3). Despite these suggested dissimilarities in relative abundance of OTUs, there was no statistically significant difference in relative proportions of pyrosequencing reads (percentage of sequences) except for *Turicibacter* (Table S3), which was found to be significantly higher in both the peach and plum groups when compared to the control group. The genus

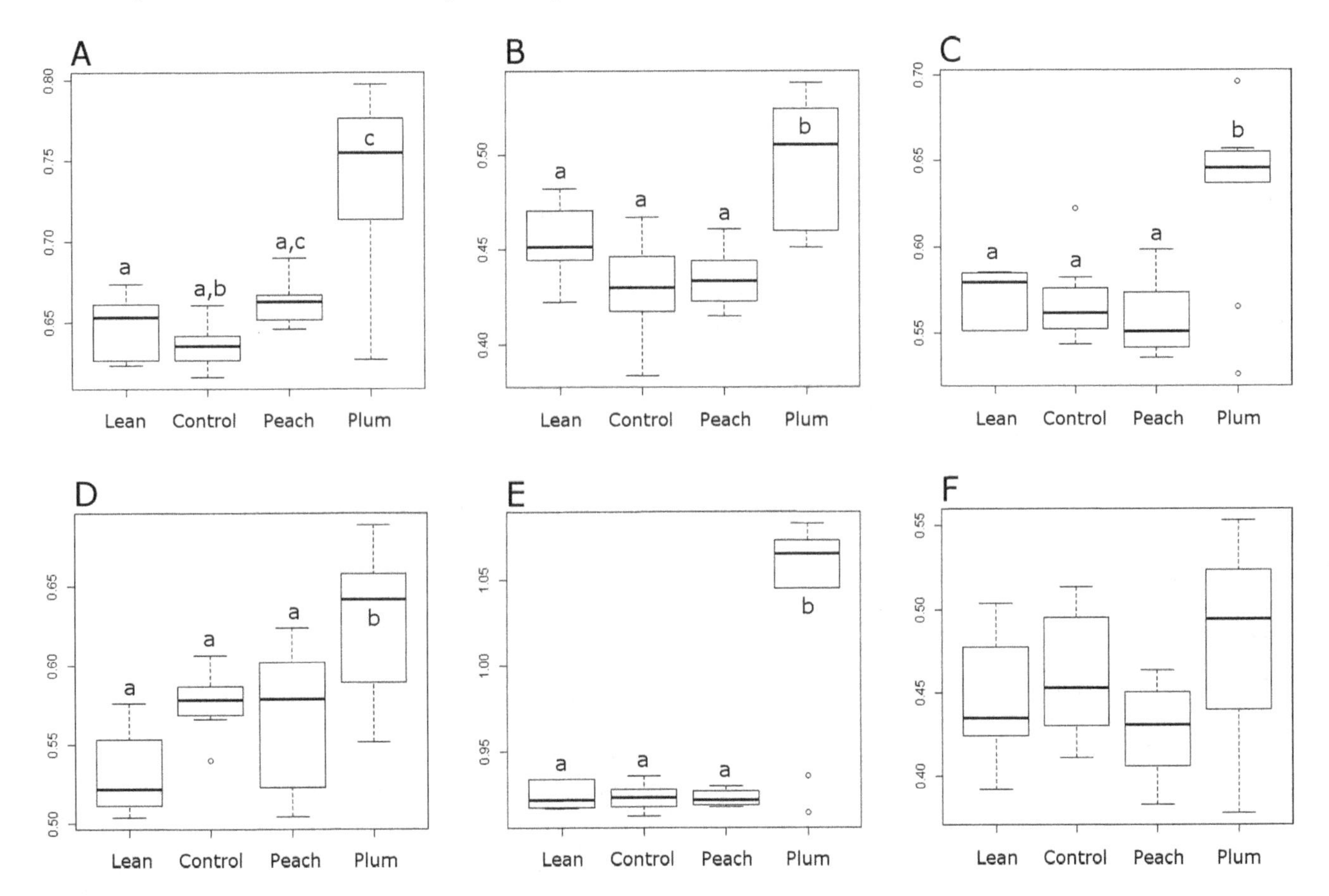

Figure 1. Quantitative real-time PCR results for Ruminococcaceae (family, A), *Faecalibacterium* (B), *Lactobacillus* (C), *Turicibacter* (D), Bacteroidetes (phylum, E) and *Bifidobacterium* (F) in the lean (n = 6), control obese (n = 8), peach (n = 7), and plum (n = 9) groups. Error bars represent the median and interquartile ranges (all results were normalized to qPCR data for total bacteria). Columns not sharing the same superscript are significantly different ($p<0.05$). *Significantly higher than all other groups.

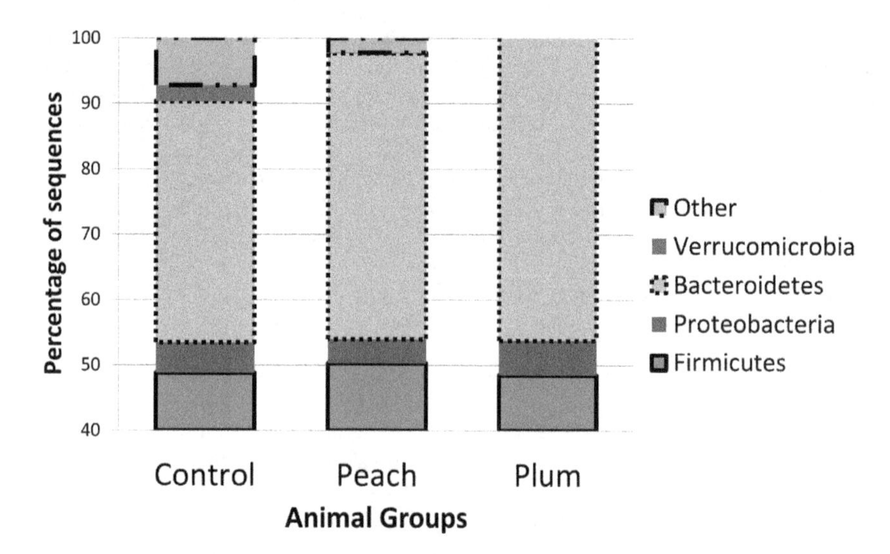

Figure 2. Composition of fecal microbiota in the control (n = 4), peach (n = 4) and plum (n = 4) groups at the phylum level. Bars represent median percentage of sequences. The y axis (percentage of sequences) was modified to also show the low abundant phyla.

Akkermansia (phylum Verrucomicrobia) was higher in the obese control group but this difference did not reach significance (p = 0.069, Table S3). Figure 4 illustrates the rarefaction curves for the control and the treatment groups. Alpha diversity indices were not significantly different among the animal groups (Table 2).

A Principal Coordinate Analysis (PCoA) analysis of the Unifrac-based unweighted distance matrices revealed useful information about the phylogenetic relationship among the fecal bacterial microbiota in the different animal groups (Figure 5). Most samples from the control obese group were separated from the peach and the plum samples in at least two of the combinations of coordinates (ANOSIM with 9999 permutations, p = 0.0012, Figure 5). It is known that when few independent factors are responsible for most of the variation, the first 2–3 coordinates explain most of the variation in the data [43]. In this study, the first three coordinates only described 41% of the variability, suggesting that many independent factors could have contributed to the observed variation in UniFrac distance values among the samples [43].

An UPGMA hierarchical clustering was created and suggested a distinctive clustering of all but one of the samples in the control group (75–100% jackknife support) (Figure 6). Expectedly, the sample from the control obese group that did not cluster with the rest of the control samples in the UPGMA hierarchical clustering was the same sample that remained independent in the PCoA analysis (Figure 5). There was not clear distinction (low jackknife support) among the samples from the peach and the plum groups (Figures 6), an observation that was also noted in the PCoA plots (Figure 5).

The analysis of all samples (control obese, treatment groups, and the one lean subject) revealed that the lean subject had higher indices of diversity and richness than any other sample analyzed (Table S2).

Fecal fatty acids analysis

Fecal fatty acids were measured in a subset of samples from the peach (n = 6), plum (n = 8), control (n = 6) and lean (n = 5) groups. The samples from the obese control group had a significantly higher concentration of acetic and propionic acid when compared to the plum and the lean group (acetic acid) and the peach and the lean group (propionic acid) ($p<0.05$), respectively (Table S4). All other fecal fatty acids, including butyric acid, were not significantly different among the animal groups (Table S4).

Body weight

Body weight at day 0 (beginning of experiment) was significantly different between the lean group and all other groups (data not shown). Animals in the plum group showed a significantly lower body weight (541.8 ±43.6) compared to control obese (644.4 ±39.3) and peach (611.1 ±39.4) group at week 11 (end of experiment, $p<0.05$, Table S1).

Discussion

There has been an increased interest in the characteristics and potential modifications of the intestinal microbiota to improve health in obese individuals. However, little information is available investigating the effect of potentially beneficial nutrients on the obese microbiota. To our knowledge, this study is the first to report the effect of peach and plum juices on the intestinal microbiota of obese rats using molecular tools, including a high-throughput sequencing technique.

Obese individuals have been reported to harbor a distinctive intestinal microbiota when compared to non-obese subjects. For example, Ley *et al.* showed a lower proportion of Bacteroidetes and a higher proportion of Firmicutes in obese mice when compared with lean mice [44]. Likewise, it has been suggested that obesity is related to phylum-level changes in the microbiota and reduced bacterial diversity [45]. However, others have found either no difference in the proportions of the main phyla or a change in proportions that seemed to contradict the original observations by Ley *et al.* [23]. In this study, qPCR analyses revealed statistically significant differences in the abundance of several fecal bacterial groups between the treatment (peach and plum) groups compared to the control and lean groups, but there was no difference between the lean and the obese control groups. The reasons for this lack of difference between lean and obese subjects are unknown but other authors have proposed a role of inter-individual differences, methods of sample preparation or methods of bacterial analysis [46].

The study of intestinal microorganisms and their relationship with fat metabolism and obesity has received increased attention

Control				Peach				Plum				
104	310	12	150	32	107	277	189	302	172	154	31	Alcaligenaceae
39	9	42	79	56	146	216	43	60	103	104	147	Bacteroidaceae
14	22	50	54	45	65	67	120	37	56	56	122	Bacteroidaceae
28	7	57	47	35	117	218	44	78	59	108	205	Bacteroidaceae
39	59	75	35	18	117	130	59	83	21	119	143	Bacteroidales
32	40	36	44	73	78	76	260	167	170	125	47	Bacteroidales
20	44	72	63	93	97	96	301	97	99	67	117	Bacteroidales
0	5	0	42	132	0	4	213	265	0	0	15	Bacteroidales
13	43	18	24	29	72	42	110	103	46	55	82	Bacteroidales
75	138	0	271	37	1155	249	353	32	0	0	150	Bacteroidales
33	40	34	21	43	66	52	159	103	70	47	103	Bacteroidales
12	26	56	11	56	52	60	29	81	39	42	69	Bacteroidales
1	8	10	13	16	67	22	42	71	108	69	95	Clostridiaceae
29	48	16	60	58	278	320	206	121	170	281	197	Clostridiaceae
22	105	78	13	120	221	35	145	87	42	78	51	Lactobacillaceae
46	370	242	45	99	153	78	103	179	215	263	22	Lactobacillaceae
97	6	1	5	0	10	115	3	103	2	127	214	Prevotellaceae
17	16	5	3	23	125	80	0	71	30	159	124	Prevotellaceae
55	30	33	31	36	34	90	99	53	44	45	68	Prevotellaceae
361	5	4	0	5	868	0	161	206	263	36	326	Ruminococcaceae
0	1	52	3	34	115	59	33	48	63	58	84	Ruminococcaceae
8	10	10	15	22	43	49	72	58	116	142	70	Turicibacteraceae
67	610	0	64	0	0	51	0	6	0	0	0	Verrucomicrobiaceae

Figure 3. Heat map showing the most abundant operational taxonomic units (OTUs, at least 500 total) in the control (n = 4), peach (n = 4) and plum (n = 4) groups. Colors represent differences in relative abundance within samples (red: higher; white: median; blue: lower).

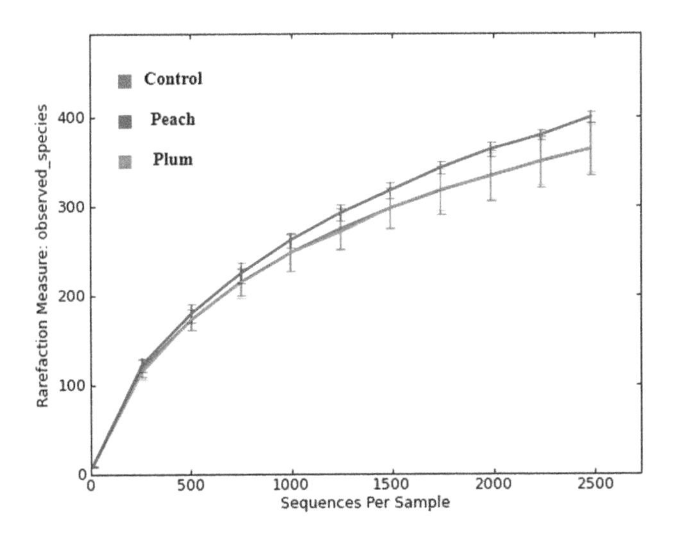

Figure 4. Rarefaction plots of 16S rRNA gene sequences obtained from fecal samples. Lines denote the average of each group; error bars represent the standard deviation. This analysis was carried out using a randomly selected 2489 sequences per sample.

over the last few years. However, little is known about how to successfully manipulate the obese gut microbiota, previous studies mainly used non-digestible carbohydrates [23]. Using an obese animal model, this study suggest that the polyphenolics in the juices played a role in the observed changes because the plum juice contained 3 times more polyphenolics and the differences in fecal microbial ecology and body weight were more marked in the plum group. For example, we found a higher abundance of *Turicibacter* in the plum group and this bacterial group has received increased attention because of its close relationship with the immune system of the host [47]. Also, we found a higher abundance of Bacteroidetes in the plum group. As mentioned above, Ley *et al.* [44] and others have shown that lean individuals generally carry a higher abundance of this group. Interestingly, in the plum group we also found a higher abundance of *Faecalibacterium* and *Lactobacillus*, important and abundant members of the phylum Firmicutes [48–49]. Moreover, we found differences in the abundance of the genus *Akkermansia* (phylum Verrucomicrobia), whose abundance has been shown to decrease in obese and type 2 diabetic mice [50]. It is important to note that our results about *Akkermansia* are somehow in disagreement with previous studies where a high abundance of this bacterial group is associated with health [50–51]. In our study, the relative abundance of *Akkermansia* was higher (although not statistically, $p = 0.069$) in obese rats and

Table 2. Median (minimum-maximum) indices of bacterial diversity (Shannon Weaver and Chao1 3%) and richness (OTUs 3%) obtained from fecal samples of the control, peach and plum groups. *P* values come from the non-parametric Kruskal-Wallis.

	Control (n = 4)	Peach (n = 4)	Plum (n = 4)	p value
Chao1	500 (434–553)	600 (547–616)	485 (463–576)	0.0592
Shannon	6.9 (6.5–7.4)	7.1 (6.6–7.5)	7.0 (6.8–7.4)	0.8741
OTUs	359 (329–370)	401 (381–413)	355 (341–401)	0.1238

These estimates are based on 2489-sequences subsamples.

the consumption of peach and plum extracts helped diminish its abundance (Table S3). This discrepancy may be explained by phenotypic differences among species within the genus or strains within the species as well as differences in the animal models utilized.

In order to obtain a better understanding of the effect of the peach and plum juices on the gut microbial ecosystem, we also measured SCFA and BCFA in fecal samples. Using an *in vitro* fecal culturing system, Bialonska *et al.* [18] showed that the inoculation of pomegranate polyphenols-rich extracts yielded significant increases in acetate, propionate and butyrate concentrations, as well as in the abundance of total bacteria, *Bifidobacterium* and *Lactobacillus* spp. Interestingly, the authors also inoculated the major pomegranate polyphenols (i.e., punicalagins) in the fecal cultures and did not observe changes in the abundance of fecal microorganisms and/or SCFA concentrations [18]. The authors of this study suggest that the effect of pomegranate extracts on fecal bacteria can be attributed to other non-punicalagins polyphenolics in pomegranate as well as glucose. Similarly, our data suggests that polyphenolics in the peach and plum juices have the potential to modify the composition of fecal SCFA concentrations *in vivo*. Moreover, the current study offers valuable information to the field of functional foods because carbohydrates were removed from the fruits. More detailed functional (metabolic) data, such as single-cell stable isotope probing, are necessary to research in more depth the complex bacterial interactions during the metabolism of polyphenolics inside the gut.

The cause of any difference in the fecal microbiota due to dietary polyphenols can be attributed to several factors. There is evidence suggesting that a proportion of dietary polyphenols can reach the large intestine in their original form [52–53], which are then subjected to microbial bioconversion [21]. Moreover, dietary polyphenols have the ability to inhibit the activity of pancreatic lipase, resulting in a reduced ability to absorb fat and consequently in a higher fecal fat content [54–56], and can promote fat oxidation and decrease lipogenesis [57]. Additionally, polyphenols are not considered as a primary energy source of microbial growth (compared to polysaccharides) [57] and possess both anti-microbial and growth-enhancing activities [15,58]. Therefore, the differences observed in this study may have arisen from the bioconversion of polyphenols by the gut microbiota, modifications of the lipid metabolism, as well as anti-microbial and growth-enhancing effects. More research, using purified polyphenols and whole extracts from polyphenolics-rich foods, is needed to understand more in depth gut microbial metabolism of polyphenols.

This study analyzed the effect of carbohydrate-free peach and plum juices on the obese fecal microbiota. However, the juices most likely contained other compounds aside the polyphenolics, such as vitamins and minerals, as peach and plum are known to contain high concentrations of these nutrients. Although it is known that several members of the intestinal microbiota are capable of utilizing and synthesizing vitamins [59–60], very little is known about the effect of these and other specific nutrients on the gut microbiota. Nonetheless, we cannot rule out the possibility that vitamins, minerals and/or other compounds in the juices could have had a contribution on the changes we observed.

The relevance of the current study to human or veterinary medicine is debatable. There are similarities in the gut microbiota of different mammals based on gut type and diet [61]. Mice and rats also share many physiological similarities with humans and other mammals, and studies in these animal species can therefore be useful to human and veterinary medicine. However, it is difficult for this and other studies to generalize about the

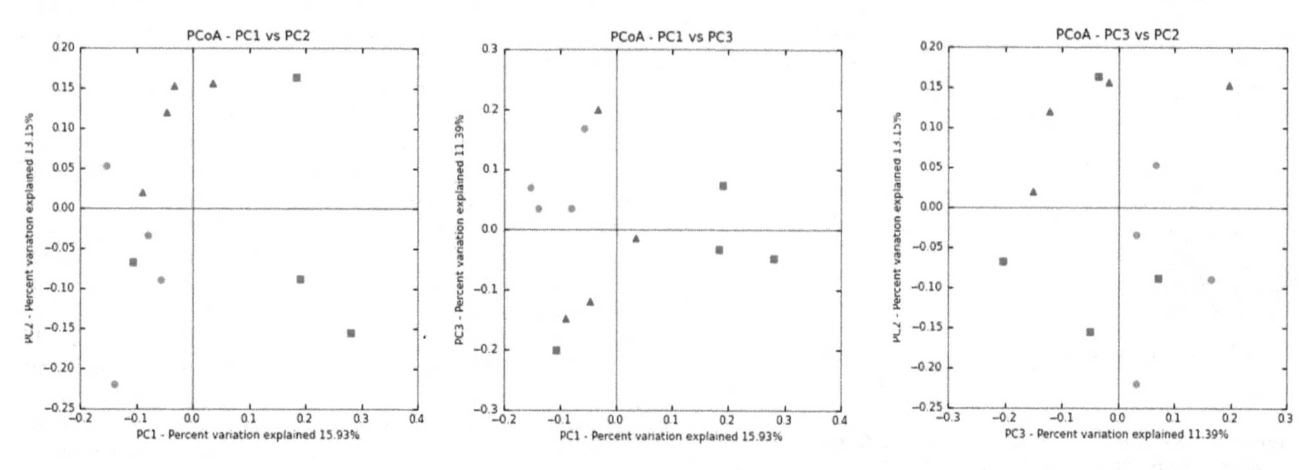

Figure 5. Principal Coordinate Analysis (PCoA) plots of the unweighted Unifrac distance matrix. The plots show each combination of the first three principal coordinates. Red (square): control; orange (circle): plum; blue (upright triangle): peach.

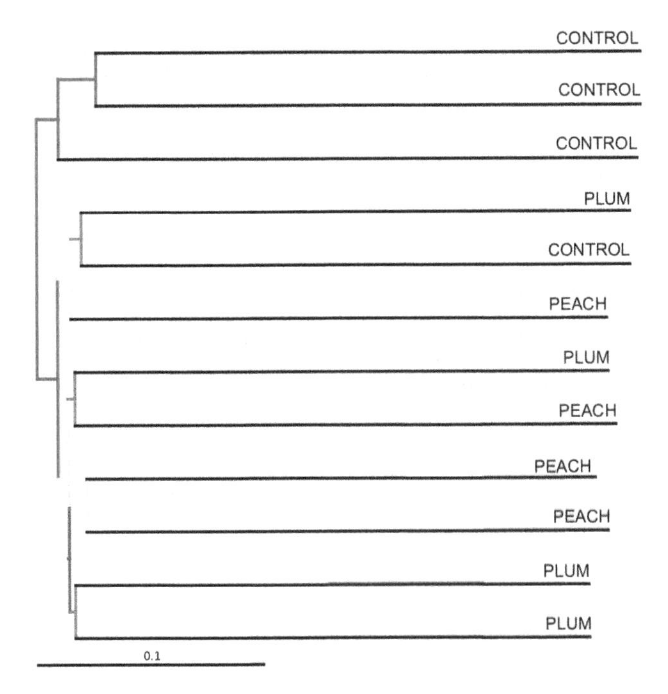

Figure 6. UPGMA hierarchical clustering using the unweighted Unifrac distance matrix. The colors represent different jackknife support: red (75–100% support); yellow (50–75%); green (25–50%); blue (<25% support). The bar represents community dissimilarity.

contribution of specific dietary nutrients to any change in the abundance or phylogenetic composition of the gut microbiota. For instance, in this study the prevention of weight gain could have been responsible for the changes in the microbiota instead or in addition to any change caused by direct microbial metabolism of the nutrients in the administered juices.

In summary, the current study suggests that the consumption of carbohydrate-free peach and plum juice has the potential to modify fecal bacterial composition in obese rats, as determined by qPCR and pyrosequencing. These changes occurred simultaneously with differences in fecal SCFA concentrations and a decrease in body weight in the plum group. Clinical research is needed to investigate the significance of our observations in preventing and treating human or veterinary patients with obesity.

Supporting Information

Figure S1 Principal Coordinate Analysis (PCoA) of the unweighted Unifrac distance matrix. The plots show each combination of the first three principal coordinates. Red (square): control; green (circle): plum; orange (horizontal triangle): peach; blue (upright triangle): lean.

Figure S2 UPGMA hierarchical clustering using the unweighted Unifrac distance matrix with the lean subject. The colors represent different jackknife support: red (75–100% support); yellow (50–75%); green (25–50%); blue (< 25% support). The bar represents community dissimilarity.

Figure S3 Heat map showing the most abundant operational taxonomic units (OTUs, at least 500 total) in one lean subject, control, peach and plum groups. Colors represent differences in relative abundance within samples (red: higher; white: median; blue: lower).

Table S1 Total body weight at the end of the study (11 weeks of consumption of peach or plum juices), juice consumption and polyphenolics content in the control, peach, plum, and lean animal groups.

Table S2 Median (minimum-maximum) indices of bacterial diversity (Shannon Weaver and Chao1 3%) and richness (OTUs 3%) obtained from fecal samples of one lean subject, control, peach and plum groups. P values come from the non-parametric Kruskal-Wallis.

Table S3 Median (minimum-maximum) relative proportions of pyrosequencing tags (percentage of sequences) for the control, peach, and plum groups. P values come from the non-parametric Kruskal Wallis test.

Table S4 Median (minimum-maximum) concentrations (µmol/g of wet feces) of short-chain fatty acids obtained from fecal samples of the control, peach, plum, and lean groups. P values come from the non-parametric Kruskal-Wallis.

Acknowledgments

The authors are grateful for the excellent support provided by the QIIME Help Forum (https://groups.google.com/forum/#!forum/qiime-forum) and to Dr. Nivedita Banerjee, Nutrition and Food Science Department at Texas A&M University, for her support with sample preparation.

Author Contributions

Conceived and designed the experiments: GDN HSM DB SUMT. Performed the experiments: GDN JFGM MM HSM YM. Analyzed the data: GDN JFGM JMS JSS SUMT. Contributed reagents/materials/analysis tools: JMS DB JSS SUMT. Wrote the paper: GDN JFGM JMS DB JSS SUMT.

References

1. Jellinger PS (2007) Metabolic consequences of hyperglycemia and insulin resistance. Clin Cornerstone 8: S30–42.
2. Holvoet P (2008) Relations between metabolic syndrome, oxidative stress and inflammation and cardiovascular disease. Verh K Acad Geneeskd Belg 70: 193–219.
3. Gallagher EJ, Leroith D, Karnieli E (2010) Insulin resistance in obesity as the underlying cause for the metabolic syndrome. Mt Sinai J Med 77: 511–523.
4. Niswender KD (2011) Basal insulin: beyond glycemia. Postgrad Med 123: 27–37.
5. Bravo L (1998) Polyphenols: chemistry, dietary sources, metabolism, and nutritional significance. Nutr Rev 56: 317–333.
6. Meydani M, Hasan ST (2010) Dietary polyphenols and obesity. Nutrients 2: 737–751.
7. Crozier A, Jaganath IB, Clifford MN (2009) Dietary phenolics: chemistry, bioavailability and effects on health. Nat Prod Rep 26: 1001–1043.
8. Chiva-Blanch G VF (2012) Polyphenols and health: moving beyond antioxidants. Journal of Berry Research 2: 63–71.
9. Hunter DC, Burritt DJ (2012) Polyamines of plant origin - An important dietary consideration for human health. In: Venketeshwer R, editor. Phytochemicals as nutraceuticals - Global approaches to their role in nutrition and health. Rijeka, Croatia: InTech. pp. 225–244.

10. Albrecht M, Jiang WG, Kumi-Diaka J, Lansky EP, Gommersall LM, et al. (2004) Pomegranate extracts potently suppress proliferation, xenograft growth, and invasion of human prostate cancer cells. J Med Food 7: 274–283.
11. Pacheco-Palencia LA, Talcott ST, Safe S, Mertens-Talcott S (2008) Absorption and biological activity of phytochemical-rich extracts from acai (Euterpe oleracea Mart.) pulp and oil in vitro. J Agric Food Chem 56: 3593–3600.
12. Mertens-Talcott SU, Rios J, Jilma-Stohlawetz P, Pacheco-Palencia LA, Meibohm B, et al. (2008) Pharmacokinetics of anthocyanins and antioxidant effects after the consumption of anthocyanin-rich acai juice and pulp (Euterpe oleracea Mart.) in human healthy volunteers. J Agric Food Chem 56: 7796–7802.
13. Lin JK, Lin-Shiau SY (2006) Mechanisms of hypolipidemic and anti-obesity effects of tea and tea polyphenols. Mol Nutr Food Res 50: 211–217.
14. Ahn YJ, Lee CO, Kweon JH, Ahn JW, Park JH (1998) Growth-inhibitory effects of Galla Rhois-derived tannins on intestinal bacteria. J Appl Microbiol 84: 439–443.
15. Lee HC, Jenner AM, Low CS, Lee YK (2006) Effect of tea phenolics and their aromatic fecal bacterial metabolites on intestinal microbiota. Res Microbiol 157: 876–884.
16. Parkar SG, Stevenson DE, Skinner MA (2008) The potential influence of fruit polyphenols on colonic microflora and human gut health. Int J Food Microbiol 124: 295–298.
17. Tzounis X, Vulevic J, Kuhnle GG, George T, Leonczak J, et al. (2008) Flavanol monomer-induced changes to the human faecal microflora. Br J Nutr 99: 782–792.
18. Bialonska D, Ramnani P, Kasimsetty SG, Muntha KR, Gibson GR, et al. (2010) The influence of pomegranate by-product and punicalagins on selected groups of human intestinal microbiota. Int J Food Microbiol 140: 175–182.
19. Tzounis X, Rodriguez-Mateos A, Vulevic J, Gibson GR, Kwik-Uribe C, et al. (2011) Prebiotic evaluation of cocoa-derived flavanols in healthy humans by using a randomized, controlled, double-blind, crossover intervention study. Am J Clin Nutr 93: 62–72.
20. Hidalgo M, Oruna-Concha MJ, Kolida S, Walton GE, Kallithraka S, et al. (2012) Metabolism of anthocyanins by human gut microflora and their influence on gut bacterial growth. J Agric Food Chem 60: 3882–3890.
21. Bolca S, Van de Wiele T, Possemiers S (2013) Gut metabotypes govern health effects of dietary polyphenols. Curr Opin Biotechnol 24: 220–225.
22. Turnbaugh PJ, Ley RE, Mahowald MA, Magrini V, Mardis ER, et al. (2006) An obesity-associated gut microbiome with increased capacity for energy harvest. Nature 444: 1027–1031.
23. Delzenne NM, Cani PD (2011) Interaction between obesity and the gut microbiota: relevance in nutrition. Annu Rev Nutr 31: 15–31.
24. Cueva C, Sanchez-Patan F, Monagas M, Walton GE, Gibson GR, et al. (2013) In vitro fermentation of grape seed flavan-3-ol fractions by human faecal microbiota: changes in microbial groups and phenolic metabolites. FEMS Microbiol Ecol 83: 792–805.
25. Queipo-Ortuno MI, Boto-Ordonez M, Murri M, Gomez-Zumaquero JM, Clemente-Postigo M, et al. (2012) Influence of red wine polyphenols and ethanol on the gut microbiota ecology and biochemical biomarkers. Am J Clin Nutr 95: 1323–1334.
26. Lacombe A, Li RW, Klimis-Zacas D, Kristo AS, Tadepalli S, et al. (2013) Lowbush wild blueberries have the potential to modify gut microbiota and xenobiotic metabolism in the rat colon. PLoS One 8: e67497.
27. Waldram A, Holmes E, Wang Y, Rantalainen M, Wilson ID, et al. (2009) Top-down systems biology modeling of host metabotype-microbiome associations in obese rodents. J Proteome Res 8: 2361–2375.
28. Kilkenny C, Browne WJ, Cuthill IC, Emerson M, Altman DG (2010) Improving bioscience research reporting: The ARRIVE guidelines for reporting animal research. PLoS Biol 8(6): e1000412.
29. Miller GL (1959) Use of dinitrosalicylic acid reagent for determination of reducing sugar. Anal Chem 31: 426–428.
30. Swain T, Hillis W (1959) The phenolic constituents of *Prinus domestica*. I. The quantitative analysis of phenolic constituents. J Sci Food Agr 10: 63–68.
31. Suchodolski JS, Xenoulis PG, Paddock CG, Steiner JM, Jergens AE (2010) Molecular analysis of the bacterial microbiota in duodenal biopsies from dogs with idiopathic inflammatory bowel disease. Vet Microbiol 142: 394–400.
32. Muyzer G, de Waal EC, Uitterlinden AG (1993) Profiling of complex microbial populations by denaturing gradient gel electrophoresis analysis of polymerase chain reaction-amplified genes coding for 16S rRNA. Appl Environ Microbiol 59: 695–700.
33. Garcia-Mazcorro JF, Suchodolski JS, Jones KR, Clark-Price SC, Dowd SE, et al. (2012) Effect of the proton pump inhibitor omeprazole on the gastrointestinal bacterial microbiota of healthy dogs. FEMS Microbiol Ecol 80: 624–636.
34. Malinen E, Rinttilä T, Kajander K, Mättö J, Kassinen A, et al. (2005) Analysis of the fecal microbiota of irritable bowel syndrome patients and healthy controls with real-time PCR. Am J Gastroenterol 100: 373–382.
35. Suchodolski JS, Markel ME, Garcia-Mazcorro JF, Unterer S, Heilmann RM, et al. (2012) The fecal microbiome in dogs with acute diarrhea and idiopathic inflammatory bowel disease. PLoS ONE 7(12): e51907.
36. Mühling M, Woolven-Allen J, Colin Murrell J, Joint I (2008) Improved group-specific PCR primers for denaturing gradient gel electrophoresis analysis of the genetic diversity of complex microbial communities. ISME J 2: 379–392.
37. Caporaso JG, Kuczynski J, Stombaugh J, Bittinger K, Bushman FD, et al. (2010) QIIME allows analysis of high-throughput community sequencing data. Nat Methods 7: 335–336.
38. Edgar RC, Haas BJ, Clemente JC, Quince C, Knight R (2011) UCHIME improves sensitivity and speed of chimera detection. Bioinformatics 27: 2194–2200.
39. Edgar RC (2010) Search and clustering orders of magnitude faster than BLAST. Bioinformatics 26: 2460–2461.
40. Cole JR, Wang Q, Cardenas E, Fish J, Chai B, et al. (2009) The Ribosomal Database Project: improved alignments and new tools for rRNA analysis. Nucleic Acids Res 37: D141–145.
41. Moreau NM, Goupry SM, Antignac JP, Monteau FJ, Le Bizec BJ, et al. (2003) Simultaneous measurement of plasma concentrations and 13C-enrichment of short-chain fatty acids, lactic acid and ketone bodies by gas chromatography coupled to mass spectrometry. J Chromatogr B Analyt Technol Biomed Life Sci 784: 395–403.
42. Hammer Ø, Harper DAT, Ryan PD (2001) PAST: Paleontological statistics software package for education and data analysis. Paleotologia Electronica 4: 9.
43. Lozupone C, Knight R (2005) UniFrac: a new phylogenetic method for comparing microbial communities. Appl Environ Microbiol 71: 8228–8235.
44. Ley RE, Backhed F, Turnbaugh P, Lozupone CA, Knight RD, et al. (2005) Obesity alters gut microbial ecology. Proc Natl Acad Sci USA 102: 11070–11075.
45. Turnbaugh PJ, Hamady M, Yatsunenko T, Cantarel BL, Duncan A, et al. (2009) A core gut microbiome in obese and lean twins. Nature 457: 480–484.
46. Duncan SH, Lobley GE, Holtrop G, Ince J, Johnstone AM, et al. (2008) Human colonic microbiota associated with diet, obesity and weight loss. Int J Obesity 32: 1720–1724.
47. Presley LL, Wei B, Braun J, Borneman J (2010) Bacteria associated with immunoregulatory cells in mice. Appl Environ Microbiol 76: 936–941.
48. Sokol H, Pigneur B, Watterlot L, Lakhdari O, Bermudez-Humaran LG, et al. (2008) *Faecalibacterium prausnitzii* is an anti-inflammatory commensal bacterium identified by gut microbiota analysis of Crohn disease patients. Proc Natl Acad Sci USA 105: 16731–16736.
49. van Baarlen P, Troost F, van der Meer C, Hooiveld G, Boekschoten M, et al. (2011) Human mucosal *in vivo* transcriptome responses to three lactobacilli indicate how probiotics may modulate human cellular pathways. Proc Natl Acad Sci U S A 108 Suppl 1: 4562–4569.
50. Everard A, Belzer C, Geurts L, Ouwerkerk JP, Druart C, et al. (2013) Cross-talk between *Akkermansia muciniphila* and intestinal epithelium controls diet-induced obesity. Proc Natl Acad Sci USA 110: 9066–9071.
51. Png CW, Linden SK, Gilshenan KS, Zoetendal EG, McSweeney CS, et al. (2010) Mucolytic bacteria with increased prevalence in IBD mucosa augment in vitro utilization of mucin by other bacteria. Am J Gastroenterol 105: 2420–2428.
52. van Dorsten FA, Grun CH, van Velzen EJ, Jacobs DM, Draijer R, et al. (2010) The metabolic fate of red wine and grape juice polyphenols in humans assessed by metabolomics. Mol Nutr Food Res 54: 897–908.
53. van Duynhoven J, Vaughan EE, Jacobs DM, Kemperman RA, van Velzen EJ, et al. (2011) Metabolic fate of polyphenols in the human superorganism. Proc Natl Acad Sci USA 108 Suppl 1: 4531–4538.
54. Ikeda I, Tsuda K, Suzuki Y, Kobayashi M, Unno T, et al. (2005) Tea catechins with a galloyl moiety suppress postprandial hypertriacylglycerolemia by delaying lymphatic transport of dietary fat in rats. J Nutr 135: 155–159.
55. Nakai M, Fukui Y, Asami S, Toyoda-Ono Y, Iwashita T, et al. (2005) Inhibitory effects of oolong tea polyphenols on pancreatic lipase in vitro. J Agric Food Chem 53: 4593–4598.
56. Lei F, Zhang XN, Wang W, Xing DM, Xie WD, et al. (2007) Evidence of anti-obesity effects of the pomegranate leaf extract in high-fat diet induced obese mice. Int J Obes (Lond) 31: 1023–1029.
57. Klaus S, Pultz S, Thone-Reineke C, Wolfram S (2005) Epigallocatechin gallate attenuates diet-induced obesity in mice by decreasing energy absorption and increasing fat oxidation. Int J Obes (Lond) 29: 615–623.
58. Coccia A, Carraturo A, Mosca L, Masci A, Bellini A, et al. (2012) Effects of methanolic extract of sour cherry (*Prunus cerasus* L.) on microbial growth. Int J Food Sci Technol 47: 1620–1629.
59. Ly NP, Litonjua A, Gold DR, Celedon JC (2011) Gut microbiota, probiotics, and vitamin D: interrelated exposures influencing allergy, asthma, and obesity? J Allergy Clin Immunol 127: 1087–1094.
60. LeBlanc JG, Milani C, de Giori GS, Sesma F, van Sinderen D, et al. (2013) Bacteria as vitamin suppliers to their host: a gut microbiota perspective. Curr Opin Biotechnol 24: 160–168.
61. Ley RE, Hamady M, Lozupone C, Turnbaugh PJ, Ramey RR, et al. (2008) Evolution of mammals and their gut microbes. Science 320: 1647–1651.

15

Metabolic Flexibility as a Major Predictor of Spatial Distribution in Microbial Communities

Franck Carbonero[1,2], Brian B. Oakley[1,3], Kevin J. Purdy[1]*

1 School of Life Sciences, University of Warwick, Coventry, United Kingdom, 2 Department of Food Science, University of Arkansas, Fayetteville, Arkansas, United States of America, 3 United States Department of Agriculture, Agricultural Research Service, Richard B. Russell Research Center, Athens, Georgia, United States of America

Abstract

A better understand the ecology of microbes and their role in the global ecosystem could be achieved if traditional ecological theories can be applied to microbes. In ecology organisms are defined as specialists or generalists according to the breadth of their niche. Spatial distribution is often used as a proxy measure of niche breadth; generalists have broad niches and a wide spatial distribution and specialists a narrow niche and spatial distribution. Previous studies suggest that microbial distribution patterns are contrary to this idea; a microbial generalist genus (*Desulfobulbus*) has a limited spatial distribution while a specialist genus (*Methanosaeta*) has a cosmopolitan distribution. Therefore, we hypothesise that this counter-intuitive distribution within generalist and specialist microbial genera is a common microbial characteristic. Using molecular fingerprinting the distribution of four microbial genera, two generalists, *Desulfobulbus* and the methanogenic archaea *Methanosarcina*, and two specialists, *Methanosaeta* and the sulfate-reducing bacteria *Desulfobacter* were analysed in sediment samples from along a UK estuary. Detected genotypes of both generalist genera showed a distinct spatial distribution, significantly correlated with geographic distance between sites. Genotypes of both specialist genera showed no significant differential spatial distribution. These data support the hypothesis that the spatial distribution of specialist and generalist microbes does not match that seen with specialist and generalist large organisms. It may be that generalist microbes, while having a wider potential niche, are constrained, possibly by intrageneric competition, to exploit only a small part of that potential niche while specialists, with far fewer constraints to their niche, are more capable of filling their potential niche more effectively, perhaps by avoiding intrageneric competition. We suggest that these counter-intuitive distribution patterns may be a common feature of microbes in general and represent a distinct microbial principle in ecology, which is a real challenge if we are to develop a truly inclusive ecology.

Editor: Douglas Andrew Campbell, Mount Allison University, Canada

Funding: This study was performed as part of an EU Commission funded Marie Curie Excellence Grant for Teams, MicroComXT (MEXT-CT-2005-024112) to KJP. The funders had no role in study design, data collection and analysis, decision to publish, or preparation of the manuscript.

Competing Interests: The authors have declared that no competing interests exist.

* E-mail: k.purdy@warwick.ac.uk

Introduction

In order to better understand the ecology of microbes and their role in the global ecosystem, it is essential to determine whether ecological ideas and theories that have been derived from studies on plants and animals are also applicable to microbes [1–4]. However, it may be that, due to differences in scale and physiologies between micro- and macroorganisms, there will be principles of ecology that are difficult to reconcile between the two. Developing an inclusive ecology represents a substantial challenge as for many years a primary assumption in microbial ecology was that while all microbes could be found in all environments, community structures were shaped by local environmental factors, exemplified in Baas-Becking's statement "*Everything is everywhere, but the environment selects*" [5]. In contrast traditional ecology, where endemism and biogeography are the norm, can be exemplified by the ecological truism that "*all species are always absent from almost everywhere*" [6]. However, microbial communities have been shown to have distinct spatial distribution patterns [7–11], yet it is unclear what processes structure microbial communities or whether ecological ideas, such as biogeography [12,13], are readily applicable to microbes.

The direct application of ecological theory to microbes is difficult because of the complexity of defining microbial taxonomic groups and the extent of microbial genetic and phenotypic diversity [14,15]. These problems particularly confound the whole bacterial assemblage analyses that are commonly used in microbial ecology. Such broad-brush studies have no equivalent in traditional ecology where studies are usually constrained within tight taxonomic boundaries. Ecologists avoid the problems common to studies in microbial ecology as they use these model species/genera to represent larger groups [16,17]. Such an approach has rarely been applied to microorganisms primarily because of the difficulty in defining homogeneous species. This issue can be avoided by studying model microbial genera if the selected genera are phylogenetically and functionally homogeneous [18]. Therefore, to study the environmental distribution of microbes we selected model genera that fitted these specific criteria and could also be defined as either specialists or generalists.

A specialist or generalist organism is most succinctly defined as having either a narrow or wide potential niche respectively. However, defining the actual extent of an organism's niche, the n-dimensional space that affects an organism's growth and survival [19], is clearly extremely challenging, if not impossible. Therefore, within traditional ecology spatial distribution has become an accepted proxy measure of niche breadth for many taxa, so it is assumed that a specialist organism with a narrow niche will have a limited spatial distribution and a generalist a wide spatial distribution [20–22]. In microbial ecology such assumptions, which underpin many ideas in ecology, have never been properly tested. There are microbes that utilise a vast array of carbon sources for energy and growth and are also capable of respiring more than one type of compound, thus are metabolically flexible with multiple electron donors and acceptors and so can confidently be defined as generalists with wide potential niches. Conversely, there are microbes that have very specific and limited metabolic needs and are clearly highly specialised with narrow potential niches. Thus, metabolic flexibility represents a fundamentally important aspect of a microbe's ecological function and, we propose, represents a reasonable proxy for niche breadth. Therefore, if the spatial distribution of metabolically generalist and specialist microbial model genera is a good proxy for niche breadth this would be a strong indicator that some rules that govern the distribution of large organisms such as plants and animals can also be applied to microbes.

We tested the idea that the extent of spatial distribution along an estuarine gradient is related to metabolic flexibility in two model microbial genera that are both distributed along the full length of the Colne estuary, Essex, UK [23,24]. Estuaries are natural environmental gradients that have been used to show spatial and evolutionary differentiation in macro-organisms [17,25] and are hotspots of biogeochemical cycling and microbial activity [26]. The model genera were the sulfate-reducing bacteria *Desulfobulbus*, a generalist genus that can respire both sulfur and nitrogen oxyanions, use fermentation for growth and metabolise a range of carbon sources [27,28] and the methanogenic archaeal genus *Methanosaeta*, a metabolic specialist that uses acetate as both electron acceptor and donor and as its sole carbon source [29]. Genotypes (at approximately the species level) of the generalist *Desulfobulbus* genus were spatially restricted to distinct regions of the estuarine salinity gradient in a manner similar to that classically described for estuarine macrofauna [17] while genotypes of the specialist *Methanosaeta* showed no such differential distribution and were monotonically distributed along the estuary [8,30–32].

These data challenge assumptions from traditional ecology that spatial distribution is a proxy measure of niche breadth leading us to propose the hypothesis that, in contrast to traditional ecological ideas, the breadth of the spatial distribution of specialist and generalist microbes is inversely related to the breadth of their potential niches. However, this hypothesis is derived from data from only two genera that are from different Domains and have very different metabolisms and so we cannot exclude the possibility that the differences seen in their distributions are due to their phylogenetic or metabolic differences and not differences in their metabolic flexibility. Therefore, to test the hypothesis above and to determine whether metabolic flexibility in general does influence the spatial distribution of microbes we expanded this study to include the most generalist methanogenic archaeal genus, *Methanosarcina*, and the highly specialised sulfate-reducing bacterial genus, *Desulfobacter*. *Methanosarcina* is the only methanogen genus able to perform all three methanogenic respiratory pathways [33] while *Desulfobacter* are SRB that respire sulfur oxyanions only, primarily whilst oxidising acetate [34]. Analysis of these additional genera produces a dataset that includes a generalist and specialist genus from both the bacterial and the archaeal Domains and that are methanogens and sulphate-reducers and so will address whether the specific metabolism or phylogeny of these four model genera or indeed their metabolic flexibility is a driver of the counter-intuitive differential distribution of *Methanosaeta* and *Desulfobulbus* detected previously.

Results

16S rRNA gene fragments, amplified by genus-specific PCR (Table S1 in File S1), were used to analyse the distribution of all four genera in sediment along the full length of the Colne estuary, Essex, UK (Figure S1) using the molecular fingerprinting method Denaturing Gradient Gel Electrophoresis (DGGE; Supplementary information). Profiles of representative bands are shown in Figure 1 (for complete profiles see Figure S3). Genotype distribution patterns from metabolic generalist (*Methanosarcina* and *Desulfobulbus*) and specialist (*Methanosaeta* and *Desulfobacter*) model genera were highly dissimilar. Both generalist models showed a restricted spatial distribution, in which particular genotypes were only found in certain regions of the estuary, while both specialist models showed no such restricted distribution, with all bands detected all along the estuary.

Cluster analysis of both DGGE band presence/absence (Jaccard) and total DGGE profile (Pearson) indicated geographically coherent clustering for the generalist models, *Desulfobulbus* and *Methanosarcina* (Figure 2A and B). The specialist models,

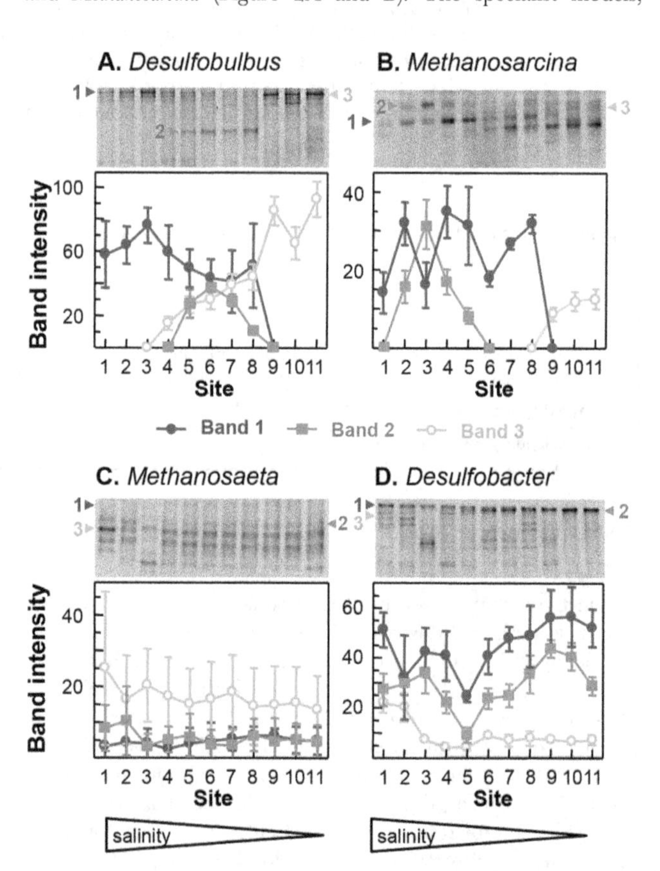

Figure 1. Corrected mean band intensities of representative bands from replicated DNA-DGGE analyses from the 11 sites (from the marine site 1 to the freshwater site 11) along the Colne estuary, Essex, UK for: A. *Desulfobulbus,* B. *Methanosarcina,* C. *Methanosaeta* and D. *Desulfobacter.* Numbered arrows on each gel image indicate band numbers. [Band 1 (line with solid circle) Band 2 (line with open square) Band 3(line with open circle)]. Error bars represent the standard error of the mean (SEM, n = 3). A complete profile is given in Figure S3.

Methanosaeta and *Desulfobacter*, showed inconsistent and geographically incoherent clustering with generally high overall similarities between sites (>59%) across the two methods (Figure 2C and D). Unconstrained Canonical Correspondence Analysis (CCA) of the DGGE fingerprints showed some clustering according to site for all genera (Figure S4). However, very low eigenvalues (scales on the CCA; 0.015–0.048) in the CCA of the specialist genera (Figure S4C and D) indicated a poor correlation between genotypic dissimilarity matrixes and environmental variables while eigenvalues (0.181–0.680) were higher for the generalist models (Figure S4A and B). Mantel and partial Mantel tests revealed a significant correlation ($p<0.05$) between the geographic distance between sites and distribution of the generalist models (Table S2A and B in File S1) whether based on Band Intensity, Jaccard or Pearson analyses. Very limited significant correlations were found for *Desulfobacter* and *Methanosaeta* (Table S2C and D in File S1), with no clear coherent correlations across all three methods of analysis.

Analysis of RNA (in this case 16S rRNA) provides an indication that the organisms detected are not only present but also active in an environment. Banding patterns in the RNA-DGGE analysis of both *Desulfobulbus* and *Methanosaeta* (Figure S5) were very similar to those seen in the DNA-DGGE (Figure 1 and Figure S3). As with DNA-DGGE *Desulfobulbus* had a restricted pattern of spatial distribution, while *Methanosaeta* genotypes were constantly detected all along the estuary [30]. Cluster analyses showed spatially coherent clustering for *Desulfobulbus* clusters (Fig 3A) and a weak but inconsistent clustering for *Methanosaeta* (Fig 3B). Unconstrained CCAs showed strong geographic clustering for *Desulfobulbus* (Eigen values 0.381–0.628; Figure S6A) and no clustering for *Methanosaeta* (Eigen values 0.075–0.138; Figure S6B). Mantel and partial Mantel tests (Table S3 in File S1) revealed a clearly significant correlation between geographic distance and the *Desulfobulbus* distribution pattern ($p<0.05$). While all Mantel tests were significant for *Methanosaeta* ($p<0.05$) no coherent correlation could be seen with partial Mantel tests. Importantly, DNA- and RNA-based *Desulfobulbus* distribution patterns were very similar, as observed previously for *Methanosaeta* [30]. This implies that detected genotypes (from analysis of DNA) are active (from analysis of RNA), and thus DNA based DGGE fingerprints represent a satisfactory representation of metabolically active populations. Pyrosequence analysis of functional genes from *Desulfobulbus* (*dsr*A) and *Methanosaeta* (*mcr*A) along the Colne estuary also shows a clear difference in distribution patterns between these two model genera, supporting the conclusions above [32].

These data show that metabolic flexibility does appear to directly affect the distribution of microbial genera; metabolic generalists (*Desulfobulbus* and *Methanosarcina*) have a specialist spatial distribution strongly correlated with environmental variables and metabolic specialists (*Methanosaeta* and *Desulfobacter*) have a generalist spatial distribution along an estuarine gradient that is not correlated to environmental factors.

Discussion

Here we show that within microbial genera metabolic flexibility appears to have a profound effect on the spatial distribution patterns of those genera, with members of metabolic generalist genera showing a narrower spatial distribution than metabolic specialist genera. These seemingly counter-intuitive distribution patterns suggest that microbes are distributed in ways that differ from most plants and animals and that these differences would have to be accounted for in a truly inclusive ecology. The comparison described here, using two specialist and two generalist genera from across both archaeal and bacterial domains and both the terminal oxidation processes of sulphate reduction and methanogenesis, suggests that neither phylogeny nor metabolism

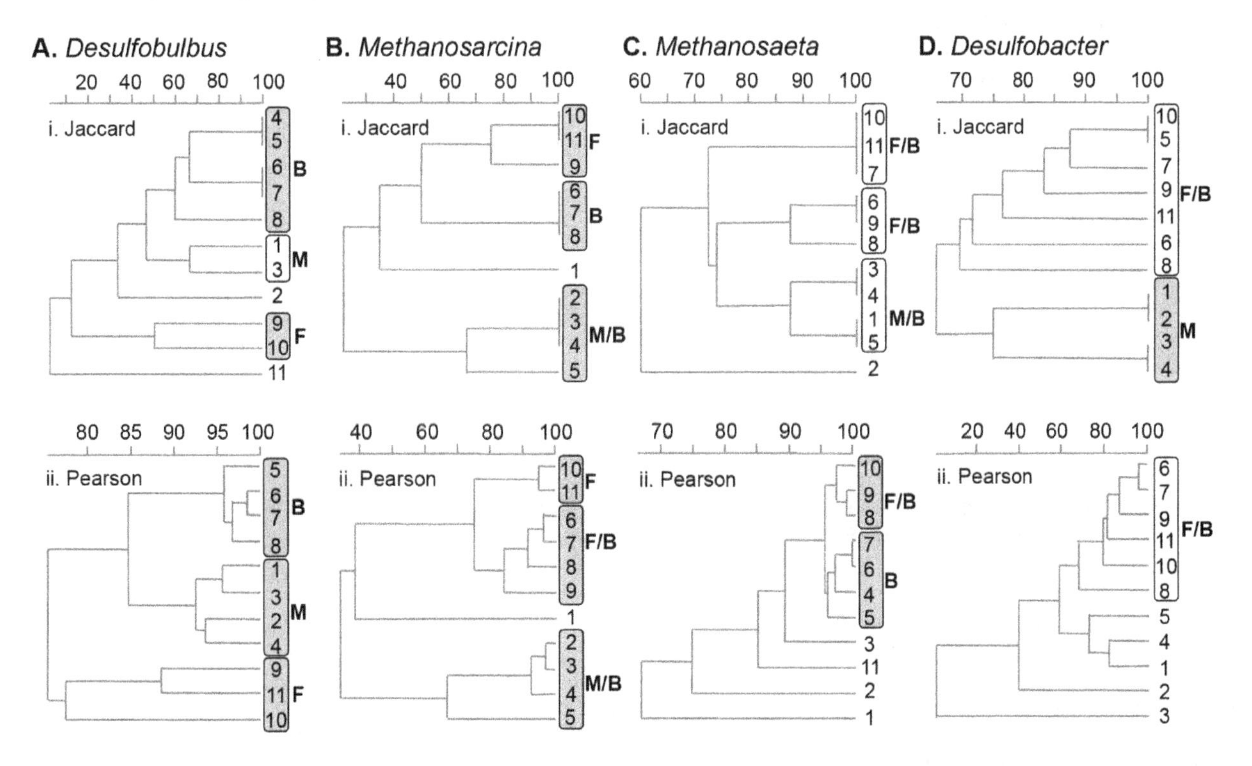

Figure 2. Cluster analyses of the DNA-DGGE profiles: i. Jaccard and ii. Pearson analyses of A. *Desulfobulbus*; B. *Methanosarcina*; C. *Methanosaeta*; D. *Desulfobacter*. Marine (M), Marine/Brackish (M/B), Brackish (B), Freshwater/Brackish (F/B) and Freshwater (F) clusters are circled. Shaded clusters are geographically continuous, unshaded clusters are not.

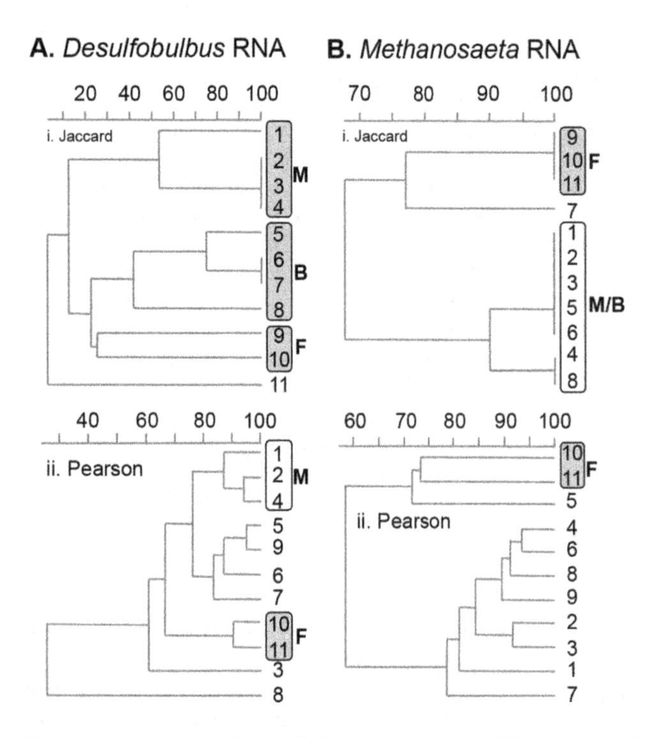

Figure 3. Cluster analyses of the RNA-DGGE profiles: i. Jaccard and ii. Pearson analyses of A. *Desulfobulbus*; B. *Methanosarcina*. Marine (M), Marine/Brackish (M/B), Brackish (B), Freshwater/Brackish (F/B) and Freshwater (F) clusters are circled. Shaded clusters are geographically continuous, unshaded clusters are not.

are major driving forces in the detected distribution of these four genera.

The distribution of the generalist methanogen genus *Methanosarcina* was assessed for the first time, to our knowledge, indicating a differential distribution pattern with genotypes restricted to marine, marine-brackish, brackish-freshwater and freshwater regions of the estuary. *Methanosarcina* strains have been isolated from various freshwater and marine environments and isolates are able to use at least two of the three methanogenic pathways (acetoclastic, hydrogenotrophic and methylotrophic methanogenesis), but with important differences in their substrate preference [33]. In an estuary, acetate concentrations are higher in freshwater regions [35] and decrease substantially due to consumption by SRB where sulfate is freely available, which is generally in all but the most freshwater sediments [36]. *Methanosarcina* are known to use one-carbon compounds, such as methylamines, that are more abundant in marine environments [37]. Thus, both the salinity gradient and the variable availability of different substrates (such as acetate and methylamines) along the estuarine gradient could be strong drivers of *Methanosarcina*'s differential spatial distribution, in a similar way to the distribution of *Desulfobulbus* along the estuary [8,31]. Thus, the distribution patterns of the generalist model genera are similar to those observed for estuarine macroorganisms, such as *Gammarus* [17,25] and comparable to classic spatial distributions seen in other systems [16].

Desulfobacter distribution was also assessed for the first time, to our knowledge. Some bands appear to fluctuate along the estuary, but no correlation was found with the environmental variables, although some weak site-related clustering was observed (Figure 2D). Thus, the only supportable conclusion is that all genotypes were present all along the estuary. Such an undifferentiated distribution pattern, also seen with *Methanosaeta*, may be a general property of metabolically specialist microbes. Thus, these data support the proposed hypothesis that, in contrast to traditional ecological ideas, the breadth of the spatial distribution of specialist and generalist microbes is inversely related to the breadth of their potential niche and indicate that the pattern of spatial distribution of a microbe is directly linked to its metabolic flexibility.

Whilst definitions of specialisation are varied and often context dependent [38] most ecologists define taxa according to the breadth of their niche; specialists have narrow niches and generalists have broad niches with spatial distribution often used as a proxy measure of niche breadth (e.g. [20,21,39]). Thus, in many cases organisms are defined as specialists or generalists based on their existing population structure but not directly on their intrinsic biological properties. Such a distribution-based definition of specialist/generalist has never been used to describe microbial populations and in this study the metabolic specialists would be classified as ecological generalists and vice versa. From an ecological perspective it is counter-intuitive that metabolic versatility should limit the spatial distribution of a species and metabolic restriction result in a broader spatial distribution [39] yet, for these four model genera in this estuary, such contrasting patterns of spatial distribution are unequivocal.

Regardless of functional group or phylogeny specific genotypes of generalists are spatially restricted to specific regions of the estuary, whereas specialists are more evenly distributed and seem generally unaffected by gross environmental variables. These data suggest that spatial differentiation in microbes occurs if individuals within a genus harbour sufficient metabolic flexibility to exploit differing conditions along an environmental gradient. Specialists, that by definition would lack the capacity to exploit differing conditions, should exhibit either a highly restricted (distribution is limited by changing environmental factors that preclude growth/survival) or a cosmopolitan distribution along a gradient (environmental factors in the gradient have little or no effect on the distribution of the specialist genera). The later seems to be the case for the specialist genera we have analysed here. This apparent paradox could be explained by the difference in terms of potential and realised niches [21,38,39]. We suggest that metabolic specialists are constrained by relatively few factors, usually the availability of one or two specific substrates (in this study, acetate for *Methanosaeta* and acetate and sulfate for *Desulfobacter*). As a consequence while metabolic specialists have a narrow potential metabolic niche they are able to occupy much of this niche space. If, as suggested, microbial specialists already occupy a great extent of their potential niche, there is little niche space that could be occupied by diverging strains. With little ecological space available for speciation intrageneric competition may be highly constrained and so species coexist along environmental gradients. For generalists the opposite is true, genotypes are spatially constrained within a narrow realised niche but have a much wider potential niche (see isolate data in Oakley et al. [32]) and so do have niche space that can be exploited by diverging strains. Therefore, we propose that all generalist microbial genotypes could potentially occupy additional niche space but are constrained from doing so by intense intrageneric competition. Such competition would be enhanced if environmental heterogeneity, such as temporal variation in the availability of electron donors and acceptors, was great, as appears to be the case on a on a microscopic level in many environments (e.g. [40]). Therefore, intrageneric competition, in concert with environmental heterogeneity, may effectively produce barriers to dispersal that facilitate localised adaptation and differentiation between strains. Thus, a generalist phenotype results in very different ecological outcomes in microbial communities compared to those seen in traditional ecological studies of plants and animals.

We further suggest that among the four ecological processes proposed by Hanson *et al.* [41] as drivers of microbial biogeography, speciation through selection by environmental variables is more relevant to generalists. In contrast, it appears that specialists are probably mainly affected by the three other processes: dispersal, mutation and drift; resulting in presumably subtle and small-scale biogeographic patterns. This challenges the primary assumption that mainly abiotic environmental factors must be accounted for to understand the ecology of microbes as there appear to be different ecological rules for different groups.

Most efforts with the aim of including microbes into our existing ecological framework has been based on using broad-brush approached to determine whether microbial communities are similar to plants and animals in their relationships and interactions with the environment. Here we show, using taxonomically focused analyses, that, in contrast to the perceived view in ecology that specialisation can be defined by spatial restriction [22], microbial metabolic specialists have a wider distribution than metabolic generalists. These results illustrate the real challenge we face in integrating microbes within ecological theory and analysis in order to build a truly unified ecology.

Materials and Methods

Ethics statement

No permissions were required to sample the estuarine sites used here as they are freely accessible to the public and not on private land.

Site description and sampling

The sample sites (see Figure S1) and sampling strategy, sediment porewater salinity, Sulfate and acetate measurements (Figure S2) and sample treatment prior to DNA and RNA extraction are described in File S1.

Nucleic acids extraction, 16S rRNA gene PCR and fingerprint analysis of the model genera

DNA and RNA were extracted from all sediment samples using the hydroxyapatite spin-column method [42,43] and 16S rRNA gene sequences amplified as described in File S1. A new set of primers specific for *Methanosarcina* 16S rRNA gene (Msc214f and Msc613r; Table S1 in File S1) was designed using the web-based primer design software PRIMER3 (http://frodo.wi.mit.edu/primer3/) and assessed for specificity and sensitivity with THERMOPHYL ([44]; http://go.warwick.ac.uk/thermophyl). Specificity was checked empirically by amplification of *Methanosarcina mazei* (DSM 2053) and *Methanosarcina acetivorans* (DSM 2834) pure culture DNA and using several other archaeal and bacterial negative controls.

Amplified 16S rRNA-gene fragments from both DNA and RNA were analysed using DGGE fingerprinting as described in Information SI. DGGE patterns corresponded extremely well with qPCR and pyrosequence analysis of genotype distribution [32,45]. In addition RNA-DGGE analysis required only a 2-step PCR but produced almost identical distribution patterns to those from the 3- or 4-step PCRs used in DNA-DGGE analyses, indicating that the additional steps do not introduce artefacts into the analysis. Dissimilarity matrixes were obtained from Pearson correlation and Jaccard analyses of the DGGE gels in GelComparII (Applied Maths, Belgium). Environmental variables were also converted into dissimilarity matrixes and correlations between the two matrices were analysed using Mantel and partial Mantel tests [46]. Canonical Correspondence Analysis (CCA) was used to assess the effect of the environmental gradient on the distribution of organisms as described in File S1.

Supporting Information

Figure S1 Map of the Colne estuary, Essex, UK showing the 11 sampling sites along the full extent of the estuary.

Figure S2 Porewater concentrations (mM) of A. chloride (salinity); B. sulphate from the 11 sampled sites along the Colne estuary. Error bars represent the standard error of the mean (SEM, n = 3).

Figure S3 Corrected mean band intensities from replicated DNA-DGGE analyses from the 11 sites (from the marine site 1 to the freshwater site 11) along the Colne estuary, Essex, UK for: A. *Desulfobulbus*, B. *Methanosarcina*, C. *Methanosaeta* and D. *Desulfobacter*. Band numbers are indicated on the left of gel images. Error bars represent the standard error of the mean (SEM, n = 3).

Figure S4 Unconstrained Canonical Correspondence Analysis tests correlating environmental variables (geographic distance (GD), chloride (Cl) and sulphate (S)) with model genera's genotypic distribution patterns from DNA-DGGE for A. *Desulfobulbus*, B. *Methanosarcina*, C. *Methanosaeta* and D. *Desulfobacter*. Geographical groupings of samples are circled and labelled.

Figure S5 Corrected mean band intensities from replicated RNA-DGGE analyses from the 11 sites (from the marine site 1 to the freshwater site 11) along the Colne estuary, Essex, UK for: A. *Desulfobulbus*, B. *Methanosaeta*. Band numbers are indicated on the left of gel images. Error bars represent the standard error of the mean (SEM, n = 3).

Figure S6 Unconstrained Canonical Correspondence Analysis tests correlating environmental variables (geographic distance (GD), chloride (Cl) and sulphate (S)) with model genera's genotypic distribution patterns from RNA-DGGE for A. *Desulfobulbus*, B. *Methanosaeta*. Geographical groupings of samples are circled and labelled.

File S1 Contains Table S1–S3 and detailed description of Materials and Methods used and basic results with associated references. Table S1: List of PCR primers and annealing temperatures used in this study. *TD - These PCRs were Touchdown PCRs. Initial annealing temperature is +10°C and reduces by 1°C per cycle for the first 10 cycles of the PCR. **Table S2. Mantel tests correlating environmental variables (geographic distance (D), chloride (C) and sulphate (S)) with model genera's DNA-DGGE based genotypic distribution patterns (based on Band intensities, Pearson and Jaccard analyses). Table S3. Mantel tests correlating environmental variables (geographic distance (GD), chloride (C) and sulphate (S)) with model genera's RNA-DGGE based genotypic distribution patterns (based on Band intensities, Pearson and Jaccard analyses).**

Author Contributions

Conceived and designed the experiments: KJP FC. Performed the experiments: FC. Analyzed the data: FC BBO KJP. Contributed reagents/materials/analysis tools: FC BBO. Wrote the paper: KJP FC BBO.

References

1. Falkowski PG, Fenchel T, Delong EF (2008) The microbial engines that drive Earth's biogeochemical cycles. Science 320: 1034–1039.
2. Prosser JI, Bohannan BJM, Curtis TP, Ellis RJ, Firestone MK, et al. (2007) The role of ecological theory in microbial ecology. Nat Rev Microbiol 5: 384–392.
3. Robinson CJ, Bohannan BJM, Young VB (2010) From structure to function: The ecology of host-associated microbial communities. Microbiol Mol Biol Rev 74: 453–476.
4. Costello EK, Stagaman K, Dethlefsen L, Bohannan BJM, Relman DA (2012) The application of ecological theory toward an understanding of the human microbiome. Science 336: 1255–1262.
5. Baas-Becking LGM (1934) Geobiologie of inleiding tot de milieukunde (in Dutch) The Hague, the Netherlands: W.P. van Stockum & Zoon.
6. Begon M, Harper JL, Townsend CR (1996) Ecology: Individuals, populations and communities. Oxford, UK: Blackwell Science Ltd.
7. Cermeno P, Falkowski PG (2009) Controls on diatom biogeography in the ocean. Science 325: 1539–1541.
8. Oakley BB, Carbonero F, van der Gast CJ, Hawkins RJ, Purdy KJ (2010) Evolutionary divergence and biogeography of sympatric niche-differentiated bacterial populations. ISME J 4: 488–497.
9. Whitaker RJ, Grogan DW, Taylor JW (2003) Geographic barriers isolate endemic populations of hyperthermophilic archaea. Science 301: 976–978.
10. Fierer N, Jackson RB (2006) The diversity and biogeography of soil bacterial communities. Proc Nat Acad Sci USA 103: 626–631.
11. Lauber CL, Hamady M, Knight R, Fierer N (2009) Pyrosequencing-based assessment of soil pH as a predictor of soil bacterial community structure at the continental scale. Appl Environ Microbiol 75: 5111–5120.
12. Gaston KJ (2000) Global patterns in biodiversity. Nature 405: 220–227.
13. Whittaker RJ, Willis KJ, Field R (2001) Scale and species richness: towards a general, hierarchical theory of species diversity. JBiogeog 28: 453–470.
14. Cohan FM (2006) Towards a conceptual and operational union of bacterial systematics, ecology, and evolution. Phil Trans Roy Soc London B 361: 1985–1996.
15. Curtis TP, Sloan WT (2004) Prokaryotic diversity and its limits: microbial community structure in nature and implications for microbial ecology. Curr Opin Microbiol 7: 221–226.
16. MacArthur RH (1958) Population ecology of some warblers of Northeastern coniferous forests. Ecology 39: 599–619.
17. Spooner GM (1947) The distribution of *Gammarus* species in estuaries. J Mar Biol Assoc 27: 1–52.
18. Philippot L, Andersson SGE, Battin TJ, Prosser JI, Schimel JP, et al. (2010) The ecological coherence of high bacterial taxonomic ranks. Nat Rev Microbiol 8: 523–529.
19. Leibold MA (1995) The niche concept revisited - Mechanistic models and community context. Ecology 76: 1371–1382.
20. Colles A, Liow LH, Prinzing A (2009) Are specialists at risk under environmental change? Neoecological, paleoecological and phylogenetic approaches. Ecol Lett 12: 849–863.
21. Julliard R, Clavel J, Devictor V, Jiguet F, Couvet D (2006) Spatial segregation of specialists and generalists in bird communities. Ecol Lett 9: 1237–1244.
22. Poisot T, Bever JD, Nemri A, Thrall PH, Hochberg ME (2011) A conceptual framework for the evolution of ecological specialisation. Ecol Lett 14: 841–851.
23. Nedwell DB, Embley TM, Purdy KJ (2004) Sulphate reduction, methanogenesis and phylogenetics of the sulphate reducing bacterial communities along an estuarine gradient. Aquat Microb Ecol 37: 209–217.
24. Purdy KJ, Munson MA, Nedwell DB, Embley TM (2002) Comparison of the molecular diversity of the methanogenic community at the brackish and marine ends of a UK estuary. FEMS Microbiol Ecol 39: 17–21.
25. Gaston KJ, Spicer JI (2001) The relationship between range size and niche breadth: a test using five species of *Gammarus* (Amphipoda). Global Ecol Biogeogr 10: 179–188.
26. Flynn AM (2008) Organic matter and nutrient cycling in a coastal plain estuary: Carbon, nitrogen, and phosphorus distributions, budgets, and fluxes. J Coastal Res 55: 76–94.
27. Laanbroek HJ, Abee T, Voogd IL (1982) Alcohol conversions by *Desulfobulbus propionicus* Lindhurst in the presence and absence of sulfate and hydrogen. Arch Microbiol 133: 178–184.
28. Kuever J, Rainey FA, Widdel F (2005) *Desulfobulbus*. In: Brenner DJ, Krieg NR and Staley JT, editors. Bergey's Manual of Systematic Bacteriology Volume 2: The *Proteobacteria*; Part C The *Alpha-, Beta-, Delta* and *Epsilonproteobacteria*. New York: Springer. pp. 988–992.
29. Patel GB (2001) *Methanosaeta*. In: Brenner DJ, Krieg NR and Staley JT, editors. Bergey's Manual of Systematic Bacteriology Volume 1: The *Archaea* and the Deeply Branching and Phototrophic *Bacteria*. New York: Springer. pp. 289–294.
30. Carbonero F, Oakley BB, Purdy KJ (2012) Genotypic distribution of a specialist model microorganism, *Methanosaeta*, along an estuarine gradient: Does metabolic restriction limit niche differentiation potential? Microb Ecol 63: 856–864.
31. Hawkins RJ, Purdy KJ (2007) Genotypic distribution of an indigenous model microorganism along an estuarine gradient. FEMS Microbiol Ecol 62: 187–194.
32. Oakley BB, Carbonero F, Dowd SE, Hawkins RH, Purdy KJ (2012) Contrasting patterns of niche partitioning between two anaerobic terminal oxidizers of organic matter. ISME J 6: 905–914.
33. Boone DR, Mah RA (2001) *Methanosarcina*. In: Brenner DJ, Krieg NR and Staley JT, editors. Bergey's Manual of Systematic Bacteriology Volume 1: The *Archaea* and the Deeply Branching and Phototrophic *Bacteria*. New York: Springer. pp. 269–276.
34. Kuever J, Rainey FA, Widdel F (2005) *Desulfobacter*. In: Brenner DJ, Krieg NR and Staley JT, editors. Bergey's Manual of Systematic Bacteriology Volume 2: The *Proteobacteria*; Part C The *Alpha-, Beta-, Delta* and *Epsilonproteobacteria*. New York: Springer. pp. 961–964.
35. Silva S (2004) Activity and diversity of sulphate-reducing bacteria, methanogenic archaea in contrasting sediments from the River Colne estuary. Colchester, UK: University of Essex.
36. Purdy KJ, Munson MA, Cresswell-Maynard T, Nedwell DB, Embley TM (2003) Use of 16S rRNA-targeted oligonucleotide probes to investigate function and phylogeny of sulphate-reducing bacteria and methanogenic archaea in a UK estuary. FEMS Microbiol Ecol 44: 361–371.
37. Fitzsimons MF, Kahni-Danon B, Dawitt M (2001) Distributions and adsorption of the methylamines in the inter-tidal sediments of an East Anglian Estuary. Environ Exp Bot 46: 225–236.
38. Devictor V, Clavel J, Julliard R, Lavergne S, Mouillot D, et al. (2010) Defining and measuring ecological specialization. J Appl Ecol 47: 15–25.
39. Kassen R (2002) The experimental evolution of specialists, generalists, and the maintenance of diversity. J Evolution Biol 15: 173–190.
40. Jørgensen BB (1977) Bacterial sulphate reduction within reduced microniches of oxidised marine sediments. Mar Biol 41: 7–17.
41. Hanson CA, Fuhrman JA, Horner-Devine MC, Martiny JBH (2012) Beyond biogeographic patterns: processes shaping the microbial landscape. Nat Rev Microbiol 10: 497–506.
42. Purdy KJ (2005) Nucleic acid recovery from complex environmental samples. Meth Enzymol 397: 271–292.
43. Purdy KJ, Embley TM, Takii S, Nedwell DB (1996) Rapid extraction of DNA and rRNA from sediments by a novel hydroxyapatite spin-column method. Appl Environ Microbiol 62: 3905–3907.
44. Oakley BB, Dowd SE, Purdy KJ (2011) ThermoPhyl: A software tool for designing thermodynamically and phylogenetically optimized quantitative-PCR assays. FEMS Microbiol Ecol 77: 17–27.
45. Purdy KJ, Hurd PJ, Moya-Laraño J, Trimmer M, Oakley BB, et al. (2011) Systems biology for ecology: From molecules to ecosystems. Adv Ecolog Res 43: 88–149.
46. Mantel N, Fleiss JL (1980) Minimum expected cell-size requirements for the Mantel-Haenszel one-degree-of-freedom chi-square test and a related rapid procedure. Amer J Epidemiol 112: 129–134.

16

Stability of Microbiota Facilitated by Host Immune Regulation: Informing Probiotic Strategies to Manage Amphibian Disease

Denise Küng[1], Laurent Bigler[2], Leyla R. Davis[1], Brian Gratwicke[3], Edgardo Griffith[4], Douglas C. Woodhams[1,5*¤]

1 Institute of Evolutionary Biology and Environmental Studies, University of Zurich, Zurich, Switzerland, **2** Institute of Organic Chemistry, University of Zurich, Zurich, Switzerland, **3** Center for Species Survival, Conservation and Science, National Zoological Park, Smithsonian Institution, Washington DC, United States of America, **4** El Valle Amphibian Conservation Center, El Valle, República de Panamá, **5** Smithsonian Tropical Research Institute, Balboa, Ancón, Panamá, República de Panamá

Abstract

Microbial communities can augment host immune responses and probiotic therapies are under development to prevent or treat diseases of humans, crops, livestock, and wildlife including an emerging fungal disease of amphibians, chytridiomycosis. However, little is known about the stability of host-associated microbiota, or how the microbiota is structured by innate immune factors including antimicrobial peptides (AMPs) abundant in the skin secretions of many amphibians. Thus, conservation medicine including therapies targeting the skin will benefit from investigations of amphibian microbial ecology that provide a model for vertebrate host-symbiont interactions on mucosal surfaces. Here, we tested whether the cutaneous microbiota of Panamanian rocket frogs, *Colostethus panamansis*, was resistant to colonization or altered by treatment. Under semi-natural outdoor mesocosm conditions in Panama, we exposed frogs to one of three treatments including: (1) probiotic - the potentially beneficial bacterium *Lysinibacillus fusiformis*, (2) transplant – skin washes from the chytridiomycosis-resistant glass frog *Espadarana prosoblepon*, and (3) control – sterile water. Microbial assemblages were analyzed by a culture-independent T-RFLP analysis. We found that skin microbiota of *C. panamansis* was resistant to colonization and did not differ among treatments, but shifted through time in the mesocosms. We describe regulation of host AMPs that may function to maintain microbial community stability. Colonization resistance was metabolically costly and microbe-treated frogs lost 7–12% of body mass. The discovery of strong colonization resistance of skin microbiota suggests a well-regulated, rather than dynamic, host-symbiont relationship, and suggests that probiotic therapies aiming to enhance host immunity may require an approach that circumvents host mechanisms maintaining equilibrium in microbial communities.

Editor: Joy Sturtevant, Louisiana State University, United States of America

Funding: Funding was provided through Swiss National Science Foundation grant 31-125099 to D.C.W. and fieldwork was supported by SCNAT (Akademie der Naturwissenschaften Schweiz). The funders had no role in study design, data collection and analysis, decision to publish, or preparation of the manuscript.

Competing Interests: The authors have declared that no competing interests exist.

* E-mail: dwoodhams@gmail.com

¤ Current address: Department of Ecology and Evolutionary Biology, University of Colorado, Boulder, Colorado, United States of America

Introduction

Environmental changes leading to disruption of the host microbiota, or dysbiosis, can lead to disease emergence [1,2,3]. Recent theory has focused on disturbance responses for microbial communities and factors important for community stability [4,5]. Besides permanent change to an altered stable state, there are four potential alternative responses to disturbance: (1) Resistance – the microbial community does not change, (2) Resilience – the microbial community is changed initially but then returns to its original composition, (3) Functional redundancy – the microbial community changes while maintaining the function of the original composition, and (4) Restoration – the microbial community recovers from a previously degraded state. Most microbial communities in the environment, for example soil communities, tend to be sensitive and not resistant to disturbance [5,6]. However, few studies have examined disturbances of host-associated microbial communities, and active host regulation of microbiota may produce greater stability than found in environmental microbial communities. Indeed, a number of host processes have been described directing the restoration or recovery of microbiota following disturbance [7].

Intentional disturbances such as antibiotic treatments intended to prevent or manage disease have the unintended effect of disrupting beneficial microbiota [8]. Transfer of microbiota from healthy to diseased hosts (e.g., fecal transplant) has proven more effective than antibiotics against *Clostridium difficile*, and has renewed interest in this type of therapy [9,10]. Mutualistic microbial communities are linked with the health of organisms across a broad taxonomic range [11,12]. In amphibian populations threatened by emerging disease, the microbial community response to potential probiotic treatments is critical for effective conservation management [13].

A diverse microbiota has been found on amphibian skin [14–20]. These cutaneous microbial communities extend host immune function and are important in the prevention or outcome of diseases such as chytridiomycosis, which is caused by the pathogenic chytrid fungus *Batrachochytrium dendrobatidis* (*Bd*) in amphibians [21,22]. The fungus is globally distributed on hundreds of amphibian species [23]. While the impacts of infection differ among species and depend on environmental context [24,25], severe outbreaks can lead to extinctions or collapse of regional amphibian faunas [26,27] and ecosystem alterations [28]. In Panama, populations of the rocket frog, *Colostethus panamansis* Dunn 1933, declined dramatically [27] and Koch's postulates were fulfilled for *Bd* as the causative agent of chytridiomycosis [29]. While this species was extremely sensitive to the disease, others, such as the glass frog *Espadarana prosoblepon* Boettger 1892, were able to persist in smaller populations [27,30].

Growth of the fungus *Bd* is often inhibited by amphibian skin microbiota [17–19,31,32], and probiotic application of antifungal bacteria is a promising tool for disease mitigation [24,33]. While sometimes effective, probiotic therapy has met with mixed results [34–36]. Probiotic screening protocols and advances in application method are under development [13]. Overcoming hurdles in effective probiotic therapy will involve a better understanding of the microbial ecology of amphibian skin, including the microbial responses to disturbance, and the host responses that maintain a functional microbiota.

Amphibian immune defenses are quite sophisticated and include most components present in mammals [37]. One immune component of particular relevance to skin infections includes release of bioactive compounds into the mucosal layer. Antimicrobial peptides (AMPs) are synthesized in dermal granular glands of many species [38–42]. AMPs are stored in the granular glands and released to the skin surface when the animal is alarmed or injured, and small amounts are constitutively expressed [43]. The response is thought to be a non-specific and fast-acting innate defense, although AMP responses can be closely linked with adaptive immunity [44]. Amphibian skin peptides can directly inhibit amphibian pathogens such as *Bd*, *Basidiobolus ranarum* or ranaviruses in vitro [45–47]. Interactions between AMPs, microbiota, and environmental conditions may be important for maintaining a functionally stable microbiota and is an ongoing area of study [48–50].

Here, we examine AMP responses to three microbial treatments of a threatened Central American amphibian. Frogs were treated with a probiotic skin bacterium, treated with skin-wash transplants from a disease-resistant species, or treated with sterile water as controls. Resistance of the microbiota to colonization by exogenous microbes would indicate that host mechanisms are maintaining homeostasis. This knowledge will aid strategies to enhance immune function and mitigate disease.

Materials and Methods

Study Species and Sites

In January 2011, we collected rocket frogs, *C. panamansis* (n = 44), from a stream at the Sierra Llorona Lodge near Colón, Panama. Glass frogs, *E. prosoblepon* (*n* = 17), were collected from Omar Torrijos National Park, Coclé Province, Panama. Frogs were transported to the El Valle Amphibian Conservation Centre (EVACC) at El Níspero Zoological Park, El Valle, Panama. Collecting permits were provided by the Autoridad Nacional del Ambiente (ANAM), and all experimental procedures were approved by the Smithsonian Tropical Research Institute (STRI) Institutional Animal Care and Use Commission. After treatments, frogs were monitored daily to record microhabitat use, behavior, and animal welfare. After the experiment, all animals were retained in captivity at EVACC and not released, according to ANAM specifications.

Animal Care

All *C. panamansis* were housed individually in 60 L plastic mesocosms situated on a shaded lawn at EVACC. The tubs were filled with 2 to 3 L of filter-sterilized water and tilted to one side to cover approximately one third of the surface, and a large rock was provided as a hide in the dry portion of the enclosure. Water was exchanged every 8 d and waste was decontaminated with bleach before disposal. Frogs were fed with small domestic crickets (*Acheta domesticus*), fruit flies (*Drosophila hydei*), or a mixture of both every other day. All *E. prosoblepon* were housed indoors in individual 2 L plastic enclosures containing water and large leaves. They were fed with *D. hydei* every other day.

Experimental Design

Frogs were captured by hand using a fresh pair of gloves for each capture. Before swabbing the frogs, they were rinsed with approximately 20 ml filter sterilized water to remove debris and transient bacteria not associated with the skin [15]. Upon capture, each *C. panamansis* was swabbed twice with a sterile rayon swab (Copan, Brescia, Italy) on the thighs, hands and feet 5 times and 10 times on the ventral surface. Swabs were then immediately stored at $-15°C$. The first swab was used for the analysis of the microbial skin community as described below, and the second swab was given to Roberto Ibáñez at the Smithsonian Tropical Research Institute in Panama City, Panama to test for *Bd* by qPCR according to Boyle et al. [51]. Rocket frogs were distributed randomly among three treatment groups. Glass frogs, *E. prosoblepon*, were collected on January 11 and 12, and rocket frogs were collected on January 13–15. After determining *Bd* infection status of all individuals, treatments of rocket frogs began January 24 (day 1). A second swab for microbial skin community analysis was obtained on day 48, at which time skin peptides were also sampled, marking the end of the experiment and a biologically relevant time point for assessing changes in skin peptides. Collection of microbes and skin peptides from *E. prosobleopon* occurred as soon as possible after use of frogs in the experiment (day 14).

Probiotic treatment. Frogs (n = 15) were exposed to the antifungal bacterium, *Lysinibacillus fusiformis*. In January 2010, *L. fusiformis* was isolated from the skin of a *C. panamansis* from Tortí, Panama and cryopreserved at STRI until use. The isolate was identified by Matthew Becker at Virginia Tech University and the 16S rDNA sequence has been deposited in the EMBL Nucleotide Sequence Database (accession number HE817768). The host frog was not infected with *Bd*, although some other amphibians at the site were *Bd* positive including two *Craugastor crassidigitus* out of 93 sampled amphibians. Some strains of the bacterium can produce tetrodotoxin [52] and have the capacity to inhibit *Bd* growth in the laboratory. Thus, Dendrobatid frogs may form symbioses with this bacterium to increase antimicrobial or anti-predator defenses. Bacteria were incubated for 96 h at room temperature on glucose-casein-KNO_3 agar plates (containing 0.5 g glucose, 0.3 g casein, 2 g KNO_3, NaCl and K_2HPO_4, 0.05 g $MgSO_4{*}7H_2O$, 0.03 g $CaCO_3$, 0.01 g $FeSO_4{*}7H_2O$ and 20 g agar per L medium). After incubation, *L. fusiformis* was washed from the agar plate with filter-sterilized water and 25 ml of the bacterial solution was poured into each of fifteen 50 ml centrifuge tubes. Rocket frogs were placed individually into the tubes for 1 hr. This treatment was performed once, on day 1 of the experiment.

Transplant treatment. Frogs (n = 15) were exposed to skin washes from *E. prosoblepon* (*n* = 16). Glass frogs appear to have an exaptation to resist *Bd* and pre-existing mucosal defenses that protect nests from pathogenic fungi [32]. Glass frogs were given daily washes in 25 ml of filter-sterilized water for 30 min. All skin washes were then mixed together and 25 ml applied to each of the 15 *C. panamansis* for 30 min in 50 ml centrifuge tubes. Starting on day 1 of the experiment, the treatment was repeated once daily for 7 d.

Control treatment. Frogs (n = 14) were held as controls with a handling regime matching that of the transplant treatment. Control frogs were given a daily wash in 25 ml filter sterilized water for 30 min each of 7 d starting on day 1 of the experiment. Any affect of stress from handling or daily treatment washes was matched in these controls.

Data Collection

Frog mass. The mass of all *C. panamansis* was measured on day 1, day 28 and day 48 of the experiment. To test for treatment effects on mass we used repeated measures ANOVA. Slope of mass change through time was compared among treatment groups by ANOVA with Tukey HSD pairwise comparisons. All statistical tests were carried out with IBM SPSS Statistics v. 19 (SPSS Inc.) unless otherwise indicated, and non-parametric tests were used when data were not normally distributed and had heterogeneous variances (Levene's test, $P>0.05$).

Microbiota. Dynamics of bacterial communities were investigated by terminal-restriction fragment length polymorphism (T-RFLP), a consistent and high-resolution culture-independent technique used to monitor microbial community changes over space or time [53–56]. DNA was extracted from swabs with the Microbial Ultra Clean DNA Kit (MO BIO) following the protocol of the manufacturer. Bacterial 16S rDNA was then amplified using the primer 27F (PET® labelled) (5′-AGA GTT TGA TCC TGG CTC AG-3′) and 1492R (5′-TAC GGY TAC CTT GTT ACG ACT-3′) (Applied Biosystems). For the 20 µl PCR reaction mixture, the thermocycling conditions were set as follows: 95°C for 5 min followed by 32 cycles of 94°C for 1 min, 52°C for 1.5 min, 72°C for 2 min and a final elongation for 10 min at 72°C. Each reaction contained the following reagents: 2 µl template DNA, 1 µl of each primer (10µM), 1 µl BSA (2 ug/µl), 2 µl dNTPs (2 mM), 4 µl buffer (5x), 1.2 µl $MgCL_2$ (25 mM), 1 µl Taq 1:10 (0.5 u/µl), and 6.8 µl water.

To minimise PCR bias, PCR reactions were run in triplicate and their products combined and checked by electrophoresis in 1% agarose with GelRED staining. In order to eliminate pseudo-terminal fragments [60],10 units of mung bean nuclease and 12 µl of 10x reaction buffer were added to the PCR product to digest single-stranded DNA. The digestion was performed with a total volume of 120 µl at 30°C for 2 h. To stop the reaction, SDS (0.1%) was added to a final concentration of 0.01%. Mung bean digested PCR products were purified with the GenElute PCR Clean-Up Kit (Sigma-Aldrich) according to the protocol provided by the manufacturer until step IV and then eluted with 33 µl of Elution Solution.

Using a spectrophotometer (NanoDrop), the absorption of samples at 260 nm was determined to quantify DNA. Samples were diluted with Milli Q water to a DNA concentration of 50 ng μl^{-1} and aliquots of the samples (200–500 ng) were digested with restriction enzymes. For the 20 µl restriction reaction, 1.5 units of either HaeIII or MspI were used with 2 µl of 10x FastDigest or 10x Tango buffer respectively.

To determine restriction fragment lengths, 2 µl of digested PCR product were run on an ABI 3730 DNA Analyser (Applied Biosystems) equipped with 36 cm capillaries filled with POP7 polymer. For each sample, we used 17.8 µl HiDi-Formamide (Applied Biosystems) and 0.2 µl GeneScan 500 LIZ size standard (Applied Biosystems). T-RF sizes and quantities, measured in fluorescence units (rfu), were determined with GeneMapper v3.7 (Applied Biosystems) using the AFLP option and the Local Southern size calling method. Peak alignment was done automatically by GeneMapper and peak parameters were set to a polynomial degree of 4, window size of 13 and a minimum of width at half maximum (base pairs) of 2. To exclude possible primer dimers or other artifacts, we analyzed peaks in the range of 80–411 bp (HaeIII) and 80–596 bp (MspI) with intensities of ≥150 RFU. Samples with less than 2 peaks were removed from the data set; in total, 6 samples digested with HaeIII and 3 samples digested with MspI were discarded. By calculating the area of each peak as a proportion of the total area, data were standardized in Microsoft Excel and the resultant data set imported into the statistical software package PAST. T-RFLP data were visualized by non-metric multidimensional scaling (nMDS) and hypotheses tested by analysis of similarity (ANOSIM) using Bray-Curtis coefficient similarity matrix, and by non-parametric multivariate analysis of variance (NPMANOVA). As a second method of microbial community analysis, the number of taxa detected in every sample was counted (assuming that each peak represents one single species) and the Simpson and Shannon indices for diversity were calculated in PAST v2.10. These measures of diversity were used to test for differences among treatment groups by ANOVA or to test for changes in microbial diversity through time with paired t-tests.

Skin peptides. On day 14 of the experiment, all *E. prosoblepon* were swabbed for microbial community analysis, and afterward, skin peptides were collected. Frogs were given a dorsal subcutaneous injection of 40 nmol norepinephrine (bitartrate salt; Sigma) per g body weight (gbw) to elicit granular gland secretions, then rinsed with 25 ml filter sterilized water and allowed to sit for 15 min. Peptide mixtures were acidified to 1% hydrochloric acid (HCl) to prevent proteolytic degradation of samples. The solution was immediately passed over C-18 Sep-Pak cartridges (Waters Corporation) previously primed with acetonitrile and rinsed, and the Sep-Paks were stored in zip-lock bags with 2 ml of 0.1% HCl. After transport to the University of Zürich, peptides bound to the Sep-Paks were eluted with 70% acetonitrile, 29.9% water, and 0.1% HCl and concentrated to dryness by centrifugation under vacuum, and weighed. The same procedure was used to collect peptides from *C. panamansis* after swabbing them at the end of the experiment for microbial community analysis on day 48.

To test the extracted skin peptides for antimicrobial activity, the growth inhibition of *Bd* zoospores was measured for a subset of samples from each *C. panamansis* treatment group and from *E. prosoblepon*. The dried peptides were dissolved in water and diluted to a concentration of 1000 µg ml^{-1} before addition to a 96-well plate in duplicate. The final peptide concentration was 500 µg ml^{-1} in the wells containing 50 µl peptide and 50 µl *Bd* zoospores in 1% tryptone broth (T broth). Preliminary tests showed that lower concentrations had no significant effects on the growth of zoospores (data not shown). To obtain *Bd*, an RIIA agar plate supplemented with 1% tryptone was inoculated with the panzootic lineage of *Bd* from the UK (generously provided by M. Fisher), grown for 7 d, and flushed with 3 ml of T broth. After a 15 min incubation, the T broth with freshly released and active zoospores at 4.2×10^6 zoospores ml^{-1} was collected in a sterile reagent reservoir. The 96-well plates were prepared: 100 µl of T broth was added to all outer wells to retain moisture in the plate. Control wells in replicates of 6 contained 50 µl water and 50 µl of *Bd* zoospores heat killed for 15 min at 60°C (negative control) or 50 µl of living *Bd* zoospores (positive growth control). The optical density at 490 nm was measured on days 0 and 7 on a multilabel counter (Victor[3], Perkin Elmer) and plates were incubated at 18°C, an optimal temperature for *Bd* growth [57]. The percentage of *Bd* growth inhibition for each peptide sample was calculated by comparison to controls. The percentage of *Bd* growth inhibition was then multiplied by the quantity of peptides produced per cm^2 surface area to calculate the peptide capacity against *Bd.* Peptide capacity is similar to the measure of peptide effectiveness used by Woodhams et al. [58] for small frogs where large quantities of peptides are not available for testing minimal inhibitory concentrations. The skin surface was estimated using the equation: surface area = 9.90*(weight in grams)$^{0.56}$ [59]. We tested for differences among treatments and species in the quantity of peptides recovered, growth inhibition of *Bd* (%), and peptide capacity by ANOVA with Tukey HSD pairwise comparisons.

To test for differences in the composition of skin peptides among the three *C. panamansis* treatment groups, skin peptides were analyzed. Ultra-high performance liquid chromatography electrospray ionization mass spectrometry (UHPLC-ESI-MS) was performed on a Waters Acquity UPLC (Waters, Milford, MA 01757, USA) connected to a Bruker maxis high-resolution quadrupole time-of-flight mass spectrometer (Bruker Daltonics, Bremen, Germany). An Acquity BEH 300 C18 column (Waters, 1.7 µm, 1×100 mm) has been used with a gradient of H_2O+0.05% TFA (A) and CH_3CN+0.05% TFA (B) at 0.2 mL/min flow rate (linear gradient from 10 to 50% B within 10 min followed by flushing with 100% B for 4 min). All solvents were purchased in best LC-MS qualities.

The mass spectrometer was operated in the positive electrospray ionization mode at 4′500 V capillary voltage, –500 V endplate offset, with a N_2 nebulizer pressure of 1.4 bar and dry gas flow of 10.0 L/min at 200°C. MS acquisitions were performed in the mass range from m/z 100 to 3′000 at 20′000 resolution (full width half maximum) and 2 scans per second. The MS instrument was optimized for maximum intensities of Bradykinin at m/z 530.8. Masses were calibrated with an electrospray calibrant solution (Fluka, Buchs, Switzerland, 20x dilution in CH_3CN) between m/z 118 and 2722.

The relative abundance (area under peak) of 16 peptides was determined by HPLC-MS for a subset of samples from each treatment group and species. The relative intensities of each peptide were ranked to produce a dataset that satisfied Box's test for homogeneity of covariance matrices. Differences in peptide profiles among treatments were analyzed by multivariate analysis of variance in PAST v2.16. Peptide dry weight was compared with the ranked intensity of the ubiquitous peptide (mass 1064) by linear regression to determine whether dry mass measurements of partially purified skin secretions enriched for hydrophobic peptides could predict the relative abundance of peptide peaks detected by HPLC. To test for a potential immune-energetic trade-off, the quantity of peptides recovered was tested for correlation with change in frog mass. Peptide composition was also tested for affect on frog mass by multiple linear regressions, overall and separately for each treatment.

Results

Survival, Bd Infection, and Body Mass

The survival rates of *C. panamansis* (N = 44) in the three different treatments were as follows: 86.7% for the probiotic treament, 85.7% for the control treatment, and 78.6% for the skin wash transplant treatment. Five frogs, randomly divided among treatments, were found to be infected with *Bd* but none of these frogs died during the experimental period. Mass loss of infected frogs was not significantly different than mass loss of uninfected frogs (Independent t-test, $t_{37} = -0.075$, P = 0.940). At the start of the experiment, mass of frogs did not differ significantly among treatments (mean +/− SD = 1.52+/−0.28 g, ANOVA, $F_{2,36} = 0.667$, P = 0.519). Treatment groups differed significantly in the change of body mass through time (Repeated measures ANOVA, Greenhouse-Geisser $F_{3.3,56.6} = 3.168$, P = 0.027). Mass loss was greatest in frogs treated with *L. fusiformis* (slope of mass loss, ANOVA, $F_{2,36} = 3.850$, P = 0.031; Fig. 1a). Mass loss was on average 11.9% for probiotic treated frogs, 0.1% for control frogs, and 7.3% for skin-wash transplant treated frogs.

Microbiota Composition

We did not detect significant differences in the composition of microbiota among the three different groups either before or after treatment (Table 1). The microbiota of all treatment groups changed over time (nMDS, Fig. 2; ANOSIM and NPMANOVA, Table 1), but the stress values of the nMDS plots were high and R values of the ANOSIM low, indicating that the changes were very small. Hence, the microbiota did not differ significantly among treatments at the beginning of the experiment, microbiota shifted to a small degree through time in the mesocosms, and microbiota did not significantly differ among treatments at the end of the experiment. The microbiota found on *E. prosoblepon* was significantly different to that found on *C. panamansis* (treatment groups combined) at all times (NPMANOVA, before: F = 4.814, P = 0.0001; after: F = 3.529, P = 0.0001).

For terminal restriction fragments (TRFs) hydrolysed with HaeIII, the number of detected taxa varied between 3 and 20 before and 2 and 14 after treatment. For MspI between 2 and 18 taxa were found both before and after treatment. There was no change in the Shannon index of diversity, and similarly the Simpson index, detected for either enzyme over time (paired t-test for combined treatments; HaeIII: t = −1.307, P = 0.200; MspI: t = 1.431, P = 0. 1617) nor among treatment groups.

Using pure culture standards and T-RFs obtained from cleavages of the 16S rDNA by several enzymes, T-RFLP can be used identify species within microbial community profiles [55]. The bacterium *L. fusiformis* was initially detected on one frog that died during the experiment (fragment lengths: 148 for Msp1 and 235 for HaeIII) and was not found on any frogs at the end of the experiment including those treated with live cultures of *L. fusiformis.*

Skin Peptides

The quantity of peptides measured at the end of the experiment differed significantly among the *C. panamansis* treatment groups (Kruskal-Wallis test, P = 0.025). Frogs treated by skin wash produced significantly less peptide per surface area than frogs treated with *L. fusiformis* (Mann-Whitney U test, P = 0.015), and control frogs were intermediate. Peptide recovery was highest in frogs treated with the probiotic, and frogs in this group also lost the most body mass during the experiment; however, there was not a significant overall correlation between peptide production and change in mass (Pearson correlation, r = −0.175, P = 0.308). Nor was there a significant affect of peptide profile on slope of mass when analyzed by multiple linear regressions overall or separately for each treatment (P's >0.05).

Bd growth inhibition caused by 500 µg ml^{-1} peptides also differed significantly among *C. panamansis* treatments (ANOVA, $F_{3,23} = 34.492$, P<0.0001) and was greatest in frogs treated with skin washes from *E. prosoblepon.* The capacity of peptide defenses

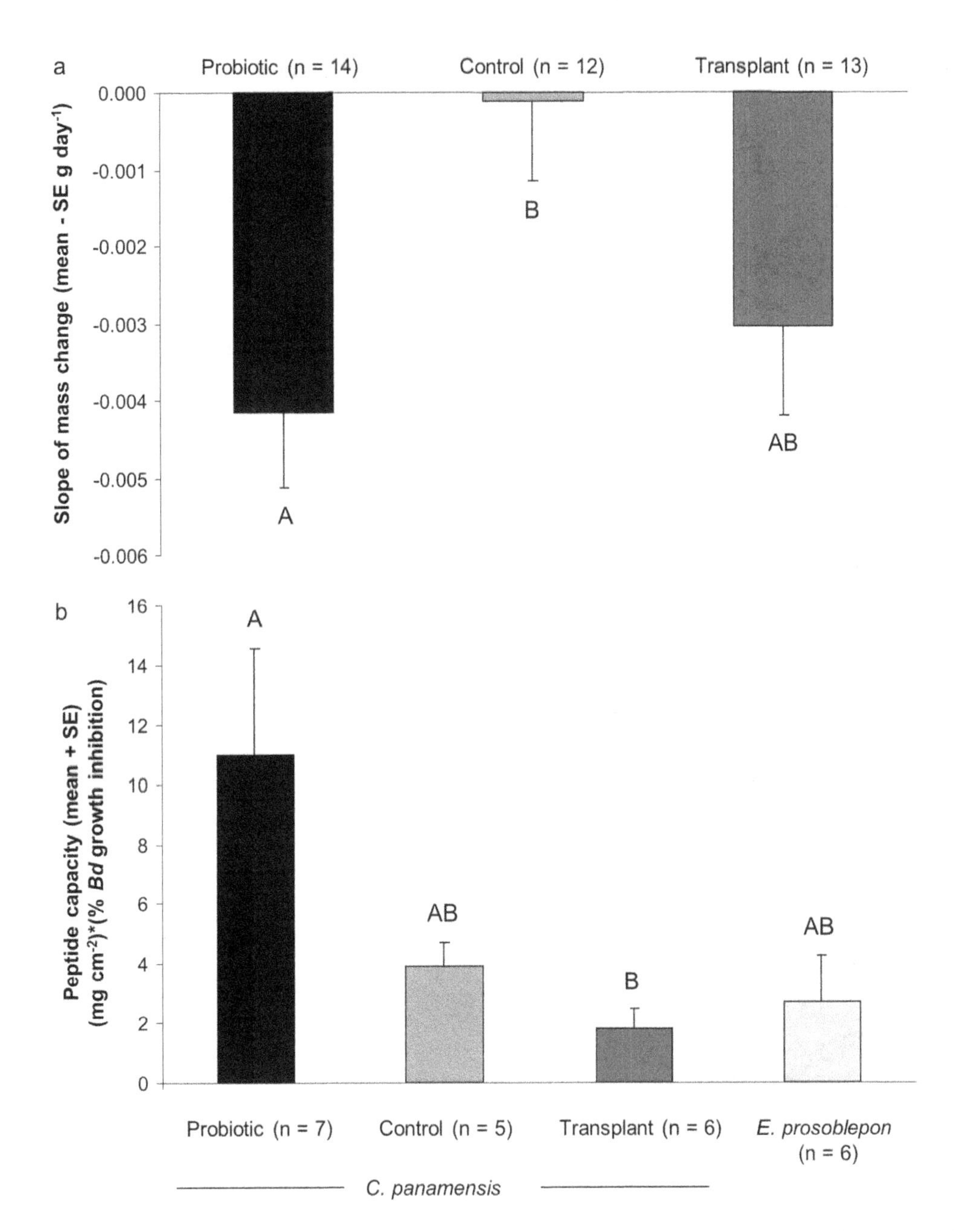

Figure 1. Body mass change and capacity of skin defense peptides against *Batrachochytrium dendrobatidis* (*Bd*) of frogs in different treatments. (a) Slope of mass change. (b) Peptide capacity calculated as peptide quantity per surface area multiplied by percent growth inhibition of *Bd* caused by 500 μg ml^{-1} peptide for each treatment or species. Letters above bars indicate homogeneous subsets based on ANOVA with Tukey HSD.

against *Bd* differed significantly among treatment groups and species (Fig. 1b). Frogs treated with *L. fusiformis* had at least twice the peptide capacity to inhibit *Bd* than frogs in any other group (Fig. 1b).

Rocket frogs, *C. panamansis*, expressed between 1 and 16 skin peptides (mean = 4.8) detected by HPLC (Table 2, Fig. 3). Common skin peptides included molecular weight 1005.6, 1064.0, and 3306.6. Glass frogs, *E. prosoblepon*, produced a different set of skin defense peptides (Table 2, Fig. 3). Primary structures have not been described. Skin peptide profiles were not significantly different among treatments of *C. panamansis* (MANOVA, Wilks' Lambda, $F_{18,50} = 0.756$, $P = 0.738$). Similar results were obtained with non-parametric tests of untransformed data. Overlapping peptide profiles suggest that peptide quantity, rather than profile, was primarily affected by treatment, with the exception of one peptide.

Peptide of mass 1064 was present in all *C. panamansis* samples, and was the only peptide that differed in relative abundance among treatment groups (ANOVA, $F_{2,33} = 3.611$, $P = 0.038$). This peptide was most abundant in transplant treated frogs, least abundant in probiotic treated frogs, and intermediate in controls. There was a significant correlation between total peptide dry mass and rank intensity of peptide mass 1064 (Fig. 4; overall Pearson correlation, $r = 0.672$, $N = 36$, $P<0.001$, $R^2 = 0.4522$). Treatment accounted for 26.4% of the total variance in rank peptide intensity controlling for the effect of peptide quantity (ANCOVA,

Table 1. Statistical analysis of treatment differences in microbial communities described by T-RFLP using either Hae3 or Msp1 enzymes.

	ANOSIM				NPMANOVA			
	Msp1		Hae3		Msp1		Hae3	
	R	p	R	p	F	p	F	p
before treatment								
P vs. C	−0.0980	0.9910	−0.0660	0.9170	0.2900	0.9980	0.5100	0.9720
P vs. T	−0.0360	0.7450	−0.0470	0.8270	0.5700	0.8960	0.6000	0.8750
C vs. T	−0.0560	0.9240	−0.0220	0.6420	0.6200	0.8820	0.7900	0.6920
after treatment								
P vs. C	−0.0300	0.7490	0.0350	0.1930	0.9600	0.4900	1.2700	0.2040
P vs. T	0.0000	0.4230	−0.0230	0.6220	0.9900	0.4350	0.7800	0.6370
C vs. T	0.0080	0.3600	0.0030	0.3880	1.2200	0.2230	1.0900	0.3240
before, after								
Treaments combined	**0.1860**	**0.0001**	**0.1420**	**0.0001**	**5.6830**	**0.0001**	**3.9450**	**0.0001**

Analysis of similarity (ANOSIM) and non-parametric multivariate analysis of variance (NPMANOVA) results are shown of the three *C. panamansis* treatments: P = probiotic treatment, C = control treatment, T = transplant treatment. Significant values identified by ANOSIM and NPMANOVA are indicated in bold.

$F_{2,32} = 17.124$, $P<0.001$, $\omega^2 = 0.264$). Thus, higher total peptide quantity was associated with lower relative abundance of peptide mass 1064 in a treatment-specific pattern (Fig. 4).

Discussion

Stability of Skin Microbiota

Probiotic therapy is a promising disease mitigation strategy currently under development as many amphibian populations

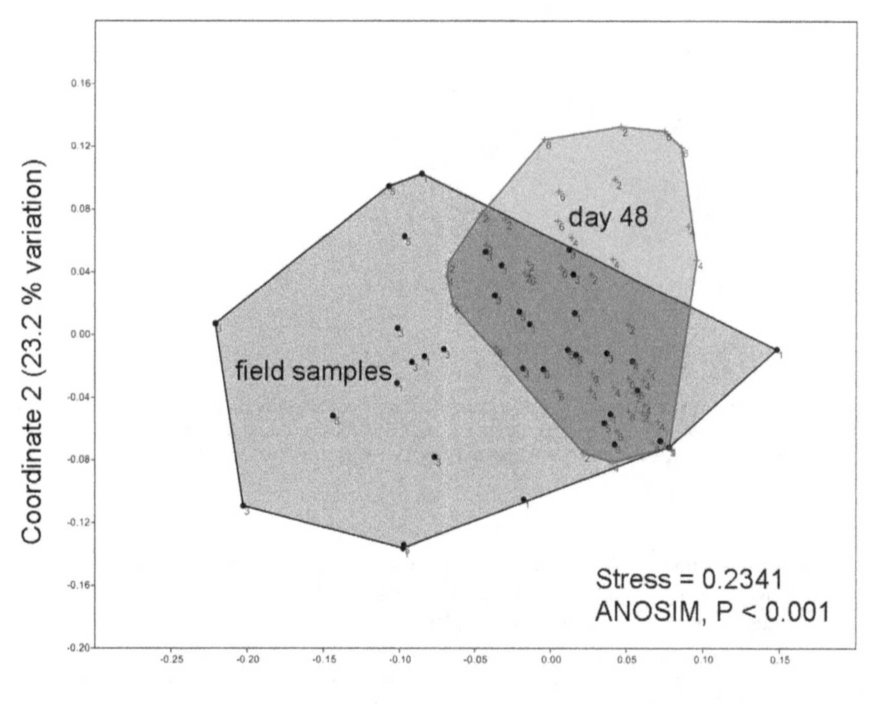

Figure 2. Skin microbial communities of *Colostethus panamansis*. Communities before (circles, field samples) and after (crosses, day 48) treatments visualized by non-metric multidimensional scaling (nMDS) of T-RFLP analysis using enzymes HaeIII and MspI. Treatments are numbered to indicate probiotic bacterium *Lysinibacillus fusiformis* (1,2), control (3,4), and skin-wash transplant from the disease-resistant glass frog *Espadarana prosoblepon* (5,6). Microbial communities were not significantly different among treatments within each time-point represented by convex hulls. Distance between objects on the plot represents relative dissimilarity (axes are in arbitrary units). Stress <0.1 indicates strong community differences and stress >0.2 indicates that differences should be interpreted with caution. Statistical analyses are presented in Table 1.

Table 2. Retention time (Rt), molecular weight (MW), prevalence, and mean relative area of each peptide based on HPLC-MS chromatograms.

Rt (min)	MW	Prevalence (%)	Mean relative area
Colostethus panamansis (N = 36)			
3.6	1041.5	61	0.14
4.45	1064.0	100	1.00
4.5	3306.6	72	0.35
4.8	1936.2	25	0.12
5.0	1512.0	42	0.28
5.3	2988.6	3	0.03
5.3	3001.5	3	0.02
5.7	2986.6	6	0.20
7.0	2974.5	17	0.65
6.8	2957.7	17	0.57
7.1	2972.6	17	0.93
7.0	2290.4	11	1.07
7.3	2231.4	6	0.63
7.6	2315.4	6	0.94
8.1	1005.6	94	1.66
9.7	1790.0	6	0.13
Espadarana prosoblepon (N = 16)			
1.45	1746.7	81	0.02
2.2	1585.8	94	0.31
3.7	2650.2	100	0.72
4.4	2698.2	100	1.14
4.55	2634.2	100	0.74
4.6	2652.2	100	0.37
4.75	1004.6	100	0.39
5.05	1004.6	100	0.51
5.45	2681.2	100	1.00
5.69	2665.1	38	0.06

Area is relative to a consistently observed peak, MW 1064.0 for *C. panamansis* and MW 2681.2 for *E. prosoblepon*.

decline worldwide [13,24,33]. Applications of probiotics may be considered a managed pulse disturbance of the microbial community, but the response to disturbance in terms of stability of host-associated microbiota has not previously been tested. We found that an amphibian species threatened by chytridiomycosis in Panama had a remarkably stable skin microbiota that was resistant to alteration by experimental treatments with skin washes from a co-occurring disease-resistant species, and with the potential probiotic bacterium *L. fusiformis*. Although *L. fusiformis* is a naturally occurring symbiont of *C. panamansis*, and may be responsible for defensive tetrodotoxin compounds found in the skin of Dendrobatids [52,61], the bacterium did not establish. We did not detect tetrodotoxin production from the bacterium grown in isolation (K. Minbiole, unpublished data). Mechanisms maintaining bacterial communities on amphibian skin have not been previously described. Skin defense peptides extracted from the skin of *C. panamansis* inhibited the growth of the pathogenic fungus *Batrachochytrium dendrobatidis* and contributed to host mechanisms maintaining the microbial composition of *C. panamansis* by limiting *L. fusiformis* and exogenous microbes from *E. prosoblepon* skin washes. Application of *L. fusiformis* led to increased peptide capacity against *Bd* in *C. panamansis*.

Although the probiotic bacterium originally isolated from *C. panamansis* did not establish under our experimental conditions, the composition of the skin microbiota of *C. panamansis* changed over the course of the 48 d experiment in all treatments. Thus, the skin-associated microbial community was resistant to experimental disturbance but showed a gradual shift through time, and was perhaps more influenced by environmental conditions than exogenous microbial exposure. The temperature in the mesocosms at El Valle over the course of the experiment (mean 22.8°C) was probably a little lower than that of the lowland rainforest at Sierra Llorona lodge were the frogs were captured. This factor could also have initiated a shift in the microbial communities on the frog skin. The degree to which amphibians depend on their environment or contact with conspecifics to maintain their microbiota long-term is unknown, but other studies have shown slight changes in microbial diversity through time in captivity [36,62]. Here, regulation of microbiota by host immune factors [63] is supported.

How amphibians acquire the microbiota on their skin remains unclear. Plausible routes of transmission include contact with conspecifics (horizontal transmission), habitat (environmental transmission), or parents (vertical transmission [32]). Colonization of *L. fusiformis* on the skin of adult *C. panamansis* after contact with high concentrations of bacteria was not successful, and there are several potential explanations. (1) Colonization may begin at early developmental stages when the microbiota reaches a stable equilibrium that then resists disturbance [64,65]. (2) Competition for resources such as nutrients or space could have prevented establishment of a new member of the skin microbiota [66,67]. (3) Resident microbiota may have prevented the invasion of *L. fusiformis* by the production of antibiotic metabolites or bacteriocins [68,69]. (4) Host immune factors in the skin including AMPs [37] may have been induced and excluded *L. fusiformis*.

Antimicrobial defense peptides extracted from the skin of *C. panamansis* differed significantly among treatments in quantity and in relative abundance of peptide mass 1064. This peptide will be targeted in future studies for primary structure determination and for testing of antifungal function. *C. panamansis* exposed to potentially beneficial bacteria and other host factors in the mucus washed from the skin of *E. prosoblepon* did not increase overall peptide quantity, but did show an increase in the relative abundance of peptide mass 1064 and a corresponding increase in *Bd* inhibition at a standardized concentration of 500 $\mu g\ ml^{-1}$ peptide. *C. panamansis* exposed to cultured *L. fusiformis* produced greater quantities of AMPs than frogs in the other two treatment groups, leading to greater defense capacity against *Bd* (Fig. 1), and suggesting a generalized induced immune response. Similarly, Schadich et al. [70] described increased peptide production in the frog *Litoria raniformis* induced by exposure to the pathogenic bacterium *Klebsiella pneumoniae*. In this study, induction of skin defense peptides likely contributed to the elimination of *L. fusiformis*, and the inability of the probiotic to establish within the host skin microbial community. At the same time, frogs in the *L. fusiformis* treatment lost significantly more weight than frogs in other treatments, indicating a potential cost to immune activation [71]. Certainly, other host responses in addition to skin peptides may have occurred simultaneously, contributing to the observed treatment effect on mass loss.

Susceptibility to Chytridiomycosis

Soon after the arrival of *Bd* at Omar Torrijos National Park in 2004, *C. panamansis* populations declined critically [27,29], whereas the frogs sampled near the Sierra Llorona lodge appeared to be

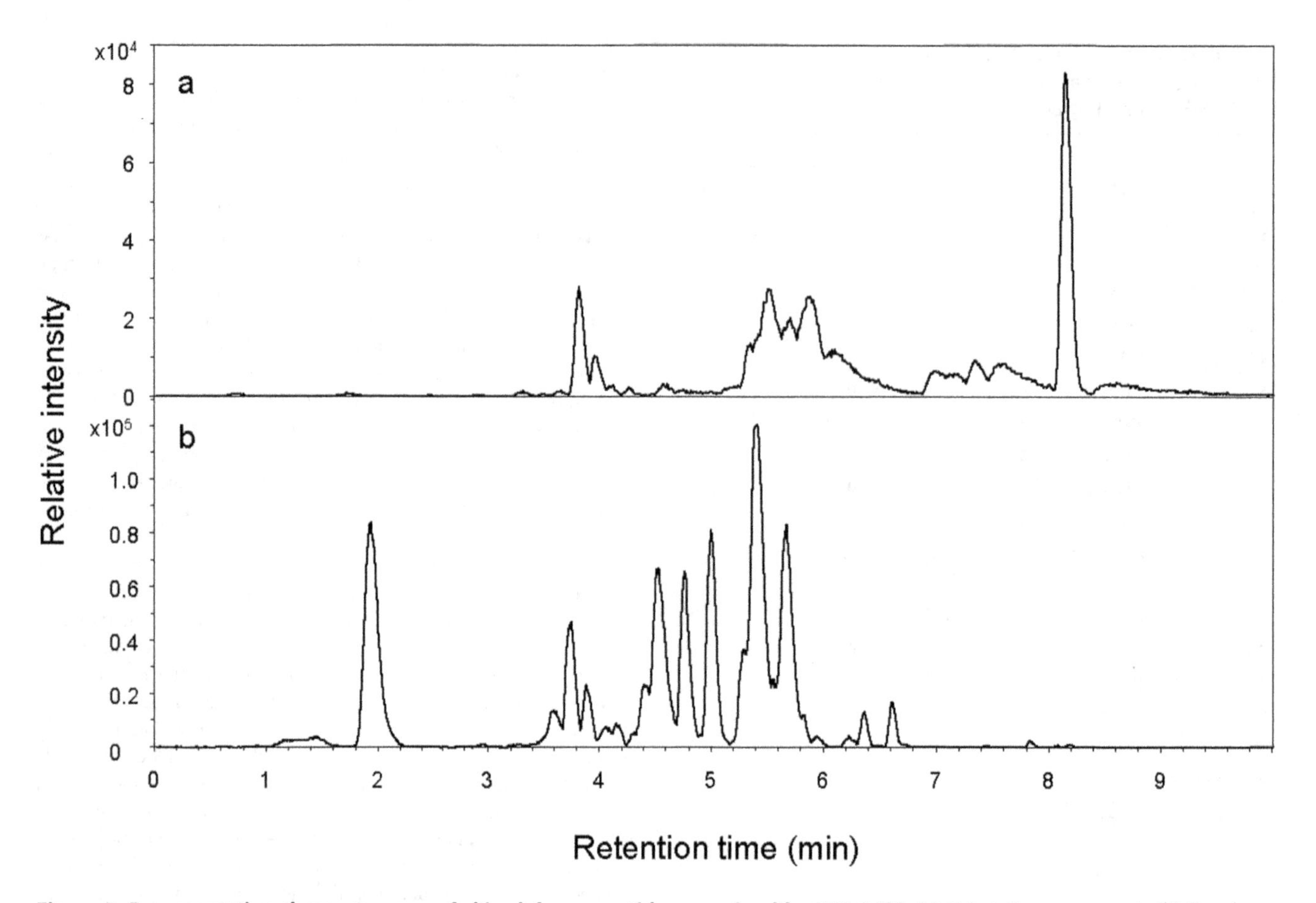

Figure 3. Representative chromatograms of skin defense peptides examined by HPLC-MS. (a) *Colostethus panamansis.* (b) *Espadarana prosoblepon.* Values of molecular weight and mean area for the detected peptides are reported in Table 2.

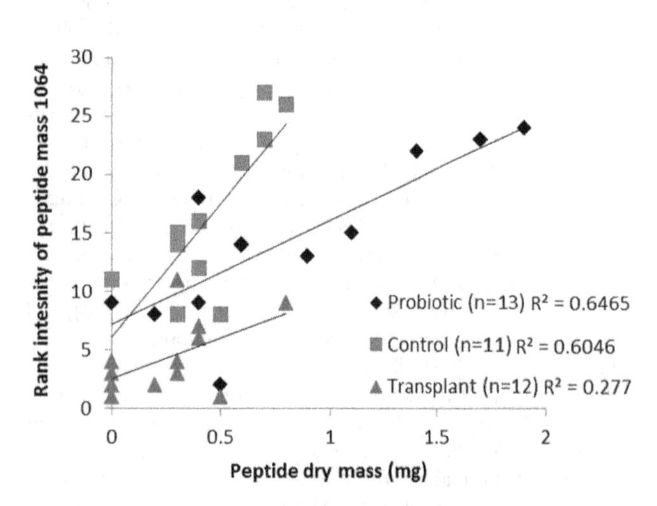

Figure 4. Peptide dry mass predicts a significant proportion of variation in peptide intensity determined by LC-MS. Overall, there was a significant correlation between peptide dry mass and rank abundance of peptide mass 1064. This relationship differed among treatments indicating a change in the relative abundance of the peptide components of skin secretions depending on microbial treatment. Transplant treated frogs had the highest relative abundance of peptide mass 1064 (lowest rank), and probiotic treated frogs the lowest relative abundance with controls intermediate. Probiotic treated frogs had significantly higher total quantity of peptides than transplant treated frogs (see text).

coexisting with *Bd*, with a prevalence of 11.4% (95% binomial confidence interval: 3.8–24.6%). That *C. panamansis* are able to persist in an area with *Bd* may be due to infection tolerance [72] or related to habitat characteristics. Temperatures in the lowland rainforest near Sierra Llorona lodge are typically higher than in the cloud forest habitat at Omar Torrijos National Park. Environmental factors such as temperature can also influence the synthesis and expression of skin defense peptides in amphibians [37,73]. While none of the frogs in this study showed clinical signs of chytridiomycosis, infection status may be an important driver of immune function, or a response to immune function including AMPs and microbiota, and thus an important target for future investigation. In particular, does microbial therapy have the same effect as a treatment of infection as it does as a prophylactic treatment?

Based on samples taken before *Bd* emergence at Omar Torrijos National Park [58], *E. prosoblepon* skin defense peptides were expected to be more effective against *Bd* growth than *C. panamansis* peptides. Thus, similar or greater *Bd* growth inhibition caused by skin peptides from all three treatment-groups of *C. panamansis* compared to *E. prosoblepon* peptides in this study was unexpected. In contrast to *C. panamansis*, *E. prosoblepon* has been able to survive for more than 8 yr at Omar Torrijos National Park, and 16 yr at Fortuna in the presence of *Bd* [27,30]. We found higher values of peptide effectiveness against *Bd* than previously reported for *C. panamansis*, and this might be explained by population origin of the frogs. Glass frogs, *E. prosoblepon*, were sampled from the same upland site as in the previous study, while *C. panamansis* were extirpated from the upland site and for this study frogs were captured from a lowland rainforest habitat. AMPs from these frogs may have been up-regulated by exposure to *Bd* or microbiota, or AMP defenses may differ among populations or habitats [74]. Stressors may also differ among sites, and long-term upregulation of stress hormones including glucocorticoids can suppress immunity [75] including AMP skin defenses in amphibians [76]. Besides the invasion of *Bd* at Omar Torrijos National Park, stressors have not been reported [29].

Considerations for Probiotic Therapy

Promoting and sustaining human health through strategies that manipulate microbial communities is a long-term goal of the Human Microbiome Project [77]. Thus, amphibians and other model vertebrate systems are important for examining host-microbiota interactions to gain a mechanistic understanding of microbial community assembly and maintenance [62]. Probiotic disease mitigation is also high on the list of conservation options available for threatened amphibians [13,24].

An intuitive strategy of reducing the biomass or diversity of resident skin microbiota may aid in the establishment of new bacteria by minimizing community interactions. However, antibiotic pre-treatment interfered with intestinal microbial community establishment in rats [78]. Becker et al. [35] first washed golden frogs, *Atelopus zeteki*, in a 1.5% solution of hydrogen peroxide to reduce microbiota before probiotic treatment. However, the bacterium *Janthinobacterium lividum* did not establish on the skin of the frogs. Pathogens can become established in hosts treated with antibiotics by exploiting the reduced competitiveness of the disturbed community [79], and intestinal disease has been linked to the outgrowth or loss of certain components of the microbiota [80–82]. Conversely, beneficial bacteria can also establish in hosts and many examples of successful probiotic use have been reported in aquaculture [83], livestock and poultry production [84], as well as in human medicine [85,86].

A recommended step for probiotic application is to use small probiotic doses and to wash bacterial cultures in a physiological solution to ensure that hosts are exposed to the living cells only, minimizing exposure to metabolic products of the bacteria including immunomodulatory toxins [34,35]. Metabolites from unwashed whole cultures may help bacteria in microbial competitive interactions; however, toxins or inordinately large probiotic doses may also elicit host immune responses. It remains unclear whether pre-treatment steps to reduce endogenous microbiota or to wash beneficial bacteria are necessary to introduce an exogenous bacterium into an existing microbial community, but this is a critical consideration for use of probiotics in disease management. The bacterium *J. lividum*, used by Harris et al. [34] on *Rana muscosa* was likely already present on many of the frogs and represents a bio-augmentation experiment. Thus, altering relative population sizes and community function within an established microbiome may be more feasible than altering community membership.

The composition of microbiota can affect host immune responses and influence disease outcome. For individuals with functional skin microbiota and immune defense, colonization resistance can be beneficial, for example in times of environmental change. On the other hand, a resistant or resilient microbiota is not desirable for enhancing host disease resistance through probiotic therapy. Probiotic therapies aim to alter the microbial community to a new stable state that is more protective than the previous state [7]. Establishment of novel microbiota may require methods to circumvent host mechanisms maintaining the microbiota. In the case of *C. panamansis* from Panamanian lowlands, the combination of microbiota that are resistant to colonization, and AMPs effective at inhibiting *Bd* growth, may favor infection tolerance and population persistence. The continuing development of probiotic strategies offers hope for populations threatened by infectious disease.

Acknowledgments

We thank Julie Ray at LaMICA biological station for logistical support and assistance with frog collection, Heidi Ross and EVACC for assistance with animal care, peptide work and permits, Matt Becker and Kevin Minbiole for identification and chemical profiling of *L. fusiformis*, and Roberto Ibáñez, Laura Reinert, and Louise Rollins-Smith for qPCR. We thank Sarah Higginbotham for support with bacteria work, El Níspero Zoo for providing space for experiments, Christian Mayer for assistance with T-RFLP, Matt Becker, Uli Reyer, Silvia Rauch, Christine Avena and the Microbiome Reading Group at the University of Colorado for comments on the manuscript.

Author Contributions

Conceived and designed the experiments: DW DK EG. Performed the experiments: DK BG EG. Analyzed the data: DK DW LB LD. Contributed reagents/materials/analysis tools: DW LB EG. Wrote the paper: DK DW BG LB.

References

1. Belden LK, Harris RN (2007) Infectious diseases in wildlife: the community ecology context. Front Ecol Environ 5: 533–539.
2. Blaser M, Falkow S (2009) What are the consequences of the disappearing human microbiota? Nat Rev Microbiol 7: 887–894.
3. Hajishengallis G, Darveau RP, Curtis MA (2012) The keystone-pathogen hypothesis. Nat Rev Microbiol 10: 717–725.
4. Fierer N, Ferrenberg S, Flores GE, Gonzalez A, Kueneman J, et al. (2012) From animalcules to an ecosystem: Application of ecological concepts to the human microbiome. Annu Rev Ecol Evol Sys 43: 137–155.
5. Shade A, Peter H, Allison AD, Baho DL, Berga M, et al. (2012) Fundamentals of microbial community resistance and resilience. Front Microbiol 3: 1–19.
6. Allison SD, Martiny JBH (2008) Resistance, resilience, and redundancy in microbial communities. Proc Natl Acad Sci U S A 105: 11512–11519.
7. Reid G, Younes JA, Van der Mei HC, Gloor GB, Knight R (2011) Microbiota restoration: natural and supplemental recovery of human microbial communities. Nat Rev Microbiol 9: 27–38.
8. Blaser M (2011) Antibiotic overuse: Stop the killing of beneficial bacteria. Nature 476: 393–394.
9. Kelly CP (2013) Fecal microbiota transplantation – an old therapy comes of age. N Engl J Med 368: 474–475.
10. van Nood E, Vrieze A, Nieuwdorp N, Fuentes S, Zoetendal EG, et al. (2013) Duodeal infusion of donor feces for recurrent *Clostridium difficile*. N Engl J Med 368: 407–415.
11. Turnbaugh PJ, Ley RE, Hamady M, Fraser-Liggett CM, Knight R, et al. (2007) The Human Microbiome Project. Nature 449: 804–810.
12. Rosenberg E, Zilber-Rosenberg I (2011) Symbiosis and development: the hologenome concept. Birth Defects Res C Embryo Today 93: 56–66.
13. Bletz MC, Loudon AH, Becker MH, Bell SC, Woodhams DC, et al. (2013) Mitigating amphibian chytridiomycosis with bioaugmentation, characteristics of effective probiotics and strategies for their selection and use. Ecol Lett 16: 807–820.
14. Taylor SK, Green DE, Wright KM, Whitaker BR (2001) Bacterial Diseases. In: Wright KM, Whitaker BR (Eds.), Amphibian Medicine and Captive Husbandry. Krieger Publishing Co., Malabar, FL, 159–179.
15. Lauer A, Simon MA, Banning JL, André E, Duncan K, et al. (2007) Common cutaneous bacteria from the eastern red-backed salamander can inhibit pathogenic fungi. Copeia 2007: 630–640.
16. Lauer A, Simon MA, Banning JL, Lam BA, Harris RN (2008) Diversity of cutaneous bacteria with antifungal activity isolated from female four-toed salamanders. ISME J 2: 145–157.
17. Flechas SV, Sarmiento C, Cárdenas ME, Medina EM, Restrepo S, et al. (2012) Surviving chytridiomycosis: Differential anti-*Batrachochytrium dendrobatidis* activity in bacterial isolates from three lowland species of *Atelopus*. PLoS ONE 7(9): e44832. doi:10.1371/journal.pone.0044832

18. Woodhams DC, Vredenburg VT, Simon MA, Billheimer D, Shakhtour B, et al. (2007) Symbiotic bacteria contribute to innate immune defences of the threatened mountain yellow-legged frog, *Rana muscosa*. Biol Cons 138: 390–398.
19. Lam BA, Walke JB, Vredenburg VT, Harris RN (2010) Proportion of individuals with anti-*Batrachochytrium dendrobatidis* skin bacteria is associated with population persistence in the frog *Rana muscosa*. Biol Cons 143: 529–531.
20. McKenzie VJ, Bowers RM, Fierer N, Knight R, Lauber CL (2012) Co-habiting amphibian species harbor unique skin bacterial communities in wild populations. ISME J 6: 588–596.
21. Berger L, Speare R, Daszak P, Green DW, Cunningham AA, et al. (1998) Chtytridiomycosis causes amphibian mortality associated with population declines in the rain forests of Australia and Central America. Proc Natl Acad Sci U S A 95: 9031–9036.
22. Fisher MC, Henk DA, Briggs CJ, Brownstein JS, Madoff LC, et al. (2012) Emerging fungal threats to animal, plant and ecosystem health. Nature 484: 186–194.
23. Global *Bd*-Mapping Project, Available: http://www.bd-maps.net/, Accessed 2013 Jul 7.
24. Woodhams DC, Bosch J, Briggs CJ, Cashins S, Davis LR, et al. (2011) Mitigating amphibian disease: Strategies to maintain wild populations and control chytridiomycosis. Front Zool 8: 8.
25. Rowley JJL, Alford RA (2013) Hot bodies protect amphibians against chytrid infection in nature. Sci Rep 3: 1515.
26. Skerratt LF, Berger L, Speare R, Cashins S, McDonald KR, et al. (2007) Spread of chytridiomycosis has caused the rapid global decline and extinction of frogs. EcoHealth 4: 125–134.
27. Crawford AJ, Lips KR, Bermingham E (2010) Epidemic disease decimates amphibian abundance, species diversity, and evolutionary history in the highlands of central Panama. Proc Natl Acad Sci U S A 107: 13777–13782.
28. Whiles MR, Lips KR, Pringle CM, Kilham SS, Bixby RJ, et al. (2006) The effects of amphibian population declines on the structure and function of Neotropical stream ecosystems. Front Ecol Environ 4: 27–34.
29. Lips KR, Brem F, Brenes R, Reeve JD, Alford RA, et al. (2006) Emerging infectious disease and the loss of biodiversity in a Neotropical amphibian community. Proc Natl Acad Sci U S A 103: 3165–3170.
30. Woodhams DC, Kilburn VL, Reinert LK, Voyles J, Meidna D, et al. (2008) Chytridiomycosis and amphibian population declines continue to spread eastward in Panama. EcoHealth 5: 268–274.
31. Harris RN, James TY, Lauer A, Simon MA, Patel A (2006) Amphibian pathogen *Batrachochytrium dendrobatidis* is inhibited by the cutaneous bacteria of amphibian species. EcoHealth 3: 53–56.
32. Walke JB, Harris RN, Reinert LK, Rollins-Smith LA, Woodhams DC (2011) Social immunity in amphibians: evidence for vertical transmission of innate defences. Biotropica 43: 396–400.
33. Vredenburg VT, Briggs CJ, Harris RN (2011) Host-pathogen dynamics of amphibian chytridiomycosis: the role of the skin microbiome in health and disease. *Fungal diseases: an emerging challenge to human, animal, and plant health* (Olsen L, Choffnes E, Relman DA & Pray L, eds), 342–355. The National Academies Press, Washington, DC.
34. Harris RN, Brucker RM, Walke JB, Becker MH, Schwantes CR, et al. (2009) Skin microbes on frogs prevent morbidity and mortality caused by a lethal skin fungus. ISME J 3: 818–824.
35. Becker MH, Harris RN, Minbiole KP, Schwantes CR, Rollins-Smith LA, et al. (2011) Towards a better understanding of the use of probiotics for preventing chytridiomycosis in Panamanian golden frogs. Ecohealth 8: 501–506.
36. Woodhams DC, Geiger CC, Reinert LK, Rollins-Smith LA, Lam B, et al. (2012) Treatment of amphibians infected with chytrid fungus: learning from failed trials with itraconazole, antimicrobial peptides, bacteria, and heat therapy. Dis Aquat Organ 98: 1–25.
37. Rollins-Smith LA, Woodhams DC (2012) Amphibian Immunity: Staying in tune with the environment. In: Demas GE, Nelson RI (Eds.), Ecoimmunology. Oxford University Press, New York, 92–143.
38. Nicolas P, Mor A (1995) Peptides as weapons against microorganisms in the chemical defense system of vertebrates. Annu Rev Microbiol 49: 277–304.
39. Rinaldi AC (2002) Antimicrobial peptides from amphibian skin: an expanding scenario. Curr Opin Chem Biol 6: 799–804.
40. Zasloff M (2002) Antimicrobial peptides of ulticellular organisms. Nature 415: 389–395.
41. Apponyi MA, Pukala TL, Brinkworth CS, Vaselli VM, Bowie JH, et al. (2004) Host-defence peptides of Australian anurans: structure, mechanisms of action and evolutionary significance. Peptides 25: 1035–1054.
42. Conlon JM, Kolodziejek J, Nowotny N (2004) Antimicrobial peptides from ranid frogs: taxonomic and phylogenetic markers and a potential source of new therapeutic agents. Biochim Biophys Acta 1696: 1–14.
43. Pask J, Woodhams DC, Rollins-Smith LA (2012) The ebb and flow of antimicrobial skin peptides defends northern leopard frogs, *Rana pipiens*, against chytridiomycosis. Glob Change Biol 18: 1231–1238.
44. Hancock REW, Nijnik A, Philpott DJ (2012) Modulating immunity as a therapy for bacterial infections. Nat Rev Microbiol 10: 243–254.
45. Chinchar VG, Wang J, Murti G, Carey C, Rollins-Smith L (2001) Inactivation of frog virus 3 and channel catfish virus by esculentin-2P and ranatuerin-2P, two antimicrobial peptides isolated from frog skin. Virology 288: 351–357.
46. Rollins-Smith LA, Doersam JK, Longcore JE, Taylor SK, Shamblin JC, et al. (2002) Antimicrobial peptide defenses against pathogens associated with global amphibian declines. Dev Comp Immunol 26: 63–72.
47. Woodhams DC, Rollins-Smith LA, Carey C, Reinert LK, Tyler MJ, et al. (2006) Population trends associated with antimicrobial peptide defenses against chytridiomycosis in Australian frogs. Oecologia 146: 531–540.
48. Easton DM, Nijnik A, Mayer ML, Hancock REW (2009) Potential of immunomodulatory host defense peptides as novel anti-infectives. Trends Biotechnol 27: 582–590.
49. Radek KA, Elias PM, Taupenot L, Mahata SK, O'Conner DT (2010) Neuroendocrine nicotinic receptor activation increases susceptibility to bacterial infections by suppressing antimicrobial peptide production. Cell Host Microbe 7: 277–289.
50. Salzman NH, Hung K, Haribhai D, Chu H, Karlsson-Sjoberg J, et al. (2010) Enteric defensins are essential regulators of intestinal microbial ecology. Nat Immunol 11: 76–83.
51. Boyle DG, Boyle DB, Olsen V, Morgan JAT, Hyatt AD (2004) Rapid quantitative detection of chytridiomycosis *Batrachochytrium dendrobatidis* in amphibian samples using real-time Taqman PCR assay. Dis Aquat Organ 60: 141–148.
52. Wang J, Fan Y, Yao Z (2010) Isolation of a *Lysinibacillus fusiformis* strain with tetrodotoxin-producing ability from puffer fish *Fugu obscurus* and the characterization of this strain. Toxicon 56: 640–643.
53. Hartmann M, Frey B, Kölliker R, Widmer F (2005) Semi-automated genetic analyses of soil microbial communities: comparison of T-RFLP and RISA based on descriptive and discriminative statistical approaches. J Microbiol Methods 61: 349–360.
54. Hartmann M, Widmer F (2006) Community structure analyses are more sensitive to differences in soil bacterial communities than anonymous diversity indices. Appl Environ Microbiol 72: 7804–7812.
55. Hartmann M, Widmer F (2008) Reliability for detecting composition and changes of microbial community by T-RFLP genetic profiling. FEMS Microbiol Ecol 63: 249–260.
56. Widmer F, Rasche F, Hartmann M, Fliessbach A (2006) Community structures and substrate utilization of bacteria in soils from organic and conventional farming systems of the DOK long-term field experiment. Appl Soil Ecol 33: 294–307.
57. Piotrowski JS, Annis SL, Longcore JE (2004) Physiology of *Batrachochytrium dendrobatidis*, a chytrid pathogen of amphibians. Mycologia 96: 9–15.
58. Woodhams DC, Voyles J, Lips KR, Carey C, Rollins-Smith LA (2006) Predicted disease susceptibility in a Panamanian amphibian assemblage based on skin peptide defenses. J Wildl Dis 42: 207–218.
59. McClanahan L, Baldwin R (1969) Rate of water uptake through the integument of the desert toad, *Bufo punctatus*. Comp Biochem Physiol 28: 381–389.
60. Lueders T, Friedrich MW (2003) Evaluation of PCR amplification bias by T-RFLP analysis of SSU rRNA and *mcrA* genes using defined template mixtures of methanogenic pure cultures and soil DNA extracts. Appl Environ Microbiol 69: 320–326.
61. Daly JW, Gusovsky F, Myers CW, Yotsu-Yamashita M, Yasumoto T (1994) First occurrence of tetrodotoxin in a dendrobatid frog *Colostehus inguinalis*, with further reports for the bufonid genus *Atelopus*. Toxicon 32: 279–285.
62. Roeselers G, Mittge EK, Stephens WZ, Parichy DM, Cavanaugh CM, et al. (2011) Evidence for a core gut microbiota in the zebrafish. ISME J 5: 1595–1608.
63. Miele R, Ponti D, Boman HG, Barra D, Simmaco M (1998) Molecular cloning of a bombinin gene from *Bombina orientalis*: Detection of NF-κB and NF-IL-6 binding sites in its promoter. FEBS Lett 431: 23–28.
64. Kanther M, Rawls JF (2010) Host-microbe interactions in the developing zebrafish. Curr Opin Immunol 22: 10–19.
65. Gonzalez A, Clemente JC, Shade A, Metcalf JL, Song S, et al. (2011) Our microbial selves: what ecology can teach us. EMBO Rep 12: 775–784.
66. Chan RCY, Reid G, Irvin RT, Bruce AW, Costerton JW (1985) Competitive exclusion of uropathogens from human uroepithelial cells by *Lactobacillus* whole cells and cell wall fragments. Infect Immun 47: 84–89.
67. Kennedy MJ, Volz PA (1985) Ecology of *Candida albicans* gut colonization: inhibition of *Candida* adhesion, colonization, and dissemination from the gastrointestinal tract by bacterial antagonism. Infect Immun 49: 654–663.
68. Boskey ER, Telsch KM, Whaley KJ, Moench TR, Cone RA (1999) Acid production by vaginal flora in vitro is consistent with the rate and extent of vaginal acidification. Infect Immun 67: 5170–5175.
69. Dobson A, Cotter PD, Ross RP, Hill C (2012) Bacteriocin production, a probiotic trait? Appl Environ Microbiol 78: 1–6.
70. Schadich E, Cole ALJ, Mason D, Squire M (2009) Effect of the pesticide carbaryl on the production of the skin peptides of *Litoria raniformis* frogs. Austral J Ecotox 15: 17–24.
71. Schmid-Hempel P, Ebert D (2003) On the evolutionary ecology of specific immune defence. Trends Ecol Evol 18: 27–32.
72. Woodhams DC, Bigler L, Marschang R (2012). Tolerance of fungal infection in European water frogs exposed to *Batrachochytrium dendrobatidis* after experimental reduction of innate immune defences. BMC Vet Res 8: 197.
73. Mattute B, Storey KB, Knoop FC, Conlon JM (2000) Induction of synthesis of an antimicrobial peptide in the skin of the freeze-tolerant frog, *Rana sylvatica*, in response to environmental stimuli. FEBS Lett 483: 135–138.

74. Tennessen JA, Woodhams DC, Chaurand P, Reinert LK, Billheimer D, et al. (2009) Variations in the expressed antimicrobial peptide repertoire of northern leopard frog *Rana pipiens* populations suggest intraspecies differences in resistance to pathogens. Dev Comp Immunol 33: 1247–1257.
75. Rhen T, Cidlowski JA (2005) Antiinflammatory action of glucocorticoids - new mechanisms for old drugs. N Engl J Med 353: 1711–1723.
76. Simmaco M, Boman A, Mangoni ML, Mignogna G, Miele R, et al. (1997) Effect of glucocorticoids on the synthesis of antimicrobial peptides in amphibian skin. FEBS Lett 416: 273–275.
77. Peterson J, Garges S, Giovanni M, McInnes P, Wang L, et al. (2009). The NIH Human Microbiome Project. Genome Res 19: 2317–2323.
78. Manichanh C, Reeder J, Gibert P, Varela E, Llopis M, et al. (2010) Reshaping the gut microbiome with bacterial transplantation and antibiotic intake. Genome Res 20: 1411–1419.
79. Brook I (2005) The role of bacterial interference in otitis, sinusitis and tonsillitis. Otolaryngol Head Neck Surg 133: 139–146.
80. Frank DN, St Amand AL, Feldman RA, Boedeker EC, Harpaz N, et al. (2007) Molecular-phylogenetic characterization of microbial community imbalances in human inflammatory bowel diseases. Proc Natl Acad Sci U S A 104: 13780–13785.
81. Sartor RB (2008) Microbial influences in inflammatory bowel diseases. Gastroenterology 134: 577–594.
82. Sokol H, Seksik P, Furet JP, Firmesse O, Nion-Larmurier I, et al. (2009) Low counts of *Faecalibacterium prausnitzii* in colitis microbiota. Inflamm Bowel Dis 15: 1183–1189.
83. Irianto A, Austin B (2002) Probiotics in aquaculture. J Fish Dis 25: 633–642.
84. Patterson JA, Burkholder KM (2003) Application of prebiotics and probiotics in poultry production. Poult Sci 82: 627–631.
85. Rastall RA, Gibson GR, Gill HS, Guarner F, Klaenhammer TR, et al. (2005) Modulation of the microbial ecology of the human colon by probiotics-prebiotics, and synbiotics to enhance human health: an overview of enabling science and potential applications. FEMS Microbiol Ecol 52: 145–152.
86. Gareau MG, Sherman PM, Walker WA (2010) Probiotics and the gut microbiota in intestinal health and disease. Nat Rev Gastroenterol Hepatol 7: 503–514.

Altered Mucus Glycosylation in Core 1 O-Glycan-Deficient Mice Affects Microbiota Composition and Intestinal Architecture

Felix Sommer[1,2], Nina Adam[1], Malin E. V. Johansson[2], Lijun Xia[3], Gunnar C. Hansson[2*], Fredrik Bäckhed[1,4*]

1 The Wallenberg Laboratory and Sahlgrenska Center for Cardiovascular and Metabolic Research, Department of Molecular and Clinical Medicine, University of Gothenburg, Gothenburg, Sweden, **2** Mucin Biology Group, Department of Medical Biochemistry, University of Gothenburg, Gothenburg, Sweden, **3** Cardiovascular Biology Research Program, Oklahoma Medical Research Foundation, Oklahoma City, Oklahoma, United States of America, **4** Novo Nordisk Foundation Center for Basic Metabolic Research, Section for Metabolic Receptology and Enteroendocrinology, Faculty of Health Sciences, University of Copenhagen, Copenhagen, Denmark

Abstract

A functional mucus layer is a key requirement for gastrointestinal health as it serves as a barrier against bacterial invasion and subsequent inflammation. Recent findings suggest that mucus composition may pose an important selection pressure on the gut microbiota and that altered mucus thickness or properties such as glycosylation lead to intestinal inflammation dependent on bacteria. Here we used TM-IEC *C1galt*$^{-/-}$ mice, which carry an inducible deficiency of core 1-derived O-glycans in intestinal epithelial cells, to investigate the effects of mucus glycosylation on susceptibility to intestinal inflammation, gut microbial ecology and host physiology. We found that TM-IEC *C1galt*$^{-/-}$ mice did not develop spontaneous colitis, but they were more susceptible to dextran sodium sulphate-induced colitis. Furthermore, loss of core 1-derived O-glycans induced inverse shifts in the abundance of the phyla Bacteroidetes and Firmicutes. We also found that mucus glycosylation impacts intestinal architecture as TM-IEC C1galt$^{-/-}$ mice had an elongated gastrointestinal tract with deeper ileal crypts, a small increase in the number of proliferative epithelial cells and thicker circular muscle layers in both the ileum and colon. Alterations in the length of the gastrointestinal tract were partly dependent on the microbiota. Thus, the mucus layer plays a role in the regulation of gut microbiota composition, balancing intestinal inflammation, and affects gut architecture.

Editor: Yolanda Sanz, Instutite of Agrochemistry and Food Technology, Spain

Funding: This study was supported by the Swedish Research Council, Torsten and Ragnar Söderbergs' foundations, IngaBritt and Arne Lundberg's foundation, Swedish Foundation for Strategic Research - The Mucus-Bacteria-Colitis Center (MBC), Knut and Alice Wallenberg foundation, National Institute of Health (U01AI095473, R01DK085691), and the regional agreement on medical training and clinical research (ALF) between Region Västra Götaland and Sahlgrenska University Hospital. The funders had no role in study design, data collection and analysis, decision to publish, or preparation of the manuscript.

Competing Interests: The authors have declared that no competing interests exist.

* E-mail: fredrik.backhed@wlab.gu.se (FB); gunnar.hansson@medkem.gu.se (GCH)

Introduction

The gastrointestinal tract harbours the most densely populated microbial ecosystem known encompassing more than 10^{14} bacteria, which outnumbers the cells of the human body by an order of magnitude. The gut microbiota resides in very close proximity to the host epithelium; however, despite this close association, the intestinal tract is normally healthy and free of inflammation primarily owing to a mechanical separation of host and microbial cells [1–3]. This separation is caused by the presence of several physical and biochemical barriers, the most important being gelatinous mucus, which covers the whole gastrointestinal tract [2]. Defects in the mucus layer are associated with intestinal inflammation [3,4].

The mucus layer in the small intestine and colon is mainly made up of multimers of the mucin MUC2, a highly O-glycosylated protein of approximately 5200 amino acids and 2.5 MDa [5]. O-glycans contribute to about 80% of its mass and therefore mainly determine the physical mucus properties. O-glycosylation of MUC2 occurs post-translationally in the Golgi apparatus starting with the addition of the initial N-acetylgalactosamine (GalNAc) to the hydroxyl groups of serine and threonine in its two PTS domains (rich in proline, threonine and serine), where the larger is a stretch of about 2300 aa centrally located in MUC2 and rich in proline, threonine and serine [6–8]. The resulting GalNAcα-O-Ser/Thr structure is known as the Tn antigen and is normally not detectable as it is extended and branched by the action of other glycosyltransferases. The primary enzymes in this process are the core 1 β1,3-galactosyltransferase (C1galt1) in mice and core 3 β1,3-N-acetylglucosaminyltransferase (C3GnT) in humans [9,10].

Mucus glycans not only protect the epithelial layer, but they also serve as an adhesion substrate for bacteria expressing adhesins and are a nutrient source for bacteria by hydrolysis through glycosidases [11,12]. Thus, glycans could be an important factor in the selection of a beneficial microbiota and homeostasis. Indeed, glycosyltransferases have been shown to impact both intestinal inflammation and microbiota composition. In humans, loss of the α-1,2-fucosyltransferase FUT2, which is involved in the formation of ABO blood group antigens on the intestinal mucosa and in body fluids, also leads to an altered microbiota and increased susceptibility to infection and inflammatory disease such as Crohn's disease [13]. In mice, another blood-group-related glycosyltransferase β-1,4-N-acetylgalactosaminyltransferase 2 (B4galnt2) has been shown to affect gut microbiota composition [14] and thereby may affect susceptibility to intestinal inflammatory diseases.

Furthermore, mice fostered with milk of mothers deficient in the α2,3-sialyltransferase St3gal4 harbour an altered gut microbiota and are more resistant to dextran sodium sulphate (DSS)-induced colitis [15]. Finally, loss of core 3-derived O-glycans results in greater susceptibility to DSS-induced colitis [16] and loss of core 1-derived O-glycans has been reported to lead to spontaneous colitis [17]. Inflammation is in both models caused by altered mucus properties that abolish the separation of host epithelium and intestinal bacteria, thereby allowing bacterial penetration and overgrowth [4]. Disease development can be ameliorated by antibiotic treatment [17].

Together, these data indicate a causal role for the gut microbiota in the induction of colitis in susceptible hosts. However, it remains to be determined if the altered mucus in these models selects for a more colitogenic microbiota that then causes inflammation. Here we used mice that carry an inducible deficiency of core 1-derived O-glycans in intestinal epithelial cells to investigate if mucus glycosylation affects gut microbial ecology and thereby host physiology as well as susceptibility to intestinal inflammation.

Materials and Methods

Mice

C1galt1$^{f/f}$;Villin-Cre-ER^{T2} (TM-IEC *C1galt1*$^{-/-}$) mice have been previously described [17] and were rederived as germ-free at Taconic. Germ-free mice were maintained in flexible film isolators. A conventional cohort was established by colonizing germ-free TM-IEC *C1galt1*$^{-/-}$ mice with caecal flora of C57BL/6 mice and bred for at least three generations in our animal facility under specific pathogen-free conditions. All mice were housed under a 12-h light cycle and fed autoclaved chow diet ad libitum (Labdiet, St Louis, Missouri, USA). TM-IEC *C1galt1*$^{-/-}$ mice were bred using heterozygous setup for Villin-Cre-ER^{T2} alleles to facilitate littermate controls. Excision of the *C1galt1* gene was induced by i.p. injection of 1 mg tamoxifen for five consecutive days. Experiments were initiated after a further ten days. All experiments were performed using protocols approved by the Gothenburg Animal Ethics Committee (339-2012, 280-2012 and 281-2012).

Histology

For histological analyses, we used 8-9-week-old female mice that were killed by cervical dislocation. Intestinal specimens were harvested and fixed in methacarn (60% dry methanol, 30% dry chloroform, 10% glacial acetic acid) for 1-2 weeks at room temperature prior to paraffin embedding and sectioning. Sections were stained with haematoxylin/eosin (HE) for morphology, Alcian blue/periodic acid-Schiff (AB-PAS) stain for glycan composition, and Mab anti-Tn antibody (clone 5F4, [18]) with rat-anti-mouse-FITC antibody (BD Pharmingen) for detection of the Tn antigen. Muscle was visualized using rabbit-anti-smooth muscle actin antibody coupled to Cy3 fluorochrome (Sigma) and proliferative cells were stained using rabbit-anti-Ki-67 (Thermo Scientific) antibody and Vectastain Elite ABC kit (Rabbit IgG, Vector labs). HE and HOECHST were used as counterstaining for enzymatic and fluorescent detection, respectively. Villus length and crypt depth were assessed using HE-stained sections. Microscopy measurements were performed using Zeiss Axiovision LE v4.8 on sections of at least five mice per genotype with each 15 individual scores. Number of lamina propria cells was counted for 5–7 mice. For ileum five complete villi and for colon ten areas between two complete crypts per mouse were counted.

DSS treatment

Colonic inflammation was induced in 9–14-week-old male and female mice by administration of 3% DSS (TdB TdB Consultancy, Uppsala, Sweden) for five days in drinking water followed by a five-day recovery period before killing the mice by cervical dislocation [19]. Weight and faeces of the mice were monitored daily. Colitis severity was assessed by calculating the disease activity index (DAI) combining the scores of weight loss, stool consistency and faecal bleeding [20]. Briefly, the scoring system was as follows: weight loss: 0 = no loss, 1 = 1–5%, 2 = 5–10%, 3 = 10–20%, 4 = >20%; stool: 0 = normal, 2 = loose stool, 4 = diarrhoea; and bleeding: 0 = no blood, 2 = cryptic blood (Hemoccult positive, Hemoccult II; Beckman Coulter), and 4 = heavy bleeding. Interleukin (IL)-1β and tumor necrosis factor (TNF)α levels were measured in protein lysates from colon tissues of TM-IEC *C1galt1*$^{-/-}$ and wild type mice using Mouse Basic Kit FlowCytomix with IL-1β and TNFα simplex (eBioscience) according to the manufacturer's instructions.

Microbiota analysis by 16S rDNA pyro-sequencing

Total genomic DNA was extracted from 100–200 mg snap frozen caecum of 8–9-week-old female mice as described previously [21–23]. The V2–V3 region of the 16S rDNA gene was amplified using barcoded primers. Purified amplicons were pooled and concentration was adjusted to 20 ng/μl for 454 pyrosequencing. Sequence data were analyzed using MacQIIME package v1.6 (http://www.wernerlab.org/software/macqiime, [24]). Briefly, sequencing reads were trimmed and mapped onto the different samples using the barcode information. Next, reads were assigned to operational taxonomic units (OTUs) using 97% identity, representative OTUs were picked and taxonomy assigned. Quality filtering was performed by removing chimeric sequences using ChimeraSlayer and by removing singletons. Relative abundance was calculated by dividing the number of reads for an OTU by the total number of sequences in the sample. Unifrac alpha and beta diversity were calculated and phylogeny constructed using UPGMA (Unweighted Pair Group Method with Arithmetic Mean). Significance of differences in abundances of various taxonomic units was calculated using t-test and false discovery-rate correction in R program.

Results

Loss of core 1-derived O-glycans in adult mice increases susceptibility to DSS-induced intestinal inflammation

In several genetic mouse models used to study intestinal inflammation conditional transgenic mice are used. However, the mice are born with the respective defect causing inflammation, which may affect microbial composition per se, and thus this strategy is suboptimal to elucidate selection effects of host structures on the gut microbiota. Here we used recently developed TM-IEC *C1galt*$^{-/-}$ mice, which carry an inducible deficiency of core 1-derived O-glycans in intestinal epithelial cells in which the floxed *C1galt1* gene is excised only after induction of the otherwise inert Cre recombinase by administration of tamoxifen. In contrast to a previous study in which TM-IEC *C1galt*$^{-/-}$ mice developed intestinal inflammation ten days after induced loss of core 1-derived O-glycans [17], TM-IEC *C1galt*$^{-/-}$ mice in our animal facility did not develop spontaneous colitis or show any signs of inflammation ten days (Fig. 1A, left panels) or even up to 26 weeks

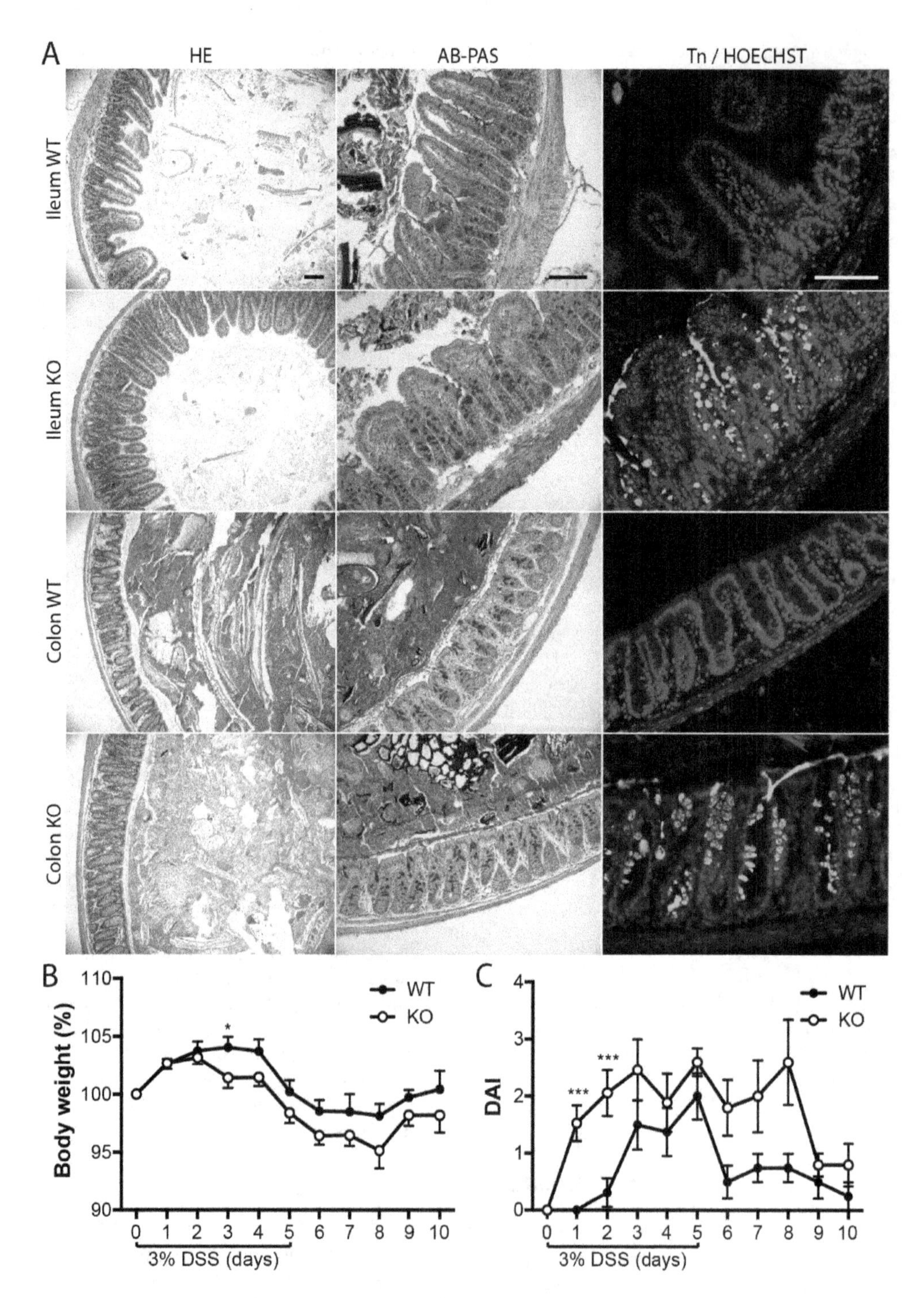

Figure 1. TM-IEC *C1galt*$^{-/-}$ mice are not spontaneously colitic but more susceptible to DSS-induced intestinal inflammation. (A) Histological analysis by HE, AB-PAS and Tn antibody staining of sections taken ten days post knockout induction from ileal and colonic tissue of wild type (WT) and TM-IEC *C1galt*$^{-/-}$ (KO) mice. Scale bars indicate 100 µm. (B) Body weight and (C) DAI during DSS treatment and recovery period (n = 16 per genotype for day 0, two mice sacrificed every day during DSS treatment; Data shows mean ± SEM; ** p<0.01, *** p<0.001).

after knockout induction. This was not caused by an inability to induce the knockout allele since the TM-IEC *C1galt*$^{-/-}$ mice showed an increased ratio of cells producing neutral versus acidic carbohydrates (Fig. 1A, middle panels) and a clear removal of core 1-derived O-glycans exposing the Tn antigen after administration of tamoxifen (Fig. 1A, right panels). Furthermore, TM-IEC *C1galt*$^{-/-}$ mice displayed an altered mucus glycosylation pattern compared with the Cre-negative wild type littermates (data not shown, in agreement with a recent report on the constitutively deleted C1galt1 [10]).

Importantly, the TM-IEC *C1galt*$^{-/-}$ mice responded more quickly and with greater severity to treatment with DSS. After three days of DSS treatment, TM-IEC *C1galt*$^{-/-}$ mice had a significantly lower body weight than the wild type littermate controls and were unable to recover from this weight difference throughout the experiment (Fig. 1B). Furthermore, disease activity

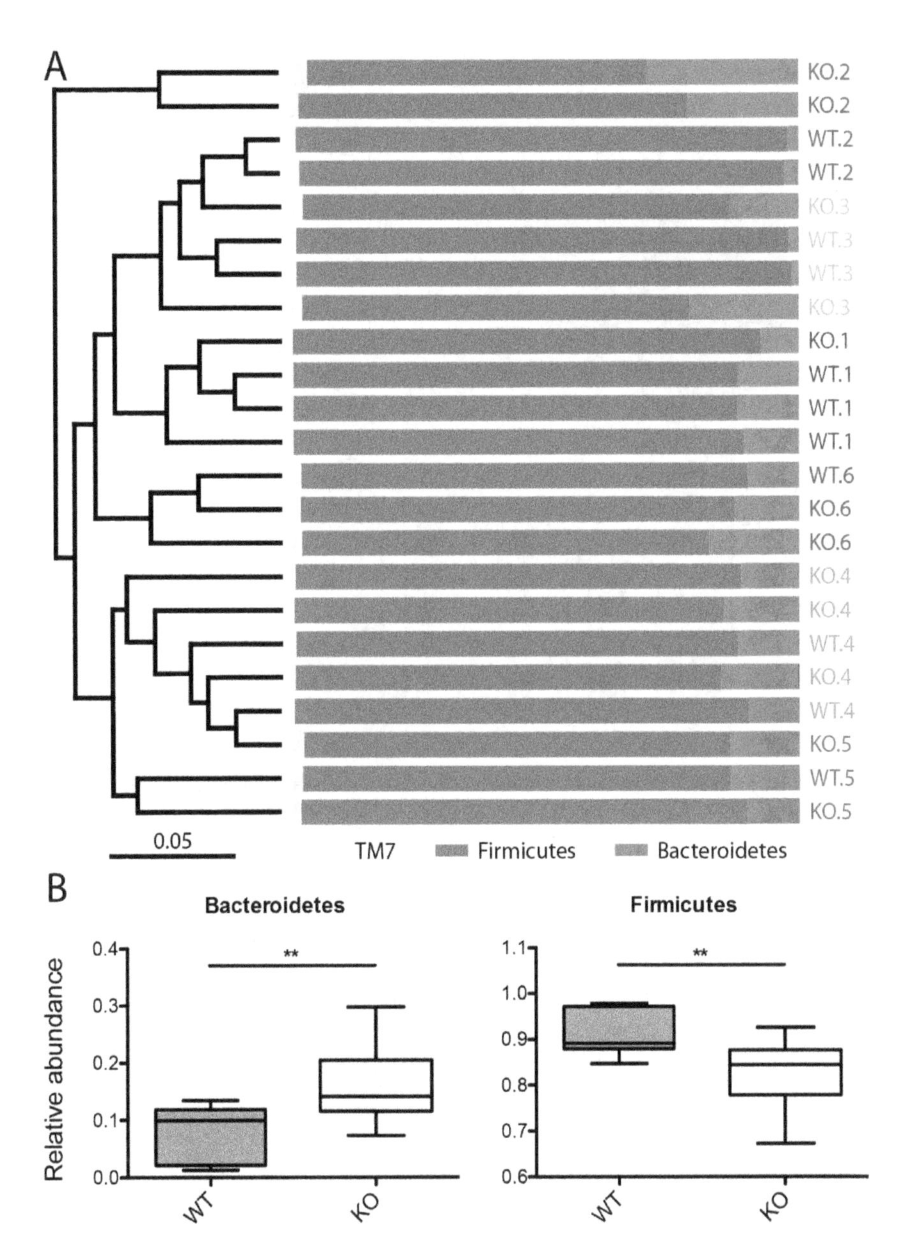

Figure 2. Mucus glycosylation influences gut microbiota composition. (A) Microbiota composition and phylogenetic tree on phylum level. Labels indicate genotype and cage number. Samples are colour coded according to cage number. (B) Relative abundance of Bacteroidetes and Firmicutes in wild type (WT) and TM-IEC *C1galt*$^{-/-}$ (KO) mice (Data shows mean ± SEM; ** $p \leq 0.01$, *** $p \leq 0.001$).

index (DAI; the combined score of weight loss, stool consistence and faecal bleeding) was higher in TM-IEC *C1galt*$^{-/-}$ mice on the first two days of the DSS treatment (Fig. 1C). Blood in the faeces was the main contributor to the DAI with only minor changes in faecal consistency. TM-IEC *C1galt*$^{-/-}$ mice seemed to recover more slowly from the DSS-induced colitis. Faecal blood could still be detected three days after the DSS treatment was stopped in TM-IEC *C1galt*$^{-/-}$ mice whereas the wild type controls recovered within one day (Fig. 1C). In addition, the colons of TM-IEC *C1galt*$^{-/-}$ mice had a slight but non-significant increase in IL-1β and a significantly elevated level of TNFα compared with those of wild type mice (Fig. S1). Taken together, these data indicate that although the TM-IEC *C1galt*$^{-/-}$ mice did not develop spontaneous colitis in our study, they were more susceptible to DSS-induced gut inflammation.

Mucus glycosylation has a minor influence on gut microbiota composition

To test the hypothesis that mucus glycosylation plays a role in selecting the microbiota and maintaining homeostasis, we analyzed the microbial composition of TM-IEC *C1galt*$^{-/-}$ mice and their wild type littermates. When comparing the overall microbial composition, we found that the mice grouped primarily according to cage and within a cage the mice grouped secondarily according to genotype, indicating that loss of core 1-derived O-glycans induced subtle alterations in the microbiota (Fig. 2A). Abundance of the phylum Bacteroidetes was increased whereas that of Firmicutes was reduced in TM-IEC *C1galt*$^{-/-}$ mice (Fig. 2B). These phylum differences were reflected by lower taxonomic levels as well with, for example, classes bacteroidia and clostridia showing the same trends (Fig. S2, S3). However, on a species level, only one operational taxonomic unit (OTU) differed significantly

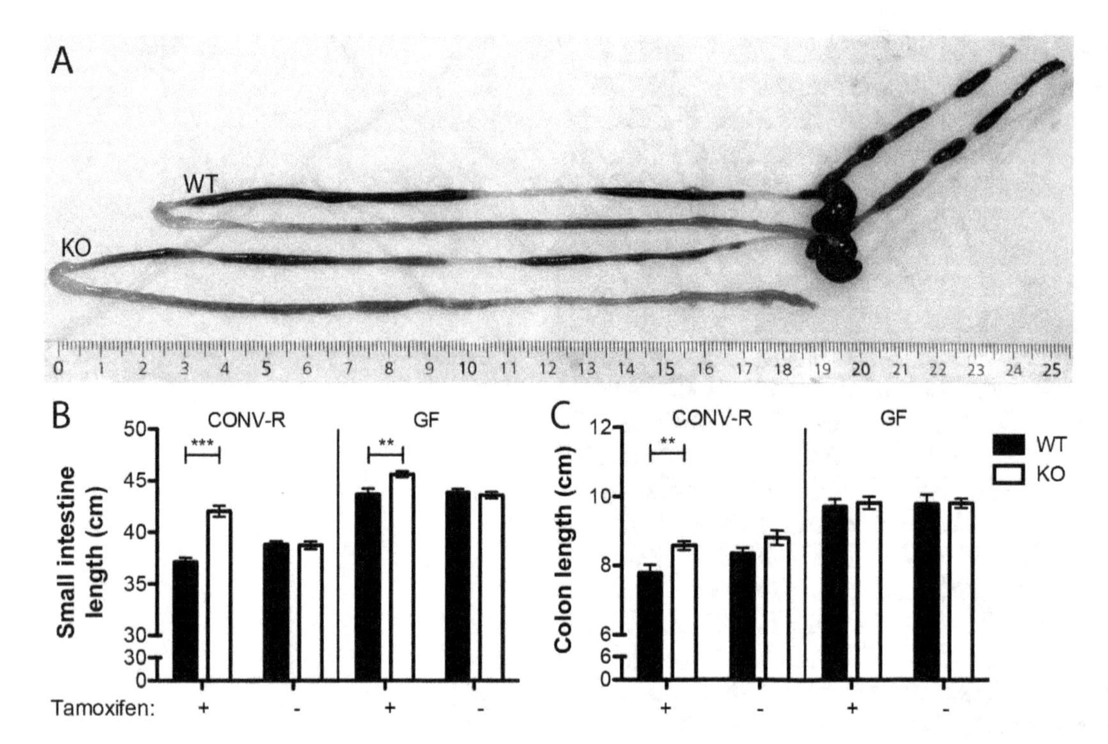

Figure 3. Lack of core 1-derived mucus O-glycosylation alters gut length. (A) Representative gastrointestinal tracts of wild type (WT) and TM-IEC *C1galt*$^{-/-}$ (KO) mice. (B) Length of the small intestine and colon of wild type (WT) and TM-IEC *C1galt*$^{-/-}$ (KO) mice raised in the presence (CONV-R) and absence of bacteria (GF) as well as with or without injection of tamoxifen to induce Cre recombinase-mediated loss of core 1-derived mucus O-glycosylation (n = 7-16 mice per group; Data shows mean ± SEM; ** $p<0.01$, *** $p<0.001$).

among TM-IEC *C1galt*$^{-/-}$ and wild type mice (Fig. S2F), indicating rather broad but subtle differences in microbial abundance caused by loss of core 1-derived O-glycans. Moreover, no differences in alpha diversity (the number of microbial species in a sample) were detected (Fig. S4). In summary, in the absence of any detectable inflammation, loss of core 1-derived O-glycans seems to exert a subtle selection effect on the gut microbiota.

Lack of core 1 mucus glycosylation alters the intestinal architecture

The most obvious phenotypic feature of TM-IEC *C1galt*$^{-/-}$ mice was an extension in the length of gastrointestinal tract. Both the small and large intestines of the TM-IEC *C1galt*$^{-/-}$ mice were about 10% longer than those of wild type controls (Fig. 3A–C). Notably, the effect on gut length was not seen in the absence of tamoxifen activation of the Cre recombinase and thus dependent on the core 1-derived O-glycans (Fig. 3B, C). Regardless of the genotype, the length of the gastrointestinal tract in germ-free mice was longer than that of conventional counterparts. Furthermore, in the small intestine the effect on gut length in the TM-IEC *C1galt*$^{-/-}$ mice seemed to be partly dependent on the microbiota as the difference in length between knockout and wild type mice decreased from 13.2% ($p<0.0001$) to 4.6% ($p = 0.0039$) in conventional and germ-free setting, respectively. In contrast, in the colon the effect on gut length seemed to be solely dependent on the microbes (decreased from 10.1%, $p = 0.006$ to 1.1%, $p = 0.72$ in the conventional and germ-free setting, respectively; Fig. 3C).

In addition to gut length, structural architecture of the gastrointestinal tract was also altered by loss of core 1-derived O-glycans. Although length of the ileal villi and depth of the colonic crypts did not differ between TM-IEC *C1galt*$^{-/-}$ and wild type mice, ileal crypts were deeper in TM-IEC *C1galt*$^{-/-}$ mice (Fig. 4A). We also observed small but significant increases in the number of Ki-67 positive cells both in ileal and colonic crypts of TM-IEC *C1galt*$^{-/-}$ mice (Fig. 4B, C), indicating slightly increased proliferation of intestinal epithelial cells. Furthermore, the circular muscle layers both in the ileum and colon were thicker in TM-IEC *C1galt*$^{-/-}$ mice (Fig. 5A, B). However, numbers of immune cells in the lamina propria were not altered in either the ileum or colon of TM-IEC *C1galt*$^{-/-}$ mice compared with wild type controls (Fig. 5C, D).

Discussion

We aimed to investigate if mucus glycosylation has a selective effect on gut microbial ecology and thereby might impact susceptibility to intestinal inflammation and host physiology. Therefore, we used TM-IEC *C1galt*$^{-/-}$ mice which carry an inducible deficiency of core 1-derived O-glycans in intestinal epithelial cells as a model since it allows us to discriminate between selection effects of host structures in the adult mouse from those present when mice are born and raised with the respective defects. We found that TM-IEC *C1galt*$^{-/-}$ mice did not show any detectable signs of inflammation, but were more susceptible to DSS-induced colitis. This is in contrast to a previous study of TM-IEC *C1galt*$^{-/-}$ mice kept in another animal house where they spontaneously developed intestinal inflammation which was dependent on the microbiota ten days after induced loss of core 1-derived O-glycans [17]. Thus, the discrepancy between our and previous findings is probably due to different housing conditions and microbiota in the animal facilities. Presumably, the gut microbiota composition of the TM-IEC *C1galt*$^{-/-}$ mice rederived and bred in our facility has a flora that is less colitogenic. This is also observed for IL-10$^{-/-}$ mice that normally have spontaneous colitis, but in our animal house only show minor inflammation [4,25]. An impact of housing on

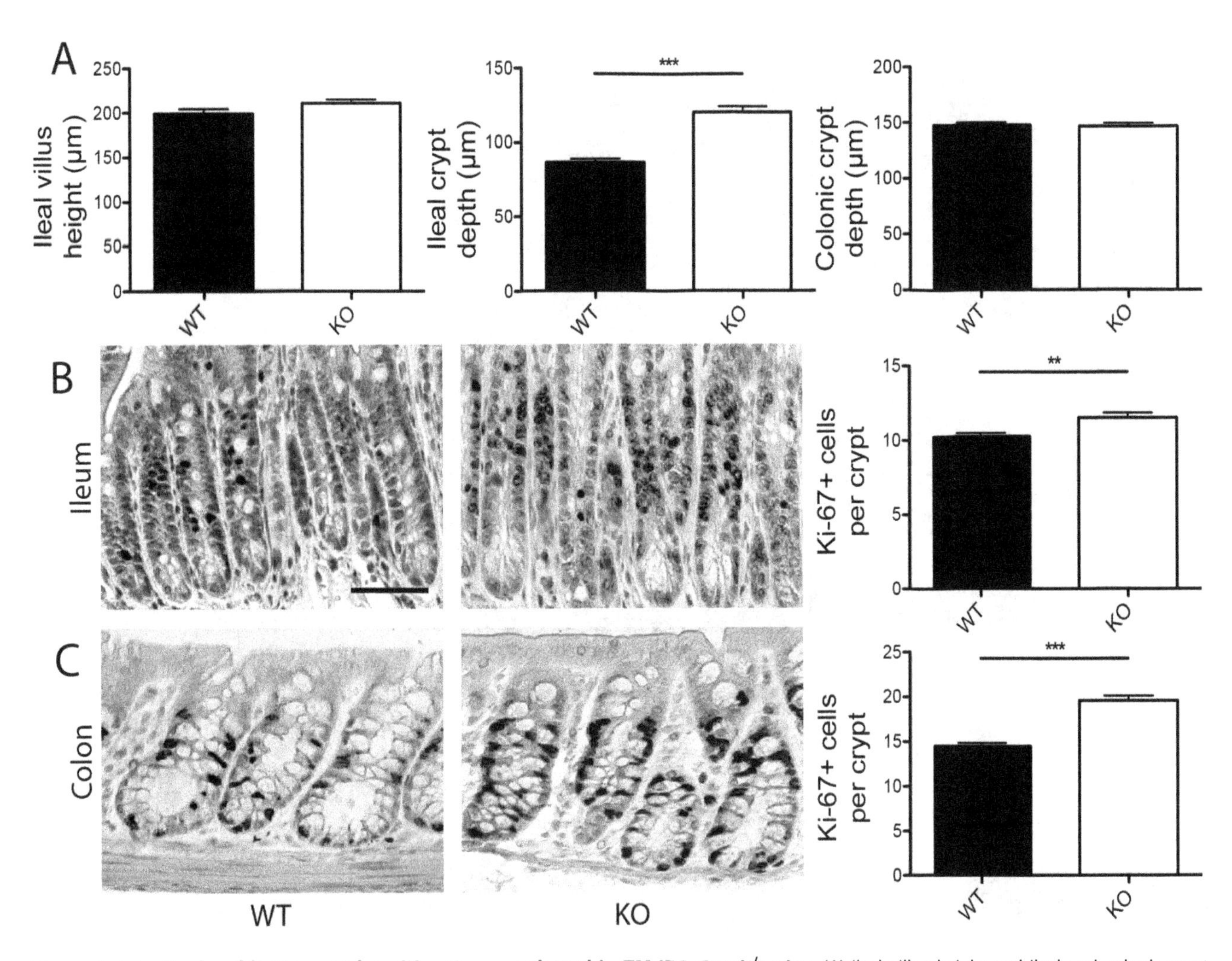

Figure 4. Intestinal architecture and proliferation are altered in TM-IEC *C1galt*$^{-/-}$ mice. (A) Ileal villus height and ileal and colonic crypt depth of wild type (WT) and TM-IEC *C1galt*$^{-/-}$ (KO) mice. (B–C) Ki-67 stainings and quantification of sections from ileal (B) and colonic (C) tissue of wild type (WT) and TM-IEC *C1galt*$^{-/-}$ (KO) mice. For quantifications n = 5–7 mice were scored. Scale bar indicates 50 µm (B–C). Data shows mean ± SEM; ** p<0.01, *** p<0.001.

experimental phenotypes has been observed previously in other models. For example, C56BL/6 mice bred by Taconic Farms harbour more lamina propria lymphocytes than those bred by Jackson Laboratory and only caecal flora from Taconic mice is able to induce lamina propria lymphocyte formation in colonized germ-free mice [26]. Similarly, differential susceptibility to streptozotocin-induced diabetes in mice from Taconic Farms, Jackson Laboratory or Charles River Laboratories has been observed [27,28]. Together, these findings highlight the importance of the microbiota in the development of several inflammatory or metabolic diseases and the need to keep experimental setups including housing conditions and microbiota as stable as possible to facilitate comparisons among studies performed in different animal facilities.

A number of recent publications highlighted the interaction between the intestinal mucus layer, its glycosylation and the microbiota for intestinal homeostasis [29–32]. Here, we made two observations that suggest a microbial role in intestinal homeostasis. (i) Loss of core 1-derived O-glycans, in the absence of inflammation, induced subtle gut microbial alterations, for example inverse shifts in the abundance of the phyla Bacteroidetes and Firmicutes, indicating that glycosylation of the intestinal mucus layer has a selective capacity on microbial ecology. Similarly, other immune components have previously been shown to modulate the microbiota [33]. (ii) The gastrointestinal tract was elongated in TM-IEC *C1galt*$^{-/-}$ mice and these alterations were partly microbially dependent. The small changes in intestinal architecture may thus potentially result from the altered microbial ecology.

We also observed other intestinal changes in TM-IEC *C1galt*$^{-/-}$ mice compared with wild type controls, namely increased ileal crypt depth, a small increase in the number of proliferating epithelial cells, and a thicker circular muscle layer in the ileum and colon of TM-IEC *C1galt*$^{-/-}$ mice. This last observation might indicate differential muscle contraction in the TM-IEC *C1galt*$^{-/-}$ mice and could potentially contribute to the observed differences in length of the gastrointestinal tract of TM-IEC *C1galt*$^{-/-}$ mice.

Notably, it has previously been shown that the gut microbiota affects several aspects of host physiology within and outside of the gastrointestinal tract including organ morphogenesis and tissue homeostasis [1]. In Drosophila, the gut microbiota modulates length of the gastrointestinal tract along with proliferation and differentiation of intestinal epithelial cells [34]. In mice, proliferation of epithelial cells is reduced in the small intestine of germ-free compared to conventionally raised mice [35–37].

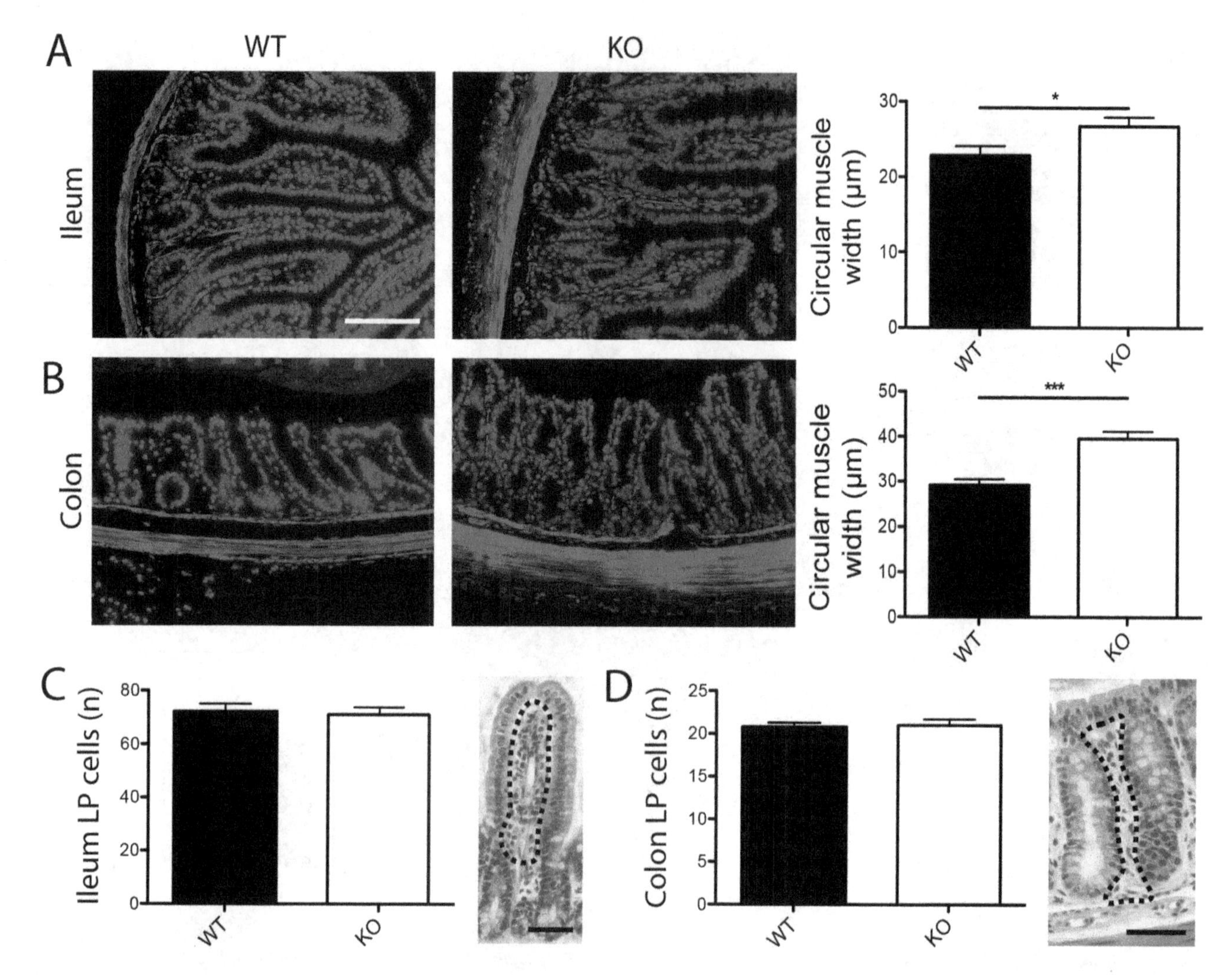

Figure 5. Intestinal muscle width is increased in TM-IEC *C1galt*$^{-/-}$ mice in absence of immune cell infiltration. (A–B) Smooth muscle actin stainings and quantification of sections from ileal (A) and colonic (B) tissue of wild type (WT) and TM-IEC *C1galt*$^{-/-}$ (KO) mice. (C–D) Quantification and representative image of ileal (C) and colonic (D) lamina propria cells. LP - lamina propria. The dotted lines represent the areas used for scoring. For quantifications n = 5–7 mice were scored. Scale bars indicate 50 µm. Data shows mean ± SEM; * $p<0.05$, *** $p<0.001$.

Furthermore, the gut microbiota influences intestinal architecture including villus and crypt morphology as well as remodelling of the intestinal vascular system [38,39]. Altogether these observations suggest that the gut microbiota represents an important environmental factor in the regulation of intestinal homeostasis and physiology.

An altered intestinal architecture without major infiltration of immune cells is reminiscent of some of the features of the human disease irritable bowel syndrome (IBS) [40]. Furthermore, both hypertrophy of the jejunal muscle layer [41] and alterations in faecal microbiota composition were reported to be associated with IBS [42]. Further studies are required to determine if TM-IEC *C1galt*$^{-/-}$ mice could represent an animal model for IBS.

Taken together, mucus glycosylation seems to be important not only for protection and lubrication of the gastrointestinal tract, but also has selective effects on the composition of the resident gut microbiota and small effects on the structure of the intestine. The small alterations in mucus properties, microbiota composition and slightly increased epithelial proliferation in TM-IEC *C1galt*$^{-/-}$ mice could indicate a skewed intestinal homeostasis with altered regenerative response and intestinal architecture in the absence of inflammation. This could potentially suggest a number of physiological defence actions preceding an overall inflammatory response with infiltrating immune cells.

Supporting Information

Figure S1 Levels of pro-inflammatory cytokines IL-1β and TNFα in colon of TM-IEC *C1galt*$^{-/-}$ and wild type mice. Proteins were isolated from colonic tissue of wild type (WT) and TM-IEC *C1galt*$^{-/-}$ (KO) mice and IL-1β and TNFα measured with n = 4 mice per group. Data shows mean ± SEM; * $p<0.05$.

Figure S2 Taxa differentially abundant among wild type and TM-IEC *C1galt*$^{-/-}$ mice. Data shows mean ± SEM; * $p<0.05$, ** $p<0.01$.

Figure S3 Abundance overview of microbial phyla on class, order, family and genus level. Labels indicate genotype and cage number. Samples are colour coded according to cage number.

Acknowledgments

We thank Caroline Jonstrand for measuring cytokines, Anna Hallén and Karin Ahlman for technical assistance and Rosie Perkins for editing the manuscript.

Author Contributions

Conceived and designed the experiments: FS MEVJ LX GCH FB. Performed the experiments: FS NA MEVJ. Analyzed the data: FS NA MEVJ GCH FB. Contributed reagents/materials/analysis tools: LX. Wrote the paper: FS GCH FB. Commented on the manuscript: FS NA MEVJ LX GCH FB.

References

1. Sommer F, Backhed F (2013) The gut microbiota - masters of host development and physiology. Nat Rev Microbiol 11: 227–238.
2. Johansson ME, Phillipson M, Petersson J, Velcich A, Holm L, et al. (2008) The inner of the two Muc2 mucin-dependent mucus layers in colon is devoid of bacteria. Proc Natl Acad Sci U S A 105: 15064–15069.
3. Johansson ME, Sjovall H, Hansson GC (2013) The gastrointestinal mucus system in health and disease. Nat Rev Gastroenterol Hepatol.
4. Johansson ME, Gustafsson JK, Holmen-Larsson J, Jabbar KS, Xia L, et al. (2013) Bacteria penetrate the normally impenetrable inner colon mucus layer in both murine colitis models and patients with ulcerative colitis. Gut.
5. Johansson ME, Ambort D, Pelaseyed T, Schutte A, Gustafsson JK, et al. (2011) Composition and functional role of the mucus layers in the intestine. Cell Mol Life Sci 68: 3635–3641.
6. Gum JR Jr, Hicks JW, Toribara NW, Siddiki B, Kim YS (1994) Molecular cloning of human intestinal mucin (MUC2) cDNA. Identification of the amino terminus and overall sequence similarity to prepro-von Willebrand factor. J Biol Chem 269: 2440–2446.
7. Jensen PH, Kolarich D, Packer NH (2010) Mucin-type O-glycosylation—putting the pieces together. FEBS J 277: 81–94.
8. Larsson JM, Karlsson H, Sjovall H, Hansson GC (2009) A complex, but uniform O-glycosylation of the human MUC2 mucin from colonic biopsies analyzed by nanoLC/MSn. Glycobiology 19: 756–766.
9. Bennett EP, Mandel U, Clausen H, Gerken TA, Fritz TA, et al. (2012) Control of mucin-type O-glycosylation: a classification of the polypeptide GalNAc-transferase gene family. Glycobiology 22: 736–756.
10. Thomsson KA, Holmen-Larsson JM, Angstrom J, Johansson ME, Xia L, et al. (2012) Detailed O-glycomics of the Muc2 mucin from colon of wild-type, core 1- and core 3-transferase-deficient mice highlights differences compared with human MUC2. Glycobiology 22: 1128–1139.
11. Derrien M, van Passel MW, van de Bovenkamp JH, Schipper RG, de Vos WM, et al. (2010) Mucin-bacterial interactions in the human oral cavity and digestive tract. Gut Microbes 1: 254–268.
12. Juge N (2012) Microbial adhesins to gastrointestinal mucus. Trends Microbiol 20: 30-39.
13. Rausch P, Rehman A, Kunzel S, Hasler R, Ott SJ, et al. (2011) Colonic mucosa-associated microbiota is influenced by an interaction of Crohn disease and FUT2 (Secretor) genotype. Proc Natl Acad Sci U S A 108: 19030–19035.
14. Staubach F, Kunzel S, Baines AC, Yee A, McGee BM, et al. (2012) Expression of the blood-group-related glycosyltransferase B4galnt2 influences the intestinal microbiota in mice. ISME J 6: 1345–1355.
15. Fuhrer A, Sprenger N, Kurakevich E, Borsig L, Chassard C, et al. (2010) Milk sialyllactose influences colitis in mice through selective intestinal bacterial colonization. J Exp Med 207: 2843–2854.
16. An G, Wei B, Xia B, McDaniel JM, Ju T, et al. (2007) Increased susceptibility to colitis and colorectal tumors in mice lacking core 3-derived O-glycans. J Exp Med 204: 1417–1429.
17. Fu J, Wei B, Wen T, Johansson ME, Liu X, et al. (2011) Loss of intestinal core 1-derived O-glycans causes spontaneous colitis in mice. J Clin Invest 121: 1657–1666.
18. Mandel U, Petersen OW, Sorensen H, Vedtofte P, Hakomori S, et al. (1991) Simple mucin-type carbohydrates in oral stratified squamous and salivary gland epithelia. J Invest Dermatol 97: 713–721.
19. Johansson ME, Gustafsson JK, Sjoberg KE, Petersson J, Holm L, et al. (2010) Bacteria penetrate the inner mucus layer before inflammation in the dextran sulfate colitis model. PLoS One 5: e12238.
20. Kim JJ, Bridle BW, Ghia JE, Wang H, Syed SN, et al. (2013) Targeted inhibition of serotonin type 7 (5-HT7) receptor function modulates immune responses and reduces the severity of intestinal inflammation. J Immunol 190: 4795–4804.
21. Yu Z, Morrison M (2004) Improved extraction of PCR-quality community DNA from digesta and fecal samples. Biotechniques 36: 808–812.
22. Salonen A, Nikkila J, Jalanka-Tuovinen J, Immonen O, Rajilic-Stojanovic M, et al. (2010) Comparative analysis of fecal DNA extraction methods with phylogenetic microarray: effective recovery of bacterial and archaeal DNA using mechanical cell lysis. J Microbiol Methods 81: 127–134.
23. Nylund L, Heilig HG, Salminen S, de Vos WM, Satokari R (2010) Semi-automated extraction of microbial DNA from feces for qPCR and phylogenetic microarray analysis. J Microbiol Methods 83: 231–235.
24. Caporaso JG, Kuczynski J, Stombaugh J, Bittinger K, Bushman FD, et al. (2010) QIIME allows analysis of high-throughput community sequencing data. Nat Methods 7: 335–336.
25. Kuhn R, Lohler J, Rennick D, Rajewsky K, Muller W (1993) Interleukin-10-deficient mice develop chronic enterocolitis. Cell 75: 263–274.
26. Ivanov II, Frutos Rde L, Manel N, Yoshinaga K, Rifkin DB, et al. (2008) Specific microbiota direct the differentiation of IL-17-producing T-helper cells in the mucosa of the small intestine. Cell Host Microbe 4: 337–349.
27. Graham ML, Janecek JL, Kittredge JA, Hering BJ, Schuurman HJ (2011) The streptozotocin-induced diabetic nude mouse model: differences between animals from different sources. Comp Med 61: 356–360.
28. Kriegel MA, Sefik E, Hill JA, Wu HJ, Benoist C, et al. (2011) Naturally transmitted segmented filamentous bacteria segregate with diabetes protection in nonobese diabetic mice. Proc Natl Acad Sci U S A 108: 11548–11553.
29. Kashyap PC, Marcobal A, Ursell LK, Smits SA, Sonnenburg ED, et al. (2013) Genetically dictated change in host mucus carbohydrate landscape exerts a diet-dependent effect on the gut microbiota. Proc Natl Acad Sci U S A 110: 17059–17064.
30. Wrzosek L, Miquel S, Noordine ML, Bouet S, Chevalier-Curt MJ, et al. (2013) Bacteroides thetaiotaomicron and Faecalibacterium prausnitzii influence the production of mucus glycans and the development of goblet cells in the colonic epithelium of a gnotobiotic model rodent. BMC Biol 11: 61.
31. Kleiveland CR, Hult LT, Spetalen S, Kaldhusdal M, Christofferesen TE, et al. (2013) The noncommensal bacterium Methylococcus capsulatus (Bath) ameliorates dextran sulfate (Sodium Salt)-Induced Ulcerative Colitis by influencing mechanisms essential for maintenance of the colonic barrier function. Appl Environ Microbiol 79: 48–56.
32. Sperandio B, Fischer N, Chevalier-Curt MJ, Rossez Y, Roux P, et al. (2013) Virulent Shigella flexneri Affects Secretion, Expression, and Glycosylation of Gel-Forming Mucins in Mucus-Producing Cells. Infect Immun 81: 3632–3643.
33. Ubeda C, Lipuma L, Gobourne A, Viale A, Leiner I, et al. (2012) Familial transmission rather than defective innate immunity shapes the distinct intestinal microbiota of TLR-deficient mice. J Exp Med 209: 1445–1456.
34. Shin SC, Kim SH, You H, Kim B, Kim AC, et al. (2011) Drosophila microbiome modulates host developmental and metabolic homeostasis via insulin signaling. Science 334: 670–674.
35. Abrams GD, Bauer H, Sprinz H (1963) Influence of the normal flora on mucosal morphology and cellular renewal in the ileum. A comparison of germ-free and conventional mice. Lab Invest 12: 355–364.
36. Crawford PA, Gordon JI (2005) Microbial regulation of intestinal radiosensitivity. Proc Natl Acad Sci U S A 102: 13254–13259.
37. Savage DC, Siegel JE, Snellen JE, Whitt DD (1981) Transit time of epithelial cells in the small intestines of germfree mice and ex-germfree mice associated with indigenous microorganisms. Appl Environ Microbiol 42: 996–1001.
38. Reinhardt C, Bergentall M, Greiner TU, Schaffner F, Ostergren-Lunden G, et al. (2012) Tissue factor and PAR1 promote microbiota-induced intestinal vascular remodelling. Nature 483: 627–631.
39. Stappenbeck TS, Hooper LV, Gordon JI (2002) Developmental regulation of intestinal angiogenesis by indigenous microbes via Paneth cells. Proc Natl Acad Sci U S A 99: 15451–15455.
40. Quigley EM, Abdel-Hamid H, Barbara G, Bhatia SJ, Boeckxstaens G, et al. (2012) A global perspective on irritable bowel syndrome: a consensus statement of the World Gastroenterology Organisation Summit Task Force on irritable bowel syndrome. J Clin Gastroenterol 46: 356–366.
41. Tornblom H, Lindberg G, Nyberg B, Veress B (2002) Full-thickness biopsy of the jejunum reveals inflammation and enteric neuropathy in irritable bowel syndrome. Gastroenterology 123: 1972–1979.
42. Jeffery IB, O'Toole PW, Ohman L, Claesson MJ, Deane J, et al. (2012) An irritable bowel syndrome subtype defined by species-specific alterations in faecal microbiota. Gut 61: 997–1006.

18

Non-Lethal Control of the Cariogenic Potential of an Agent-Based Model for Dental Plaque

David A. Head[1]*, Phil D. Marsh[2,3], Deirdre A. Devine[3]

1 School of Computing, University of Leeds, Leeds, United Kingdom, **2** Microbiology Services, PHE Porton, Salisbury, United Kingdom, **3** Department of Oral Biology, School of Dentistry, University of Leeds, United Kingdom

Abstract

Dental caries or tooth decay is a prevalent global disease whose causative agent is the oral biofilm known as plaque. According to the ecological plaque hypothesis, this biofilm becomes pathogenic when external challenges drive it towards a state with a high proportion of acid-producing bacteria. Determining which factors control biofilm composition is therefore desirable when developing novel clinical treatments to combat caries, but is also challenging due to the system complexity and the existence of multiple bacterial species performing similar functions. Here we employ agent-based mathematical modelling to simulate a biofilm consisting of two competing, distinct types of bacterial populations, each parameterised by their nutrient uptake and aciduricity, periodically subjected to an acid challenge resulting from the metabolism of dietary carbohydrates. It was found that one population was progressively eliminated from the system to give either a benign or a pathogenic biofilm, with a tipping point between these two fates depending on a multiplicity of factors relating to microbial physiology and biofilm geometry. Parameter sensitivity was quantified by individually varying the model parameters against putative experimental measures, suggesting non-lethal interventions that can favourably modulate biofilm composition. We discuss how the same parameter sensitivity data can be used to guide the design of validation experiments, and argue for the benefits of *in silico* modelling in providing an additional predictive capability upstream from *in vitro* experiments.

Editor: Zezhang Wen, LSU Health Sciences Center School of Dentistry, United States of America

Funding: DAH was funded by the BHRC, University of Leeds. The funders had no role in study design, data collection and analysis, decision to publish, or preparation of the manuscript.

Competing Interests: The authors have declared that no competing interests exist.

* Email: d.head@leeds.ac.uk

Introduction

Dental caries is a common disease that reduces quality of life globally and represents a substantial economic burden for health organisations [1]. It is caused by acids (primarily lactic) in the oral cavity which initiate demineralisation of tooth enamel, resulting in lesions that can develop into cavities. These acids are produced as a by-product of the glycolysis of dietary sugars by the bacterial community that comprises the oral biofilm known as dental plaque. However, there is no single pathogenic species responsible for caries. Although *Streptococcus mutans* has been widely implicated and studied in this context, carious lesions can exist in the absence of *S. mutans* and be absent in its presence [1,2]. It is instead thought that any acid-producing species that can metabolise sugars at low pH, *i.e.* those that are both *acidogenic* and *aciduric*, can contribute to the caries process. This description includes *S. mutans*, but also other mutans streptococci, lactobacilli, bifidobacteria, and others. Clinical trials have confirmed the correlation between biofilm composition and caries progression, with healthy and diseased sites associated with distinct subpopulations of bacteria [3–5].

The bacterial composition of dental plaque, and biofilms in general, is the result of a dynamic interplay between microbial physiology and external perturbations from the environment and host, and in this sense can be meaningfully regarded as ecosystems [6–8]. For oral biofilms, this has been formalised into the *ecological plaque hypothesis* [9]. Varying external conditions therefore alter biofilm composition, potentially increasing or decreasing the fraction of cariogenic species, and *in vitro* studies have shown that the low pH resulting from pulsing glucose into mixed species biofilms causes population shifts favouring *S. mutans* and lactobacilli [10,11]. This ecological perspective suggests alternative therapeutic strategies: rather than eradication of bacteria, which is both difficult and undesirable as many oral bacteria are also beneficial to the host, it may be possible to instead modulate the biofilm composition to favour bacterial communities with a lower fraction of acidogenic, aciduric species. Indeed, *in vitro* studies have demonstrated that fluoride below lethal concentrations can reduce the fraction of putative cariogenic organisms [12,13]. The mechanism is thought to be a reduction of their aciduricity and acidogenicity [14–17], and reducing their competitiveness with respect to non-pathogenic species such as *S. gordonii* [18–20].

Identifying targets to modulate biofilm composition to the benefit of the host is challenging due to the significant complexity of the system, with many coupled mechanisms driving population changes. Systematically varying the many candidate factors in *in vitro* experiments is costly both in time and expense. It is desirable

Table 1. Physical and biological parameters that were systematically varied in this study.

Label	Meaning	Primary value	Range
$t^{\rm durn}$	Duration of pulse cycle	6 h	2–10 h
$K_{\rm acid}^{A}$	Half concentration for acid inhibition for A	2×10^{-5} mol/L	$5\times10^{-6}-10^{-4}$ mol/L
$K_{\rm acid}^{NA}$	Half concentration for acid inhibition for NA	2×10^{-7} mol/L	$5\times10^{-8}-10^{-6}$ mol/L
$K_{\rm nut}^{A}$	Half concentration for nutrient uptake for A	5 g/L	0.1–20 g/L
$K_{\rm nut}^{NA}$	Half concentration for nutrient uptake for NA	20 mg/L	5–40 mg/L
$K_{\rm a}$	Effective dissociation constant	10^{-9} mol/L	$10^{-10}-10^{-8}$ mol/L
$h_{\rm plaque}$	Thickness of the plaque layer	150 μm	50–250 μm
$h_{\rm saliva}$	Thickness of the saliva layer	100 μm	25–350 μm
$[{\rm Gl}]_{\rm inter}$	Concentration of sugar between pulses	5 mg/L	1–15 mg/L
$D_{\rm acid}$	Diffusion coefficient for the acid (uniform)	10^3 μm^2/s	140–2700 μm^2/s
$Y_{\rm rel}^{\rm EPS}$	Relative yield factor for EPS	0.4	0.2–1.5
$k_{\rm death}$	Linear factor in the kill rate	$10^{-4}/\mu$m h	$2\times10^{-5}-5\times10^{-4}/\mu$m h
$d^{\rm max}$	Threshold diameter for cell division	5 μm	4 μm–10 μm
$\kappa^{\rm char}$	Characteristic stiffness	50 pN/μm	30–200 pN/μm
$\sigma^{\rm div}$	Width of daughter mass ratios after division	0.1	0.05–0.2

When one parameter was varied over the range given in the fourth column, the remaining parameters took the values given in the third column. See *Methods* for further details.

to introduce an additional predictive layer before *in vitro* modelling to highlight promising targets, and this can be realised by *in silico* modelling, *i.e.* computational simulation of mathematical models. Such models generate quantitative predictions over broad ranges of parameter space in relatively short time. In addition, they are not restricted to specific species of bacteria but can systematically incorporate strain and sub-strain variation by continuously varying parameters related to cellular physiology. They are thus well suited to probing populations of species defined by their function, rather than their genetic identity. Early mathematical models of dental plaque by Dibdin *et al.* adopted a continuum approach in which concentrations of various dispersed phases varied smoothly with distance from the enamel surface [21–26]. This approach has recently been advanced by Ilie *et al.* who included numerous coupled fields to more realistically represent acid buffering, polyglucose storage *etc.* [27]. These studies did not consider the changes in biofilm composition necessary to study population response to perturbations. Such questions can be naturally addressed using agent based modelling, which is established in biofilm research [28–36] but has not yet been applied to plaque.

The aim of this paper is to describe findings from an agent based model of *supragingival* plaque, *i.e.* that component of the oral biofilm above the gumline that is responsible for dental caries, developed to probe the relationship between evolving biofilm composition and cariogenic potential. The model consists of two competing populations of bacteria, one that is pathogenic in that it is both acidogenic and aciduric, and a second non-pathogenic population which, while acidogenic, cannot metabolise sugars at low pH and thus is non-aciduric. These populations are labelled A and NA respectively, for 'aciduric' and 'non-aciduric'. It is found that one population or the other dominates at late times, with a 'tipping point' between the two outcomes that depends on a range of parameters relating to both intrinsic physiological processes and external factors such as the frequency of sugar intake. Treatments intended to modify these parameters could therefore drive plaque composition towards a healthy, non-cariogenic state. In addition, sensitivity analysis reveals the relative importance of each model parameter on putative experimental measurements which, as well as highlighting the most important mechanisms relevant to plaque function, can also guide validation experiments by identifying parameter-measurement pairings with high sensitivity, suitable for fixing input parameters from *in vitro* data.

Results

Interspecific competition

Snapshots of a section of a biofilm for the primary parameters in Table 1 are given in Fig. 1 for the initial conditions at $t=0d$, at an early time $t=10d$, and at the final time point $t=200d$. This time scale is relevant to biofilm accumulation at stagnant sites, such as the approximal surfaces between teeth. Even for the early time point there is visible aggregation of each distinct population compared to the initial condition, and these aggregates evolve into the depth-spanning domains visible at the later time. This is primarily a consequence of daughter particles remaining localised to their mother during division, driving the aggregation through shared proximity of descendants from the same progenitor. Note also that there is a clear gradient in the concentration of acid produced by the biofilm, with higher concentrations near the enamel surface and lower concentrations as one moves through the biofilm into the saliva layer, as seen in real plaque [37]. Lateral gradients in acid are not visible in these snapshots but are also present, in particular at late times when the large domains of A produce more acid than NA. Larger, colour images and movies are available in Figure S1 and Movies S1 and S2 of the Supplementary Information.

The two populations A and NA are not competing for the sole carbon source (glucose), since there is no mass transfer limitation of glucose as the model is defined. They are however competing for space, in that the fixed system size imposed by the plaque thickness $h_{\rm plaque}$ restricts the total number of particles of either

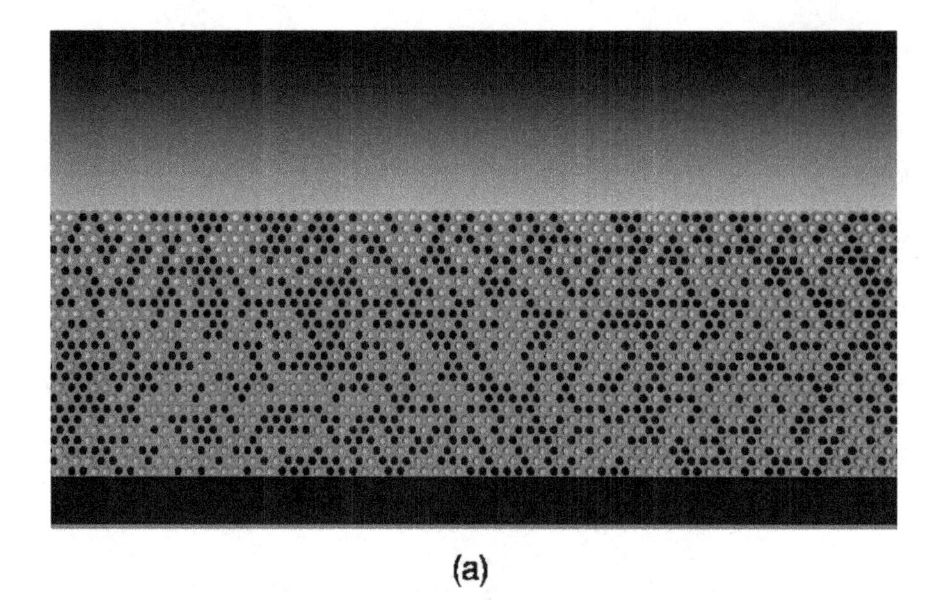

(a)

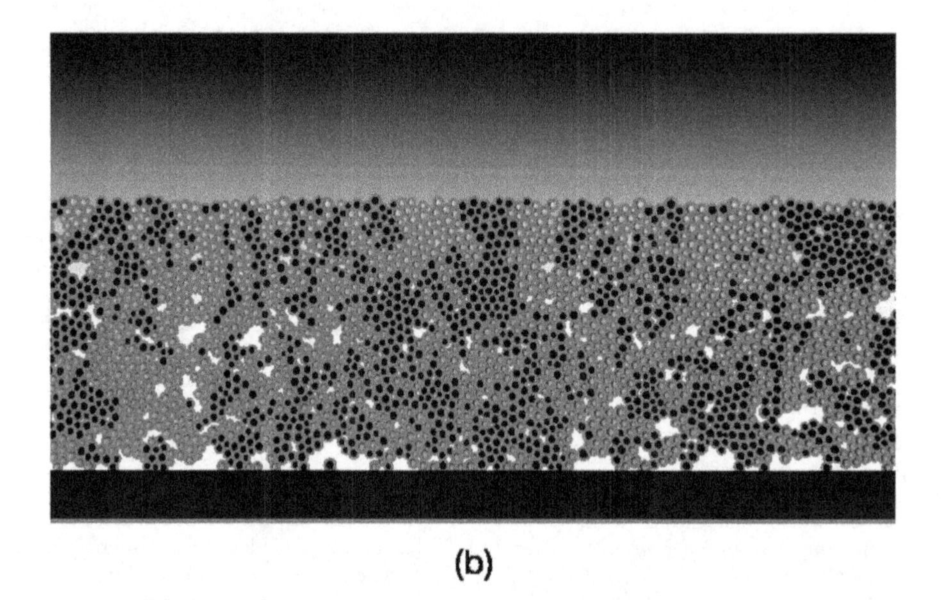

(b)

(c)

Figure 1. Snapshots of a section of a biofilm taken at time points (a) $t=0d$, (b) $t=10d$ and (c) $t=200d$ for the primary parameters in Table 1 (the full system size is roughly 10 times wider). Light grey (black) discs denote aggregates of A (NA), respectively, encased in EPS shells (grey). The shaded field in the background corresponds to the acid, with light (dark) regions for high (low) concentrations. The white regions near the base correspond to high concentrations of acid, visible in the voids created by cell death (see text). Larger, colour images and movies for $t^{\mathrm{durn}}=4h$ and $t^{\mathrm{durn}}=8h$ are available in Figure S1 and Movies S1 and S2 of the Supplementary Information.

population, $N_{\mathrm{tot}}=N_{\mathrm{A}}+N_{\mathrm{NA}}$, to remain roughly constant in time. Therefore, if one population exhibits a net growth rate faster than the other, it will increase its fraction within the biofilm at the expense of the other, whose fraction will contract. The only other form of direct interaction between the two populations is the acid produced which, particularly during the glucose pulse, severely reduces the growth of NA relative to A. Since A produces more acid that NA, this drives the formation of lateral pH gradients mentioned above.

Alternating differential growth

The dominant population at late times is that with the greater net growth rate when averaged over both the inter and intra-pulse periods. If one population exhibited the greater rate in both periods, it would clearly outgrow the other. However, for this study we have chosen parameters for which population A exhibits the greater growth rate during the pulse when the pH is low, but population NA grows the fastest between glucose pulses when the pH is high. This situation is representative of oral bacteria as determined in chemostat experiments, for instance *S. mutans* is more aciduric that *S. gordonii*, but exhibits lower growth around neutral pH [38,39]. Although the parameters were varied to cover a broad range of values, encompassing functionally similar species and strains, they were limited to ensure this basic alternation between the most competitive population remained true.

An example of this alternation is given in Fig. 2, which shows both the pH measured at the enamel surface, and the fraction of the biofilm that belongs to population A, *i.e.* $N_{\mathrm{A}}/N_{\mathrm{tot}}$, during three consecutive glucose pulses. For clarity of presentation the long inter-pulse periods have been compressed in this diagram, but the behaviour of both quantities can be inferred as explained in the caption. It is evident that the fraction $N_{\mathrm{A}}/N_{\mathrm{tot}}$ increases during the glucose pulse, but decreases again before the start of the subsequent pulse, when it again increases. Since A is more aciduric than NA, more acid should be produced as the fraction $N_{\mathrm{A}}/N_{\mathrm{tot}}$ increases, and this is also evident in the figure where there is a clear correlation between pH and biofilm composition.

The example shown in Fig. 2 is for a somewhat frequent glucose pulsing with $t^{\mathrm{durn}}=4h$. In this case, the increase in $N_{\mathrm{A}}/N_{\mathrm{tot}}$ during the pulse is greater than the decrease between pulses. The long-term trend is therefore for A to dominate, and correspondingly the pH during pulses to drop. This can be regarded as a *ratchet* effect whereby each full pulse cycle increases the fraction of A by a small amount. The small shifts in composition and pH in the diagram lead to biologically significant changes when integrated over many such cycles. It might be expected that a longer t^{durn}, and hence a longer period of time spent in the high-pH environment, will result in a larger drop in $N_{\mathrm{A}}/N_{\mathrm{tot}}$ between pulses and a net decrease of $N_{\mathrm{A}}/N_{\mathrm{tot}}$, resulting in a slow drift towards a NA-dominated state. This is indeed observed, but instead we now consider data averaged over whole pulses for which ratcheting is not visible, to better focus on long-term trends.

Transition between homeostatic and pathogenic biofilms

Fig. 3 shows the fraction of population A versus time for pulse duration t^{durn} increasing from 2 h to 10 h, where each data point corresponds to the averaged value over a 2 d period so that the sawtooth variation in Fig. 2 is not visible. Starting from a 50:50 population of A:NA, the biofilm composition becomes increasingly A-dominated with time for low t^{durn} corresponding to frequent glucose pulses, *i.e.* frequent acid challenges. By contrast, for infrequent pulses with high t^{durn}, $N_{\mathrm{A}}/N_{\mathrm{tot}}$ becomes small, signifying the biofilm becoming predominantly of type NA. Varying the frequency of glucose pulses in the biofilm environment therefore determines the late-time fate of the biofilm, *i.e.* whether it is A-dominated (pathogenic) or NA-dominated (homeostatic).

It is evident from the figure that the biofilm composition $N_{\mathrm{A}}/N_{\mathrm{tot}}$ at any given time point continuously increases as the frequency of the glucose pulses increase. There is an intermediate value $t^{\mathrm{durn}}\approx 6h$ where the variation in biofilm composition is not discernible over the available data window. It can therefore be hypothesised that there is a transition value of t^{durn} about which $N_{\mathrm{A}}/N_{\mathrm{tot}}$ remains at 50:50 for all times, and neither population comes to dominate the other. This is supported by quantitative analysis of the late-time variation of $N_{\mathrm{A}}/N_{\mathrm{tot}}$, which can be shown to exponentially increase towards unity for low t^{durn}, and exponentially decrease towards zero for high t^{durn}. This is demonstrated by semi-logarithmic plot in Fig. 4, which shows examples of the exponential decay of A ($t^{\mathrm{durn}}>6h$) or NA ($t^{\mathrm{durn}}<6h$). The fits in these figures are to a simple exponential,

$$\frac{N_{\mathrm{A/NA}}(t)}{N_{\mathrm{tot}}(t)}=A\exp^{-kt}\ , \qquad (1)$$

where k is the rate at which the minority fraction decays. The inset to the figure shows that this rate decreases continuously to zero as t^{durn} approaches a critical point close to $6h$. The exponential fit (1) is only successful if an initial transient of roughly 10 days is removed prior to fitting; no simple fit, including logistic growth, was found to fit the entire data range.

Multi-factorial modulation of pathogenicity

The duration of glucose pulses t^{durn} is not the only factor that determines the relative competitiveness of the two populations. Any of the model parameters listed in Tables 1 and 2 can, in principle, affect the growth rates of one or both populations over the course of a complete pulse, and therefore influence the selection of the dominant population. This is expected for the metabolic parameters such as the half-concentrations, but equally holds true for environmental factors. For example, increasing the biofilm thickness h_{plaque} with all other parameters held fixed, results in an increased production of acid and a lower pH, reducing the metabolic rate of population A to a lesser degree than NA. This shift in competitiveness might be enough to make A the dominant population at late times, even when t^{durn} is greater than 6h.

Confirmation that parameters other than t^{durn} can promote a transition between pathogenic and homeostatic biofilms is presented in Fig. 5. This shows the variation of both the intra-pulse pH and the biofilm composition as a function of time, for a

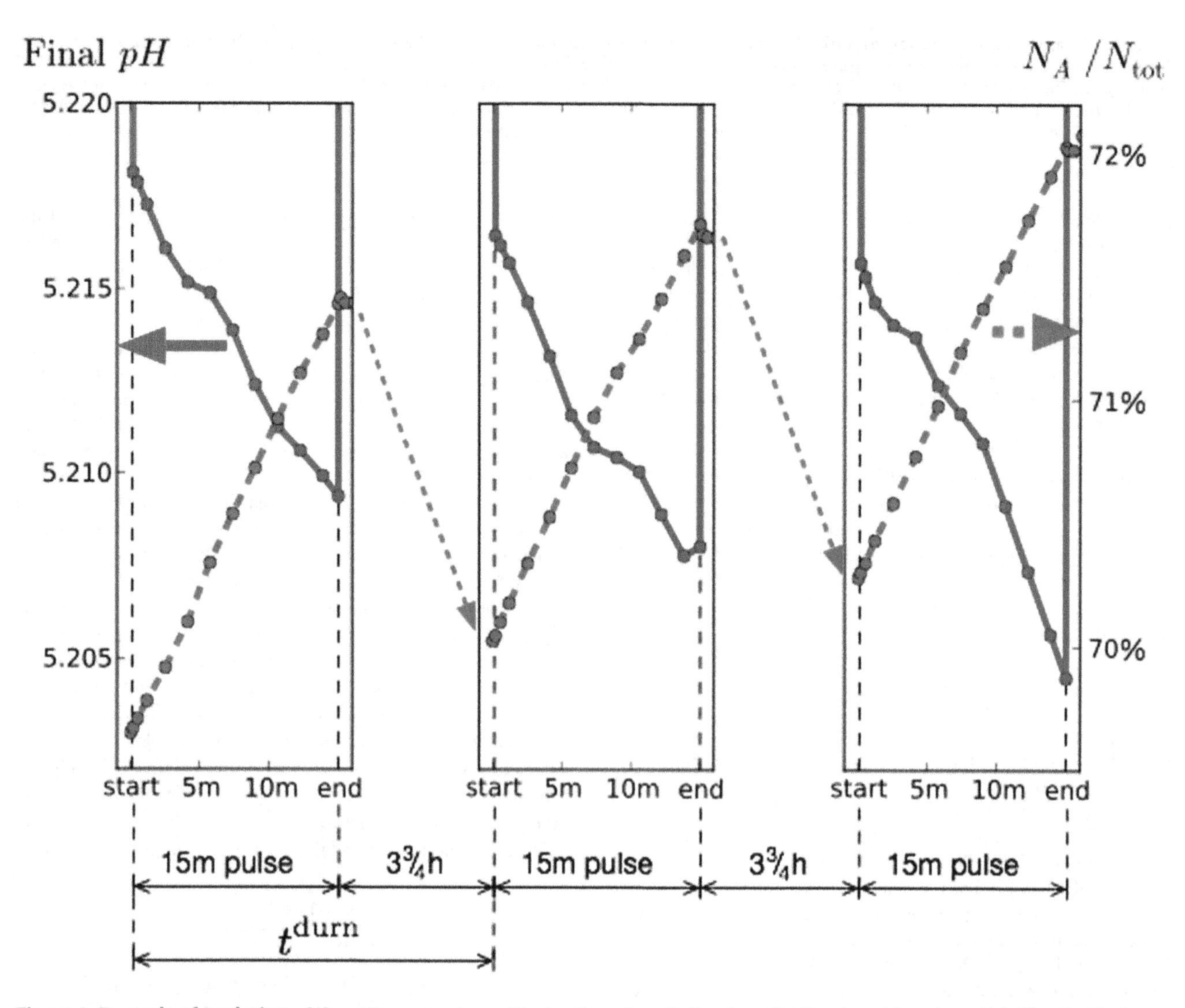

Figure 2. Example of evolution of the pH averaged over the tooth surface (left axis; solid lines) and fraction of biofilm that belongs to population A (right axis; broken lines) for 3 consecutive glucose pulses. This example is for the primary parameters of Table 1 except the total pulse cycle duration $t^{durn} = 4h$ here. The pH is observed to decrease for successive pulses, concomitant with an increase in the fraction of A. During the inter-pulse period, which has been compressed for clarity, the fraction of A decreases slowly (shown schematically by the diagonal arrows), and the pH takes much higher values around 6.0 (not shown).

series of independent runs that differ only in one of the aciduricity parameters, K^{A}_{acid} and K^{NA}_{acid}. There is a clear trend from a late-time biofilm composition that is A-dominated with a low pH, arising for high K^{A}_{acid} or low K^{A}_{acid}, to a NA-dominated state with a higher pH for low K^{A}_{acid} or high K^{A}_{acid}. Thus, increasing the aciduricity of one population enhances its competitiveness with respect to the other. Note that in Fig. 5(b) the pH at the final time point first increases with the parameter K^{NA}_{acid} as it is increased, but then decreases with K^{NA}_{acid} for the higher values considered. The initial increase is because of the increasing dominance of the NA population as just discussed. The subsequent decrease arises because, although the NA remain dominant, they become increasingly aciduric as K^{NA}_{acid} is increased, resulting in increased acid production. Indeed, increasing K^{NA}_{acid} until it equals K^{A}_{acid} would result it two populations of equal aciduricity, rendering our labels A and NA meaningless.

Systematically varying each of the parameters in Table 1 reveals that most of them can also promote this transition. This is summarised in Fig. 6, which shows the biofilm composition and pH for 10 of the parameters between the extremes of the ranges given in the table. In all cases the observed variation is intuitive. For instance, increasing the diffusion D of acid results in more rapid dispersal from the system according to the boundary conditions of *Methods*, resulting in a higher pH and a lower fraction of A. Conversely, reducing the death rate k_{death} increases acid production, but this effect is weak as demonstrated below. The parameters d^{max}, κ and σ^{div} have been omitted from this figure since the signal-to-noise ratio was too small to discern any trend.

Parameter sensitivity

It is not readily apparent from Fig. 6 which parameters are the most important for determining the late-time biofilm composition, making it unsuitable for identifying potential targets for controlling biofilm fate. In addition, the time point for the predicted quantities (pH and composition at $t=200d$) are not suitable for *in vitro* validation, for which a much shorter time scale is convenient. Both

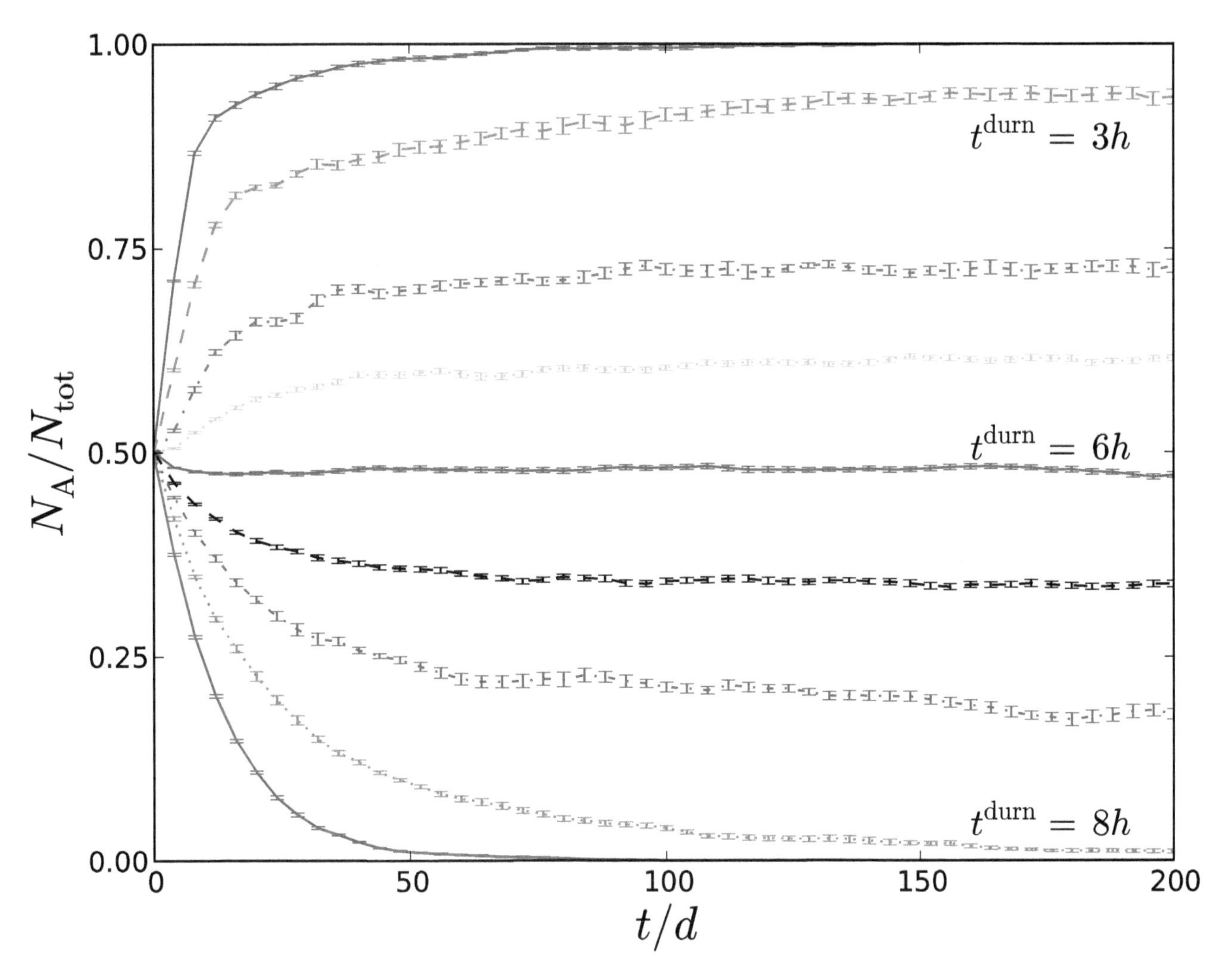

Figure 3. Fraction of system that belong to population A versus time for (from top to bottom) t^{durn} = 2 h, 3 h, 4 h, 5 h, 5.5 h, 6 h, 6.5 h, 7 h, 8 h, 10 h, respectively. Error bars show scatter over at least 10 independent runs.

of these issues can be resolved by tabulating the changes in a range of putative experimental measurements with respect to each input parameter, allowing pairings that exhibit high or low sensitivity to be immediately identified. The sensitivity heat map for this model is Fig. 7, where each entry shows the predicted percentage change in a range of measurable outcomes for a 1% change in each input parameter. The measurable outcomes include 4 predictions for single-species biofilms, *i.e.* the concentration of H^+ ions during and between pulses for systems comprised purely of A or NA, which reach dynamical steady states in the order of days. In addition, 4 outcomes for mixed-species films were considered. This includes the variation in the critical pulse duration t^{durn} separating pathogenic from homeostatic biofilms, and 3 quantities measured at a time point of $t=20d$ starting from an initial 50:50 composition of A:NA, namely the fraction of N_A/N_{tot}, and the concentration of H^+ (again during and between pulses).

Inspection of the table immediately reveals that the quantity $N_A/N_{tot}|_{t=20d}$ is the most sensitive to a number of parameters, suggesting this would be a useful quantity to measure in experiments. Single-species experiments may be the best to measure each species' metabolic half-constants, although environmental factors are clearly also important. Reading across rows rather than down columns suggests that the final 5 parameters in the diagram give weak or no change to the measurable outcomes, suggesting these mechanisms are not worthy of further investigation. The variations are not necessarily zero, as seen by inspection of the raw data given in Table S1 of the Supplementary Information, but their influence is evidently weak.

Discussion

Mathematical modelling represents a powerful tool for biofilm investigation, providing the capability for quantitative prediction upstream from *in vitro* models and *in vivo* trials. All conceivable measures of biomass composition, structure, and associated chemical gradients can be extracted non-invasively from an *in silico* biofilm. In addition, such data are acquired in an accelerated time frame, which for the results presented here translates into 200 days real time in approximately 10 hours of simulation time, and this ratio could be further improved with additional numerical optimisation and parallelisation. This rapidity allows ranges of each input parameter to be systematically assayed and the corresponding effect on the growing biofilm to be determined, both qualitatively in terms of the nature of the changes, and quantitatively in terms of if these changes are significant or slight.

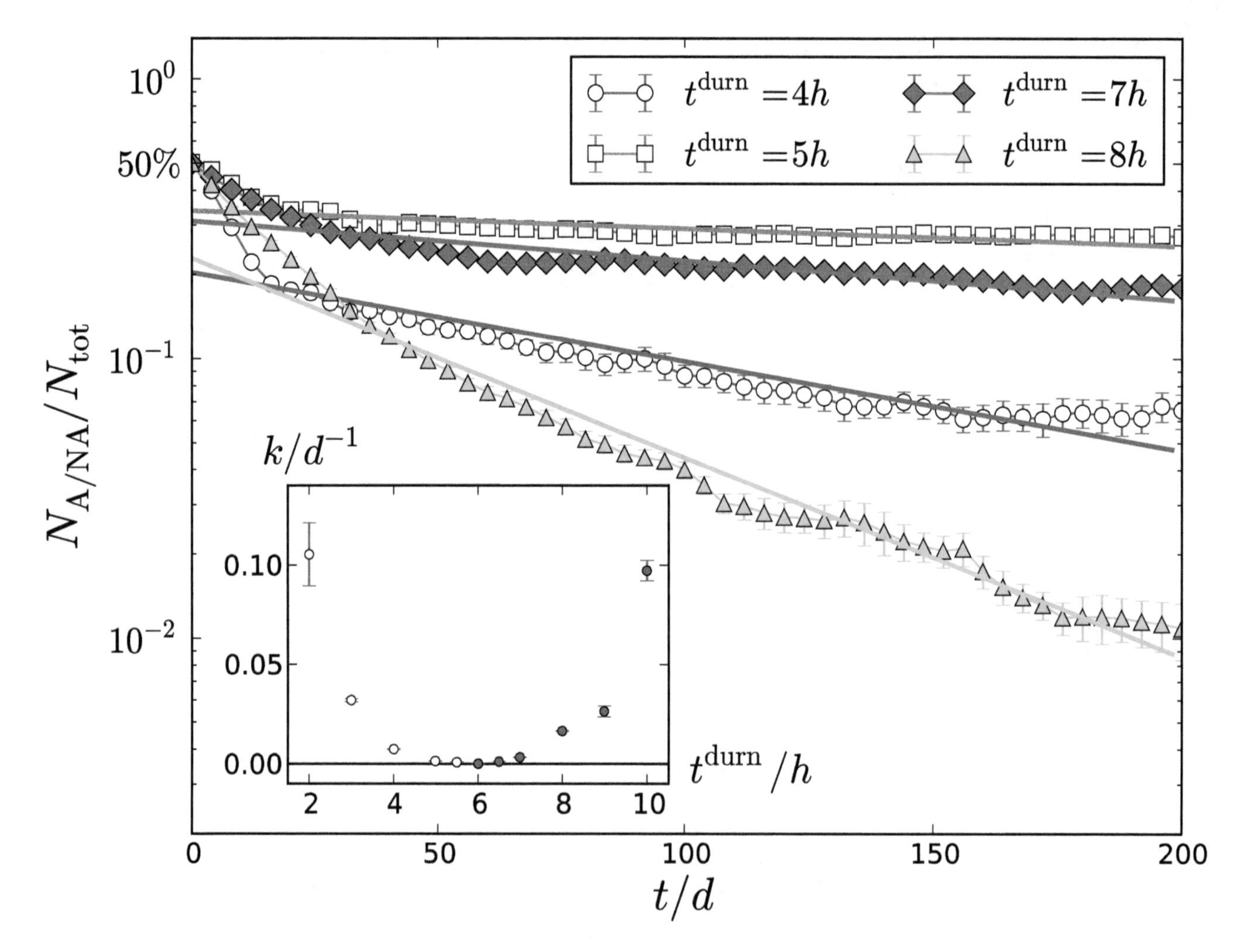

Figure 4. Plot of the fraction of the minority population (NA for open symbols, A for closed symbols) versus time, for the pulse cycle durations t^{durn} shown in the legend. Solid straight lines are the fit to exponential decay for times $t > 10d$. (Inset) Fitted rates versus t^{durn}.

The freedom with which parameters can be varied makes this form of modelling well suited to studying oral biofilms. The microbial composition of these complex multi-species communities varies greatly between human hosts, with phylogenetically distant organisms fulfilling overlapping functional roles. Focussing on microbial function rather than genetic identity is therefore desirable when developing clinical treatments, and mathematical modelling facilitates this by permitting the rapid assaying of parameters that vary between functionally-similar species. This should be contrasted with the equivalent *in vitro* experiments, which would need many repeats with different species, strains and sub-strains to sample equivalent ranges of metabolic activity.

When applied to competing acidogenic populations varying in their rates of nutrient uptake $K_{nut}^{A/NA}$ and aciduricity $K_{acid}^{A/NA}$ as in

Table 2. Physical and biological parameters that were not varied in this study and kept at the values shown.

Symbol	Description	Value	Reference
t^{intra}	Duration of carbohydrate pulse	15 m	[41]
L	Longitudinal film width	2 mm	-
$[Gl]_{intra}$	Concentration of sugar during a pulse	50 g/L	[38,39,47–50]
ρ^c	Cell density (excluding water)	0.2 pg/μm^3	[31]
ρ^e	EPS density (excluding water)	4×10^{-2} pg/μm^3	[30]
q_{max}	Base reaction rate	5/h	[38,39]
Y^c	Yield factor for cell mass	0.1	[38,39]
M^{acid}	Molecular weight of (lactic) acid	90.08 g/mol	-

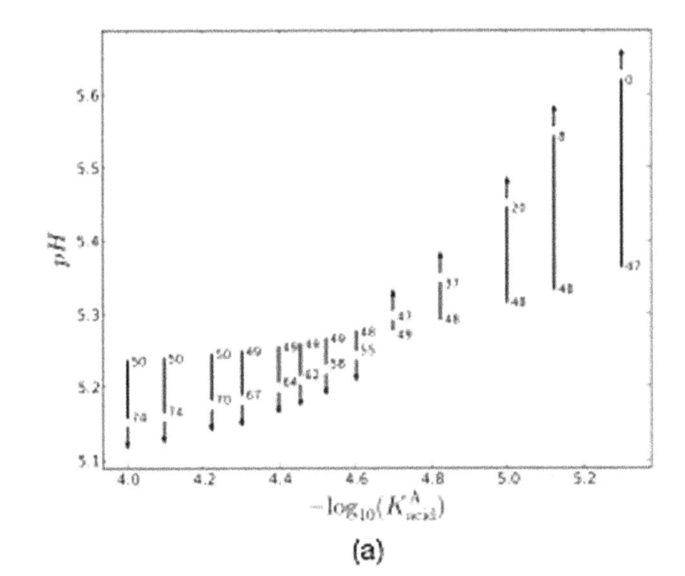

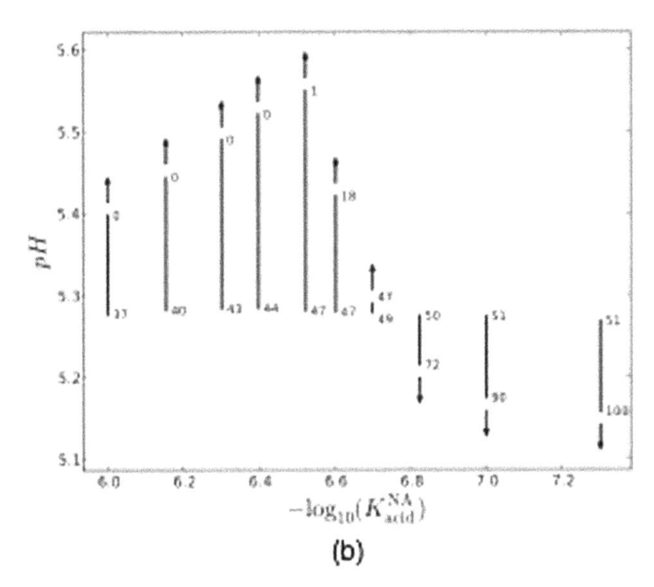

Figure 5. Variation of biofilm composition and pH during the glucose pulse as a single parameter is varied, for (a) K^{A}_{acid} and (b) K^{NA}_{acid}. These parameters are plotted as $-\log_{10}(K^{A/NA}_{acid})$, corresponding to the pH at which cell metabolism is 50% inhibited by acidity. Each line segment shows the variation in pH, starting from 2 d and finishing on 200 d, with the arrow showing the direction of increasing time. The numbers show the percentage of biofilm occupied by A (*i.e.* N_A/N_{tot}) at the start and end points.

this work, this approach demonstrates that these parameters measurably affect the cariogenic potential of (supragingival) oral biofims; see Fig. 7. This insight can be used to suggest treatments for modulating biofilm composition towards a benign homeostatic state, and indeed reducing aciduricity is one mechanism by which fluoride promotes improved oral health [12,13]. This highlights the potentially controlling role of sublethal treatments in modulating population dynamics. It also highlights an additional advantage of mathematical modelling, in that it allows us to isolate this one mechanism from many other postulated roles for fluoride, simply because these alternatives were not included in the model.

Confirmation of the predictive capability of this or any model requires model validation. The parameter sensitivity map in Fig. 7 can serve as a framework for model validation, in two respects. Firstly, it helps identify experiments that can be used to estimate specific model parameters. If the sensitivity of any single experimental measure with respect to a given parameter is high, corresponding to bright entries in the table, the resulting estimate will have a low signal-to-noise ratio and thus should be reliable. Secondly, parameters for which the sensitivity is low need not be determined to any degree of precision for reliable model predictions. It is clear from Fig. 7 that the parameters for cell division, stiffness and death rates have little effect on any of the postulated experimental measures, suggesting that rough approximation of their values should suffice.

Parameter sensitivity can also be used to suggest directions for further model development. It is apparent from Fig. 7 that the parameters for inter-pulse glucose concentration $[\mathrm{Gl}]_{\mathrm{inter}}$ and acid buffering K_a have a measurable influence on the resulting predictions. Both of these mechanisms were incorporated in an approximate manner in this work, and this sensitivity suggests that further modelling would benefit from expanding each mechanism to include more detail, albeit with the overhead of an increased number of parameters. Indeed, series of reactions for both glucose storage and acid buffering have already been specified for the non-growing biofilms of Ilie *et al.* [27], and these could be incorporated in an agent-based model such as ours.

The real plaque ecosystem maintains a dynamic equilibrium between multiple species whose relative fractions vary with environmental conditions, but never vanish entirely [1]. By contrast, here one species type is progressively removed from the system, as long as the frequency of glucose intake and low-pH challenges remains constant. This most likely represents the simplicity of the model, and many model extensions are likely to permit subpopulations of functionally dissimilar species to be perpetually maintained. Additional community complexity in the form of more than two distinct species types interacting *via* a range of interactions, mutualistic, antagonistic and otherwise, should permit a dynamically stable biofilm composition. Heterogenous biofilm composition leads to chemical gradients and an extended habitat range for species that would not be able to persist in a homogenous biofilm, features that can be included in agent based modelling. Other possibilities for maintaining a dynamic equilibrium have been discussed elsewhere [6–8]. Finally, we notice that clonal variation was not included in this version of the model, so all members of a population are phenotypically identical. It is, however, straightforward to introduce such a feature into agent-based modelling, at the expense of additional parameter fitting.

Analysis

The mathematical model employed here is based on the Individual based Model (IbM), an established agent-based model for biofilms that has been applied to a range of bacterial communities and environments [28–36]. Such models consist of two coupled phases, a particulate phase where each particle represents a bacterium or bacterial aggregate, and a series of overlapping continuous phases representing the concentration fields of one or more dissolved species, *e.g.* nutrients or metabolic products. The variant here admits biomechanics in that the particles are interconnected *via* springs representing adhesion by the extracellular polymeric substances (EPS). Here we summarise only those features of the model relevant to the subsequent discussion, and direct the reader to [40] for further details.

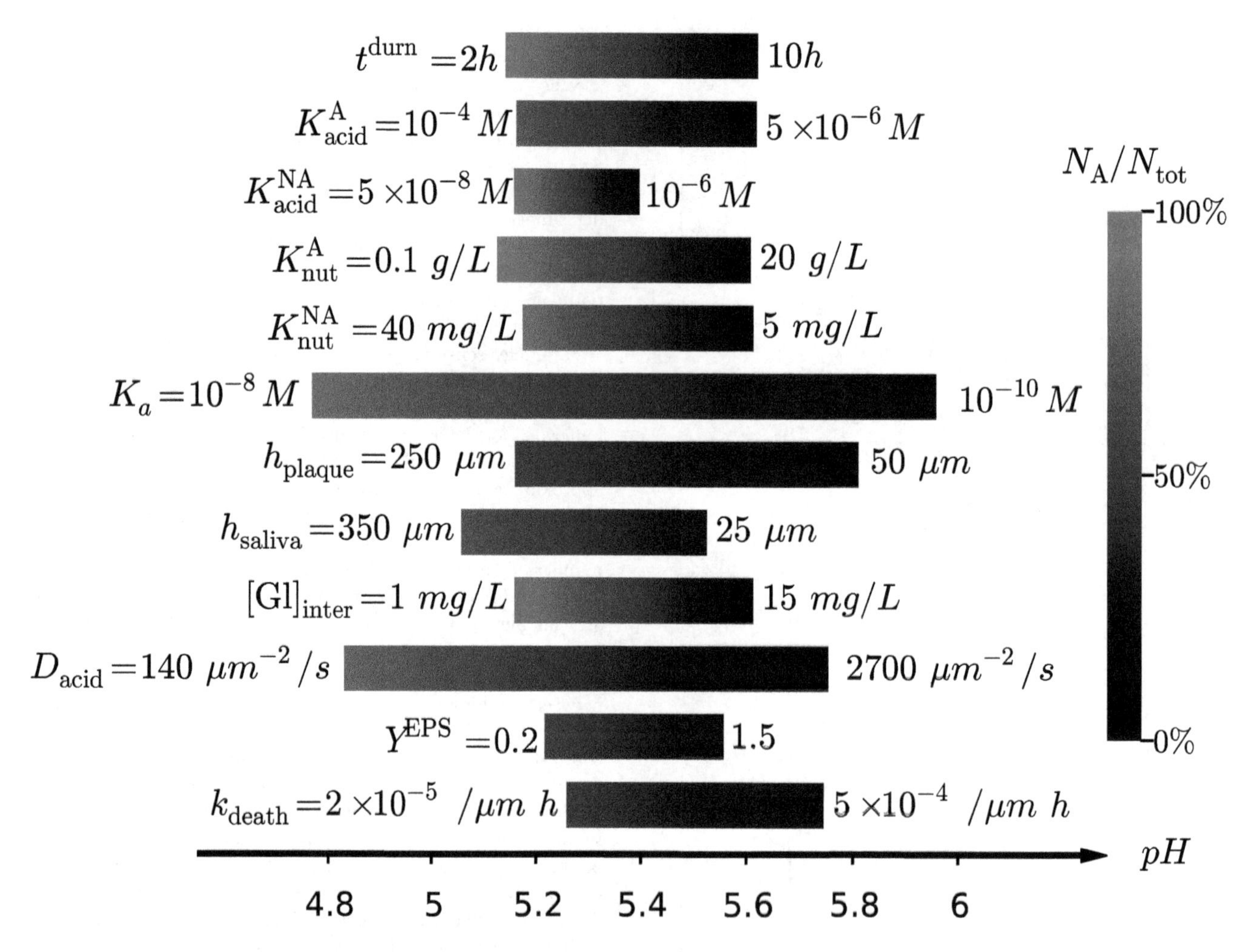

Figure 6. Variation of pH during the pulse at $t=200d$ as the first 12 parameters of Table 1 are varied. Each bar corresponds to a single parameter being varied with the remaining held fixed at their primary values. The shading corresponds to the fraction of the biofilm occupied by population A as in the calibration bar.

Model overview

In this study we consider a mixed film consisting of two microbial populations that differ in their ability to metabolise sugars in the presence of low pH. These are referred to as A for aciduric and NA for non-aciduric, and parameters relating to each species are labelled A or NA accordingly. Additionally there is a single scalar field representing the concentration of lactic acid produced by each particle's glycolysis of dietary sugar. The system domain is schematically shown in Fig. 8(a). A two-dimensional geometry has been employed as this permits biofilms of lateral extent far exceeding their thickness to be simulated within a reasonable timeframe. Periodic boundaries have been assumed in the direction parallel to the enamel surface.

The nutrient (glucose) is not represented as a spatially-varying field, but is instead assumed to have a uniform concentration with no gradients. The concentration does however vary in time as shown in Fig. 8(b). This follows a feast-famine protocol representing the dietary intake of fermentable carbohydrates, whereby the concentration $[Gl]$ alternates between short periods at a high value $[Gl]_{intra}$, interspersed with extended periods at a lower value $[Gl]_{inter}$. $[Gl]_{intra}$ is taken to be far above the half-concentrations for nutrient uptake (see below), so the metabolism for both populations is saturated during the pulse. The duration of the pulse has been fixed at 15 minutes as this represents a typical removal time of acid from the oral cavity [1,41]. The duration of a complete pulse cycle (*i.e.* inter plus intra periods), t^{durn}, is a key parameter of the model.

The model parameters are listed in two separate tables, where the free parameters that were systematically varied in this study are listed in Table 1, and the fixed parameters whose values were estimated from the literature and not varied are given in Table 2. Additional numerical parameters, such as the convergence parameters for the chemical and mechanical relaxation, were tested to be sufficiently small to not affect the results and are not quoted here.

Cell metabolism

Acid buffering within plaque results in a lower concentration of H^+ ions, and therefore a higher pH, than for the same concentration of lactic acid in aqueous solution [41]. However, empirical titration curves cannot be easily incorporated into models as discussed elsewhere [23]. Rather than include a series of coupled reactions as in [27], which would slow down our simulations and introduce additional parameters, we instead treat the dissociation of lactic acid to H^+ as a single-step process with an *effective* dissociation constant K_a that is far lower in value than aqueous dissociation. K_a thus becomes a free parameter which we systematically vary. The concentration of glucose between pulses,

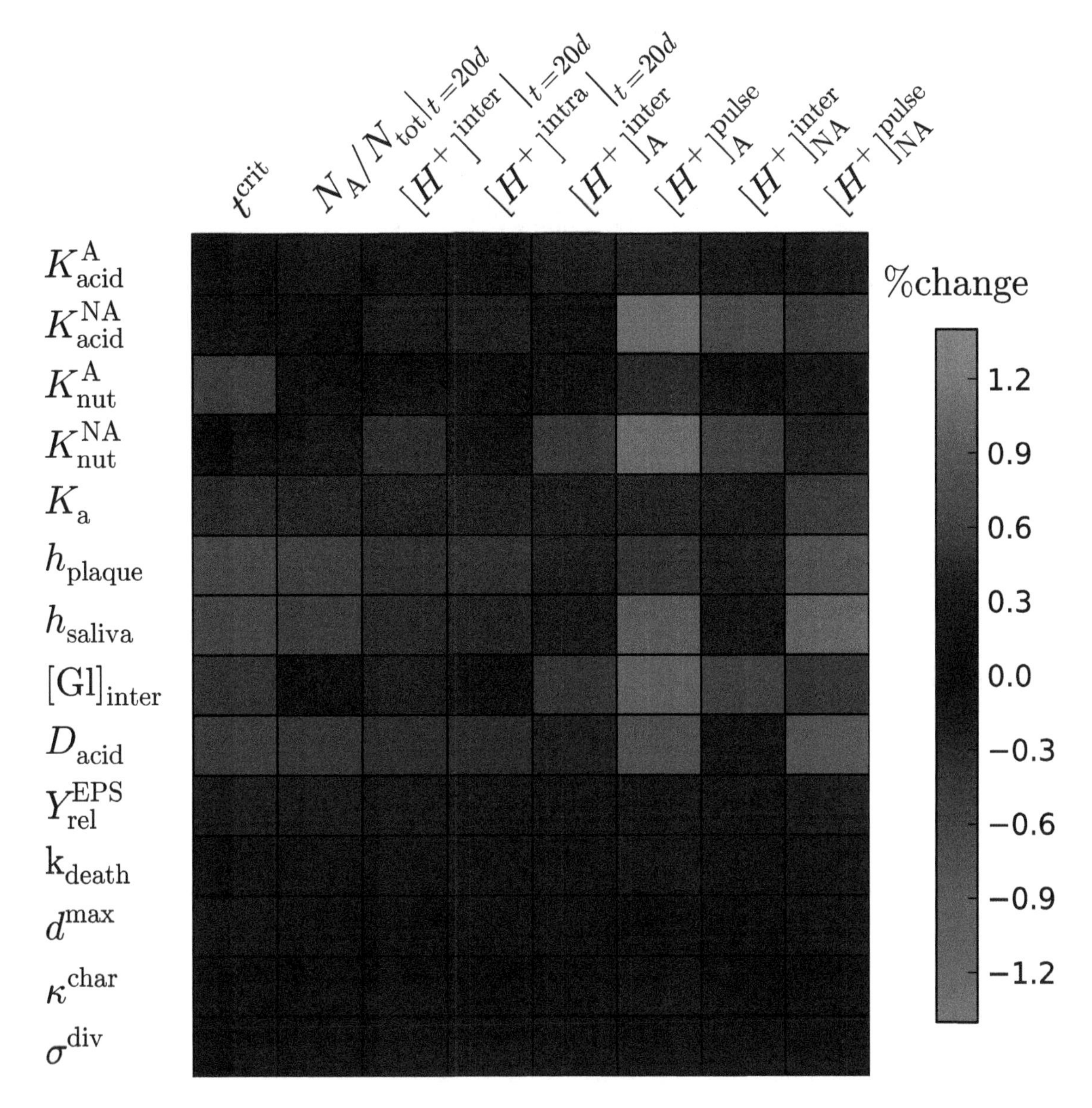

Figure 7. (Colour online) Sensitivity heat map between input parameters (rows) and measurable outputs (columns). Brightness corresponds to the relative change (increase or decrease) in the output quantity as the parameter is increased by 1%, as indicated by the colour bar. Actual values (with error bars) provided in Table S1 of the Supplementary Information.

which depends on the balance of storage and conversion reactions, is also simplified to a single free parameter $[Gl]_{inter}$.

The glycolysis of glucose to lactic acid is assumed to depend on two factors, the concentration of glucose $[Gl]$ and the local acidity $[H^+]$. Both quantities modulate the overall reaction rate as independent Monod factors with half-concentrations K_{nut} and K_{acid} for nutrient uptake and acid inhibition respectively, as shown in Fig. 8(c). The full expression for the reaction rate r_i (mass per unit time) for the particle with label i and mass m_i is

$$r_i = m_i q_{max} \left(\frac{[Gl]}{[Gl] + K_{nut}} \right) \left(\frac{K_{acid}}{[H^+] + K_{acid}} \right), \qquad (2)$$

where $[Gl]$ is in units of mass per unit volume, and $[H^+]$ in molarity. Each half-concentration has the same units as its corresponding concentration, and have separate values for A and NA, *e.g.* K^A_{nut} and K^{NA}_{nut}.

The spatial distribution of lactic acid, which is both produced by (2) and modulates it by determining $[H^+]$, obeys the standard reaction-diffusion equation in which local production of acid is balanced by diffusion away from the source. The primary value for the diffusion coefficient D_{acid} is taken to be that for lactic acid in water, but was also systematically varied. The reaction-diffusion equation was numerically solved using geometric multi-grid on a rectangular mesh [40], with no-flux boundary conditions at the enamel surface and the requirement that the acid at the upper surface of the saliva layer is zero.

Once the r_i for each particle i is determined, the resulting rate of increase in mass is computed as $Y^c r_i$, where Y^c is the dimensionless yield factor; here we fix $Y^c = 0.1$ for both bacterial populations, comparable to representative oral bacteria [38,39]. EPS is assumed to be produced at a rate proportional to the cell mass, *i.e.* the mass of the EPS increases at a rate $Y^{EPS}_{rel} Y^c r_i$, where Y^{EPS}_{rel} is a free parameter that we systematically vary. Particle and

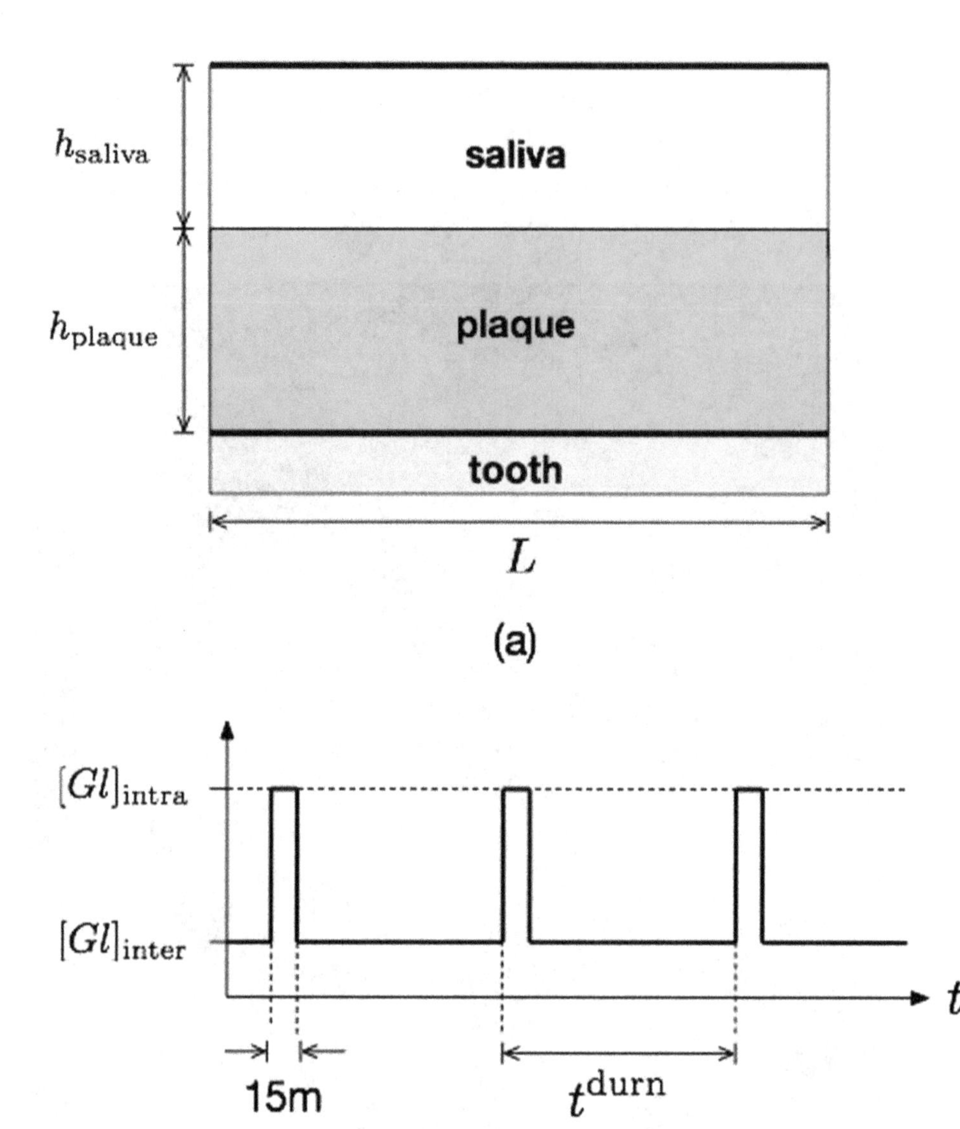

h_{saliva}
saliva
h_{plaque}
plaque
tooth
L
(a)
$[Gl]_{\mathrm{intra}}$
$[Gl]_{\mathrm{inter}}$
t
15m
t^{durn}
(b)

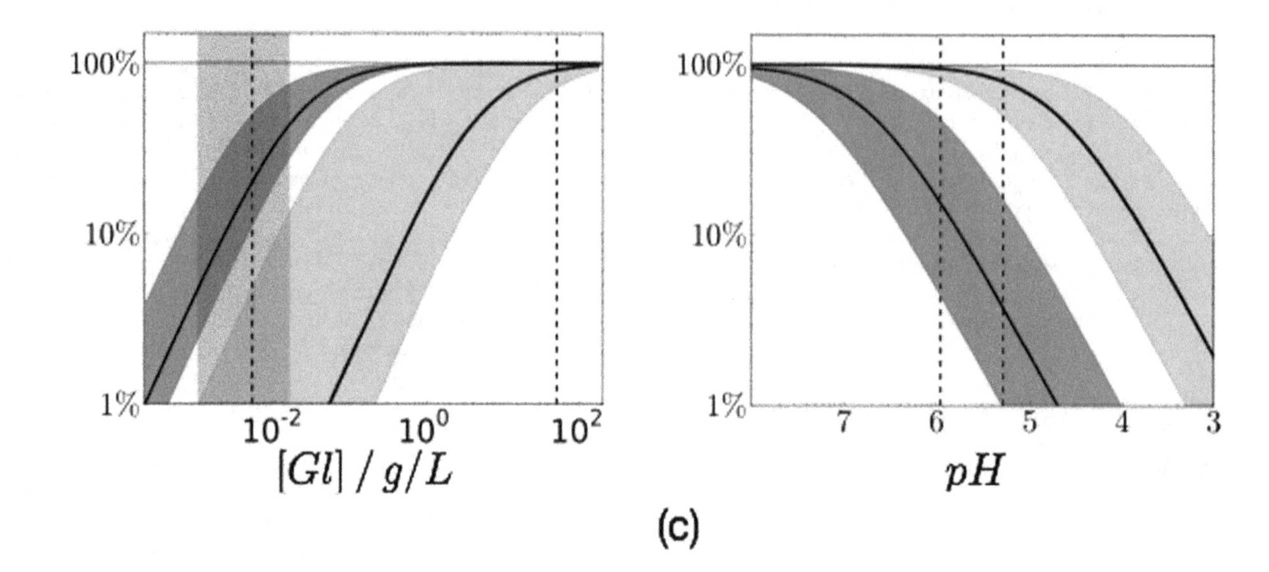

100%
10%
1%
10^{-2}
10^{0}
10^{2}
$[Gl]\ /\ g/L$
7
6
5
4
3
pH
(c)

Figure 8. Schematic of the model. (a) The system geometry. (b) The temporal variation of the environmental glucose follows a feast-famine protocol with a total cycle time t^{durn}, including the 15 minute pulse. (c) Relative metabolic rates due to nutrient (left) and acid inhibition (right) as per (2), so 100% means saturation and 1% means a significant reduction. For both plots, the leftmost solid curve corresponds to the primary value for NA, and the right to A. Each curve also has a shaded region giving the total range considered for each parameter. For the left plot, the parameters for $[Gl]_{\text{inter}}$ and $[Gl]_{\text{intra}}$ are shown as vertical dashed lines, with the shaded region for the former showing the range considered. For the right plot, the vertical dashed curves show typical pH values during (right) and in between (left) glucose pulses for the primary values in Table 1.

EPS masses are converted to physical size by assuming fixed densities of both components as listed in Table 2.

When a particle diameter exceeds the critical value $d^{\max}$ it divides into two daughter particles. The mass of the mother particle, m_{m}, is conserved but is distributed asymmetrically to the daughters, *i.e.* with masses m_{d}^1 and m_{d}^2 obeying

$$m_{\text{d}}^1, m_{\text{d}}^2 = f m_{\text{m}}, (1-f) m_{\text{m}}, \quad (3)$$

where f is a random number drawn from a Normal distribution with mean 0.5 and width σ^{div}. The mass of the EPS is distributed similarly, with the same f. The two parameters $d^{\max}$ and σ^{div} are treated as free and systematically varied.

Redistribution and removal of biomass

After particle growth and division, the particles are rearranged so as to ensure the biofilm as a whole is mechanically stable. This procedure, described in detail in [40], involves constructing a network of springs connecting nearby particles. Each of these springs has a stiffness that is proportional to the local EPS mass and the model parameter κ^{char}. In addition, springs between particles and the enamel surface have a stiffness that is also proportional to κ^{char}. The free parameter κ^{char} is here varied around values suggested by AFM experiments [42] as shown in Table 1. The numerical procedure involves converting the requirement of mechanical equilibrium to a sparse matrix equation and solving using the conjugate gradient method. This matrix approach has also been adopted for plant biofilms, motivated by numerical performance [43].

The plaque biofilm is limited in thickness to h_{plaque}, in that any particles whose centre exceeds this height is removed from the system. This can be thought of as a simplistic representation of biomass removal due to fluid shear by flowing saliva [32], and can be achieved *in vitro* by a constant depth film fermenter [13,44–46]. Above the plaque biofilm is the saliva layer of thickness h_{saliva} in which there is no biomass, but the concentration of lactic acid still continuously varies until vanishing at the upper saliva surface.

Additionally, since bacteria deep within a biofilm exhibit a lower viability than those near the surface, we include a second mechanism for particle removal, namely that cells are 'killed' at a rate that depends on their distance from the exposed biofilm surface, with the highest death rate near the enamel. The death rate per unit depth per unit time is denoted k_{death} and is a free parameter that is varied here. This parameter is the same for both populations.

Parameter sampling

Given the large number of parameters, a systematic investigation of all combinations was not feasible. Instead, each free parameter was assigned a primary value based on chemostat experiments of oral bacteria [38,39], known environmental conditions for oral biofilms [1,7,8] and related models [27]. The primary value for the effective dissociation constant K_{a} was selected to give a realistic pH during the glucose pulse, and that for $[Gl]_{\text{inter}}$ to give a critical value of t^{durn} (see below) around 6 h. Each parameter was then varied over the range specified in Table 1 with all other parameters held fixed.

The sensitivity heat map discussed in *Results* refers to linear response, *i.e.* small changes in the input parameters for which the corresponding change in each output parameter was proportional. For each model parameter, a range of variations was tested to ensure the corresponding variation in the outcome was linear (unlike the data in Fig. 6, which is concerned with non-linear trends over a broad variation of parameters), which was typically the case for a 5–10% change in the model parameter. The sensitivity was then scaled to a 1% value. The raw data (with errors) are given in Table S1 of the Supplementary Information.

Supporting Information

Figure S1 Snapshots of examples biofilms. (a) $t^{\text{durn}} = 8h$ at a time point of 10 days, (b) the same after 200 days, and (c) $t^{\text{durn}} = 4h$ after 200 days. Green (blue) discs correspond to populations of A (NA), respectively. The red in the background corresponds to the lactic acid, with red (black) corresponding to high (low) concentrations respectively. All other parameters are the same as Fig. 1 in the main article.

Table S1 Raw data for the sensitivity heap map. Values correspond to the percentage change in the measurable outcome (columns) for a 1% change in the model parameter (rows). Error bars are either given explicitly, or as variation in the final 2 digits (shown in brackets).

Movie S1 Movie corresponding to Fig. S1(b). Extends from $t=0$ to $t=200$ days. Colour coding and parameters as per Fig. S1(b).

Movie S2 Movie corresponding to Fig. S1(c). Extends from $t=0$ to $t=200$ days. Colour coding and parameters as per Fig. S1(c).

Author Contributions

Analyzed the data: DAH. Wrote the paper: DAH DD PM. Designed the software used in the modelling and analysis: DAH.

References

1. Marsh PD, Martin MV (2009) Oral Microbiology. Elsevier.
2. Marsh PD, Featherstone A, McKee AS, Hallsworth AS, Robinson C, et al. (1989) A microbiological study of early caries of approximal surfaces in schoolchildren. Journal of Dental Research 68: 1151–1154.
3. Gross EL, Beall CJ, Kutsch SR, Firestone ND, Leys EJ, et al. (2012) Beyond Streptococcus mutans: Dental Caries Onset Linked to Multiple Species by 16S rRNA Community Analysis. PLoS ONE 7: e47722.
4. Wolff D, Frese C, Maier-Kraus T, Krueger T, Wolff B (2013) Bacterial Biofilm Composition in Caries and Caries-Free Subjects. Caries Research 47: 69–77.

5. Xu H, Hao W, Zhou Q, Wang W, Xia Z, et al. (2014) Plaque Bacterial Microbiome Diversity in Children Younger than 30 Months with or without Caries Prior to Eruption of Second Primary Molars. PLoS ONE 9: e89269.
6. Allison DG, Gilbert P, Lappin-Scott HM, Wilson W, editors (2000) Community structure and cooperation in biofilms. Cambridge: Cambridge University Press.
7. Filoche S, Wong L, Sissons C (2010) Oral Biofilms: Emerging Concepts in Microbial Ecology. Journal of Dental Research 89: 8–18.
8. Marsh PD, Devine D, Moter A (2011) Dental plaque biofilms: Communities, conflict and control. Periodontology 2000 55: 16–35.
9. Marsh PD (1994) Microbial ecology of dental plaque and its significance in health and disease. Advances in Dental Research 8: 263–271.
10. Bradshaw DJ, McKee AS, Marsh PD (1989) Effects of carbohydrate pulses and pH on population shifts within oral microbial communities in vitro. Journal of Dental Research 68: 1298–1302.
11. Hoeven JS, Jong MH, Camp PJM, Kieboom CWA (1985) Competition between oral Streptococcusspecies in the chemostat under alternating conditions of glucose limitation and excess. FEMS Microbiology Letters 31: 373–379.
12. Bradshaw DJ, McKee AS, Marsh PD (1990) Prevention of population shifts in oral microbial communities in vitro by low fluoride concentrations. Journal of Dental Research 69: 436–441.
13. Bradshaw DJ, Marsh PD, Hodgson RJ, Visser JM (2002) Effects of glucose and fluoride on competition and metabolism within in vitro dental bacterial communities and biofilms. Caries Research 36: 81–86.
14. Eisenberg AD, Bender GR, Marquis RE (1980) Reduction in the aciduric properties of the oral bacterium Streptococcus mutans GS-5 by fluoride. Archives of Oral Biology 25: 133–135.
15. McDermid AS, Marsh PD, Keevil CW, Ellwood DC (1985) Additive Inhibitory Effects of Combinations of Fluoride and Chlorhexidine on Acid Production by Streptococcus-Mutans and Streptococcus-Sanguis. Caries Research 19: 64–71.
16. Hata S, Iwami Y, Kamiyama K, Yamada T (1990) Biochemical mechanisms of enhanced inhibition of fluoride on the anaerobic sugar metabolism by Streptococcus sanguis. Journal of Dental Research 69: 1244–1247.
17. Phan TN, Reidmiller JS, Marquis RE (2000) Sensitization of Actinomyces naeslundii and Streptococcus sanguis in biofilms and suspensions to acid damage by fluoride and other weak acids. Archives of microbiology 174: 248–255.
18. Kreth J, Merritt J, Shi W, Qi F (2005) Competition and Coexistence between Streptococcus mutans and Streptococcus sanguinis in the Dental Biofilm. Journal Of Bacteriology 187: 7193–7203.
19. Liu J, Wu C, Huang IH, Merritt J, Qi F (2011) Differential response of Streptococcus mutans towards friend and foe in mixed-species cultures. Microbiology 157: 2433–2444.
20. Tanzer JM, Thompson A, Sharma K, Vickerman MM, Haase EM, et al. (2012) Streptococcus mutans Out-competes Streptococcus gordonii in vivo. Journal of Dental Research 91: 513–519.
21. Dawes C, Dibdin GH (1986) A Theoretical Analysis of the Effects of Plaque Thickness and Initial Salivary Sucrose Concentration on Diffusion of Sucrose into Dental Plaque and its Conversion to Acid During Salivary Clearance. Journal of Dental Research 65: 89–94.
22. Dawes C (1989) An Analysis of Factors Influencing Diffusion from Dental Plaque into a Moving Film of Saliva and the Implications for Caries. Journal of Dental Research 68: 1483–1488.
23. Dibdin GH (1990) Plaque Fluid and Diffusion: Study of the Cariogenic Challenge by Computer Modeling. Journal of Dental Research 69: 1324–1331.
24. Dibdin GH (1992) A finite-difference computer model of solute diffusion in bacterial films with simultaneous metabolism and chemical reaction. Bioinformatics 8: 489–500.
25. Pearce EIF, Dibdin GH (1995) The Diffusion and Enzymic Hydrolysis of Monofluorophosphate in Dental Plaque. Journal of Dental Research 74: 691–697.
26. Dibdin GH, Dawes C (1998) A Mathematical Model of the Influence of Salivary Urea on the pH of Fasted Dental Plaque and on the Changes Occurring during a Cariogenic Challenge. Caries Research 32: 70–74.
27. Ilie O, van Loosdrecht M, Picioreanu C (2012) Mathematical modelling of tooth demineralisation and pH profiles in dental plaque. Journal Of Theoretical Biology 309: 159–175.
28. Kreft JU, Picioreanu C, Wimpenny J, van Loosdrecht M (2001) Individual-based modelling of biofilms. Microbiology 147: 2897.
29. Picioreanu C, Kreft JU, van Loosdrecht M (2004) Particle-Based Multidimensional Multispecies Biofilm Model. Applied and Environmental Microbiology 70: 3024.
30. Xavier JB, Picioreanu C, van Loosdrecht M (2005) A framework for multidimensional modelling of activity and structure of multispecies biofilms. Environmental Microbiology 7: 1085–1103.
31. Xavier JB, Foster KR (2007) From the Cover: Cooperation and conflict in microbial biofilms. Proceedings Of The National Academy Of Sciences Of The United States Of America 104: 876–881.
32. Chambless JD, Stewart PS (2007) A three-dimensional computer model analysis of three hypothetical biofilm detachment mechanisms. Biotechnology And Bioengineering 97: 1573–1584.
33. Laspidou C, Kungolos A, Samaras P (2010) Cellular-automata and individual-based approaches for the modeling of biofilm structures: Pros and cons. Desalination 250: 390–394.
34. Wang Q, Zhang T (2010) Review of mathematical models for biofilms. Solid State Communications 150: 1009–1022.
35. Nadell CD, Foster KR, Xavier JB (2010) Emergence of spatial structure in cell groups and the evolution of cooperation. PLoS Computational Biology 6: e1000716.
36. Lardon LA, Merkey BV, Martins S, Dötsch A, Picioreanu C, et al. (2011) iDynoMiCS: nextgeneration individual-based modelling of biofilms. Environmental Microbiology 13: 2416–2434.
37. Dirksen TR, Little MF, Bibby BG (1963) The pH of carious cavities-II. The pH at different depths in isolated cavities. Archives of Oral Biology 8: 91–97.
38. Hamilton IR, Phipps PJ, Ellwood DC (1979). Effect of Growth Rate and Glucose Concentration on the Biochemical Properties of Streptococcus mutans Ingbritt in Continuous Culture. Available: http://iai.asm.org/content/26/3/861.abstract.
39. Marsh PD, McDermid AS, Keevil CW, Ellwood DC (1985). Environmental Regulation of Carbohydrate Metabolism by Streptococcus sanguis NCTC 7865 Grown in a Chemostat. Available: http://mic.sgmjournals.org/content/131/10/2505.full.pdf+html.
40. Head DA (2013) Linear surface roughness growth and flow smoothening in a three-dimensional biofilm model. Physical Review E (Statistical, Nonlinear, and Soft Matter Physics) 88: 032702.
41. Strålfors A (1948) Studies of the microbiology of caries; the buffer capacity of the dental plaques. Journal of Dental Research 27: 587–592.
42. Lau PCY, Dutcher JR, Beveridge TJ, Lam JS (2009) Absolute Quantitation of Bacterial Biofilm Adhesion and Viscoelasticity by Microbead Force Spectroscopy. Biophysical Journal 96: 2935–2948.
43. Rudge TJ, Steiner PJ, Phillips A, Haseloff J (2012) Computational modeling of synthetic microbial biofilms. ACS Synthetic Biology 1: 345–352.
44. Kinniment SL, Wimpenny JWT, Adams D, Marsh PD (1996) Development of a Steady-State Oral Microbial Biofilm Community using the Constant-Depth Film Fermenter. Microbiology 142: 631–638.
45. Metcalf D, Robinson C, Devine D, Wood S (2006) Enhancement of erythrosine-mediated photodynamic therapy of Streptococcus mutans biofilms by light fractionation. Journal of Antimicrobial Chemotherapy 58: 190–192.
46. He Y, Peterson BW, Jongsma MA, Ren Y, Sharma PK, et al. (2013) Stress Relaxation Analysis Facilitates a Quantitative Approach towards Antimicrobial Penetration into Biofilms. PLoS ONE 8: e63750.
47. Ellwood DC, Phipps PJ, Hamilton IR (1979). Effect of Growth Rate and Glucose Concentration on the Activity of the Phosphoenolpyruvate Phosphotransferase System in Streptococcus mutans Ingbritt Grown in Continuous Culture. Available: http://iai.asm.org/content/23/2/224.abstract.
48. Hamilton IR, Ellwood DC (1983). Carbohydrate metabolism by Actinomyces viscosus growing in continuous culture. Available: http://iai.asm.org/content/42/1/19.abstract.
49. Ellwood DC, Hamilton IR (1982). Properties of Streptococcus mutans Ingbritt growing on limiting sucrose in a chemostat: repression of the phosphoenolpyruvate phosphotransferase transport system. Available: http://iai.asm.org/content/36/2/576.abstract.
50. Marsh PD, McDermid AS, Keevil CW, Ellwood DC (1985) Effect of Environmental-Conditions on the Fluoride Sensitivity of Acid Production by S-Sanguis Nctc-7865. Journal of Dental Research 64: 85–89.

19

Architectural Design Drives the Biogeography of Indoor Bacterial Communities

Steven W. Kembel[1,2,3⁹], James F. Meadow[2,3*⁹], Timothy K. O'Connor[2,3,4], Gwynne Mhuireach[2,5], Dale Northcutt[2,5], Jeff Kline[2,5], Maxwell Moriyama[2,5], G. Z. Brown[2,5,6], Brendan J. M. Bohannan[2,3], Jessica L. Green[2,3,7]

1 Département des sciences biologiques, Université du Québec à Montréal, Montréal, Québec, Canada, **2** Biology and the Built Environment Center, University of Oregon, Eugene, Oregon, United States of America, **3** Institute of Ecology and Evolution, University of Oregon, Eugene, Oregon, United States of America, **4** Department of Ecology and Evolutionary Biology, University of Arizona, Tucson, Arizona, United States of America, **5** Energy Studies in Buildings Laboratory, University of Oregon, Eugene, Oregon, United States of America, **6** Department of Architecture, University of Oregon, Eugene, Oregon, United States of America, **7** Santa Fe Institute, Santa Fe, New Mexico, United States of America

Abstract

Background: Architectural design has the potential to influence the microbiology of the built environment, with implications for human health and well-being, but the impact of design on the microbial biogeography of buildings remains poorly understood. In this study we combined microbiological data with information on the function, form, and organization of spaces from a classroom and office building to understand how design choices influence the biogeography of the built environment microbiome.

Results: Sequencing of the bacterial 16S gene from dust samples revealed that indoor bacterial communities were extremely diverse, containing more than 32,750 OTUs (operational taxonomic units, 97% sequence similarity cutoff), but most communities were dominated by Proteobacteria, Firmicutes, and Deinococci. Architectural design characteristics related to space type, building arrangement, human use and movement, and ventilation source had a large influence on the structure of bacterial communities. Restrooms contained bacterial communities that were highly distinct from all other rooms, and spaces with high human occupant diversity and a high degree of connectedness to other spaces via ventilation or human movement contained a distinct set of bacterial taxa when compared to spaces with low occupant diversity and low connectedness. Within offices, the source of ventilation air had the greatest effect on bacterial community structure.

Conclusions: Our study indicates that humans have a guiding impact on the microbial biodiversity in buildings, both indirectly through the effects of architectural design on microbial community structure, and more directly through the effects of human occupancy and use patterns on the microbes found in different spaces and space types. The impact of design decisions in structuring the indoor microbiome offers the possibility to use ecological knowledge to shape our buildings in a way that will select for an indoor microbiome that promotes our health and well-being.

Editor: Bryan A. White, University of Illinois, United States of America

Funding: This research was funded by a grant to the Biology and the Built Environment Center from the Alfred P. Sloan Foundation Microbiology for the Built Environment Program (http://www.sloan.org/major-program-areas/basic-research/microbiology-of-the-built-environment/). The funders had no role in study design, data collection and analysis, decision to publish, or preparation of the manuscript.

Competing Interests: The authors have declared that no competing interests exist.

* E-mail: jfmeadow@gmail.com

⁹ These authors contributed equally to this work.

Introduction

Biologists and designers are beginning to collaborate in a new field focused on the microbiology of the built environment [1,2]. These collaborations, which integrate perspectives from ecology and evolution, architecture, engineering and building science, are driven by a number of interrelated observations. First, it is increasingly recognized that buildings are complex ecosystems comprised of microorganisms interacting with each other and their environment [3–5]. Second, the built environment is the primary habitat of humans; humans spend the majority of their lives indoors where they are constantly coming into contact with the built environment microbiome (the microbial communities within buildings) [6]. Third, evidence is growing that the microbes living in and on people, the human microbiome, play a critical role in human health and well-being [7–9]. Together, these observations suggest that it may be possible to influence the human microbiome and ultimately human health, by modifying the built environment microbiome through architectural design.

Despite this potential, we remain in the very early stages of understanding the link between design and the microbiology of the indoor environment. A comprehensive understanding of the mechanisms that shape indoor ecosystems will entail disentangling the relative contributions of biological processes including

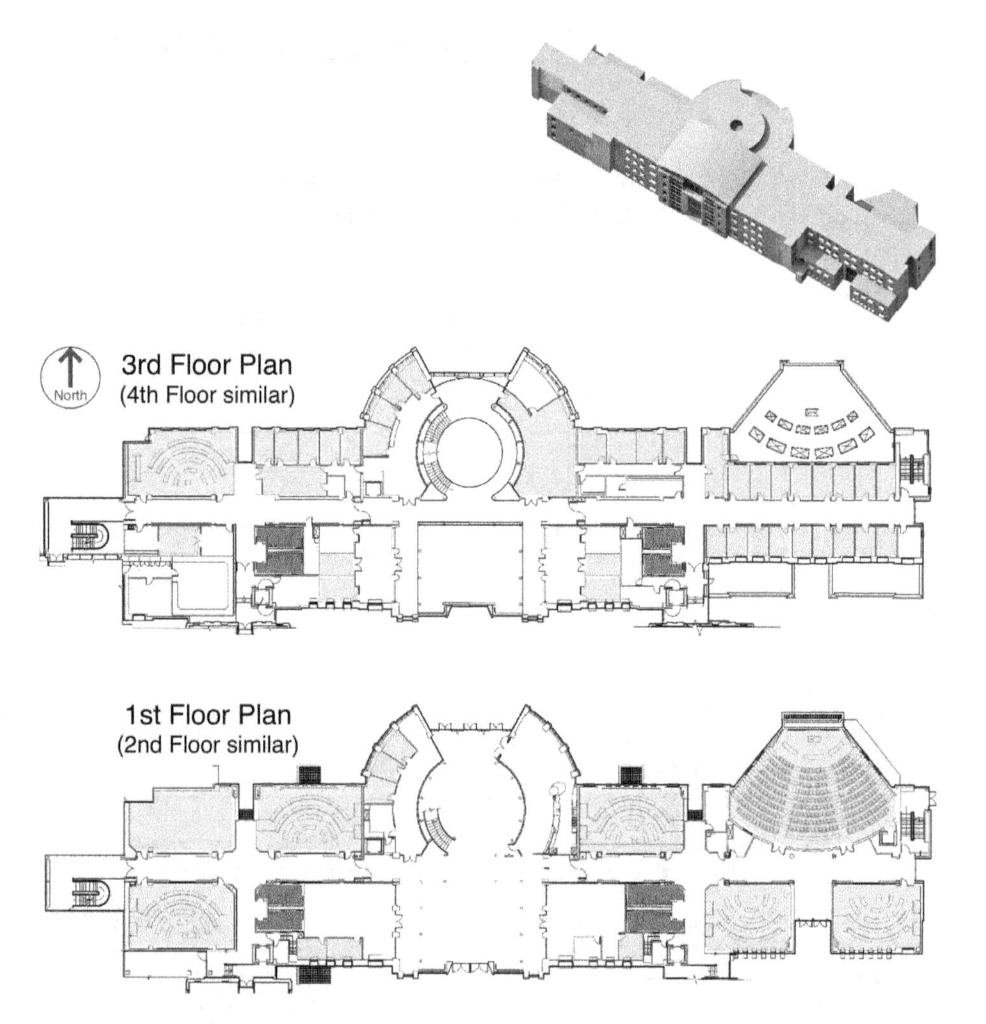

Figure 1. Architectural layout for two of four floors in Lillis Hall. Restrooms (brown), offices (blue) and classrooms (yellow) are shown to illustrate space type distribution throughout Lillis. The first two floors of the building are primarily devoted to classrooms and share a similar floor-plan. The 3rd and 4th floors contain most offices in the building and also share a similar floor-plan. The building has a basement and penthouse spaces; these are largely building support spaces, including mechanical rooms and storage.

environmental selection, dispersal, diversification, and ecological drift [10]. To date, most research has focused on understanding the influence of environmental selection and dispersal on the built environment microbiome. Environmental conditions including humidity and air temperature have been shown to influence the growth rate and survival of many microbial taxa [3,5,11] and correlate with the composition of bacterial communities indoors [4]. Many bacteria and fungi exhibit strong microhabitat associations and increased growth under conditions of higher humidity and in the presence of water sources, such as in kitchens and restrooms [12,13]. The dispersal of microbes into and within the built environment also appears to have a significant influence on indoor ecosystems. The sources of microbes include those from outdoor habitats such as air and soil brought into the building via ventilation systems or carried into the building by macroorganisms [4,14–16], microbes from indoor sources such as water, carpets and other surfaces within a building [13,17], and microbes emitted from macroorganisms within the building including humans, pets and plants [18,19]. The relative importance of these different sources of microbes indoors is not well understood, but is likely to differ as a function of space (e.g. geographic location [20]), time (e.g. year and season of sampling [15]), and building design and operation [4].

The biological processes described above can be fundamentally altered by building design. However many questions remain unanswered regarding how design aspects – such as the *function*, *form* and *organization* of a building - shape the indoor microbiome. *Function* refers to the collection of activities and uses that a building and its spaces serve. Functional requirements are translated into the variety and number of space types within a building – for example offices, restrooms, and hallways. Function is also a key determinant of the design criteria for environmental conditions including temperature, relative humidity, and light levels. *Form* refers to geometry of a building and the spaces within it, while *organization* refers to the spatial relationships among indoor spaces. Form and organization are highly interrelated and both involve design choices that influence human circulation (the source, variation and movement of people), air circulation (the source, variation and movement of air), and environmental conditions throughout a building.

To understand how design choices influence the biogeography of indoor bacterial communities, we collected microbiological, architectural, and environmental data in 155 rooms throughout a multi-use classroom and office building (Lillis Hall; Fig. 1). We focus on the bacterial communities in settled dust, because it represents an integrative record of microbial biodiversity in indoor spaces [21]. Our study addresses two overarching questions. First, at the scale of the entire building, do function, form and organization predict variation in the built environment microbiome? Second, for rooms that serve the same function (rooms that are of the same space type), which aspects of form and organization most influence the built environment microbiome?

Methods

Study Location

We analyzed bacterial communities in dust collected from 155 spaces in the Lillis Hall, a four-story classroom and office building on the University of Oregon campus in Eugene, Oregon, USA. This building was chosen as a study site for several reasons. Architecturally, Lillis Hall was designed to accommodate natural ventilation for both fresh air and cooling; the building is thin, allowing most rooms access to the building skin for supplying outside air directly through windows and louvers, and it has a central atrium used for exhausting air through stack ventilation. From a study design perspective, diverse space types, occupancy levels, and building management strategies were located in close proximity within the same building, making it possible to compare their relative influences on indoor biogeography.

Architectural Design Data

Data on architectural design attributes of each space including function, form, and organization were obtained using architectural plans, field observation, and a building information model (Fig. 1). Spaces in the building were classified into one of seven *space types*. This classification system was developed for the present study based on the Oregon University System's space type codes and definitions [40]. These categories are based on the overall architectural design and intended human use pattern for each space, and include *circulation* (e.g. hallways, atria), *classrooms*, *classroom support* (e.g. reading and practice rooms), *offices*, *office support* (e.g. most storage spaces, conference rooms), *building support* (e.g. mechanical equipment rooms, janitor closets), and *restrooms*. We measured numerous spatial and architectural attributes of each space including *level* (floor), *wing* (east versus west), *size* (net floor area), *air handling unit* (AHU) (13 different AHUs supply air to different rooms, so AHU is a categorical variable with 15 levels, one for each AHU as well as a 'none' category for rooms without mechanically supplied air, and a 'multiple' category for circulation spaces fed by multiple supply sources), and a separate binary variable denoting whether the space was only capable of being *naturally ventilated* by unfiltered outside air (e.g. via windows or louvers; 41 rooms) or by dedicated mechanical AHU supply (114 rooms).

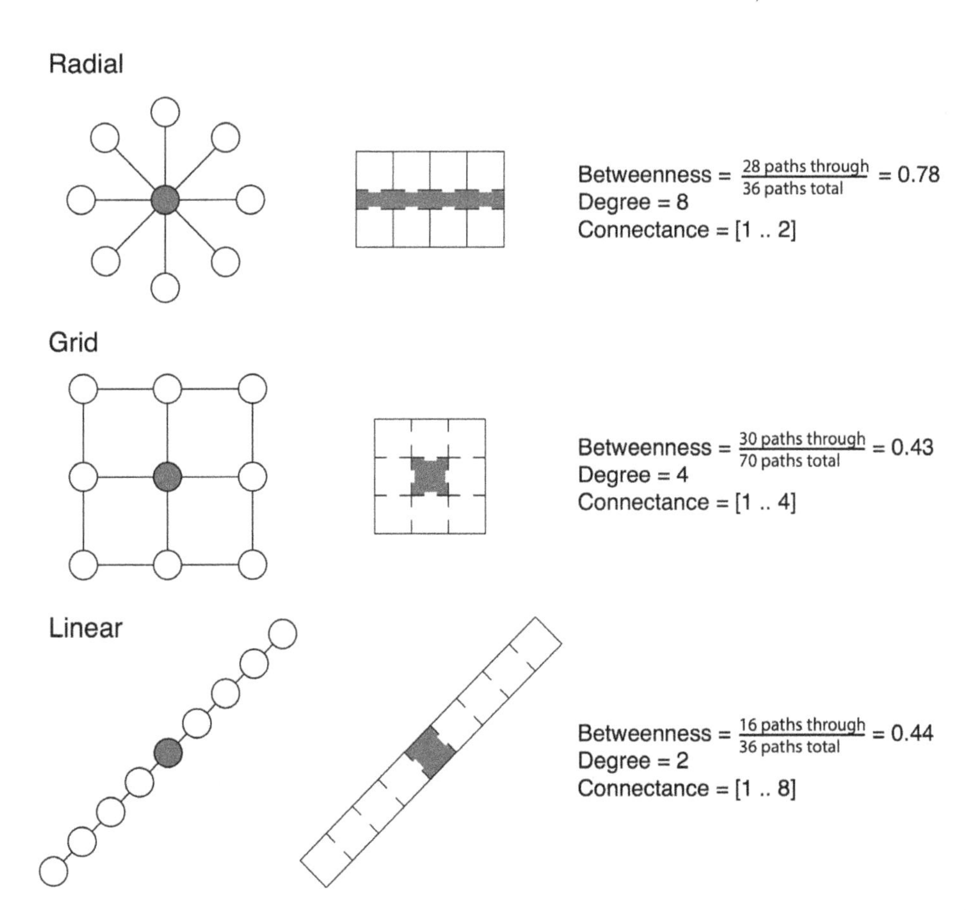

Figure 2. Network analysis metrics used to quantify spatial arrangement of spaces within Lillis Hall. Examples in the left column follow classic network representation, while those in the right column embody the architectural translation of networks. Shaded nodes and building spaces correspond to centrality measures [22] of *betweenness* (the number of shortest paths between all pairs of spaces that pass through a given space over the sum of all shortest paths between all pairs of spaces in the building) and *degree* (the number of connections a space has to other spaces); *connectance distance* (the number of doors between any two spaces) is a pairwise metric, shown here as the range of connectance distance values for each complete network/building. Since *betweenness* and *degree* strongly co-vary and are both measures of network centrality [22], they are considered together in some analyses.

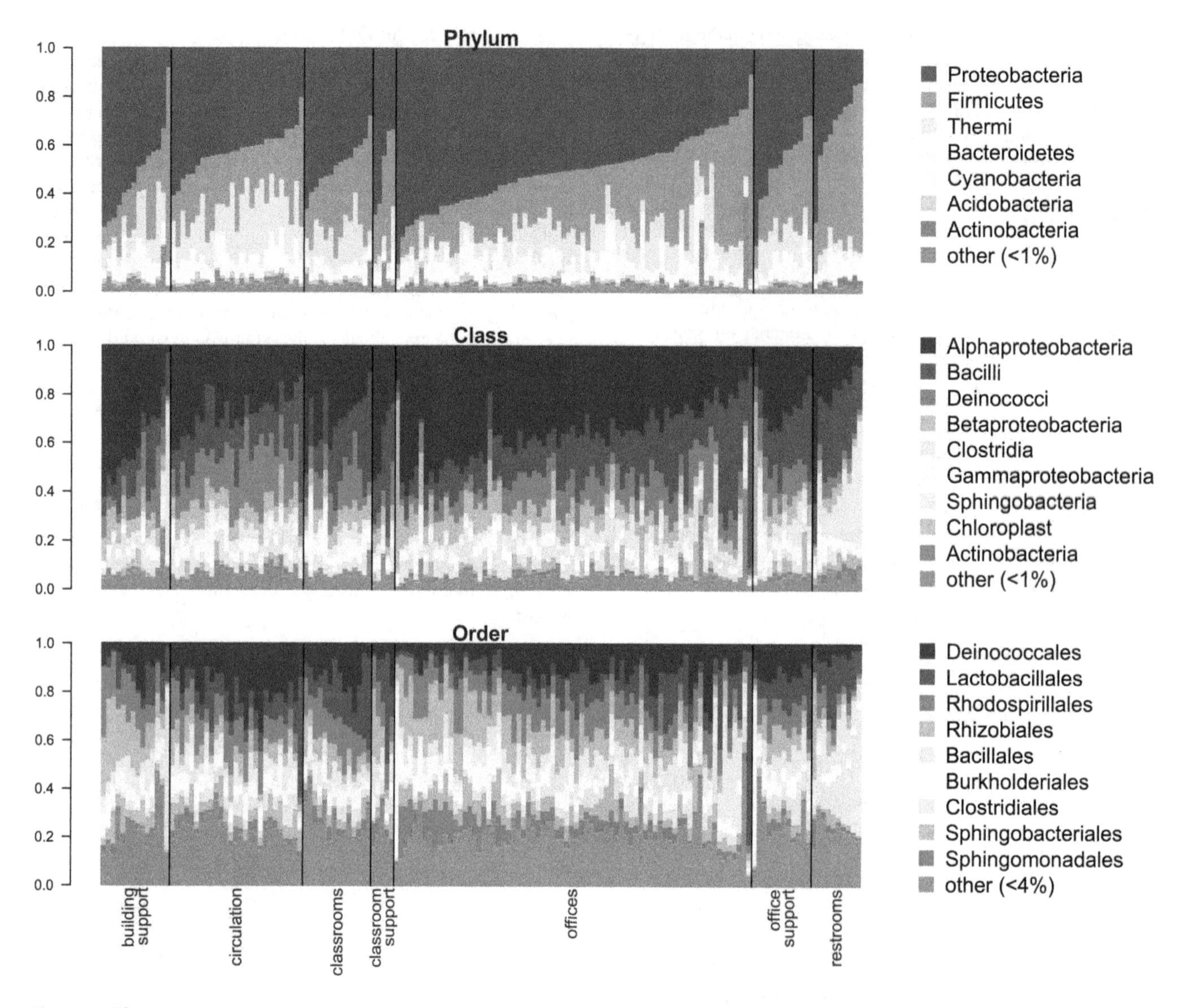

Figure 3. The taxonomic composition of bacterial communities sampled from dust in Lillis Hall. Samples are organized by space type, and relative abundances are shown for groups comprising more than 1% (for phylum and class level) and 4% (for order level).

Metrics related to *form* and *organization* were quantified using network analysis (Fig. 2) and information from building construction drawings. Spaces were considered to be spatially connected if they shared a doorway or other physical connection that would permit a person to move directly between the two spaces. The network of spatial connections among spaces was used to calculate two measures of network centrality [22,41] for each space in the building: *betweenness*, a measure of the fraction of shortest paths among all spaces in the building that would pass through a space, and *degree*, the number of connections a space has to other spaces. The network of spatial connections between spaces was also used to define a *connectance distance* between all pairs of spaces in the building, defined as the minimum number of spaces a person would need to travel through to move between two spaces. We considered using ventilation-based distance (how much duct length separates two connected spaces) as a connectance distance, however preliminary investigation indicated that connectance distance and ventilation distance were strongly correlated.

Human use patterns are a product of functional classification, but they also dictate form and organizational attributes of building design. In this study, human use patterns for each space were estimated based on a qualitative assessment of the expected patterns of *human diversity* and *annual occupied hours* in each space. Briefly, human diversity was defined on a three-point scale, ranging from low human diversity (spaces likely to be occupied by at most a single individual during a typical day; e.g. a closet) to high human diversity (spaces likely to be occupied by numerous different individuals during a typical day; e.g. a hallway). Annual occupied hours (person-hours per year) were similarly defined along a three-point scale from low (spaces that are typically vacant or occupied at low density; e.g. a mechanical support space) to high (spaces that are frequently occupied at relatively high density; e.g. administrative offices). Both of these human occupancy variables are explained in more detail in Table S1.

At the time of microbial community sampling, ambient air temperature and relative humidity measurements were taken from each space. Relative humidity measurements were detrended using daily mean values to account for temporal changes over the sampling period.

Table 1. Variance in biological dissimilarity among bacterial communities from all spaces, as well as just offices, (Canberra distance) explained by different variables in Lillis Hall.

Room types	Explanatory variable	R^2	*P-value*
all rooms	Space type	0.06	0.001
	Air source - air handling unit (AHU)	0.13	0.001
	Building floor	0.01	0.001
	Space size	0.01	0.001
	Building wing - East/West	0.01	0.341
	Building side - North/South	0.01	0.001
	Occupant diversity	0.01	0.001
	Annual occupied hours	0.01	0.015
	Centrality (betweenness)	0.01	0.001
	Centrality (degree)	0.01	0.001
	Temperature	0.01	0.024
	Relative Humidity*	0.01	0.001
	Natural ventilation capability	0.01	0.001
offices	Air source - air handling unit (AHU)	0.07	0.001
	Building floor	0.07	0.001
	Space size	0.02	0.025
	Building wing - East/West	0.01	0.541
	Centrality (betweenness)	0.02	0.005
	Centrality (degree)	0.02	0.016
	Temperature	0.02	0.002
	Relative Humidity*	0.01	0.786
	Natural ventilation capability	0.02	0.001

Variance explained (R^2) and statistical significance (*P-value*) quantified with a PERMANOVA test; since *P-values* are from permutational tests involving 999 permutations, they are only reported down to 0.001. All variables and their respective units are described in the methods section and Table S1.
*detrended using daily averages.

Biological Sampling

Sampling of dust was carried out with a Shop-Vac® 9.4L Hang Up vacuum (www.shopvac.com; #215726) fitted with a Dustream™ Collector vacuum filter sampling device (www.inbio.com/dustream.html). Dust samples were collected by vacuuming an area of approximately 2m^2 on horizontal surfaces above head level for 2 minutes in each space. We preferentially chose these surfaces for sampling since they minimized the frequency of disturbance by cleaning, and thus likely serve as a long-term sample of airborne particles in each space [21]. All samples were collected during June 22–24, 2012. Building construction was completed in 2003, and dust has presumably been accumulating in some sampled spaces since that time.

Dust samples were stored at −80°C until DNA extraction. Dust was manually extracted from filters, and used for DNA extraction. Whole genomic DNA was isolated from samples using MO BIO PowerLyzer™ PowerSoil® DNA Isolation Kit (MO BIO, Carlsbad, CA) according to manufacturer's instructions with the following modifications: bead tubes were vortexed for 10 min; solutions C4 and C5 were substituted for PW3 and PW4/PW5 solutions from the same manufacturer's PowerWater® DNA isolation kit. Bacterial communities were profiled by sequencing a ~420 bp fragment of the V4 region of the bacterial 16S rRNA gene using a custom library preparation protocol [24]. Briefly, the protocol consisted of two PCRs. The first amplified the V4/V5 region using the primers 5′-AYTGGGYDTAAAGNG-3′ and 5′-CCGTCAATTYYTTTRAGTTT-3′ [42,43] and appended a 6 bp barcode and partial Illumina sequencing adaptor. Forward and reverse strands were labeled with different barcodes, and the unique combination of these barcodes was used to pool samples in post-processing.

All extracted samples were amplified in triplicate for PCR1 and triplicates were pooled before PCR2. PCR1 (25 µL total volume per reaction) consisted of the following ingredients: 5 µL 5x HF buffer (Thermo Fisher Scientific, U.S.A.), 0.5 µL dNTPs (10 mM), 0.25 µL Phusion Hotstart II polymerase (Thermo Fisher Scientific, U.S.A.), 13.25 µL certified nucleic-acid free water, 0.5 µL forward primer (10 uM), 0.5 µL reverse primer (10 uM), and 5 µL template DNA. The PCR1 conditions were as follows: initial denaturation for 30 s at 98°C; 20 cycles of 20 s at 98°C, 30 s at 50°C and 30 s at 72°C; and 72°C for 10 min for final extension. After PCR1, the triplicate reactions were pooled and cleaned with the QIAGEN Minelute PCR Purification Kit according to the manufacturers protocol (QIAGEN, Germantown, MD). Amplified products from PCR1 were eluted in 11.5 µL of Buffer EB. For PCR2, a single primer pair was used to add the remaining Illumina adaptor segments to the ends of the concentrated amplicons of PCR1. The PCR2 (25 µL volume per reaction) consisted of the same combination of reagents that was used in PCR1, along with 5 µL concentrated PCR1 product as template. The PCR 2 conditions were as follows: 30 s denaturation at 98°C; 15 cycles of 10 s at 98°C, 30 s at 64°C and 30 s at 72°C; and 10 min at 72°C for final extension.

Amplicons were size-selected by gel electrophoresis: gel bands at c. 500bp were extracted and concentrated, using the ZR-96 Zymoclean Gel DNA Recovery Kit (ZYMO Research, Irvine, CA), following manufacturer's instructions, quantified using a Qubit Fluorometer (Invitrogen, NY), and pooled in equimolar concentrations for library preparation for sequencing. Resulting libraries were sequenced in two multiplexed Illumina MiSeq lanes (paired-end 150 base pair sequencing) at the Dana Farber Cancer Institute (Boston, MA). All sequence data and metadata have been deposited in the open-access data repository Figshare (http://figshare.com/articles/Lillis_Dust_Sequencing_Data/709596).

Sequence Processing

We processed raw sequence data with the *FastX_Toolkit* (http://hannonlab.cshl.edu/fastx_toolkit) and *QIIME* [44] software pipelines to eliminate low-quality sequences and de-multiplex sequences into samples. Sequences were trimmed to a length of 200 bp (100 bp from each paired end). We retained sequences with an average quality score of 30 over 97% of the sequence length after trimming. After trimming, quality filtering and rarefaction of each sample to 2,100 sequences to ensure equal sampling depth across samples, 329,700 sequences from 155 samples remained and were included in all subsequent analyses. We binned sequences into operational taxonomic units (OTUs) at a 97% sequence similarity cutoff using *UCLUST* [45] and assigned taxonomy to each OTU using the BLAST taxon assignment algorithm and Greengenes version *4feb2011* core set [46] as implemented in *QIIME* version 1.4. We inferred phylogenetic relationships among all bacterial OTUs using a maximum likelihood GTR+Gamma phylogenetic model in *FastTree* [47].

Data Analysis

Statistical analysis was performed in *R* [48]. Pairwise community dissimilarity was calculated using the quantitative, taxonomy-based Canberra distance metric, implemented in the *vegan* package

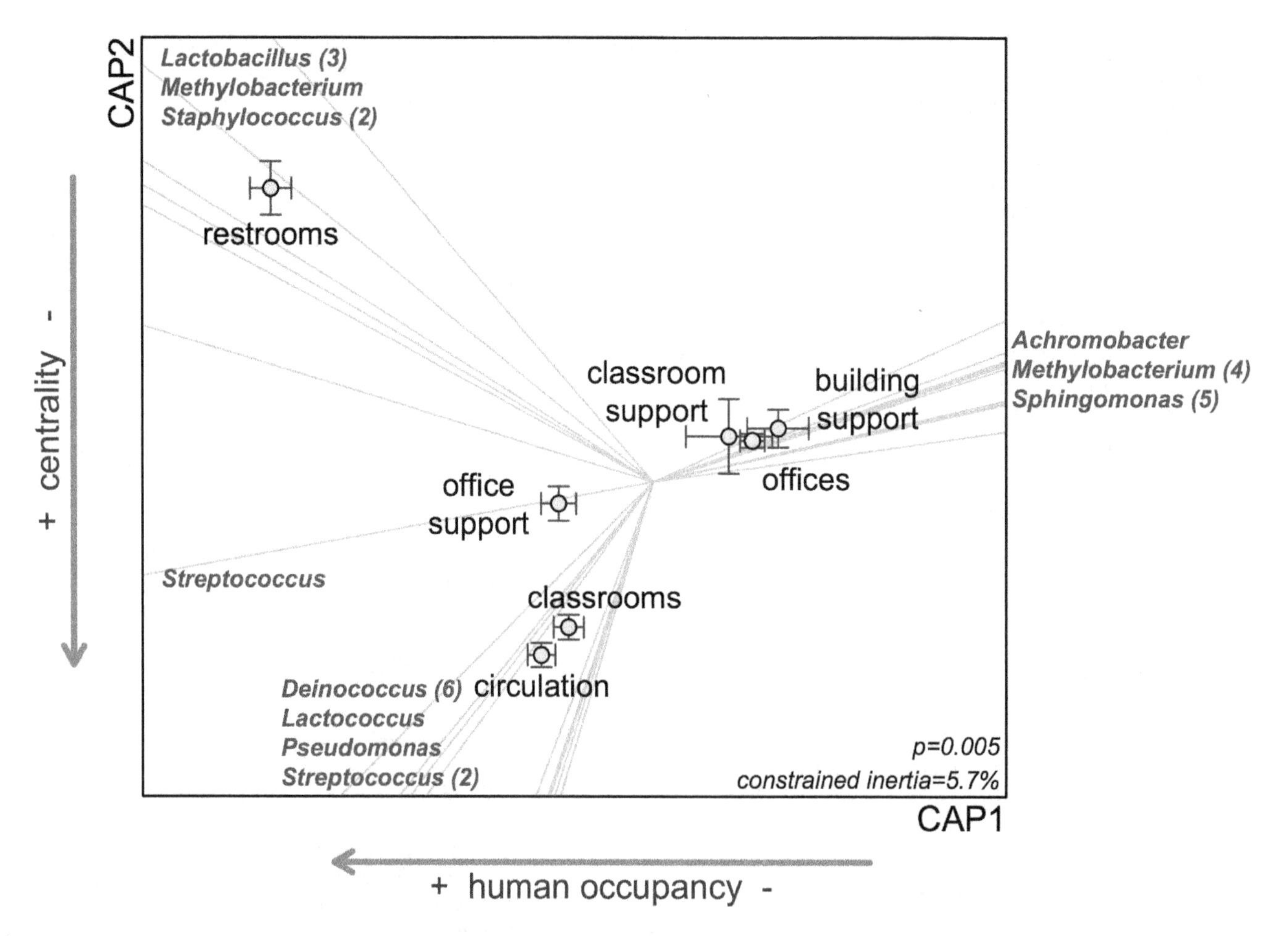

Figure 4. Dust communities within a building cluster by space type and are strongly correlated with building centrality and human occupancy. Points represent centroids (±SE) from distance based redundancy analysis (DB-RDA). Space types hold significantly different communities ($P = 0.005$), though this is driven primarily by restrooms. Bacterial OTUs that have the strongest influence in sample dissimilarities are shown at the margins; numbers in parentheses indicate multiple OTUs in the same genus. Centrality (along y-axis) represents network betweenness and degree; human occupancy (along x-axis) represents annual occupied hours and human diversity. All four correlates (simple linear models as a factor of ordination axis) are significant along their respective axes (all $P < 0.001$).

[49] in *R*. We also assessed the consequences of beta-diversity metric choice on our results; correlations between potential metrics are included as Fig. S1. Constrained ordinations (distance-based redundancy analysis; DB-RDA) were created utilizing the *capscale* function in *vegan*. Correlations reported on ordination axes, indicated by arrows, are based on simple linear models of environmental variables against ordination axes. Indicator taxa analysis [50] was performed using the *indval* function in the *labdsv* package [51]. Mantel and partial mantel tests were used to investigate the correlations between community and environmental distance matrices, including a distance-decay comparison, using the *mantel* function in *vegan*. Permutational multivariate analysis of variance (*PERMANOVA*) was used to test community differences between groups of samples as a way to identify drivers of variation in community structure, using the *adonis* function in *vegan*. All permutational tests were conducted with 999 permutations, and thus *p-values* are reported down to, but not below, 0.001.

Results

Building-scale Design Influences on the Built Environment Microbiome

Bacterial communities in dust from Lillis Hall were highly diverse. Using barcoded Illumina sequencing of 16S rRNA genes, we detected 32,964 operational taxonomic units (OTUs; defined at a 97% sequence similarity cut-off) in 791,192 sequences from 155 samples (19,403 OTUs and 325,500 sequences after rarefaction to 2,100 sequences per sample). Most of these OTUs were rare, occurring in one (49.9%) or two (13.3%) samples, and at low relative abundance (61.1% of OTUs were singletons or doubletons). However, OTUs from several taxonomic groups including Alpha-, Beta-, and Gamma-Proteobacteria, Firmicutes, and Deinococci were abundant and common in almost all dust samples we collected (Fig. 3 and Fig. S2). There were 58 OTUs belonging to these taxonomic groups that were present in 95% or more of all samples we collected. These ubiquitous OTUs were also abundant, representing 0.1% of the OTU richness but >28% of all sequences.

Spaces differing in their architectural design characteristics contained distinctive bacterial communities. Analysis of the variance in bacterial community composition explained by

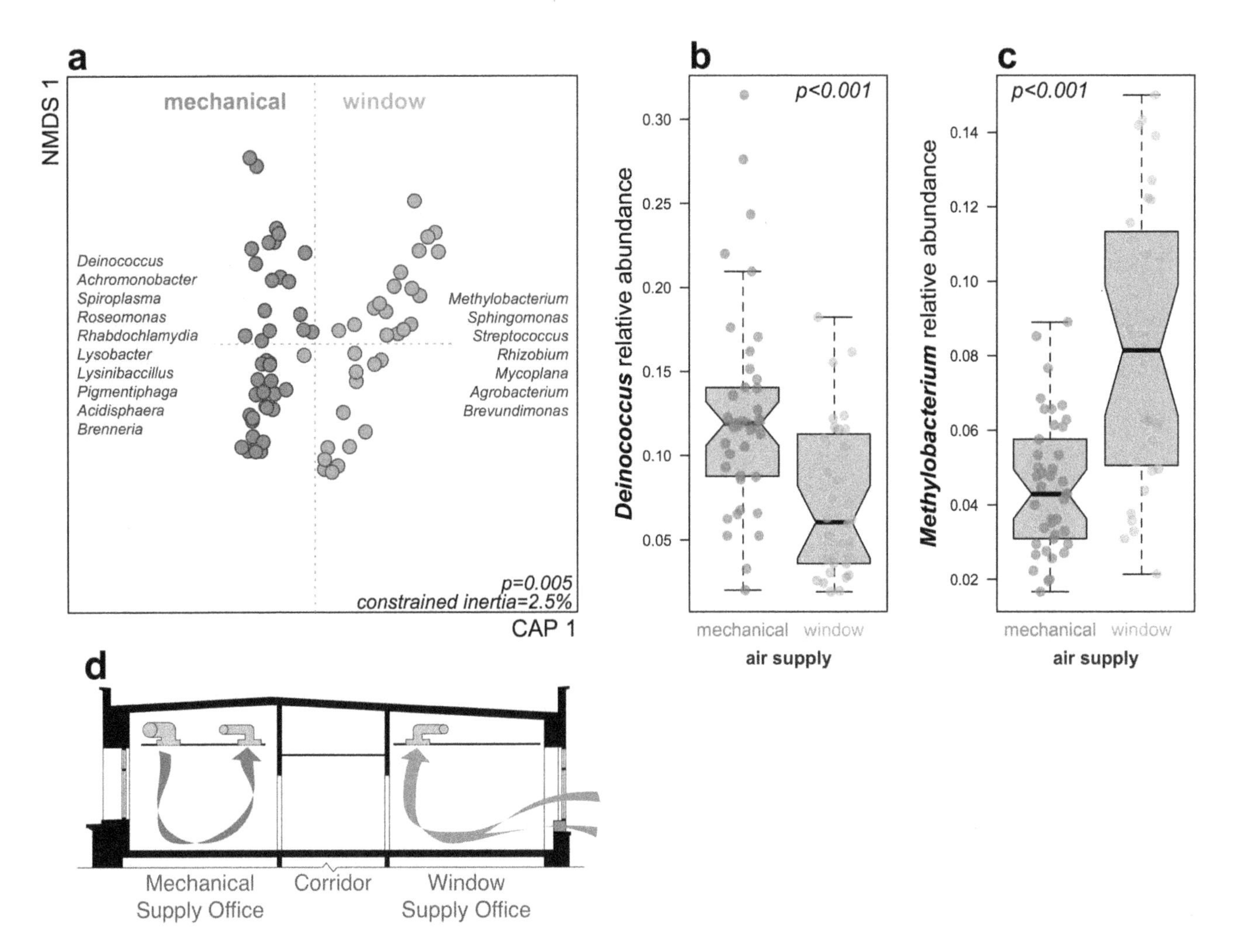

Figure 5. Offices contain significantly different dust microbial communities depending on ventilation source. a) The first axis is constrained by whether or not offices have operable window louvers (blue) or not (red). Taxon names on either side are grouped from the 25 strongest weighting OTUs in either direction. **b)** *Deinococcus* were 1.7 times more abundant in mechanically ventilated offices compared to window ventilated offices. **c)** The opposite pattern was observed for *Methylobacterium* OTUs, which were 1.8 times more abundant in window ventilated offices. Boxplots delineate (from bottom) minimum, Q1, median, Q3, and maximum values; notches indicate 95% confidence intervals. **d)** Cross-sectional view of representative Lillis Hall offices. Offices on the south side of the building (left) received primarily mechanically ventilated air, while offices on the north side of the building (right) are equipped with operable windows as a primary ventilation air source.

different factors (Table 1; PERMANOVA on Canberra distances) indicated that space type and air handling unit (AHU) explained the greatest proportion of variance ($R^2 = 0.06$ & 0.13, respectively; both $P = 0.001$). Nearly all other variables considered in this study (Table 1) were significantly correlated with biological variation as well, but explained a far smaller portion of the overall variance in microbial community structure at the scale of the building. Thus Table 1 can be seen as a potential list of building features that can, in the future, be targeted when attempting to account for microbiological variation in architectural design.

Restrooms explained a substantial amount of the variation observed between space types; bacterial communities in restrooms were compositionally distinct from other space types ($R^2 = 0.06$; $P = 0.001$; from PERMANOVA on Canberra distances). In addition to serving a distinct function, restrooms were characterized architecturally by relatively low network centrality (quantified as network betweenness and degree [22]; network terminology outlined in Fig. 2). This is because in Lillis hall, restrooms generally only have a single door and are rarely or never on a path between any two other spaces. Restrooms also had a high diversity of human occupants (defined as a high number of different occupants throughout the day; explicit definitions of occupancy variables provided in Table S1). Indicator taxa analysis detected numerous OTUs that were associated with restrooms, predominantly belonging to taxa that are commonly associated with the human gut and skin microbiome including *Lactobacillus*, *Staphylococcus*, and *Streptococcus*. Taxa including *Lactobacillus*, *Staphylococcus* and Clostridiales were also more abundant in restrooms compared with other space types, while *Sphingomonas* were relatively less abundant in restrooms (Fig. 4).

Aside from restrooms, bacterial communities in Lillis hall tended to vary with both human occupancy and room centrality (Fig. 4). For instance hallways, which had high human occupancy and high occupant diversity (e.g., relatively many occupants and many *different* occupants throughout the day) as well as high centrality (hallways often serve as a pathway between rooms), were distinct from spaces such as mechanical support rooms and faculty offices with the opposite set of attributes (Fig. 4). While there were few statistically significant indicator taxa from individual space types other than restrooms, there was variation in the abundance

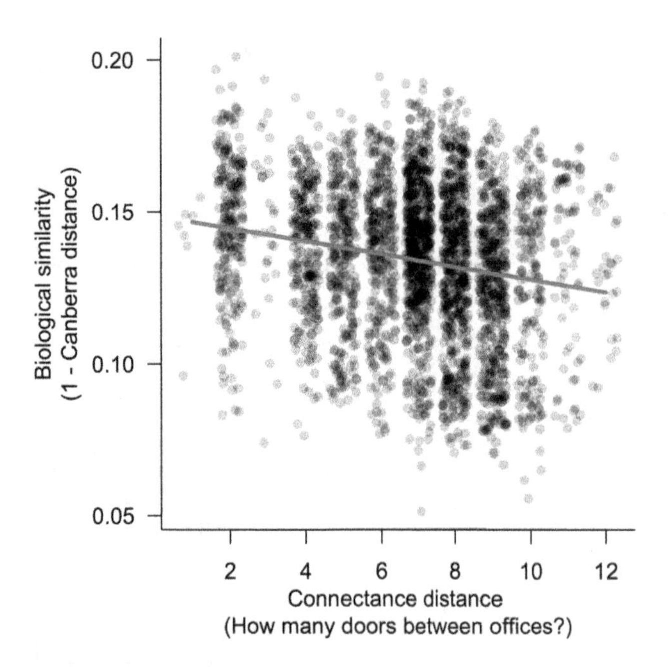

Figure 6. Offices in Lillis Hall show a strong distance-decay pattern. When only considering a single space type, biological similarity (y-axis; 1 - Canberra distance) decreases with connectance distance (number of intermediate space boundaries [e.g., doors] one would walk through to travel the shortest distance between any two spaces) (Mantel test; $R = 0.189$; $P = 0.002$). The same pattern was also observed at the whole-building scale (not shown; Mantel test; $R = 0.112$; $P = 0.001$).

of major bacterial taxa among these spaces. Taxa including *Lactococcus*, *Pseudomonas*, and *Streptococcus* were more abundant in the centrally located and highly-occupied spaces (Fig. 4), while *Achromobacter* and *Methylobacterium* were more abundant in the less central and less occupied spaces. Space types did not vary significantly in terms of their overall bacterial OTU richness or diversity (ANOVA using rarefied OTU richness and Shannon diversity; $P = 0.2$ & 0.9, respectively).

Design Influences on the Built Environment Microbiome within a Space Type

The large number of office spaces (73 offices) made it possible to test for drivers of microbial community variation among offices. Using a single space type also allowed us to hold relatively constant several building parameters. Specifically, parameters including space size, relative humidity, and occupancy varied less across offices than across all rooms at the building-scale. Variation in bacterial community structure among faculty offices was largely explained by the ventilation source in offices, with mechanically ventilated faculty offices containing a distinctive set of bacterial taxa when compared with window ventilated faculty offices (Fig. 5; $R^2 = 0.025$; $P = 0.005$). Taxa including *Deinococcus*, *Achromonobacter*, and *Roseomonas* were associated with mechanically ventilated faculty offices, while *Methylobacterium*, *Sphingomonas*, and *Streptococcus* were more closely associated with window ventilated faculty offices. Two of the most abundant of these strongly weighting taxa, *Deinococcus* and *Methylobacterium*, when grouped by genus, show consistent abundance differences between offices with different ventilation strategies. We found a strong association between the spatial connectance distance of offices (the number of doors through which one must walk between any two spaces) versus the microbial community similarity of offices (Fig. 6; $R = 0.19$; $P = 0.002$; from a Mantel test of Canberra distance vs. spatial connectance distance). This association was also significant at the building scale, regardless of space type ($R = 0.11$; $P = 0.001$).

Discussion

In this paper we first asked: at the scale of the entire building, do *function*, *form* and *organization* predict variation in the built environment microbiome? Our data suggest that the answer is yes. In architecture, function translates to space type, which in Lillis Hall was the strongest predictor of microbiome variation throughout the building. Due to the integrative nature of architectural design, function often drives patterns in the form and organization of spaces throughout a building, and form and organization are necessarily difficult to disentangle. Although form and organization are distinct aspects of architectural design, we did not attempt to draw a distinction between them in our analyses, since nearly every building variable herein relates to both. In Lillis Hall, design choices resulted in distinct space types that greatly differed in terms of their architectural characteristics, which were related to variation in microbial community composition at the building-scale. We also focused our analyses on the most common space type in Lillis Hall: offices. Specifically, we asked which aspects of form and organization most influenced the built environment microbiome in offices. We found that network betweenness, building floor, space size, and ventilation source were the strongest predictors for microbiome variation, even after holding function constant.

Despite the microbiome variation across space types, we detected a core built environment microbiome [23] of bacterial taxa that were present in nearly every indoor space we sampled. This core microbiome was dominated by taxa including members of the Proteobacteria and Firmicutes that are commonly found in indoor dust [15], although other common indoor dust taxa such as Actinobacteria were rare in this building (c. 1% of sequences). Many of the common taxa in the indoor dust microbiome were also detected in air and surface samples from the same building [24], suggesting that resuspension and settling of microbes from these pools of potential colonists are contributing to the communities detected in dust. The synchrony among these three microbial pools (air, surfaces and dust) within Lillis Hall suggests a conserved core building microbiome. Likely sources of this core microbiome include humans, soils and plants. We found that several of the bacterial taxa most strongly associated with restrooms as well as with high occupant diversity space types, such as classrooms, are also known to be associated with the human microbiome (e.g. *Lactobacillus* and *Staphylococcus*), while bacteria in low occupant diversity space types such as faculty offices and mechanical support spaces were more indicative of outdoor environments such as soils and the phyllosphere (e.g. *Methylobacterium*).

There has been a recent debate regarding the relative importance of dispersal from outdoor sources versus the conditions within buildings for determining the structure of indoor microbial communities [16,24,25]. We found evidence for the importance of both types of processes: the potential for dispersal from outdoor sources (e.g. ventilation air source, natural ventilation capacity) and conditions within the building (e.g. space type, building floor, temperature and relative humidity) influenced microbial community structure. This suggests that dispersal- and niche-based explanations will be required to understand the dynamics of the built environment microbiome. As in any ecological community, the spatial and temporal scale used to define indoor communities will have a large impact on the processes that give rise to patterns of diversity [26], as will the organisms being studied (e.g. bacteria vs. fungi), and this could explain differences between our findings and those of other recent studies [16]. For example, in a large multi-use building with high occupant density such as Lillis Hall, variation in human activities and uses among space types may be the main driver of microbial community structure. In smaller buildings with lower occupant density and stronger connections to outdoor air sources (such as greater reliance on natural ventilation), dispersal from outdoors may be the more important driver of indoor microbial community structure [16].

Our study highlights network analysis as a potentially powerful tool for applying indoor ecology and biogeography to the future of building design. Our network analyses quantified patterns in the form and organizations among spaces throughout Lillis Hall. From an architectural standpoint, room arrangement within Lillis Hall follows a double-loaded corridor design, where highly-central circulation spaces (e.g. hallways) connect most rooms in the building together with few intermediate spaces (*radial* design in Fig. 2). This *radial* design strategy, compared to *linear* or *grid* designs, reduces the range of connectance distances while increasing the centrality (betweenness and degree) of circulation spaces. We found that centrality was strongly correlated with variation in microbial communities. Since increased centrality of a space inherently increases human traffic through that space, and both of these attributes predicted microbial community composition in the present study, our findings suggest that the arrangement of spaces within a building is one promising way to influence microbial community composition.

We found that design decisions can influence the ecology of microbes within a space type - for example, in faculty offices the source of ventilation air (window- or louver-supplied versus mechanically-supplied ventilation) had a large impact on bacterial community structure and the abundance of some common taxa (e.g. *Deinococcus* and *Methylobacterium*; Fig. 5). While neither of these genera are known to influence human health, the unusually high abundance of the former in building dust, and particularly in mechanically ventilated offices, gives us insight into potential selective pressures within the built environment. *Deinococcus* is a genus best known to microbiologists for the extreme oxidative stress-, desiccation- and UV-tolerance of *Deinococcus radiodurans* [27,28]. Members of this genus are commonly found in soils, on plants, on humans, and have been detected previously in building dust and bioaerosols, but at far lower frequency than in our study [21,29]. It is plausible that consistently low relative humidity in mechanically ventilated offices, as well as UV light from windows, created indoor environmental conditions that selected for *Deinococcus* in dust assemblages, while window ventilated offices received more frequent inputs from airborne phyllosphere and soil microbial communities, leading to higher abundances of *Methylobacterium*.

As our understanding of the drivers of indoor microbiology improve, it may be possible to design spaces that foster or inhibit the growth and accumulation of different microbial taxa in order to promote a healthier indoor microbiome. But promoting a healthy indoor microbiome will require improved information about the human microbiome and health. At this point our understanding of the drivers of microbial ecology indoors has outpaced our understanding of related health implications [1,2,4,12,13,15,17,24,30,31]. Microbial biodiversity in the surrounding environment has been linked to human health and well-being [32–34], but for the vast majority of microbial taxa, we have no idea if their impact on our health is positive, negative, or neutral. Considering that the indoor microbiome represents a major potential source of microbes colonizing the human microbiome [1,12,13], as our knowledge about commensal microbiota expands [35–39], it is foreseeable that we will be able to target beneficial groups of indoor microbial taxa. Thus, while future studies will be needed to understand the health implications of indoor microbial communities, our results give clear evidence that design choices can influence the biogeography of microbial communities indoors, and thereby influence the interactions between the human microbiome and the built environment microbiome.

Conclusion

Churchill famously stated that *"[w]e shape our buildings, and afterwards our buildings shape us."* Humans help to direct microbial biodiversity patterns in buildings – not only as building occupants, but also through architectural design strategies. The impact of human design decisions in structuring the indoor microbiome offers the possibility to use ecological knowledge to shape our buildings in a way that will select for an indoor microbiome that promotes our health and well-being.

Supporting Information

Figure S1 High degree of correlation between three beta-diversity metrics. Multivariate community analysis was carried out with the Canberra taxonomic metric; this choice results in de-emphasis of the most abundant species (as opposed to using the Bray-Curtis dissimilarity metric), and also ignores nuanced evolutionary relationships between bacterial OTUs (as opposed to using the phylogenetic Weighted UniFrac distance). While the choice of a beta-diversity metric can impact results, the three potential candidates that we explored resulted in largely the same distance between samples in multivariate space. All three metrics are bounded between 0 and 1. Pearson's correlations (*r*) are given in the upper right panels.

Figure S2 The taxonomic composition of bacterial communities sampled from dust in the Lillis Business Complex. The relative abundance of sequences assigned to taxa at different taxonomic levels is indicated by the relative width of categories at each level. Bacterial taxonomy was visualized using Krona (http://sourceforge.net/projects/krona/; Ondov et al. 2011).

Acknowledgments

We would like to thank Daniel Aughenbaugh, Laura Cavin, Keith Herkert, Kate Laue, Alyssa Phanitdasack, Iman Rajaie and Jake Reichman for their help during sampling. Cornelis de Kluyver, Stephanie Bosnyk, Gordon Burke, Greg Haider, George Hecht, Don Neet, Pete Rocksvold, Frank Sharpy and Del Smith were instrumental in assisting with building operations and access during the study.

Author Contributions

Conceived and designed the experiments: SWK JLG BJMB DN JK. Performed the experiments: SWK TKO GM DN MM. Analyzed the data: SWK JFM DN JK MM. Contributed reagents/materials/analysis tools: GZB BJMB JLG. Wrote the paper: SWK JFM JLG.

References

1. Corsi RL, Kinney KA, Levin H (2012) Microbiomes of built environments: 2011 symposium highlights and workgroup recommendations. Indoor Air 22: 171–172.
2. Kelley ST, Gilbert JA (2013) Studying the microbiology of the indoor environment. Genome Biol 14: 202.
3. Tang JW (2009) The effect of environmental parameters on the survival of airborne infectious agents. J Roy Soc Interface 6 Suppl 6: S737–46.
4. Kembel SW, Jones E, Kline J, Northcutt D, Stenson J, et al. (2012) Architectural design influences the diversity and structure of the built environment microbiome. ISME J: 1–11.
5. Frankel M, Bekö G, Timm M, Gustavsen S, Hansen EW, et al. (2012) Seasonal variations of indoor microbial exposures and their relation to temperature, relative humidity, and air exchange rate. Appl Environ Microbiol 78: 8289–8297.
6. Klepeis NE, Nelson WC, Ott WR, Robinson JP, Tsang AM, et al. (2001) The National Human Activity Pattern Survey (NHAPS): a resource for assessing exposure to environmental pollutants. J Expo Anal Env Epid 11: 231–252.
7. Sternberg EM (2009) Healing Spaces: the Science of Place and Well-Being. Cambridge, MA: Belknap Press.
8. Ulrich R, Zimring C, Zhu X, DuBose J, Seo H, et al. (2008) A review of the research literature on evidence-based healthcare design. HERD J 1: 61–125.
9. Robinson CJ, Bohannan BJM, Young VB (2010) From structure to function: the ecology of host-associated microbial communities. Microbiol Mol Biol R 74: 453–476.
10. Vellend M (2010) Conceptual synthesis in community ecology. Q Rev Biol 85: 183–206.
11. Arundel AV, Sterling EM, Biggin JH, Sterling TD (1986) Indirect health effects of relative humidity in indoor environments. Environ Health Persp 65: 351–361.
12. Flores GE, Bates ST, Caporaso JG, Lauber CL, Leff JW, et al. (2013) Diversity, distribution and sources of bacteria in residential kitchens. Environ Microbiol 15: 588–596.
13. Flores GE, Bates ST, Knights D, Lauber CL, Stombaugh J, et al. (2011) Microbial biogeography of public restroom surfaces. PloS ONE 6: e28132.
14. Lee T, Grinshpun SA, Martuzevicius D, Adhikari A, Crawford CM, et al. (2006) Relationship between indoor and outdoor bio-aerosols collected with a button inhalable aerosol sampler in urban homes. Indoor Air 16: 37–47.
15. Rintala H, Pitkaranta M, Toivola M, Paulin L, Nevalainen A (2008) Diversity and seasonal dynamics of bacterial community in indoor environment. BMC Microbiol 8: 56.
16. Adams RI, Miletto M, Taylor JW, Bruns TD (2013) Dispersal in microbes: fungi in indoor air are dominated by outdoor air and show dispersal limitation at short distances. ISME J. DOI: 10.1038/ismej.2013.28.
17. Qian J, Hospodsky D, Yamamoto N, Nazaroff WW, Peccia J (2012) Size-resolved emission rates of airborne bacteria and fungi in an occupied classroom. Indoor Air 22: 339–351.
18. Fujimura KE, Johnson CC, Ownby DR, Cox MJ, Brodie EL, et al. (2010) Man's best friend? The effect of pet ownership on house dust microbial communities. J Allergy Clin Immun 126: 410–412.
19. Hospodsky D, Qian J, Nazaroff WW, Yamamoto N, Bibby K, et al. (2012) Human occupancy as a source of indoor airborne bacteria. PLoS ONE 7: e34867.
20. Amend AS, Seifert KA, Samson R, Bruns TD (2010) Indoor fungal composition is geographically patterned and more diverse in temperate zones than in the tropics. P Natl Acad Sci USA 107: 13748–13753.
21. Rintala H, Pitkäranta M, Täubel M (2012) Microbial communities associated with house dust. Adv Appl Microbiol 78: 75–120.
22. Freeman LC (1978) Centrality in social networks conceptual clarification. Soc Networks 1: 215–239.
23. Shade A, Handelsman J (2012) Beyond the Venn diagram: the hunt for a core microbiome. Environ Microbiol 14: 4–12.
24. Meadow JF, Altrichter AE, Kembel SW, Kline J, Mhuireach G, et al. (2013) Indoor Airborne Bacterial Communities Are Influenced By Ventilation, Occupancy, and Outdoor Air Source. Indoor Air. doi:DOI: 10.1111/ina.12047.
25. Martiny JBH, Bohannan BJM, Brown JH, Colwell RK, Fuhrman JA, et al. (2006) Microbial biogeography: putting microorganisms on the map. Nat Rev Microbiol 4: 102–112.
26. Levin SA (1992) The Problem of Pattern and Scale in Ecology: The Robert H. MacArthur Award Lecture. Ecology 73: 1943.
27. Slade D, Radman M (2011) Oxidative stress resistance in *Deinococcus radiodurans*. Microbiol Mol Biol R 75: 133–191.
28. Battista JR, Earl AM, Park MJ (1999) Why is *Deinococcus radiodurans* so resistant to ionizing radiation? Trends Microbiol 7: 362–365.
29. Shade A, McManus PS, Handelsman J (2013) Unexpected Diversity during Community Succession in the Apple Flower Microbiome. mBio 4: e00602–12.
30. Meadow JF, Bateman AC, Herkert KM, O'Connor TK, Green JL (2013) Significant changes in the skin microbiome mediated by the sport of roller derby. PeerJ 1: e53.
31. Bowers RM, McLetchie S, Knight R, Fierer N (2011) Spatial variability in airborne bacterial communities across land-use types and their relationship to the bacterial communities of potential source environments. ISME J 5: 601–612.
32. Hanski I, Von Hertzen L, Fyhrquist N, Koskinen K, Torppa K, et al. (2012) Environmental biodiversity, human microbiota, and allergy are interrelated. Proc Nat Acad Sci USA 109: 8334–8339.
33. Heederik D, Von Mutius E (2012) Does diversity of environmental microbial exposure matter for the occurrence of allergy and asthma? J Allergy Clin Immun 130: 44–50.
34. Ege MJ, Mayer M, Normand A, Genuneit J, Cookson WOCM, et al. (2011) Exposure to Environmental Microorganisms and Childhood Asthma. New Engl J Med 364: 701–709.
35. Pflughoeft KJ, Versalovic J (2012) Human Microbiome in Health and Disease. Annu Rev Pathol-Mech 7: 99–122.
36. Human Microbiome Project Consortium (2012) Structure, function and diversity of the healthy human microbiome. Nature 486: 207–214.
37. Claesson MJ, Jeffery IB, Conde S, Power SE, O'Connor EM, et al. (2012) Gut microbiota composition correlates with diet and health in the elderly. Nature 488: 178–184.
38. Shendell DG, Mizan SS, Yamamoto N, Peccia J (2012) Associations between quantitative measures of fungi in home floor dust and lung function among older adults with chronic respiratory disease: a pilot study. J Asthma 49: 502–509.
39. Low SY, Dannemiller K, Yao M, Yamamoto N, Peccia J (2011) The allergenicity of *Aspergillus fumigatus* conidia is influenced by growth temperature. Fungal Biol 115: 625–632.
40. Oregon University System Physical Facilities Inventory Manual (2000) Available: http://ous.edu/sites/default/files/dept/capcon/files/imanual.pdf.Accessed May 2013. Accessed 27 May 2013.
41. Csardi G, Nepusz T (2006) The igraph software package for complex network research. InterJournal Complex Sy: 1695.
42. Claesson MJ, Wang Q, O'Sullivan O, Greene-Diniz R, Cole JR, et al. (2010) Comparison of two next-generation sequencing technologies for resolving highly complex microbiota composition using tandem variable 16S rRNA gene regions. Nucleic Acids Res 38: e200.
43. Caporaso JG, Lauber CL, Walters WA, Berg-Lyons D, Huntley J, et al. (2012) Ultra-high-throughput microbial community analysis on the Illumina HiSeq and MiSeq platforms. ISME J: 1–4.
44. Caporaso JG, Kuczynski J, Stombaugh J, Bittinger K, Bushman FD, et al. (2010) QIIME allows analysis of high-throughput community sequencing data. Nat Methods 7: 335–336.
45. Edgar RC (2010) Search and clustering orders of magnitude faster than BLAST. Bioinformatics 26: 2460–2461.
46. DeSantis TZ, Hugenholtz P, Larsen N, Rojas M, Brodie EL, et al. (2006) Greengenes, a chimera-checked 16S rRNA gene database and workbench compatible with ARB. Appl Environ Microbiol 72: 5069–5072.
47. Price MN, Dehal PS, Arkin AP (2010) FastTree 2 - approximately maximum-likelihood trees for large alignments. PloS ONE 5: e9490.
48. R Development Core Team (2010) R: A Language and Environment for Statistical Computing. Available: http://cran.r-project.org.
49. Oksanen J, Kindt R, Legendre P, O'Hara B, Simpson GL, et al. (2007) Vegan: community ecology package. R package version 1.17. Available: http://cran.r-project.org/package = vegan.
50. Dufrêne M, Legendre P (1997) Species assemblages and indicator species:the need for a flexible asymmetrical approach. Ecol Monogr 67: 345–366.
51. Roberts DW (2012) labdsv: Ordination and Multivariate Analysis for Ecology. R package version 1.5-0. Available: http://cran.r-project.org/package = labdsv.

The Value of Patch-Choice Copying in Fruit Flies

Shane Golden, Reuven Dukas*

Animal Behaviour Group, Department of Psychology, Neuroscience & Behaviour, McMaster University, Hamilton, Ontario, Canada

Abstract

Many animals copy the choices of others but the functional and mechanistic explanations for copying are still not fully resolved. We relied on novel behavioral protocols to quantify the value of patch-choice copying in fruit flies. In a titration experiment, we quantified how much nutritional value females were willing to trade for laying eggs on patches already occupied by larvae (social patches). Females were highly sensitive to nutritional quality, which was positively associated with their offspring success. Females, however, perceived social, low-nutrition patches (33% of the nutrients) as equally valuable as non-social, high-nutrition ones (100% of the nutrients). In follow-up experiments, we could not, however, either find informational benefits from copying others or detect what females' offspring may gain from developing with older larvae. Because patch-choice copying in fruit flies is a robust phenomenon in spite of potential costs due to competition, we suggest that it is beneficial in natural settings, where fruit flies encounter complex dynamics of microbial communities, which include, in addition to the preferred yeast species they feed on, numerous harmful fungi and bacteria. We suggest that microbial ecology underlies many cases of copying in nature.

Editor: Johan J. Bolhuis, Utrecht University, Netherlands

Funding: This research has been funded by the Natural Sciences and Engineering Research Council of Canada, Canada Foundation for Innovation, and Ontario Innovation Trust. The funders had no role in study design, data collection and analysis, decision to publish, or preparation of the manuscript.

Competing Interests: The authors have declared that no competing interests exist.

* Email: dukas@mcmaster.ca

Introduction

In many animal species, individuals copy the choices of others. Examples include choices of feeding sites [1–3], territories [4,5], egg laying substrates [6–8] and mates [9–11]. Depending on the system, copying can have substantial effects on organismal ecology and evolution. For example, conspecific aggregation at feeding and egg laying sites can promote species coexistence [12,13] and mate choice copying can influence the intensity and direction of sexual selection [14–16].

While it is widely agreed that copying can influence animal ecology and evolution, it is often unclear how the possible fitness benefits from copying outweigh the likely costs. For example, patch-choice copying typically involves a focal individual choosing a feeding or egg laying site that is either occupied by other individuals (models) or contains products left by these individuals. There are probably only two non-mutually exclusive explanations for such copying. The first explanation involves pure information: a focal can either find a satisfactory patch faster, or locate a better patch among the available alternatives by copying others than by exploring on its own [4,17–19]. That is, the first explanation focuses on two related difficulties that animals have in locating optimal resource patches. Either the patches are hidden, so it takes time to find them, or it is difficult and time consuming to assess the multitude of features that determine patch quality. Given individuals' limited time horizon, focals that copy others can shorten the time devoted to exploration and hence increase the time spent exploiting without compromising on the quality of the patch utilized. This proposition, of course, is based on the tenuous assumption that the models indeed have chosen the optimal patch.

The other explanation for patch-choice copying involves material benefits that focals can gain from joining others, which include reduced per capita risk of attack by predators and parasitoids, and enhanced foraging efficiency and thermoregulation [20–25]. It is worth noting that, when patch-choice copying involves joining others, focals and models might face asymmetric payoffs: while a focal can gain more from joining than from settling alone, the models might lose from having another individual joining [26]. The obvious costs from joining others are competition for resources and reduced patch quality caused by accumulating waste products [7,19,20]. Competition can cause another possible asymmetric payoff that is size dependent. For example, newly hatched larvae may lose more from competition than the older resident larvae.

While there are numerous reports of copying in a wide variety of species and contexts, the value of copying has been rarely quantified. We have recently developed protocols for quantifying patch-choice copying in fruit flies (*Drosophila melanogaster*). Larvae and adults from both established laboratory strains and recently caught wild populations copy the choices of others: adult females prefer the egg laying substrates chosen by other females [27,28], both male and female adults are attracted to volatiles emanating from conspecific larvae, females show a strong preference for laying eggs in patches with larvae over unoccupied alternatives, and larvae also show significant attraction to patches already occupied by larvae [29–31]. The establishment of fruit flies as a model system for research on patch-choice copying offers new opportunities. First, the fruit fly system allows one to conduct highly controlled experiments assessing the factors that influence patch choice copying. Second, findings from the behavioral analyses of patch-choice copying can be extended to research on the genetics and neurobiology of such behavior in a highly amenable model system. Indeed there has recently been increased interest in establishing simple model systems for research on the mechanisms that control social behavior as well as behavioral decisions in general [32–35].

To elucidate the value of patch-choice copying in fruit flies, we conducted a series of experiments. We began with a titration experiment designed to quantify the perceived value that females assign to food occupied by larvae. This involved testing female preferences between reference patches and test patches of varying food qualities, which were either occupied or unoccupied by larvae. In follow-up experiments, we compared larval success on occupied and unoccupied patches of relevant food qualities. This allowed us to translate patch-choice copying by females into the consequent success of their offspring. Because females showed strong patch-choice copying even when nutritionally superior patches were readily available and in spite of the expected costs owing to larval competition, we wished to assess whether females would moderate their strong tendency to copy when the occupied patches either contain numerous larvae or have already experienced heavy consumption by larvae. Finally, to assess possible informational benefits to females, we tested whether larvae were better than adult females at assessing food quality.

Materials and Methods

Nutritional Titration

We maintained two population cages of several hundred *Drosophila melanogaster Canton-S* following standard protocol [27]. To quantify the value that females assign to patches already occupied by larvae, we placed each of 192 recently mated female inside a 60 mm Petri dish. The bottom of the dish contained agar, which provided moisture. On top of the agar, we placed two discs cut from a thin layer of fly medium. Both discs were 1.1 cm in diameter and each contained 0.5 ml food (Fig. 1a). The reference disc always had standard food in which 1 litre contained 60 g dextrose, 30 g sucrose, 32 g yeast, 75 g cornmeal, 20 g agar, 2 g methyl-paraben and water. The test disc was either fresh (non-social) or contained five early second instar larvae that had fed on that disc for 24 h (social). The test disc had standard food or one of two lower food concentrations containing either 33% or 11% of the nutrients (dextrose, sucrose, yeast, and cornmeal) available in the standard food and a larger proportion of water. The reference and test discs were 3 cm apart with the central 2 cm being a trough filled with fine sand (Fig. 1a) to prevent larvae located on the social discs from crossing to the reference discs. We housed all dishes in a chamber kept at 25°C and 90% RH and allowed the females in the Petri dishes to lay eggs overnight for 14 h. Then we discarded the females and counted the number of eggs laid on each disc. We used a generalized linear model with a Tweedie distribution and identity link function and conducted pairwise comparisons with Bonferroni corrections and 95% Wald confidence intervals [36]. See Data S1 for the raw data for all experiments.

Larval Success on Social vs. Non-Social Food

Our nutritional titration experiment indicated that females perceive social food with about one third the nutrients as equally valuable as the non-social reference food (Fig. 1b). We thus wished to quantify the success of females' eggs on social vs non-social food discs of distinct nutritional concentrations. To assess the value of laying eggs on currently versus previously occupied patches, we also included a previously social treatment. We had a total of 6 treatments involving 2 food concentrations, 100% and 33%, and 3 social treatments, non-social, social and previously social. We omitted the 11% food concentration because females in the titration experiment mostly avoided it even when it was social (Fig. 1b). The food discs were identical in constitution and volume to the 100% and 33% food discs in the titration experiment.

The non-social discs contained unmodified food. To generate the social and previously social discs, we placed on each disc 5 24-hour old first instar larvae and allowed these larvae to feed for 24 hours. In the social disc treatment, we kept the now second instar, 48 h old larvae on each disc. In the formerly social disc, we removed the larvae. That is, both the social and previously social discs were equally modified by the five larvae prior to the placement of focal eggs. Then the focal larvae emerging on the formerly social disc could reap potential benefits from such previous food modification without experiencing competition with the older larvae. Thus the formerly social disc gave us a greater power for quantifying possible benefits of prior food modification by larvae.

We placed each food disc inside a 35 mm Petri dish lined with agar, added to each disc five focal eggs and housed all the dishes in a chamber kept at 25°C and 90% RH. When the five older larvae in the social dishes pupated, we removed these pupae. We then monitored the number of focal larvae reaching pupation and calculated the larval developmental rate as the cumulative proportion of larvae reaching pupation while taking the final pupal number as 1. We counted all eclosing adults and calculated the proportion of eggs that produced adults. Because females are heavier than males, we sexed the adults, dried them in an oven at 70°C for 3 days and weighed groups of five flies of the same sex on a microbalance.

Because no larvae survived in the social 33% treatment, we conducted two separate analyses. First, we omitted the social treatment and compared larval performance in the four treatments of non-social and formerly social on 33% and 100% food. Second, we compared larval performance in all three treatments of non-social, formerly social and social on the 100% food.

We analyzed larval development rate and the proportion of eggs surviving to adulthood using a generalized estimating equation with a gamma distribution and log link function [29]. We had sufficient sample sizes for analyzing adult dry mass only for the 100% food (Fig. 2E, F). These data met ANOVA assumptions and we thus used a two-way ANOVA with a Tukey HSD. We conducted all post-hoc pairwise comparisons using the sequential Bonferroni method adjusting for multiple comparisons.

Larval Success on Abundant Food

In our previous larval success experiment, larvae were reared on 0.5 ml of food. Because the results indicated strong effects of competition, we tested larval success on social and non-social discs each containing 2.5 ml of 100% food. As a reference, fruit fly laboratories typically rear a few dozen flies per vial containing 5 ml of similar food [37,38]. By providing abundant food, we wished to maximize our ability to detect possible benefits that larvae may gain from developing on social food. All other protocol details were similar to those detailed above. That is, The social food contained 5 larvae and the non-social food had no larvae.

Females' Patch Choice When the Social Patches Have Had High Larval Densities

In our titration experiment (Fig. 1B), females showed a strong preference for laying eggs near larvae even though this reduced their offspring success in our laboratory settings (Figs 2, 3). Because larval crowding and the consequent lower larval success

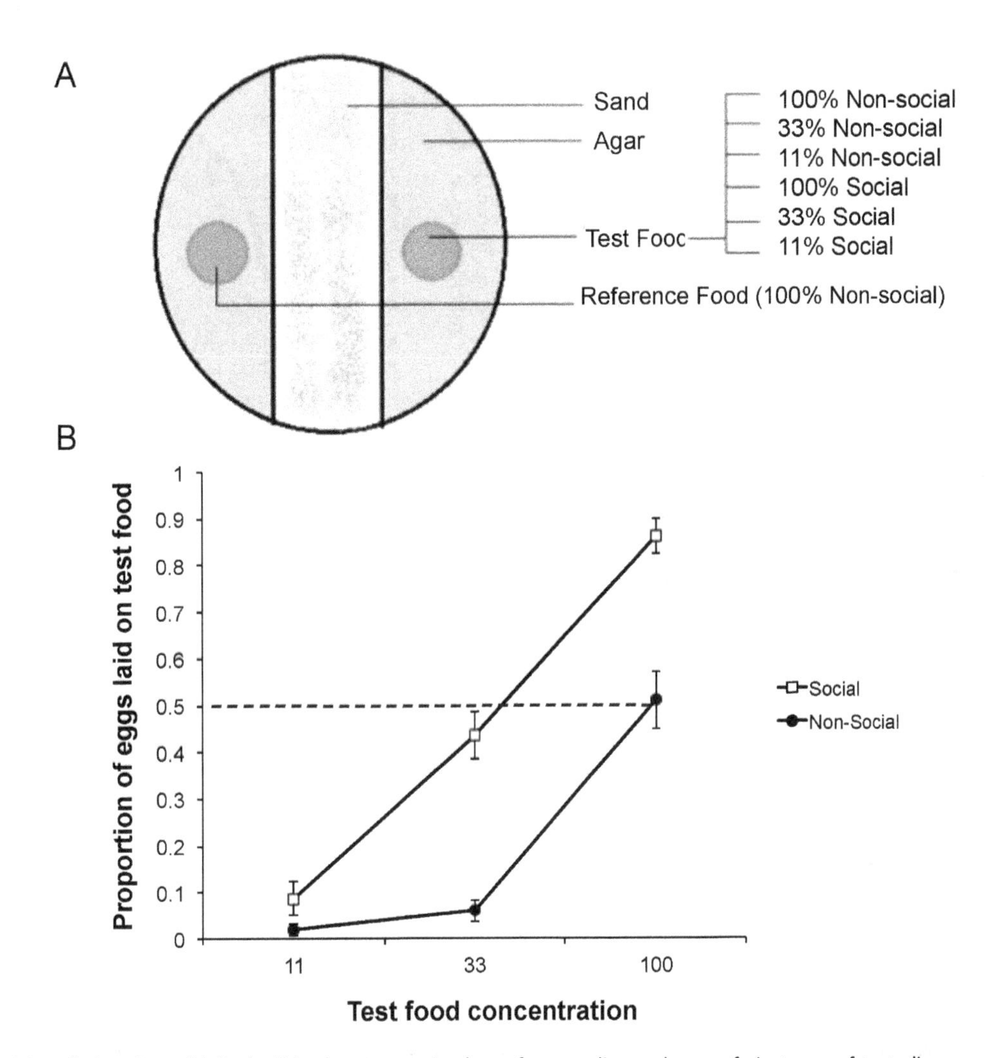

Figure 1. Nutritional titration. (A) Each dish always contained a reference disc and one of six types of test discs varying in nutritional concentration and larval presence. Sand at the centre of the dish prevented larval crawling to the reference food. (B) The average proportion of eggs (±1 SE) laid on the test disc as a function of its nutrient concentration and presence or absence of larvae (social or non-social). The horizontal dashed line indicates random choice. N = 30 replicates per treatment. Females laid more eggs on the test food in the presence than absence of larvae.

are prevalent in nature as well [39,40], we expected females to make egg laying decisions that balance their perceived benefit from laying next to larvae versus the expected cost due to larval overcrowding. We thus allowed females to choose between either a non-social patch and a social patch occupied by 5 larvae, or a non-social patch and a social patch occupied by 20 larvae. We predicted that females would lay a lower proportion of eggs on the social food when it was more crowded.

We used a protocol modified from Durisko et al [30]. We placed each recently mated female inside a plastic cage (15 cm wide, 30 cm long, and 15 cm high), which contained two 35 mm Petri dishes placed at the opposite far corners of each cage. One dish was non-social and the other was social. Both dishes contained 0.5 ml food discs composed of 100% standard lab diet. The dishes were lined with a layer of agar to prevent desiccation. Non-social food discs were unoccupied. The social food discs had either 5 or 20 middle second instar larvae, which we had added 6 h before the addition of females.

We allowed the females to lay eggs overnight. In the following morning, we removed the females from the cages, counted the number of eggs on each food disc and analyzed the proportion of eggs laid in the social dish out of the total number of eggs that a female laid. Based on preliminary data indicating effects of larval density on egg location, we also counted the number of eggs laid on the agar layer within 1 cm of the food disc and calculated the proportion of eggs laid on agar versus food in the social dish. We analyzed the data using a generalized linear model with a Tweedie distribution and log link function.

The experiment above tested females' sensitivity to larval density. It is possible however, that females are more sensitive to the condition of food as indicated by the microbial community and waste products rather than to the number of larvae already on the food. To test this possibility, we allowed females to choose between either a non-social patch and a social patch that had been previously occupied by 5 larvae, or a non-social patch and a social food patch that had been previously occupied by 20 larvae. Again, we predicted that females would lay a lower proportion of eggs on the social food that had been more crowded.

Forty-eight hours before the experiment, we transferred groups of either 5 or 20 middle second instar larvae to social food discs and kept them in 35 mm Petri dishes lined with agar. We also kept unoccupied food discs in Petri dishes lined with agar. All food discs contained 0.5 ml of 100% standard lab diet. By the day of the experiment, all larvae on the social discs had pupated. We then

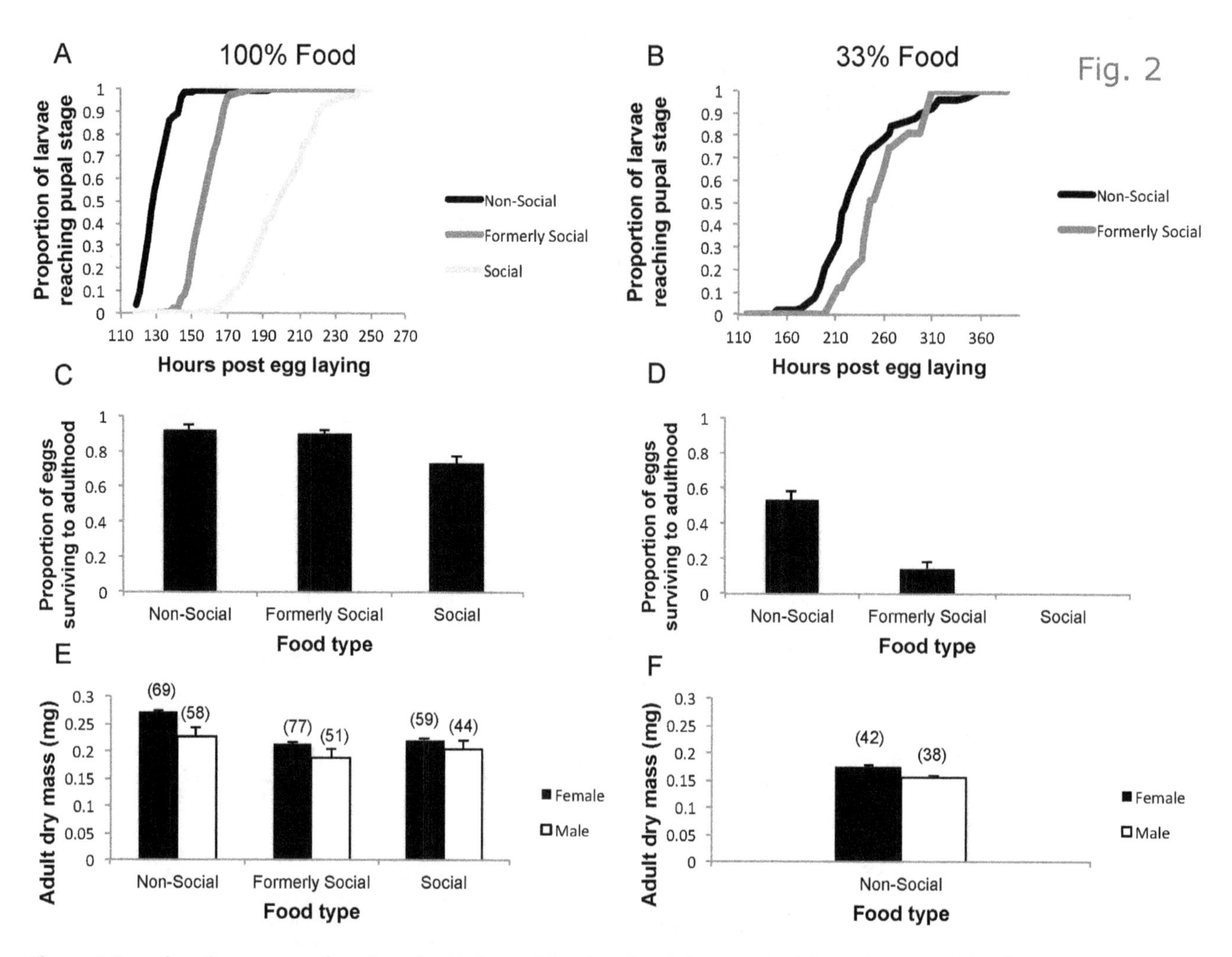

Figure 2. Larval performance as function of a disc's nutritional and social status. The left panels (A, C, and E) refer to the 100% nutrients while the right panels (B, D, and F) refer to the 33% nutrients. (A) and (B) show the time it takes for the larvae to develop from eggs into pupae. (C) and (D) show the proportions of eggs that survived to adulthood (mean+SE). In (B) and (D), survival in the social treatment was 0. (E) and (F) show the adult dry mass (mean+SE). N= 30 replicates for each treatment. The number of eclosing adults is shown above the bars in panels E and F.

placed one non-social food disc and one social disc in 60 mm Petri dishes lined with agar. The social disc had been previously consumed by either 5 or 20 larvae but was free of larvae and pupae by the time of the test. Discs were placed 2 cm apart. We then added a recently mated female to each 60 mm dish through a hole in the lid, which was then plugged with foam. We allowed the females to lay eggs overnight. In the morning, we removed the females from the dishes and counted the number of eggs on the social and non-social food discs. We analyzed the proportion of eggs on each type of social food using a generalized linear model with a Tweedie distribution and log link function.

Adult vs. Larval Abilities to Detect Differences in Yeast Concentration of Food

Because we documented a lower larval success of eggs laid at social patches, we wished to test whether the benefit of patch choice copying is related to information rather than to joining. To this end, we tested whether larvae could detect pertinent patch characteristics that adult females could not. We had two treatments testing larval and adult females' abilities to detect differences in yeast content between adjacent patches. We focused on yeast rather than sugar because larval and adult perception of sweetness is well documented [41–44]. One test involved a reference 100% standard fly medium vs standard medium with only 33% of the yeast content, and the other test involved a reference 100% standard medium vs standard medium with only 50% of the yeast content. All other medium ingredients were identical.

We added either one recently mated adult female or five mid-second instar larvae to Petri dishes containing one reference food disc and one food disc with lower yeast concentration (either 33% or 50%) placed 2 cm apart. We added the adults and focal larvae in the evening at an identical location 1 cm between the food discs. We gave them 14 hours to decide where to lay eggs or feed. In the following morning, we counted the number of eggs laid on each food disc in the adult female treatments and counted the number of larvae on each food disc in the larval treatments. We then calculated the proportion of eggs laid and the proportion of larvae on the reference 100% disc and analyzed the data with a generalized linear model with a Tweedie distribution and identity link function.

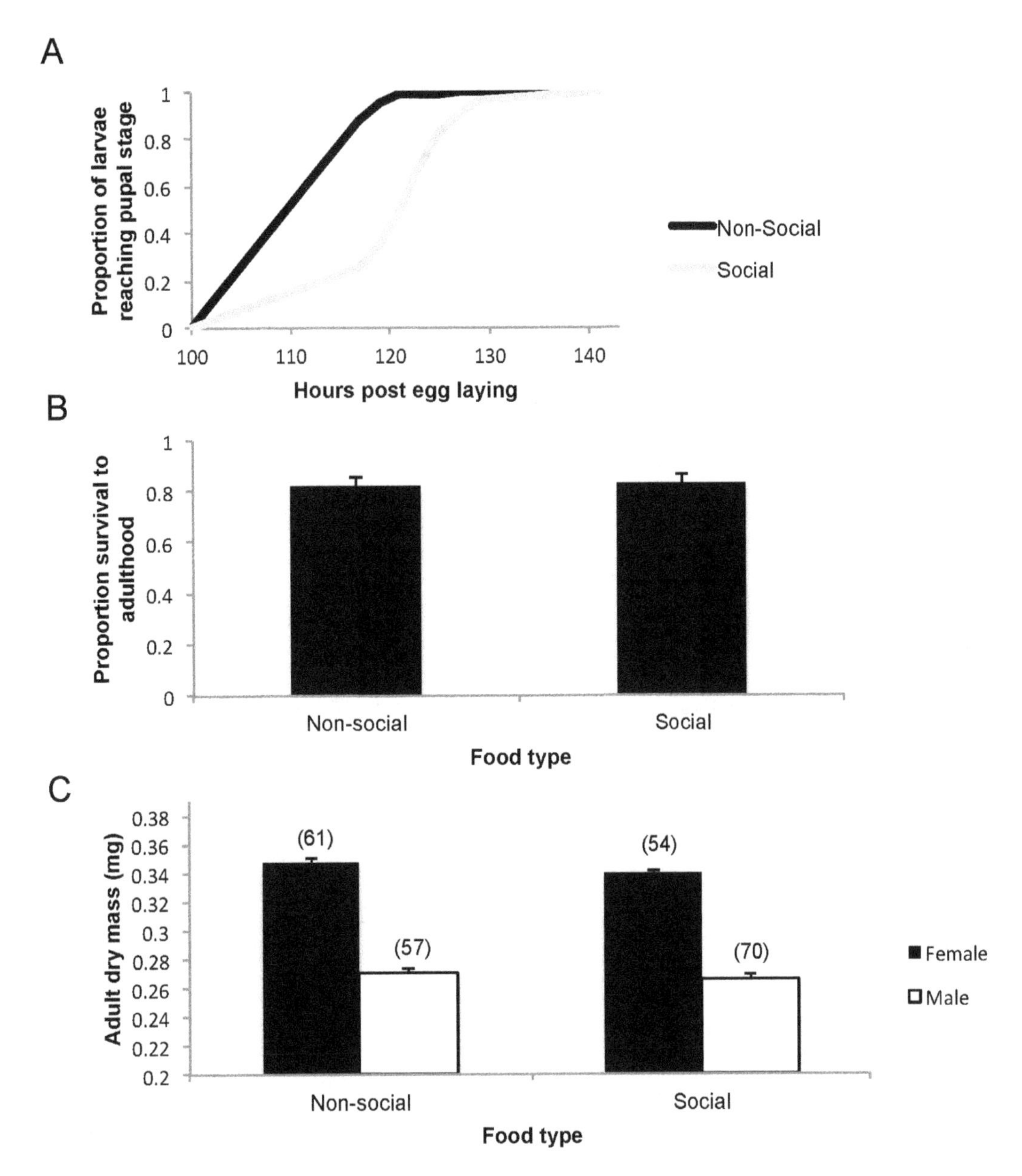

Figure 3. Performance measures of focal larvae on abundant food. Discs were either social or non-social (n = 30 replicates per treatment). (A) Time from egg laying to pupal formation (B) The proportion of eggs surviving to adulthood (mean+SE). (C) The adult dry mass of females and males in both conditions (mean+SE). Numbers in brackets above the bars indicate the number of adults in each group.

Results

Nutritional Titration

Females laid significantly higher proportions of eggs on the test food when it was social than non-social at all three food concentrations (Wald $\chi^2_1 = 49$, $P<0.001$ for the main effect and $P<0.01$ for the three pairwise comparisons with Bonferroni corrections, Fig. 1).

Larval Success on Social vs. Non-Social Food

Larval performance across food qualities. Owing to 100% mortality in the social 33% food treatment, we could compare larval performance across food qualities only for the non-social and previously social treatments. Larvae developed much faster (Wald $\chi^2_1 = 474.74$, $P<0.001$; Fig. 2A,B) and had higher survival rates on the 100% than 33% food (Wald $\chi^2_1 = 75.6$, $P<0.001$; Fig. 2C,D). Similarly, larvae developed much faster (Wald $\chi^2_1 = 33.361$, $P<0.001$; Fig. 2A,B) and had higher survival rates in the non social than formerly social treatments (Wald $\chi^2_1 = 75.769$, $P<0.001$; Fig. 2C,D).

Because survival rates in the 33% food treatment were low, we could only compare adult body mass across food qualities in the non-social treatments. Adults in the 100% food quality were much heavier than those in the 33% food quality (Wald $\chi^2_1 = 512.96$, $P<0.001$; Fig. 2E,F).

Larval performance across social treatments. This analysis could include only the 100% food owing to 100% mortality in the social 33% food treatment. Larvae developed significantly faster in the non-social treatment, intermediate in the formerly social treatment, and slowest in the social treatment (Wald $\chi^2_2 = 1700$, $P<0.001$; Fig. 2A). Post-hoc pairwise comparisons showed that each treatment was significantly different from the other two ($P<0.001$).

Survival to adulthood was significantly affected by the social treatment (Wald $\chi^2_2=13.9$, $P=0.001$; Fig. 2C). Survival was similar in the non-social and formerly social treatment (post-hoc pairwise comparison, $P=0.709$) but higher in each of these treatments than in the social treatment (post-hoc pairwise comparisons, $P=0.002$ and 0.005 for the non-social and formerly social treatment respectively).

Adult mass was significantly affected by the social treatment ($F_{2,61}=85.2$, $P<0.001$; Fig. 2E). In both males and females, adults of the non social treatment were heavier than those of the social and formerly social treatments (Tukey HSD, $P<0.001$). While males of the formerly social treatment were lighter than those in the social treatment ($P=0.007$), females of the formerly social and social treatments had similar masses ($P=0.438$).

Larval Success on Abundant Food

Larvae developed faster in the non-social condition than in the social condition (Wald $\chi^2_1=34.683$, $P<0.001$; Fig. 3A). However, the same proportion of focal eggs survived to adulthood (Wald $\chi^2_1=0.014$, $P=0.905$; Fig. 3B). Adult flies in the non-social condition were heavier than adults in the social condition (Wald $\chi^2_1=4.515$, $P=0.034$; Fig. 3C).

Females' Patch Choice When the Social Patches Have Had High Larval Densities

Females laid similar proportions of eggs in the social dishes occupied by 5 and 20 larvae (Wald $\chi^2_1=0.204$, $P=0.651$; Fig. 4A). However, females placed a greater proportion of their eggs on the agar in the social dishes with 20 than 5 larvae (Wald $\chi^2_1=4.649$, $P=0.031$; Fig. 4B). When females had a choice between non-social and previously occupied social discs, they laid a similar proportion of their eggs on the social disc regardless of the number of larvae that had previously occupied it (Wald $\chi^2_1=0.472$, $P=0.492$; Fig. 4C).

Adult vs. Larval Abilities to Detect Differences in Yeast Concentration Of Food

The proportion of eggs that females laid on the 100% food and the proportion of larvae choosing the 100% food were similar when the alternative had only 33% of yeast concentration (Wald $\chi^2_1=0.227$, $P=0.634$; Fig. 5). When the alternative was 50% yeast concentration, females showed a greater preference than larvae for the higher quality food (Wald $\chi^2_1=3.835$, $P=0.05$; Fig. 5).

Discussion

Our titration experiment (Fig. 1) indicated that, while females were highly sensitive to the nutritional values of alternative patches, they perceived low-nutrition patches occupied by larvae (social patches with 33% of the nutrients) as suitable as the reference, unoccupied patches (non-social patches with 100% of the nutrients). The larval success experiment (Fig. 2) indicated that the females' sensitivity to nutrient concentration was highly justified: their larvae developed significantly faster, had higher survival rates and produced larger adults on the non-social 100% than non-social 33% patches. Because females were willing to trade the nutritional quality of patches for the opportunity to lay eggs at patches already occupied by larvae, we expected that such choice would translate into some larval benefit. However, we did not find such an advantage. First, in all cases, larval success on the social patches was lower than that on non-social patches (Fig. 2). Second, in the previously social treatment, we removed the larvae that had occupied the patches before placing focal eggs. This allowed us to test for possible benefits that females could gain from laying eggs at patches that have been occupied by larvae while eliminating the negative effects of competition from such larvae. Even in this case, however, we found a cost rather than benefit from laying on previously occupied patches (Fig. 2). Finally, one could argue that our larval to food-volume ratio was too high so that larval competition obscured a gain occurring when food is abundant. To address this possibility, we repeated the larval success experiment with a much lower larval to food-volume ratio. Even in this case, however, larvae performed better under the non-social then social treatment (Fig. 3). The mechanism underlying this negative social effect is unknown and will require close examination in the future.

To further assess the egg laying decisions by females, we wished to quantify females' responses to clear signs of competition in social patches due to either the previous or current presence of many larvae. Although we expected females to reduce their preferences for the social patches when they were either crowded or heavily exploited, we found no such moderation (Fig. 4). Finally, although the sense of taste provides important information about the nutritional quality of food, it is insufficient for assessing whether all nutrients required for optimal larval development are available [41,43–45]. We thus proposed that the presence of feeding larvae is the best cue indicating to females that a substrate is nutritionally sufficient. First, the substrate is adequate for sustaining the larvae as indicated by the fact that they are alive. Second, the larvae are highly mobile and are adept at exploring and settling at the best locally available food [29,46]. Contrary to our expectation, however, we found in two experiments that adult females were as sensitive as larvae to realistic variations in nutritional qualities (Fig. 5).

To summarize our key results, we have strong evidence that females assign high values to patches already occupied by larvae as we quantified by titrating the nutritional quality of the patches (Fig. 1) and we could translate these values into the relevant currency of larval success (Figs 2, 3). Our data, however, indicated neither informational gain (Fig. 5) nor direct benefits from patch choice copying (Figs 2, 3). How can this puzzle be resolved? We propose four non-mutually exclusive explanations related to fruit flies' ecology under natural settings. The first three explanations deal with microbial ecology while the last one focuses on fruit fly parasitoids, which, alongside microbes, are the prominent natural enemies of fruit fly larvae. While the third explanation (microbial information) pertains to the informational benefits of patch choice copying, all other three explanations relate to the direct benefits to larvae from joining other larvae.

Competition with Microbes

While fruit flies feed on yeast species growing on fallen fruit [47], such fruit are also consumed by numerous other fungi as well as bacteria. This means that the other microbes can adversely impact yeast through exploitation competition. Furthermore, microbial interference competition involves a rich arsenal of compounds toxic to other microbes as well as to animals. That is, such compounds can either hamper yeast growth, thus reducing the amount of food available to larvae, or have direct negative effects on larval survival and growth [48–53]. Although highly pertinent for our understanding of the behavior of larval and adult fruit flies, the microbial ecology relevant to fruit flies remains mostly unexplored. A notable exception is work by Rohlfs and colleagues [54,55], which quantified negative effects of three mold species on fruit fly larvae and indicated that groups of five and 10

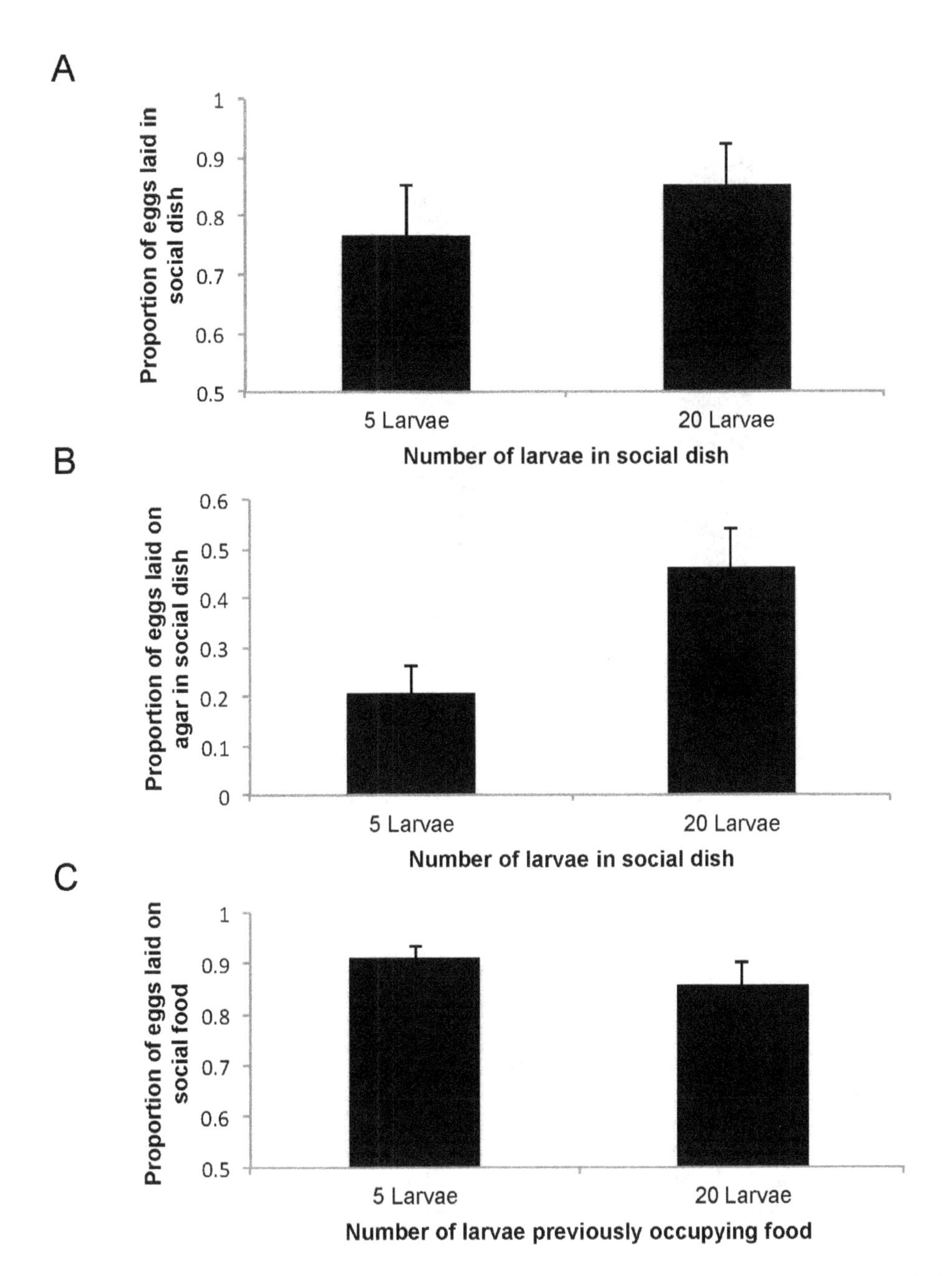

Figure 4. Social patch choice under high larval densities. The proportion (mean+SE) of eggs laid at the social disc, which currently (A, B) or previously (C) contained either five or 20 larvae. In each case, females could choose between laying at a social or non-social disc. (A) The proportion of eggs laid in the social dish out of all eggs laid. (B) The proportion of eggs laid on agar rather than on the food disc out of the eggs laid in the social dish. N = 24 replicates per treatment. (C) The proportion of eggs laid on the social disc, which had been previously consumed by either 5 or 20 larvae, out of all eggs laid. No larvae were present on the food at the time of egg laying. N = 28 replicates per treatment.

larvae were more effective at suppressing mold growth than single larvae. Another relevant observation is that fruit flies possess a dedicated olfactory circuit tuned to geosmin. Fruit flies rely on this circuit to avoid feeding and egg laying on substrates containing geosmin-producing microbes, which are harmful to fruit flies [56]. This indicates that fruit flies are sensitive to the constitution of microbial communities at prospective egg laying sites. It is thus likely that, by preferring to lay eggs at patches already occupied by larvae over unoccupied patches, females in natural settings ensure that their newly hatched larvae will be better protected from microbes harmful either to their larvae or to their larval yeast-food.

Group Enhancement of Favourable Yeasts

There appear to be mutualistic interactions between some yeast species and fruit flies. Adults and larvae inoculate fruit with yeast and larval activity promotes the growth of certain yeast species [57–59]. While some of the positive effects of larvae on yeast can be modulated through churning of the substrate, the larval gut bacteria also produce antifungals, which could selectively suppress mold and thus enhance the growth of the preferred yeast food [31,60–62]. Intriguingly, adult and larval fruit fly attraction to food inhabited by larvae is mediated by volatiles emitted from gut bacteria [31]. Hence it is likely that females in nature lay eggs in

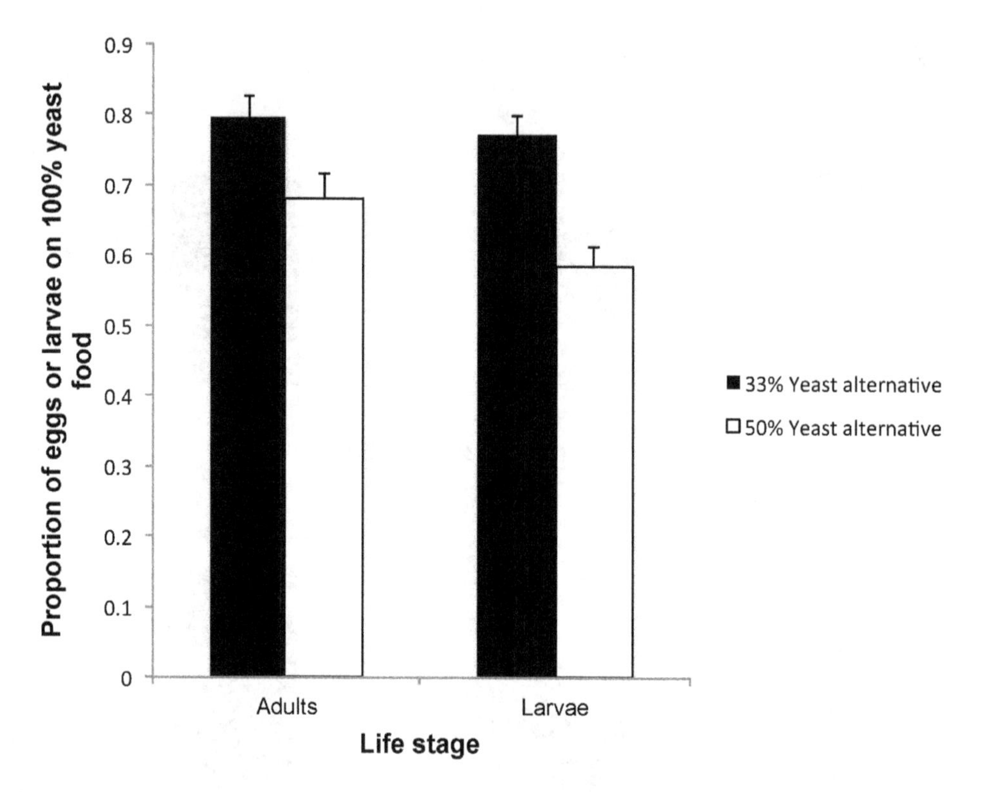

Figure 5. Patch choice by adult females versus larvae. In one experiment (black bars), adult females or larvae had a choice between a disc containing the regular yeast concentration (100%) or a disc containing 33% of the regular yeast concentration. In the other experiment (white bars), adult females or larvae had a choice between a disc containing the regular yeast concentration (100%) or a disc containing 50% of the regular yeast concentration. N = 80 replicates per nutrition treatment for larvae, and N = 60 replicates per nutrition treatment for the adult females.

occupied patches because such patches are more favourable for further growth of yeast food than are unoccupied patches.

Microbial Information

While we found no evidence that larval presence provides superior nutritional information about patch quality that females cannot readily assess, the discussion above suggests that larval presence is the best indicator that the microbial ecology is favourable to larval growth. That is, it is likely that different fruit patches allow for the optimal growth of different microbial species with only some of them being hospitable to fruit flies. For example, substrates may vary in their ability to sustain the growth of harmful mold and bacteria versus the yeast species favoured by fruit flies. Assuming that females cannot assess all the relevant ecological settings that would influence fungal growth, the presence of thriving larvae may be the best cue indicating that a patch is providing the appropriate microbial environment.

Parasitoid Avoidance

Larval parasitoids are a major source of fruit fly mortality in natural settings and fruit flies possess a suite of behavioral and physiological adaptations for reducing parasitoid success [63–66]. One way by which larvae can avoid parasitism is through hiding in micro-sites inaccessible to parasitoids. Although newly hatched larvae are not proficient at burrowing, older larvae, especially ones in the third instar stage, have stronger and larger mandibular hooks containing several teeth [67] and they spend much of their time tunnelling deep inside the substrate [34]. It is thus possible that, by laying eggs close to larvae, females ensure that their hatching offspring can hide in burrows dug by the older larvae. Limited evidence indeed indicates that larvae hidden deep in natural fruit experience lower rates of parasitoid attacks [24].

Patch Choice Copying in Other Species

Our work on the value of patch choice copying in fruit flies can inform and be informed by research on copying in other species. Perhaps the best studied and most relevant system involves the economically important bark beetles (Scolytidae), which aggregate at host trees. While there are many species of bark beetles, we focus here on obligate parasites, which attack and kill trees [68]. Long-distance attraction to host trees in bark beetles is mediated by pheromones. Early colonizers benefit from attracting others because a critical mass of beetles and perhaps associated fungi are necessary for overcoming the massive defence mounted by the host tree [68–70]. Because prospective females gain from joining patch occupiers, the adaptive function of patch choice copying is clear.

There are at least two major differences between the fruit fly and bark beetle systems. First, in the bark beetles, there is active recruitment by early colonizers, which is crucial for their success [69,71]. In fruit flies, cis-vaccenyl acetate (cVA), has been referred to as an aggregation pheromone [57,72]. However, cVA is produced only by males, who transfer it during copulation to females [73], in which it signals to prospective males that the females are recently mated and unreceptive. Indeed females emitting cVA are much less attractive to males than females with no cVA [74–76]. It is thus likely that cVA has a relatively negligible role in long-distance attraction compared to the dominant role of microbial volatiles [31,77,78]. That is, there is no critical evidence indicating active recruitment of conspecifics in fruit flies.

The second and somewhat related difference between the bark beetle and fruit fly systems is the change in patch attractiveness with density. In the bark beetle system, there is a clear decline in tree attractiveness once a threshold beetle density has been reached. Such decline can readily be explained. Functionally, the occupiers no longer require further individuals once the tree is dying. Mechanistically, the occupiers can readily modulate patch attractiveness by ceasing to emit the aggregation pheromone [69,71]. In the fruit fly system, we failed to identify the predicted lower patch attractiveness under higher density. It is likely, however, that, in natural settings, cues from microbes associated with high density could decrease patch attractiveness or even repel females, as does geosmin discussed above.

Most other systems in which patch choice copying occurs are not as well studied as bark beetles. We suggest, however, that fruit flies can serve as an excellent general model system for further research on the topic owing to their amenability to research in the ecological, evolutionary and mechanistic domains. Our work so far suggests that direct benefits from joining others are likely in many systems even when such benefits are not observed under controlled settings. The most likely reason for such discrepancies is an involvement of harmful microbes in natural settings, which a group is more likely to overcome than an individual. Similarly, because the microbial ecology and dynamics is complex, prospective individuals probably gain the best available information from relying on others, because the others' presence indicates a suitable microbial setting. Our proposition about the central importance of microbes will require extensive experimental work in collaboration with microbial ecologists.

Acknowledgments

We thank J. Holmes, V. Mavandadi, and T. Padilla for assistance and K. Abbott L. Dukas and two anonymous referees for comments on the manuscript.

Author Contributions

Conceived and designed the experiments: SG RD. Performed the experiments: SG. Analyzed the data: SG. Wrote the paper: SG RD.

References

1. Thorpe W (1963) Learning and Instinct in Animals. London: Methuen and Co. 56–81 p.
2. Waite RK (1981) Local enhancement for food finding by Rooks (*Corvus frugilegus*) foraging on grassland. Zeitschrift fur Tierpsychologie 57: 15–36.
3. Krebs JR (1973) Social learning and the significance of mixed-species flocks of chickadees (*Parus* spp.). Canadian Journal of Zoology 51: 1275–1288.
4. Stamps JA (1987) Conspecifics as cues to territory quality: a preference of juvenile lizards (*Anolis aeneus*) for previously used territories. The American Naturalist 129: 629–642.
5. Betts MG, Hadley AS, Rodenhouse N, Nocera JJ (2008) Social information trumps vegetation structure in breeding-site selection by a migrant songbird. Proceedings of the Royal Society B: Biological Sciences 275: 2257–2263.
6. Fletcher RJ, Miller CW (2008) The type and timing of social information alters offspring production. Biology Letters 4: 482–485.
7. Prokopy RJ, Roitberg BD (2001) Joining and avoidance behavior in nonsocial insects. Annual Review of Entomology 46: 631–665.
8. Raitanen J, Forsman JT, Kivelä SM, Mäenpää MI, Välimäki P (2014) Attraction to conspecific eggs may guide oviposition site selection in a solitary insect. Behavioral Ecology 25: 110–116.
9. Dugatkin LA (1992) Sexual selection and imitation: females copy the mate choice of others. American Naturalist 139: 1384–1389.
10. Galef BG, White DJ (1998) Mate-choice copying in Japanese quail, *Coturnix coturnix japonica*. Animal Behaviour 55: 545–552.
11. Alonzo SH (2008) Female mate choice copying affects sexual selection in wild populations of the ocellated wrasse. Animal Behaviour 75: 1715–1723.
12. Krijger CL, Sevenster JG (2001) Higher species diversity explained by stronger spatial aggregation across six neotropical *Drosophila* communities. Ecology Letters 4: 106–115.
13. Shorrocks B, Sevenster JG (1995) Explaining Local Species Diversity. Proceedings of the Royal Society of London Series B: Biological Sciences 260: 305–309.
14. Wade MJ, Pruett Jones SG (1990) Female Copying Increases the Variance in Male Mating Success. Proceedings of the National Academy of Sciences of the United States of America 87: 5749–5753.
15. Kirkpatrick M, Dugatkin LA (1994) Sexual selection and the evolutionary effects of mate copying. Behavioral Ecology and Sociobiology 34: 443–449.
16. Agrawal AF (2001) The evolutionary consequences of mate copying on male traits. Behavioral Ecology and Sociobiology 51: 33–40.
17. Valone TJ, Templeton JJ (2002) Public information for the assessment of quality: a widespread social phenomenon. Philosophical Transactions of the Royal Society of London Series B-Biological Sciences 357: 1549–1557.
18. Danchin E, Giraldeau L-A, Valone TJ, Wagner RH (2004) Public information: from nosy neighbors to cultural evolution. Science 305: 487–491.
19. Danchin E, Wagner RH (1997) The evolution of coloniality: the emergence of new perspectives. Trends in Ecology & Evolution 12: 342–347.
20. Allee WC (1931) Animal Aggregations. A Study in General Sociology. Chicago: University of Chicago Press.
21. Arnold W (1988) Social thermoregulation during hibernation in alpine marmots (*Marmota marmota*). Journal of Comparative Physiology B 158: 151–156.
22. Willis CR, Brigham RM (2007) Social thermoregulation exerts more influence than microclimate on forest roost preferences by a cavity-dwelling bat. Behavioral Ecology and Sociobiology 62: 97–108.
23. Beauchamp G (2014) Social Predation: How Group Living Benefits Predators and Prey. London: Elsevier Academic Press.
24. Rohlfs M, Hoffmeister TS (2004) Spatial aggregation across ephemeral resource patches in insect communities: an adaptive response to natural enemies? Oecologia 140: 654–661.
25. Wertheim B, van Baalen E-JA, Dicke M, Vet LEM (2005) Pheromone-mediated aggregation in nonsocial arthropods: an evolutionary ecological perspective. Annual Review of Entomology 50: 321–346.
26. Pulliam HR, Caraco T (1984) Living in groups: is there an optimal group size? In: Krebs JR, Davies NB, editors. Behavioural Ecology. 2nd ed. Oxford: Blackwell. pp. 122–147.
27. Sarin S, Dukas R (2009) Social learning about egg laying substrates in fruit flies. Proceedings of the Royal Society of London B-Biological Sciences 276: 4323–4328.
28. Battesti M, Moreno C, Joly D, Mery F (2012) Spread of social information and dynamics of social transmission within *drosophila* groups. Current Biology 22: 309–313.
29. Durisko Z, Dukas R (2013) Attraction to and learning from social cues in fruit fly larvae. Proceedings of the Royal Society of London B-Biological Sciences 280: 20131398.
30. Durisko Z, Anderson B, Dukas R (2014) Adult fruit fly attraction to larvae biases experience and mediates social learning. Journal of Experimental Biology 217: 1193–1197.
31. Venu I, Durisko Z, Xu JP, Dukas R (2014) Social attraction mediated by fruit flies' microbiome. Journal of Experimental Biology 217: 1346–1352.
32. Sokolowski MB (2010) Social interactions in "simple" model systems. Neuron 65: 780–794.
33. Robinson GE, Fernald RD, Clayton DF (2008) Genes and social behavior. Science 322: 896–900.
34. Durisko Z, Kemp B, Mubasher A, Dukas R (2014) Dynamics of social interactions in fruit fly larvae. PLoS ONE 9: e95495.
35. Yang C-H, Belawat P, Hafen E, Jan LY, Jan Y-N (2008) *Drosophila* egg-laying site selection as a system to study simple decision-making processes. Science 319: 1679–1683.
36. IBM-Corp. (2011) IBM SPSS Statistics for Windows, Version 21.0. Armonk, NY: IBM Corp.
37. Ashburner M (1989) *Drosophila* a Laboratory Handbook. Cold Spring Harbor: Cold Spring Harbor Laboratory Press.
38. Roberts DB, editor (1998) *Drosophila*: a Practical Approach. 2nd ed. Oxford ; New York: IRL Press at Oxford University Press. xxiv, 389 p.
39. Atkinson WD (1979) A field investigation of larval competition in domestic *Drosophila*. Journal of Animal Ecology 48: 91–102.
40. Grimaldi D, Jaenike J (1984) Competition in natural populations of mycophagous *Drosophila*. Ecology 65: 1113–1120.
41. Masek P, Scott K (2010) Limited taste discrimination in *Drosophila*. Proceedings of the National Academy of Sciences 107: 14833–14838.

42. Burke CJ, Waddell S (2011) Remembering nutrient quality of sugar in *Drosophila*. Current Biology 21: 746–750.
43. Vosshall LB, Stocker RF (2007) Molecular architecture of smell and taste in *Drosophila*. Annual Review of Neuroscience 30: 505–533.
44. Yarmolinsky DA, Zuker CS, Ryba NJP (2009) Common sense about taste: from mammals to insects. Cell 139: 234–244.
45. Stafford JW, Lynd KM, Jung AY, Gordon MD (2012) Integration of taste and calorie sensing in *Drosophila*. The Journal of Neuroscience 32: 14767–14774.
46. Schwarz S, Durisko Z, Dukas R (2014) Food selection in larval fruit flies: dynamics and effects on larval development. Naturwissenschaften 101: 61–68.
47. Begon M (1982) Yeasts and *Drosophila*. In: Ashburner M, Carson HL, Thompson JN, editors. The Genetics and Biology of *Drosophila*. London: Academic Press. pp. 345–384.
48. Janzen DH (1977) Why fruits rot, seeds mold, and meat spoils. The American Naturalist 111: 691–713.
49. Demain A, Fang A (2000) The natural functions of secondary metabolites. In: Fiechter A, editor. History of Modern Biotechnology I: Springer Berlin Heidelberg. pp. 1–39.
50. Janisiewicz WJ, Korsten L (2002) Biological control of postharvest diseases of fruits. Annual Review of Phytopathology 40: 411–441.
51. Sharma R, Singh D, Singh R (2009) Biological control of postharvest diseases of fruits and vegetables by microbial antagonists: A review. Biological Control 50: 205–221.
52. Lacey LA, Shapiro-Ilan DI (2008) Microbial control of insect pests in temperate orchard systems: potential for incorporation into IPM. Annual Review of Entomology 53: 121–144.
53. Arndt C, Cruz MC, Cardenas ME, Heitman J (1999) Secretion of FK506/FK520 and rapamycin by *Streptomyces* inhibits the growth of competing *Saccharomyces cerevisiae* and *Cryptococcus neoformans*. Microbiology 145: 1989–2000.
54. Rohlfs M (2005) Density-dependent insect-mold interactions: effects on fungal growth and spore production. Mycologia 97: 996–1001.
55. Rohlfs M, Obmann B, Petersen R (2005) Competition with filamentous fungi and its implication for a gregarious lifestyle in insects living on ephemeral resources. Ecological Entomology 30: 556–563.
56. Stensmyr Marcus C, Dweck Hany KM, Farhan A, Ibba I, Strutz A, et al. (2012) A conserved dedicated olfactory circuit for detecting harmful microbes in *Drosophila*. Cell 151: 1345–1357.
57. Wertheim B, Dicke M, Vet LEM (2002) Behavioural plasticity in support of a benefit for aggregation pheromone use in *Drosophila melanogaster*. Entomologia Experimentalis Et Applicata 103: 61–71.
58. Stamps JA, Yang LH, Morales VM, Boundy-Mills KL (2012) *Drosophila* regulate yeast density and increase yeast community similarity in a natural substrate. PLoS ONE 7: e42238.
59. Wertheim B, Marchais J, Vet LEM, Dicke M (2002) Allee effect in larval resource exploitation in *Drosophila*: an interaction among density of adults, larvae, and micro-organisms. Ecological Entomology 27: 608–617.
60. Crowley S, Mahony J, van Sinderen D (2012) Comparative analysis of two antifungal *Lactobacillus plantarum* isolates and their application as bioprotectants in refrigerated foods. Journal of Applied Microbiology 113: 1417–1427.
61. Mauch A, Dal Bello F, Coffey A, Arendt EK (2010) The use of *Lactobacillus brevis* PS1 to in vitro inhibit the outgrowth of *Fusarium culmorum* and other common *Fusarium* species found on barley. International Journal of Food Microbiology 141: 116–121.
62. Schnürer J, Magnusson J (2005) Antifungal lactic acid bacteria as biopreservatives. Trends in Food Science & Technology 16: 70–78.
63. Carton Y, Bouletreau M, Alphen JJMv, Lenteren JCv (1986) The *Drosophila* parasitic wasps. In: Ashburner M, Carson HL, Thompson JN, editors. The Genetics and Biology of *Drosophila*. London: Academic Press. pp. 347–934.
64. Fleury F, Ris N, Allemand R, Fouillet P, Carton Y, et al. (2004) Ecological and genetic interactions in Drosophila–parasitoids communities: a case study with D. *Melanogaster, D. Simulans* and their common *Leptopilina* parasitoids in south eastern France. Genetica 120: 181–194.
65. Kacsoh BZ, Lynch ZR, Mortimer NT, Schlenke TA (2013) Fruit flies medicate offspring after seeing parasites. Science 339: 947–950.
66. Hwang RY, Zhong L, Xu Y, Johnson T, Zhang F, et al. (2007) Nociceptive neurons protect *Drosophila* larvae from parasitoid wasps. Current Biology 17: 2105–2116.
67. Bodenstein D (1950) The postembryonic development of *Drosophila*. In: Demerec M, editor. Biology of *Drosophila*. Cold Spring Harbor: Cold Spring Harbor Laboratory Press. pp. 275–367.
68. Paine TD, Raffa KF, Harrington TC (1997) Interactions among Scolytid bark beetles, their associated fungi, and live host conifers. Annual Review of Entomology 42: 179–206.
69. Wood DL (1982) The role of pheromones, kairomones, and allomones in the host selection and colonization behavior of bark beetles. Annual Review of Entomology 27: 411–446.
70. Raffa KF, Berryman AA (1983) The role of host plant resistance in the colonization behavior and ecology of bark beetles (Coleoptera: Scolytidae). Ecological Monographs 53: 27–49.
71. Raffa KF, Aukema BH, Bentz BJ, Carroll AL, Hicke JA, et al. (2008) Cross-scale drivers of natural disturbances prone to anthropogenic amplification: the dynamics of bark beetle eruptions. BioScience 58: 501–517.
72. Bartelt RJ, Schaner AM, Jackson LL (1985) *cis*-vaccenyl acetate as an aggregation pheromone in *Drosophila melanogaster*. Journal of Chemical Ecology 11: 1747–1756.
73. Brieger G, Butterworth FM (1970) *Drosophila melanogaster*: identity of male lipid in reproductive system. Science 167: 1262.
74. Ejima A, Smith BPC, Lucas C, van der Goes van Naters W, Miller CJ, et al. (2007) Generalization of courtship learning in *Drosophila* is mediated by cis-vaccenyl acetate. Current Biology 17: 599–605.
75. Dukas R, Dukas L (2012) Learning about prospective mates in male fruit flies: effects of acceptance and rejection. Animal Behaviour 84: 1427–1434.
76. Keleman K, Vrontou E, Kruttner S, Yu JY, Kurtovic-Kozaric A, et al. (2012) Dopamine neurons modulate pheromone responses in *Drosophila* courtship learning. Nature 489: 145–149.
77. Becher PG, Flick G, Rozpędowska E, Schmidt A, Hagman A, et al. (2012) Yeast, not fruit volatiles mediate *Drosophila melanogaster* attraction, oviposition and development. Functional Ecology 26: 822–828.
78. Stökl J, Strutz A, Dafni A, Svatos A, Doubsky J, et al. (2010) A deceptive pollination system targeting Drosophilids through olfactory mimicry of yeast. Current Biology 20: 1846–1852.

21

Identifying Keystone Species in the Human Gut Microbiome from Metagenomic Timeseries Using Sparse Linear Regression

Charles K. Fisher, Pankaj Mehta*

Department of Physics, Boston University, Boston, Massachusetts, United States of America

Abstract

Human associated microbial communities exert tremendous influence over human health and disease. With modern metagenomic sequencing methods it is now possible to follow the relative abundance of microbes in a community over time. These microbial communities exhibit rich ecological dynamics and an important goal of microbial ecology is to infer the ecological interactions between species directly from sequence data. Any algorithm for inferring ecological interactions must overcome three major obstacles: 1) a correlation between the abundances of two species does not imply that those species are interacting, 2) the sum constraint on the relative abundances obtained from metagenomic studies makes it difficult to infer the parameters in timeseries models, and 3) errors due to experimental uncertainty, or mis-assignment of sequencing reads into operational taxonomic units, bias inferences of species interactions due to a statistical problem called "errors-in-variables". Here we introduce an approach, Learning Interactions from MIcrobial Time Series (LIMITS), that overcomes these obstacles. LIMITS uses sparse linear regression with boostrap aggregation to infer a discrete-time Lotka-Volterra model for microbial dynamics. We tested LIMITS on synthetic data and showed that it could reliably infer the topology of the inter-species ecological interactions. We then used LIMITS to characterize the species interactions in the gut microbiomes of two individuals and found that the interaction networks varied significantly between individuals. Furthermore, we found that the interaction networks of the two individuals are dominated by distinct "keystone species", *Bacteroides fragilis* and *Bacteroided stercosis*, that have a disproportionate influence on the structure of the gut microbiome even though they are only found in moderate abundance. Based on our results, we hypothesize that the abundances of certain keystone species may be responsible for individuality in the human gut microbiome.

Editor: Bryan A. White, University of Illinois, United States of America

Funding: PM and CKF were supported by a Sloan Research Fellowship (to PM), National Institutes of Health Grants K25GM086909 (to PM), and internal funding from Boston University. The funders had no role in study design, data collection and analysis, decision to publish, or preparation of the manuscript.

Competing Interests: The authors have declared that no competing interests exist.

* Email: pankajm@bu.edu

Introduction

Metagenomic sequencing technologies have revolutionized the study of the human-associated microbial consortia making up the human microbiome. Sequencing methods now allow researchers to estimate the relative abundance of the species in a community without having to culture individual species [1–3]. These studies have shown that microbial cells vastly outnumber human cells in the body, and that symbiotic microbial communities are important contributors to human health [1]. For example, a recent study by Ridaura et al [4] demonstrated that transplants of gut microbial consortia are sufficient to induce obesity in previously lean mice or to promote weight loss in previously obese mice, suggesting an intriguing hypothesis that the composition of the gut microbiome may also contribute to obesity in humans. Many other studies have found significant links between the composition of human-associated microbial consortia and diseases including cancer and austim spectrum disorder [4–10]. Despite the recent revelations highlighting the importance of the microbiome to human health, relatively little is known about the ecological structure and dynamics of these microbial communities.

A microbial community consists of a vast number of species, all of which must compete for space and resources. In addition to competition, there are also many symbiotic interactions where certain species benefit from the presence of other microbial species. For example, a small molecule that is secreted by one species can be metabolized by another [11]. These species interactions provide a window with which to view the ecology of a microbial community, and allow one to make predictions about the effect of perturbations on a population [12]. For example, removing a species that engages in mutualistic interactions may diminish the abundance of other species that depend on it for survival. Given their utility for understanding the ecology of a community, there is tremendous interest in developing techniques to infer interactions between species from metagenomic data [12–14].

There are two approaches to inferring dependencies between microbial species from metagenomic studies: cross-sectional analysis, and timeseries analysis [12–16]. Cross-sectional studies pool samples of the relative abundances of the microbial species in a particular environment (e.g. the gut) from multiple individuals and utilize correlations in the relative abundances as proxies for effective interactions between species. By contrast, timeseries analysis follows the relative abundances of the microbial species in a particular environment, for a single individual, over time and utilizes dynamical modeling (e.g. ordinary differential equations) to understand dependencies between species.

Any methods for making reliable inferences about species interactions from metagenomic studies must overcome three major obstacles. First, as shown below, a correlation between the abundances of two species does not imply that those species are interacting. Second, metagenomic methods measure the relative, not absolute, abundances of the microbial species in a community. This makes it difficult to infer the parameters in timeseries models. Finally, errors due to experimental measurement errors and/or mis-assignment of sequencing reads into operational taxonomic units (OTUs), bias inferences of species interactions due to a statistical problem called "errors-in-variables" [17]. We will show that each of these obstacles can be overcome using a new method we call LIMITS (Learning Interactions from MIcrobial Time Series). LIMITS obtains a reliable estimate for the topology of the directed species interaction network by employing sparse linear regression with bootstrap aggregation ("Bagging") to learn the species interactions in a discrete-time Lotka-Volterra (dLV) model of population dynamics from a time series of relative species abundances [18,19].

Results

Correlation does not imply interaction

Many previous works use the correlation between the relative abundances of two microbial species in an environment (e.g. the gut) as a proxy for how much the species interact. In particular, a high degree of correlation between the abundances of two species is often taken as a proxy for a strong mutualistic interaction, and large anti-correlations, as indicative of a strong competitive interactions. Using correlations as a proxy for interactions suffers from several drawbacks. First, there are important subtleties involved in calculating correlations between species from relative abundances, but previous studies have presented algorithms (e.g. SparCC) to mitigate these problems [14]. More importantly, the abundances of two species may be correlated even if those species do not directly interact. For example, if species A directly interacts with species B, and species B directly interacts with species C, the abundances of species A and C are likely to be correlated even though they do not directly interact. Finally, since correlation matrices are necessarily symmetric, all interactions learned using correlations must also be symmetric.

The problems with using correlations in species abundances as proxies for species interactions can be illustrated with a simple numerical simulation. We used the dLV model (Eq. 2) to simulate timeseries of the absolute abundances of 10 species for 1000 timesteps, starting from 100 different initial conditions, for two arbitrary species interaction matrices (see Materials and Methods). See Figure S1 for example time series. Figure 1 compares the Pearson correlation matrices calculated from the absolute species abundances obtained from the dLV simulations and the true interaction matrices. It is clear from Fig. 1 that there is no obvious relationship between the correlations and the species interactions. That is, correlations in species abundances are actually very poor proxies for species interactions.

In general, the relationship between the interaction coefficients (c_{ij}) and the correlations in the species abundances is described by a complicated non-linear function that is difficult to compute or utilize. This can be seen by linearizing the dynamics of $\ln x_i(t)$ (Eq. 4) around their equilibrium values, $\ln\langle x_i(t)\rangle$. This stochastic process is a first-order autoregessive model described by

$$\ln x(t+1) = \omega + J \ln x(t) + \zeta(t), \quad (1)$$

where $\omega_i = -\sum_j c_{ij}\langle x_j\rangle$, and $J_{ij} = \delta_{ij} + c_{ij}\langle x_j\rangle$ is the Jacobian obtained by linearizing around the equilibrium species abundances, and $\zeta(t) = \ln \eta(t)$. If V is the covariance matrix with elements $V_{ij} = \langle(\ln x_i(t) - \langle \ln x_i(t)\rangle)(\ln x_j(t) - \langle \ln x_j(t)\rangle)\rangle$ and Σ is the covariance matrix of the Gaussian noise $\zeta(t)$, then $\mathrm{vec}(V) = (I_{n^2} - J \otimes J)^{-1}\mathrm{vec}(\Sigma)$, where $\otimes$ is the Kronecker product and vec is the matrix vectorization operator [20]. Thus, the interaction matrix is related to the covariance matrix by complex relation even in the linear regime of the dynamics. For this reason, it is very difficult, if not impossible, to determine the interaction coefficients using only knowledge of the correlations and equilibrium abundances.

Cross-sectional studies that pool data across individuals and utilize the correlations between the abundances of different species as proxies for species dependencies are especially affected by this problem. This suggests that time-series data is likely to be more suited for inferring ecological interactions than cross-sectional data.

Timeseries inferrence with relative species abundances

Even though the species interaction coefficients cannot be inferred from the correlations in species abundances, it is possible to reliably infer the interaction matrix using timeseries models. To do so, one utilizes a discrete time Lotka-Volterra Model (dLV) that relates the abundance of species i at a time $t+1$ ($x_i(t+1)$) to the abundances of *all* the species in the ecosystem at a time t ($\vec{x} = \{x_1(t), \ldots, x_N(t)\}$). These interactions are encoded in the dLV through a set of interaction coefficients, c_{ij}, that describe the influence species j has on the abundance of species i [19], and inferring these interaction coefficients is the major goal of this work. The effect of species j on species i can be beneficial ($c_{ij} > 0$), competitive ($c_{ij} < 0$), or the two species may not interact ($c_{ij} = 0$).

As shown in the Materials and Methods, given a time-series of the absolute abundances of the microbes in an ecosystem, one can learn the interaction coefficients by performing a linear regression of $\ln x_i(t+1) - \ln x_i(t)$ against $\vec{x}(t) - \langle\vec{x}\rangle$, where $\langle x\rangle$ is the vector of the equilibrium abundances. It is important to note that each of these linear regressions can be performed independently for species $i = 1, \ldots, N$. In the following, we assume that the population dynamics are stable and that the equilibrium $\langle x\rangle$ (or $\langle\tilde{x}\rangle$) can be estimated by taking the median species abundances over the time series.

Recall that most modern metagenomic techniques can only measure the relative abundances of microbes, not absolute abundances. This introduces additional complications into the problem of inferring species interactions using timeseries data. Although it is straight forward to infer species interactions by applying linear regression to a timeseries of absolute abundances, it is not *a priori* clear that linear regression still works when applied to a timeseries of the relative abundances. An important technical problem that arises when using relative abundances is that the design matrix for the regression is singular because of the sum constraint on relative abundances of species ($\sum_i \tilde{x}_i(t) = 1$). As a result, there is no unique solution to the ordinary least squares problem applied to timeseries of relative species abundances. Nevertheless, the design matrix can be made to be invertible if one, or more, of the species are not included as variables in the regression.

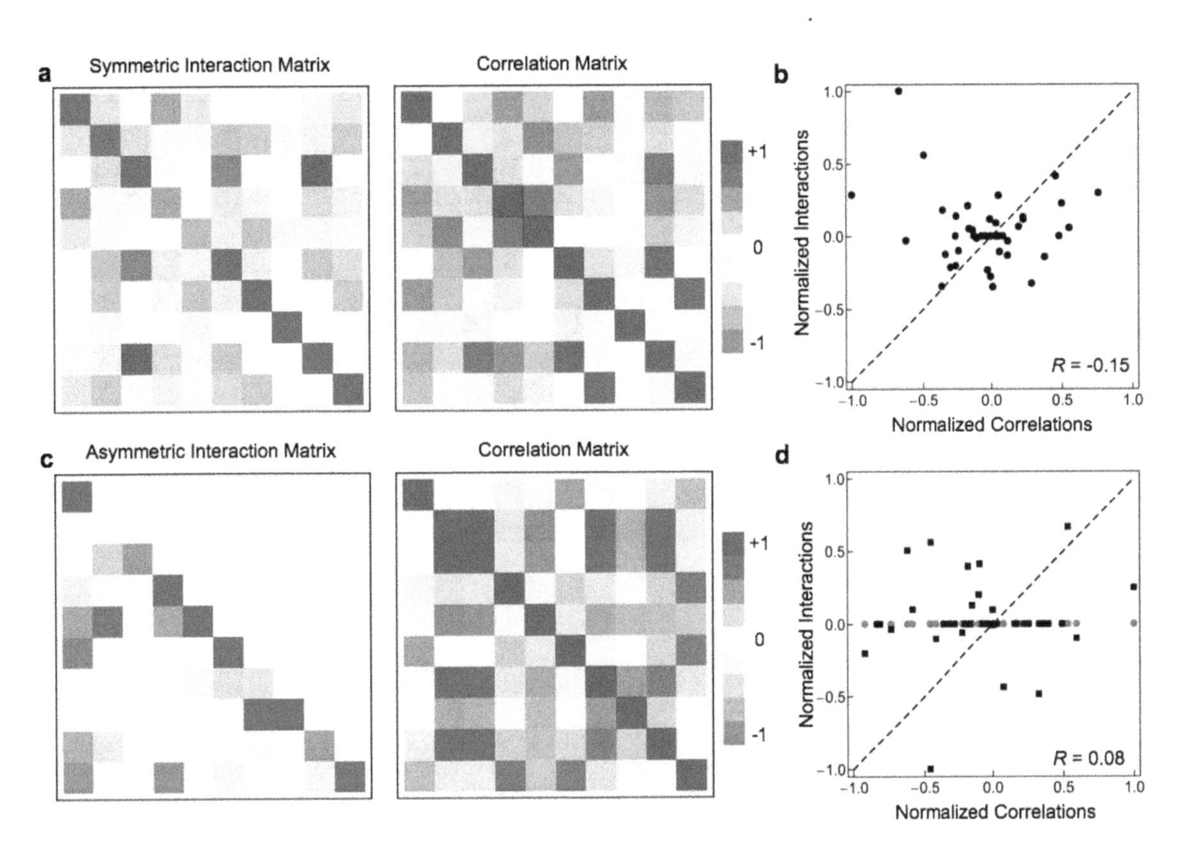

Figure 1. There is no simple relation between interaction coefficients and correlations in abundance. a) A symmetric interaction matrix and the corresponding correlation matrix. b) There is no relation between the interaction parameters and the correlations in abundance for the symmetric interaction matrix. c) An asymmetric interaction matrix and the corresponding correlation matrix. d) There is no relation between the interaction parameters and the correlations in abundance for the asymmetric interaction matrix. Points from above the diagonal in the interaction matrix are gray circles, whereas points from below the diagonal are black squares. In a and c, matrix elements have been scaled so that the smallest negative element is -1, the largest positive element is $+1$, and all elements retain their sign. In b and d, interaction coefficents were scaled so that the largest element by absolute value has $|c_{ij}| = 1$.

This insight motivates the use of a forward stepwise regression for selecting the covariate species that explain the changes in abundance of species i. In such a procedure, interactions and species are added sequentially to the regression as long as they improve the predictive power of model (see Fig. 2a). Since the design matrix now only contains a sub-set of all possible species, it is never singular and the linear regression problem is well-defined. Furthermore, the goal of forward selection is to include only the strongest, most important species interactions in the model. Therefore, the resulting interaction networks are sparse and, hence, easily interpretable.

The procedure for forward stepwise regression is illustrated in Fig. 2a. We know that each species must interact with itself, so c_{ii} is the only interaction coefficient allowed to be nonzero in the first iteration. In each subsequent iteration, one additional interaction c_{ij} is included in the model by scanning over all other species and choosing the one that produces the lowest error at predicting a test dataset. This is repeated as long as the prediction error decreases by a pre-specifed percentage that controls the sparsity of the model. A larger required improvement in prediction results in more sparse solutions. Note that the threshold defining the required improvement in prediction is a relative improvement of one model relative to another (expressed as a percent), not a measure of the absolute error in out of sample prediction.

Forward stepwise regression is a greedy algorithm, which results in a well-known instability [18]. This instability can be 'cured' using a method called bootstrap aggregation, or "Bagging" (Fig. 2b,c) [18]. To bag forward stepwise regression, the data are randomly partitioned into a training set used for the regression and a test set used for evaluating the prediction error. The required improvement in prediction is a percentage that refers to how much the mean squared error evaluated on the test dataset must decrease in order to include an additional variable in the regression. The random partitioning of the data into training and test sets is repeated many times, each one resulting in a different estimate for the interaction matrix. The classical approach to Bagging calls for averaging these different estimates, but this destroys the sparsity of the solution. For this reason, we use the median of the estimates instead of averages. This still greatly improves the stability of the inferred interaction matrix but preserves its sparsity. We call our algorithm LIMITS (Learning Interactions from MIcrobial Time Series).

Figure 3 presents the results from applying LIMITS to infer the same interaction matrices discussed in Figure 1. The data consist of either absolute or relative abundances from timeseries with 500 timesteps and 10 different initial conditions. The sample sizes are quite large, but not as large as used for the calculation of the correlations. The inferred parameters match the true interaction coefficients very accurately – the smallest R^2 between the inferred and true parameters is 0.82 – for both the symmetric (Fig. 3a–b) and asymmetric (Fig. 3c–d) interaction matrices using either absolute or relative species abundances.

To ensure that the exceptional performance of our sparse linear regression approach to inferring species interactions was not a

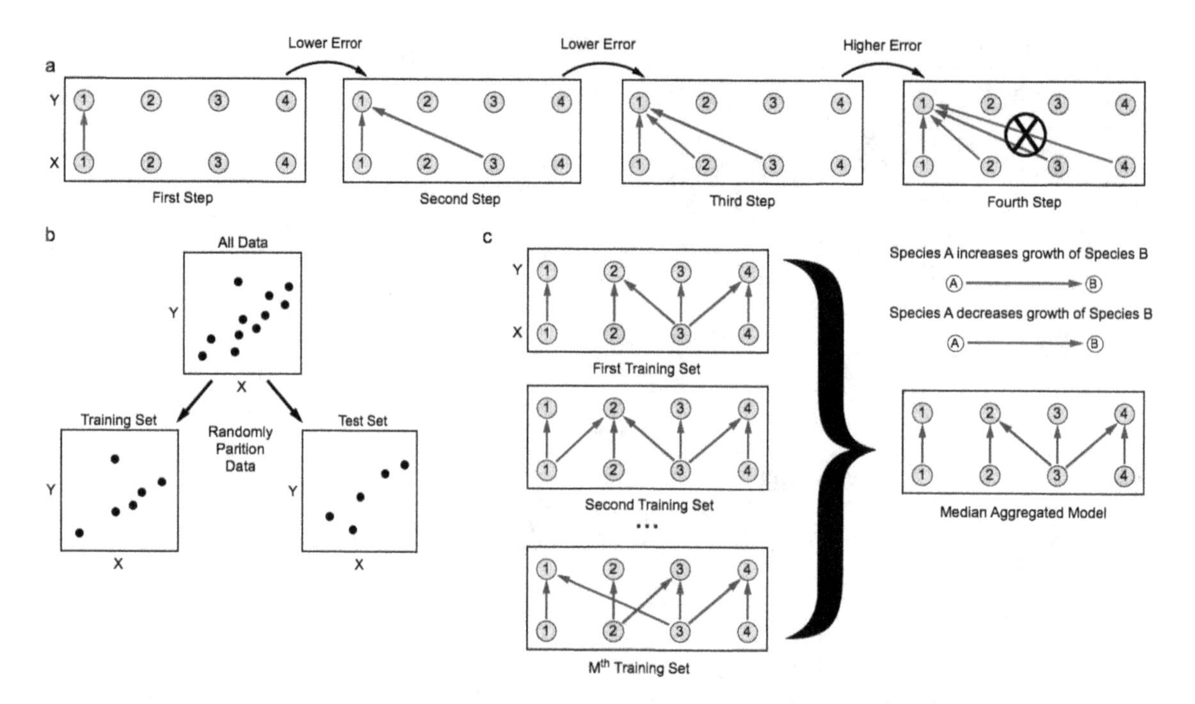

Figure 2. Schematic illustrating forward stepwise regression and median bootstrap aggregating. a) In forward stepwise regression, interactions are added to the model one at a time as long as including the additional covariate lowers the prediction error by a pre-defined threshold. b) The prediction error used for variable selection is evaluated by randomly partitioning the data into a training set used for the regression and a test used to evaluate the prediction error. c) Multiple models are built by repeatedly applying forward stepwise regression to random partitions of the data, each containing half the data points. The models are aggregated, or "bagged", by taking the median, which improves the stability of the fit while preserving the sparsity of the inferred interactions.

fluke due to a particular choice of interaction matrices, we calculated the correlation between the true and inferred parameters for many randomly generated interaction matrices (see Fig. 4 and Materials and Methods). Note that the average magnetidue of c_{ij} relative to the noise in these simulations is fixed, and that these simulations have no measurement errors (see next section for more

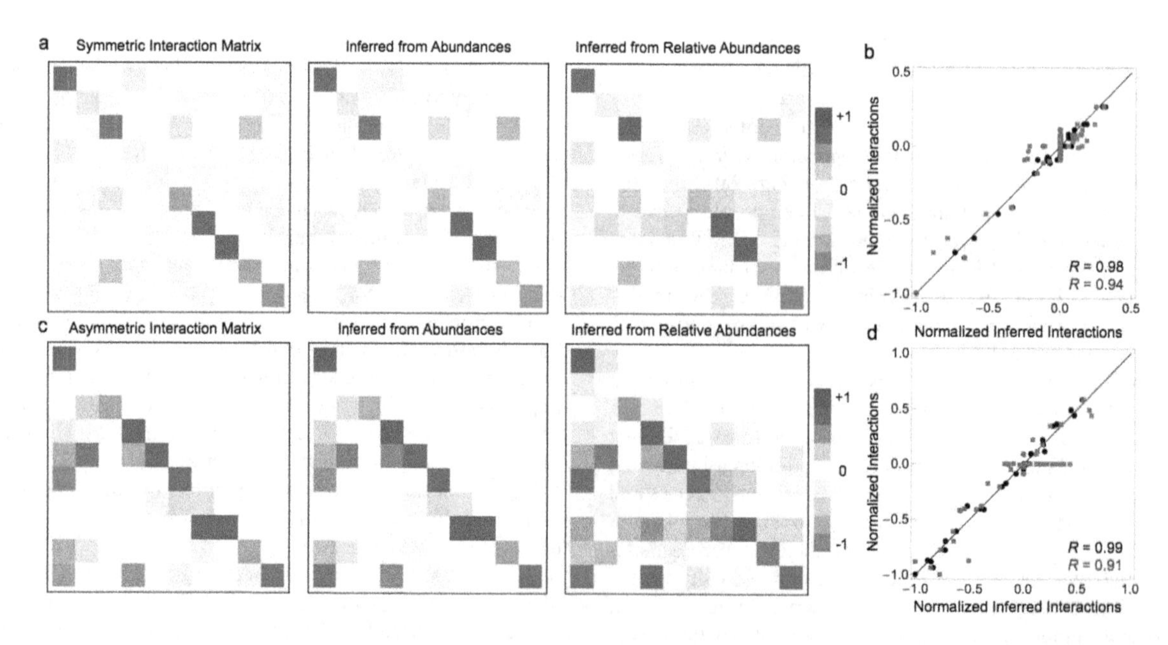

Figure 3. Example fits of interaction parameters using sparse linear regression. a) A symmetric interaction matrix (left), the corresponding matrix inferred from absolute abundance data (middle), and the corresponding matrix inferred from relative abundance data (right). b) There is good aggreement between the true and inferred interactions, from both absolute (black) and relative (gray) abundances, for the symmetric interaction matrix. c) An asymmetric interaction matrix (left), the corresponding matrix inferred from absolute abundance data (middle), and the corresponding matrix inferred from relative abundance data (right). d) There is good aggreement between the true and inferred interactions, from both absolute (black) and relative (gray) abundances, for the asymmetric interaction matrix. The prediction error threshold was set to 5% in for all fits.

on this). The performance of the algorithm obviously depends on sample size, but LIMITS generally perform admirably at inferring ecological interactions (see Figs. 4a–b for symmetric and asymmetric matrices, respectively). Furthermore, Figs. 4c–d show that our results do *not* strongly depend on the required improvement in prediction over most reasonable choices of threshold (0–5%). These results demonstrate that in the absence of measurement noise, LIMITS can successfully learn the interaction parameters from both absolute and relative abundances.

Inferring interactions in the presence of measurement errors

Up to this point, our analyses have ignored the impact of "measurement noise" on the inferred species interactions. There are two important sources of measurement noise in metagenomic data. The first source is experimental noise introduced by sequencing errors. The second, and perhaps larger source of noise, is the mis-classification of sequencing reads into operational taxonomic units (OTUs). Most metagenomic studies rely on the sequencing of 16S rRNA to estimate species composition and diversity in a community. These 16S sequences are binned into groups, or OTUs, that contain sequences with a predetermined degree of similarity. By comparing the sequences in an OTU to known sequences in an annotated database, it is often possible to assign OTUs to particular species or strains. In general, this is an extremely difficult bioinformatics problem [21] and is likely to be a significant source of measurement errors. Thus, any algorithm for inferring species interactions must be robust to measurement errors.

At first glance, it is tempting to assume that measurement noise, which we assume is multiplicative, simply adds to the stochastic ($\ln \eta_i(t)$) term that acts on the dependent variable ($\ln x_i(t+1) - \ln x_i(t)$) and, therefore, should have little impact on the inferred interactions (see Methods). However, as we discuss below, this is not the case since the $x_i(t)$ also act as the independent variables in the regression. Standard regression techniques assume that the independent variables are known exactly, and violation of this assumption results in biased parameter estimates even for asymptotically large sample sizes [17]. For example, in the simplest case of a 1-dimensional regression $Y = \alpha + \beta X$ the estimator $\hat{\beta}$ is always less than the real β, i.e. $\hat{\beta} = \mathrm{COV}(X,Y)/\mathrm{VAR}(X) \leq \beta$ with equality only if there is no measurement error on X. The bias induced by using noisy

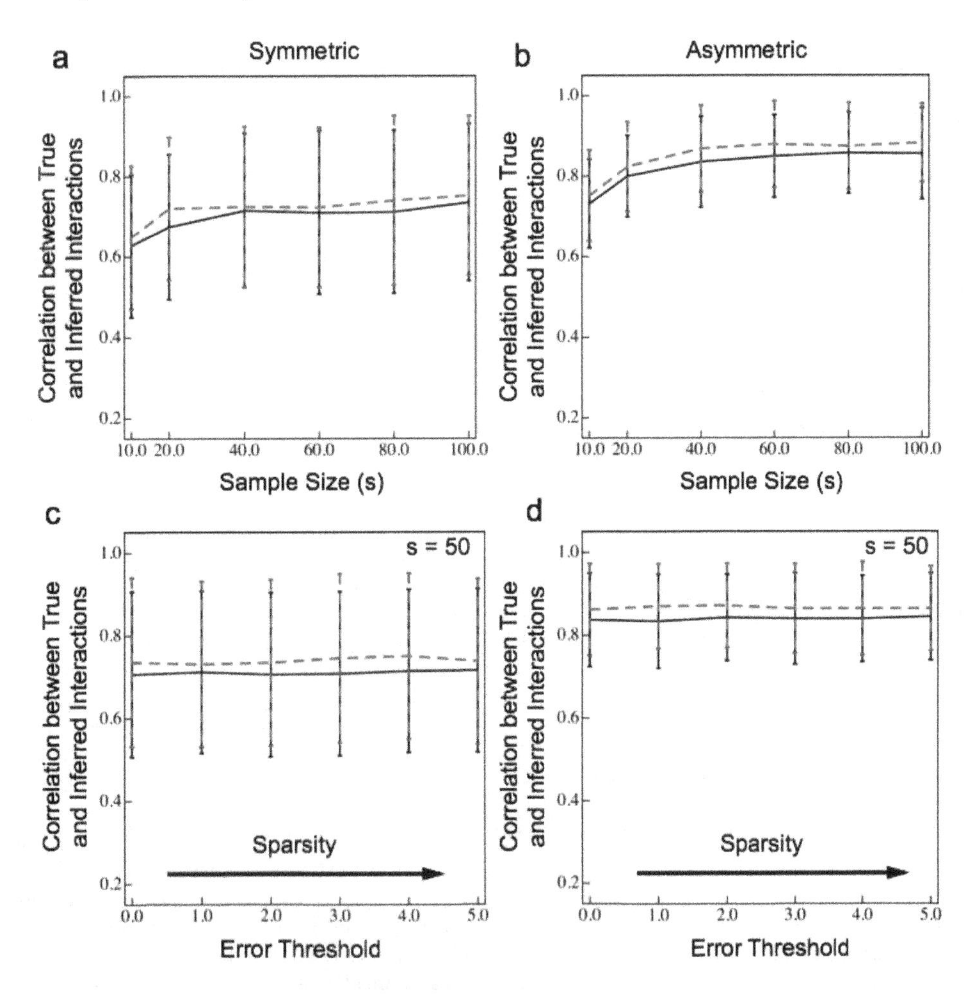

Figure 4. Performance of sparse linear regression as a functon of sample size and the prediction error threshold. a) Performance on absolute (red) and relative (black) abundances as a function of sample size for symmetric interaction matrices. b) Performance on absolute (red) and relative (black) abundances as a function of sample size for asymmetric interaction matrices. c) Performance on absolute (red) and relative (black) abundances as a function of the out-of-bag error threshold for symmetric interaction matrices. d) Performance on absolute (red) and relative (black) abundances as a function of the out-of-bag error threshold for symmetric interaction matrices. Error bars correspond to $\pm$ one standard deviation, and lines connect the means.

indepedent variables in regression is known as the "errors-in-variables" problem in the statistical literature [17]. Analysis of the errors-in-variables bias for multivariate regression is more complicated than the 1-dimensional example; nevertheless, it can be stated quite generally that the interaction parameters inferred in the presence of signficant measurement errors will be incorrect. The most reliable method for mitigating the errors-in-variables bias is to measure additional data on some "instrumental variables" that provide information on the true values of the relative species abundances [17]. Unfortunately, in most cases we often do not have access to such additional data.

Although the errors-in-variables bias cannot be eliminated, the topology of the interaction network can still be reliably inferred using our sparse linear regression approach even when the measurements of the relative species abundances are very noisy. Knowledge of which interactions are beneficial ($c_{ij} > 0$), competitive ($c_{ij} < 0$), or zero ($c_{ij} = 0$) defines the topology of the interaction network. In the following simulations, we focus only on whether a given interaction is zero ($c_{ij} = 0$) or not ($c_{ij} \neq 0$) because we found that errors in the signs of the interactions were rare (Fig. S2). The accuracy of the interaction topologies inferred from noisy relative abundance data were assessed using simulations with randomly chosen (asymmetric) interaction matrices. We computed the specificity – the fraction of species pairs correctly identified as non-interacting – and the sensitivity – the fraction of species pairs correctly identified as interacting – of the inferred topologies (Fig. 5) [22]. Figure 5a shows that LIMITS produced specificities between 60% and 80% for different choices of the prediction error threshold, and that the performance was relatively insenstive to multiplicative measurement noise up to, and even beyond, 10%. By contrast, Fig. 5b shows that applying forward variable selection to the entire dataset, i.e. without Bagging, produced results that were very senstive to the choice of the prediction error threshold and even produced specificities as low as 0%. The sensitivities for detecting interacting species with (Fig. 5c) or without (Fig. 5d) Bagging are both quite good, ranging between 70% and 80% for measurement noise up to 10%. These results demonstrate that both the forward stepwise regression and the median bootrap aggregation are crucial components of the LIMITS algorithm. Moreover, LIMITS reliably infers the topology of the species interaction network even when there are significant errors-in-variables.

Keystone species in the human gut microbiome

Emboldened by the success of our algorithm on synthetic data, we applied LIMITS to infer the species interactions in the gut microbiomes of two individuals. The data from Caporaso et al [3] were obtained from the MGRAST database [23] in a pre-processed form; i.e. as relative species abundances instead of raw sequencing data. These data consisted of approximately a half-year of daily sampled relative species abundances for individual (a) and a full year of daily sampled relative species abundances for individual (b). Note that an interaction matrix has of order n^2 free parameters for n species. Also, Figures 4 and S2 show that the performance of LIMITS plateaus around n^2 data points. Thus, in general, we require at least n^2 timepoints in order to infer the interactions from n species. The number of available timepoints was $O(100)$, so we considered only the 10 most abundant species from individuals (a) and (b). Because the most abundant species were not entirely the same in individuals (a) and (b), we studied 14 species obtained by taking the union of the 10 most abundant species from individual (a) with the 10 most abundant species from individual (b). The 14 species are listed in Fig. 6.

The species interaction network of the gut microbiome of individual (a) (shown in Fig. 6a) is dominated by the species *Bacteroides fragilis* even though it is found only in moderate abundances. *Bacteroides fragilis* has 6 outgoing interactions in individual (a), in contrast to the other 13 species that have 0–3 outgoing interactions. The species interaction network of the gut microbiome of individual (b) (shown in Fig. 6b) is also dominated by a single species, *Bacteroides stercosis*, which is also found only in moderate abundaces. *Bacteroides stercosis* has 4 outgoing interactions in individual (b), in contrast to the other 13 species that have 0–2 outgoing interactions. In addition, many of the interactions involving *Bacteroides fragilis* and *Bacteroides stercosis* are beneficial interactions. Based on these results, we refer to *Bacteroides fragilis* and *Bacteroides stercosis* as "keystone species" of the human gut microbiome because these two species exert tremendous influence on the structure of the microbial communities, even though they have lower median abundances than some other species.

Additionally, we observed that the species interaction topology of the gut microbiome of individual (a) differs substaintially from the species interaction topology of the gut microbiome of individual (b), as is clear from Fig. 6. In individual (a), *Bacteroides fragilis* is much more abundant than *Bacteroides stercosis* and, in turn, it is *Bacteroides fragilis* that dominates the interaction network of individual (a). Likewise, in individual (b), *Bacteroides stercosis* is much more abundant than *Bacteroides fragilis* and, in turn, it is *Bacteroides stercosis* that dominates the interaction network of individual (b). This observation motivates us to propose an intriguing hypothesis, that the abundances of certain keystone species are responsible for individuality of the human gut microbiome. Of course, much more data, from a larger population, will be required to confirm or reject this hypothesis.

Discussion

Metagenomic methods are providing an unprecedented window into the composition and structure of micriobial communities. They are revolutionizing our knowledge of microbial ecology and highlight the important roles played by the human microbiome in health and disease. Nevertheless, it is important to carefully consider the tools used to analyze these data and to address their associated challenges. We have highlighted three major obstacles that must be addressed by any study designed to use metagenomic data to analyze species interactions: 1) a correlation between the abundances of two species does not imply that those species are interacting, 2) the sum constraint on the relative abundances obtained from metagenomic studies makes it difficult to infer the parameters in timeseries models, and 3) errors due to experimental uncertainty, or mis-assignment of sequencing reads into operational taxonomic units (OTUs), bias inferrences of species interactions due to a statistical problem called "errors-in-variables".

To overcome these obstacles, we have introduced a novel algorithm, LIMITS, for inferring species interaction coefficients that combines sparse linear regression with bootstrap aggregation (Bagging). Our method provides reliable estimates for the topology of the species interaction network even when faced with significant measurement noise. The interaction networks constructed using our approach are sparse, including only the strongest ecological interactions. Regularizing the inference of the interaction network by favoring sparse solutions has the benefit that the results are easily interpretable, enabling the identification of keystone species with many important interactions. Furthermore, our work suggests that it is difficult to learn species interactions from cross-sectional

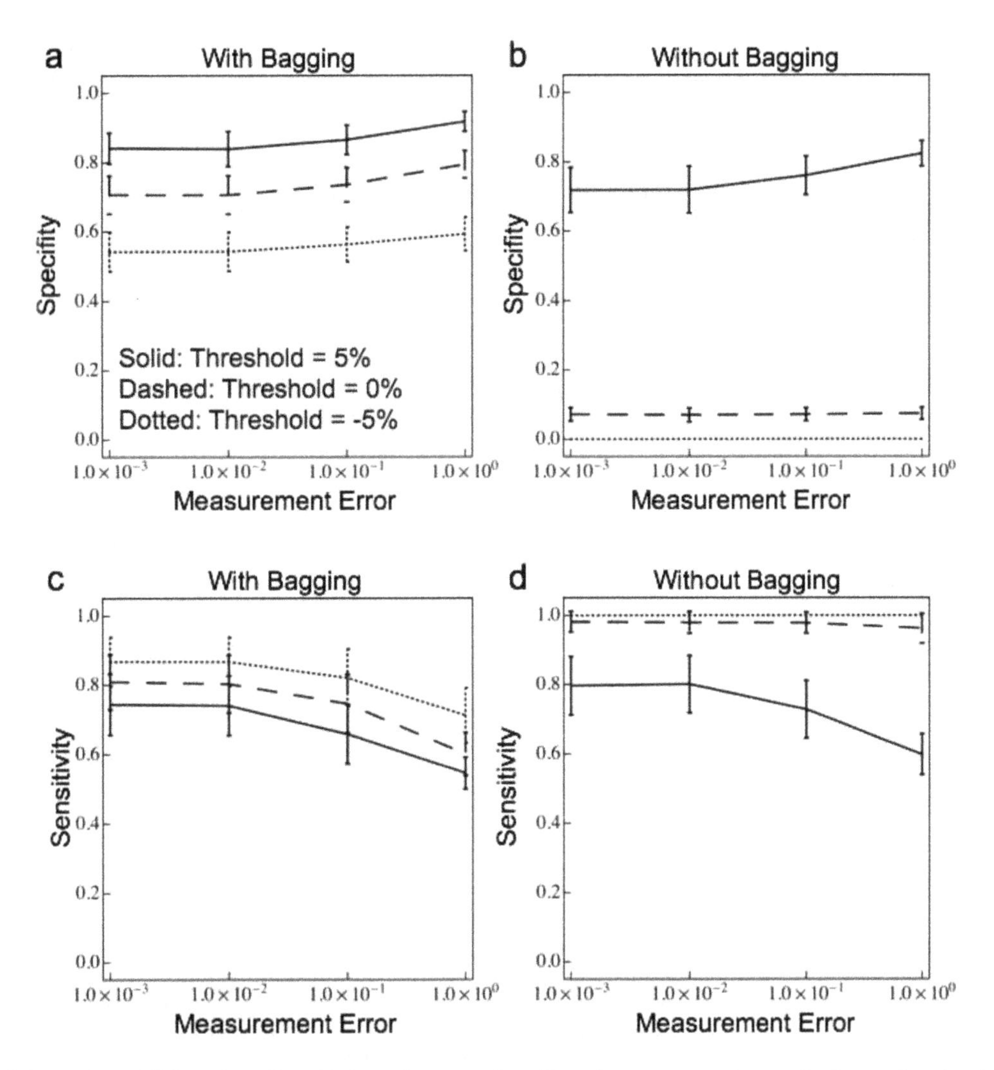

Figure 5. Sensitivity and specificity of predicted interactions as a function of measurement error for bagged and unbagged models. Specificity refers to the fraction of species pairs correctly identified as non-interacting, while sensitivity refers to the fraction of species pairs correctly identified as interaction. Both measures range from 0 (poor performance) to 1 (good performance). a) Specifity of sparse linear regression with Bagging as a function of measurement error for different prediction error thresholds. b) Specificity of sparse linear regression trained on the entire data set without Bagging as a function of measurement error for different prediction thresholds. c) Sensitivity of sparse linear regression with Bagging as a function of measurement error for different prediction error thresholds. d) Sensitivity of sparse linear regression trained on the entire data set without Bagging as a function of measurement error for different prediction error thresholds. Notice that without bagging, model performance is extremely sensitive to choice of the threshold for the required improvement in prediction for adding new interactions.

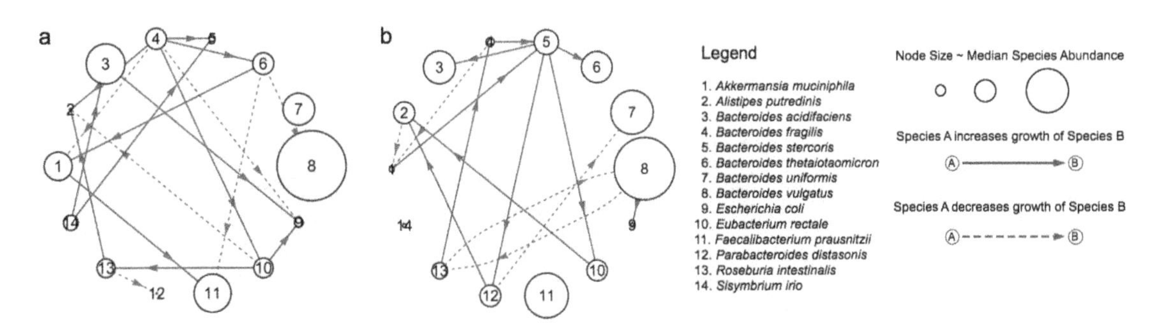

Figure 6. Interaction topologies of abundant species in the guts of two individuals. The size of a node denotes the median relative species abundance, beneficial interactions are shown as solid red arrows, and competetive interactions are shown as dashed blue arrows. In individual a) species 4 *Bacteroides fragilis* acts as a keystone species with 6 outgoing interactions, compared to a median number of outgoing interactions of 1. In individual b) species 5 *Bacteroides stercosis* acts as a keystone species with 4 outgoing interactions, compared a median number of outgoing interactions of 1. The 14 species included in the model were obtained by taking the union of the top 10 most abundant species from individuals a and b. The required improvement in prediction was set to 3%, graphs obtained using other prediction thresholds are shown in the Supporting Information.

studies that pool samples of the relative abundances of the microbial species from multiple individuals. This highlights the importance of collecting extended time-series data for understanding microbial ecological dynamics.

We applied LIMITS to time-series data to infer ecological interaction networks of two individuals and found that the interaction networks are dominated by distinct keystone species. This motivated us to propose a hypothesis: that the abundances of certain keystone species are responsible for individuality of the human gut microbiome. While more data will be required to confirm or reject this hypothesis, it is intriguing to examine its potential consequences for the human microbiome. The keystone species hypothesis implies that even small perturbations to an environment can have a large impact on the composition of its resident microbial consortia if those perturbations affect a small number of important "keystone" species. Moreover, relatively small differences in individual diets, or minor differences in the interaction between the host immune system and the gut microbiota, that affect keystone species may be sufficient to organize gut microbial consortia into distinct types of communities, or "enterotypes" [24,25].

Our analysis identified the closely related species *Bacteroides fragilis* and *Bacteroides stercosis* as potential keystone species of the gut microbiome [26]. Previous studies have suggested that the abundance of *Bacteroides fragilis* modulates the levels of several metabolites and, in turn, the composition of the gut microbiome in a mouse model of gastrointestinal abnormalities associated with autism spectrum disorder [9]. Abundance of both *Bacteroides fragilis* and *Bacteroides stercosis* are associated with an increased risk of colon cancer [5,6,10], and previous authors even suggest that *Bacteroides fragilis* acts as a critical "keystone pathogen" in the development of the disease [8]. Classical ecological models of species interaction demonstrate that the manner of the interaction between two species is not solely a function of their identity, but is highly dependent on the environment in which the interaction takes place [27,28]. Increased abundance of *Bacteroides* species is associated with high fat diets, including typical Western diets with a high consumption of red meat that are associated with increased cancer risk [5]. It is possible that *Bacteroides fragilis* and *Bacteroides stercosis* act as keystone species in individuals consuming high fat diets due to their ability to convert bile into metabolites that are used by other members of the microbial community [5,9].

The keystone species hypothesis can be experimentally tested by perturbing the abundance of individual species in a microbial consortium and observing the effect on the composition of the community. Our prediction is that most perturbations will have little impact on the overall structure of the microbial community, but perturbations applied to a small number of keystone species will have a large impact on the structure of the community. Due to ethical concerns, it is difficult to envision a direct experimental test of the keystone species hypothesis in human microbiota and, therefore, to test our specific predictions in regards to the keystone species *Bacteroides fragils* and *Bacteroides stercosis*. Nevertheless, experimental tests could be performed in animal models, or even in culture if a large enough microbial consortia can be assembled.

Materials and Methods

Discrete Time Lotka-Volterra Model for absolute abundances

Metagenomic sequencing methods have made it possible to follow the time evolution of a microbial population by determining the relative abundances of the species in a community in discrete intervals (e.g. one day). Given the discrete nature of these data, it is most sensible to use a discrete-time model of population dynamics. The discrete-time Lotka-Volterra (dLV) model of population dynamics (sometimes called the Ricker model) relates the abundance of species i at a time $t+\delta t$ ($x_i(t+\delta t)$) to the abundances of *all* the species in the ecosystem at a time t ($\vec{x}=\{x_1(t)\ldots x_N(t)\}$). These interactions are encoded in the dLV through a set of interaction coefficients, c_{ij}, that describe the influence species j has on the abundance of species i [19], and inferring these interaction coefficients is the major goal of this work. The abundance of a species can also change due to environmental and demographic stochastic effects. The dLV can be generalized to include stochasticity by including a log-normally distributed multiplicative noise, $\eta_i(t)$. Specifically, the dynamics are modeled by the equations

$$x_i(t+\delta t)=\eta_i(t)x_i(t)\exp\left(\delta t\sum_j c_{ij}(x_j(t)-\langle x_j\rangle)\right), \quad (2)$$

where $\langle x_j\rangle$ is equilibrium abundance of species j and is set by the carrying capacity of the environment. In writing these equations in this form, we have assumed that in the absence of noise the dLV equations have a unique steady-state solution with the abundances given by $\langle x_j\rangle$. Notice that in the absence of multiplicative noise and the limit $\delta t\to 0$, the dLV model reduces to the usual Lotka-Volterra differential equations:

$$\frac{1}{x_i(t)}\frac{dx_i(t)}{dt}=\sum_j c_{ij}(x_j(t)-\langle x_j\rangle). \quad (3)$$

In what follows, without loss of generality, we set $\delta t=1$. This is equivalent to measuring time in units of δt. To fit microbial data, it is actually helpful to work with the logarithm of Equation (2). Furthermore, we assume that the sampling time and the update time are both equal to 1. Taking the logarithm of Eq. 2 yields,

$$\ln x_i(t+1)-\ln x_i(t)=\zeta_i(t)+\sum_j c_{ij}(x_j(t)-\langle x_j\rangle), \quad (4)$$

where by construction $\zeta_i(t)=\ln\eta_i(t)$ is a normally distributed variable. This logarithmic form of the dLV model is especially convenient for inferring species interactions from time series of species abundances because the inference problem reduces to standard linear regression (as discussed below).

Thus, far we have assumed that it is possible to directly measure the absolute abundances $x_i(t)$. However, in practice, metagenomic sequencing studies typically provide relative abundances $\tilde{x}_i=Z^{-1}x_i$, where $Z=\sum_j x_j$. Provided that the number of species is large ($N\gg 1$), the fluctuations in the total population size $Z(t)=\sum_i x_i(t)$ around its mean value $\langle Z\rangle=\sum_i\langle x_i\rangle$ will be small. In this case, the dynamics of the relative abundances ($\tilde{x}_i(t)$) are well-described by a modified dLV model:

$$\tilde{x}_i(t+1)\approx\eta_i(t)\tilde{x}_i(t)\exp\left(\sum_j\tilde{c}_{ij}(\tilde{x}_j(t)-\langle\tilde{x}_j\rangle)\right), \quad (5)$$

where we have defined the new interaction coefficients $\tilde{c}_{ij}=\langle Z\rangle c_{ij}$ which are related to the true interaction coefficients c_{ij} by the average population size. Thus, relative species abundance data can be modeled using the dLV model, but the interaction coefficients are known only up to an arbitrary multiplicative constant.

However, as discussed in the main text, the design matrix for relative species abundances is singular so simple linear regression fails.

In all of the simulations discussed in the main text the stochasticity was set to $\ln \eta_i(t) \sim \mathcal{N}(0,0.1)$.

Linear Regression

Suppose we are given data consisting of the absolute (or relative) abundances of the species in a population of N species in the form of timeseries of length T starting from M initial conditions. We infer each row ($\vec{c}_i = \{c_{ij}\}_{j=1}^N$) of the interaction coefficient matrix (C) separately. The equilibrium population $\langle x \rangle$ (or $\langle \tilde{x} \rangle$) is assumed to correspond to the population median. The design matrix is an $M(T-1) \times N$ matrix with rows $X_l = \{x_1^{(k)}(t) - \langle x_1 \rangle, ..., x_N^{(k)}(t) - \langle x_N \rangle\}$. The data vector has length $M(T-1)$ and is given by $\vec{v}_i = \{\ln x_i^{(k)}(1) - \ln x_i^{(k)}(0), ..., \ln x_i^{(k)}(T) - \ln x_i^{(k)}(T-1)\}$. Note that any of the timepoints with $x_i^{(k)}(t) = 0$ were left out of the regression because the logarithm of zero is undefined. The least squares estimate for the interaction coefficients is $\hat{c}_i = X^+ \vec{v}_i$, where the $+$ denotes the psuedo-inverse.

Outline of the LIMITS Algorithm

Here, we present a high-level outline of the LIMITS algorithm (see Fig. 2). An implementation of the LIMITS algorithm written in Mathematica (Wolfram Research, Inc.) is available in the Supporting Information and on the author's webpage at http://physics.bu.edu/~pankajm. Since all of the regressions are performed independently for each species, we will only describe the algorithm for inferring a row of the interaction matrix ($\vec{c}_i = \{c_{ij}\}_{j=1}^N$). One simply loops over i to obtain the full interaction matrix. Moreover, bootstrap aggregating simply involves performing the whole proceedure L times, thereby constructing multiple estimates $\vec{c}_i^{(1)}, \ldots, \vec{c}_i^{(L)}$ and taking their median. Thus, the only thing that takes some effort to explain is how to construct one of estimate of $\vec{c}_i$.

1. First, we randomly partition the data into a training set and a test set, each containing half the data points.
2. A set of active coefficients is initialized to $\text{ACTIVE} = \{c_{ii}\}$ and a set of inactive coefficients is initialized to $\text{INACTIVE} = \{c_{ij}\}_{j \neq i}$.
3. A linear regression including only species i is performed on the training set, and the inferred coefficient is used to calculate a prediction error (called ERROR) for the test dataset.
4. For each coefficient c in INACTIVE create $\text{TEST} = \text{ACTIVE} \bigcup \{c\}$ and perform a linear regression using the coefficients in TEST against the training dataset.
5. Next, the inferred coefficients are used to calculate the prediction errors for the test dataset. The particular species j with the smallest prediction error is retained, and we call this error ERROR(j).
6. If $100 \times (\text{ERROR} - \text{ERROR}(j))/\text{ERROR}$ is greater than a pre-specified required improvement in prediction then we set $\text{ERROR} = \text{ERROR}(j)$, $\text{ACTIVE} = \text{ACTIVE} \bigcup \{c_{ij}\}$, and c_{ij} is deleted from INACTIVE, otherwise we terminate the loop and return an estimate for the interactions $\vec{c}_i = \{c_{ij}\}_{j=1}^N$.

Generating Test Interaction Matrices

Interaction matrices (of size 10×10 for 10 species) were randomly generated to ensure that they were sparse, and that they resulted in a stable internal equilibrium. First, the equilibrium abundances were drawn from a lognormal distribution with median 1 and scale parameter 0.1. The diagonal elements of the matrix were randomly drawn from a uniform distribution over the interval $c_{ii} \sim (-1.9/\bar{x}_i, -0.1/\bar{x}_i)$, where $\bar{x}_i$ is the randomly chosen equilibrium abundance of species i. Next, random c_{ij} parameters were added to the matrix one at a time as long as the Lotka-Volterra system remained stable until 10 off diagonal interactions had been added. The algorithm we used for generating random interaction matrices is provided in Code S1.

Supporting Information

Code S1 Code for the LIMITS algorithm and an example implemented in Mathematica (Wolfram Research, Inc.) Version 8.0.

Figure S1 Example Lotka-Volterra Time Series. Representative simulations of Lotka-Volterra time series with a) a symmetric interaction matrix and b) an asymmetric interaction matrix.

Figure S2 Errors in Interaction Topology as a Function of Sample Size. The errors in inferred interaction topologies as a function of sample size for a) a symmetric interaction matrices and b) an asymmetric interaction matrices. The black line represents the true positive rate, the dashed blue line represents the false negative rate, and the red line represents the rate of sign errors.

Figure S3 Interaction topologies of abundant species in the guts of two individuals using different prediction error thresholds. The size of a node denotes the median relative species abundance, beneficial interactions are shown as solid red arrows, and competetive interactions are shown as dashed blue arrows. The 14 species included in the model were obtained by taking the union of the top 10 most abundant species from individuals a and b.

Acknowledgments

We would like to acknowledge useful conversations with Sara Collins and Daniel Segré and thank Alex Lang and Javad Noorbakhsh for comments on the mansuscript.

Author Contributions

Conceived and designed the experiments: CKF PM. Analyzed the data: CKF. Wrote the paper: CKF PM.

References

1. Turnbaugh PJ, Ley RE, Hamady M, Fraser-Liggett CM, Knight R, et al. (2007) The human microbiome project. Nature 449: 804–810.
2. Yatsunenko T, Rey FE, Manary MJ, Trehan I, Dominguez-Bello MG, et al. (2012) Human gut microbiome viewed across age and geography. Nature 486: 222–227.
3. Caporaso JG, Lauber CL, Costello EK, Berg-Lyons D, Gonzalez A, et al. (2011) Moving pictures of the human microbiome. Genome Biol 12: R50.
4. Ridaura VK, Faith JJ, Rey FE, Cheng J, Duncan AE, et al. (2013) Gut microbiota from twins discordant for obesity modulate metabolism in mice. Science 341: 1241214.

5. Moore W, Moore LH (1995) Intestinal oras of populations that have a high risk of colon cancer. Applied and Environmental Microbiology 61: 3202–3207.
6. Ahn J, Sinha R, Pei Z, Dominianni C, Wu J, et al. (2013) Human gut microbiome and risk for colorectal cancer. Journal of the National Cancer Institute 105: 1907–1911.
7. Turnbaugh PJ, Hamady M, Yatsunenko T, Cantarel BL, Duncan A, et al. (2008) A core gut microbiome in obese and lean twins. Nature 457: 480–484.
8. Hajishengallis G, Darveau RP, Curtis MA (2012) The keystone-pathogen hypothesis. Nature Reviews Microbiology 10: 717–725.
9. Hsiao EY, McBride SW, Hsien S, Sharon G, Hyde ER, et al. (2013) Microbiota modulate behavioral and physiological abnormalities associated with neurodevelopmental disorders. Cell 155: 1451–1463.
10. Wu S, Rhee KJ, Albesiano E, Rabizadeh S, Wu X, et al. (2009) A human colonic commensal promotes colon tumorigenesis via activation of t helper type 17 t cell responses. Nature medicine 15: 1016–1022.
11. Klitgord N, Segrè D (2010) Environments that induce synthetic microbial ecosystems. PLoS computational biology 6: e1001002.
12. Faust K, Raes J (2012) Microbial interactions: from networks to models. Nature Reviews Microbiology 10: 538–550.
13. Stein RR, Bucci V, Toussaint NC, Buffie CG, Rätsch G, et al. (2013) Ecological modeling from timeseries inference: Insight into dynamics and stability of intestinal microbiota. PLoS computational biology 9: e1003388.
14. Friedman J, Alm EJ (2012) Inferring correlation networks from genomic survey data. PLoS Computational Biology 8: e1002687.
15. Ruan Q, Dutta D, Schwalbach MS, Steele JA, Fuhrman JA, et al. (2006) Local similarity analysis reveals unique associations among marine bacterioplankton species and environmental factors. Bioinformatics 22: 2532–2538.
16. Xia LC, Ai D, Cram J, Fuhrman JA, Sun F (2013) Efficient statistical significance approximation for local similarity analysis of high-throughput time series data. Bioinformatics 29: 230–237.
17. Fuller WA (1980) Properties of some estimators for the errors-in-variables model. The Annals of Statistics: 407–422.
18. Breiman L (1996) Bagging predictors. Machine learning 24: 123–140.
19. Hofbauer J, Hutson V, Jansen W (1987) Coexistence for systems governed by difference equations of lotka-volterra type. Journal of Mathematical Biology 25: 553–570.
20. Enders W (2008) Applied econometric time series. John Wiley & Sons.
21. Wooley JC, Godzik A, Friedberg I (2010) A primer on metagenomics. PLoS computational biology 6: e1000667.
22. Bishop CM, Nasrabadi NM (2006) Pattern recognition and machine learning, volume 1. springer New York.
23. Glass EM, Wilkening J, Wilke A, Antonopoulos D, Meyer F (2010) Using the metagenomics rast server (mg-rast) for analyzing shotgun metagenomes. Cold Spring Harbor Protocols 2010: pdb–prot5368.
24. Arumugam M, Raes J, Pelletier E, Le Paslier D, Yamada T, et al. (2011) Enterotypes of the human gut microbiome. Nature 473: 174–180.
25. Turnbaugh PJ, Ridaura VK, Faith JJ, Rey FE, Knight R, et al. (2009) The effect of diet on the human gut microbiome: a metagenomic analysis in humanized gnotobiotic mice. Science translational medicine 1: 6ra14.
26. Shah H, Collins M (1989) Proposal to restrict the genus bacteroides (castellani and chalmers) to bacteroides fragilis and closely related species. International journal of systematic bacteriology 39: 85–87.
27. MacArthur R (1970) Species packing and competitive equilibrium for many species. Theoretical population biology 1: 1–11.
28. Chesson P, Kuang JJ (2008) The interaction between predation and competition. Nature 456: 235–238.

Environmental Correlates of H5N2 Low Pathogenicity Avian Influenza Outbreak Heterogeneity in Domestic Poultry in Italy

Lapo Mughini-Gras[1¤*], Lebana Bonfanti[1], Paolo Mulatti[1], Isabella Monne[1], Vittorio Guberti[2], Paolo Cordioli[3], Stefano Marangon[1]

1 Istituto Zooprofilattico Sperimentale delle Venezie (IZSVe), Legnaro, Padua, Italy, **2** Institute for Environmental Protection and Research (ISPRA), Ozzano dell'Emilia, Bologna, Italy, **3** Istituto Zooprofilattico Sperimentale della Lombardia e dell'Emilia Romagna (IZLER), Brescia, Italy

Abstract

Italy has experienced recurrent incursions of H5N2 avian influenza (AI) viruses in different geographical areas and varying sectors of the domestic poultry industry. Considering outbreak heterogeneity rather than treating all outbreaks of low pathogenicity AI (LPAI) viruses equally is important given their interactions with the environment and potential to spread, evolve and increase pathogenicity. This study aims at identifying potential environmental drivers of H5N2 LPAI outbreak occurrence in time, space and poultry populations. Thirty-four environmental variables were tested for association with the characteristics of 27 H5N2 LPAI outbreaks (i.e. time, place, flock type, number and species of birds affected) occurred among domestic poultry flocks in Italy in 2010–2012. This was done by applying a recently proposed analytical approach based on a combined non-metric multidimensional scaling, clustering and regression analysis. Results indicated that the pattern of (dis)similarities among the outbreaks entailed an underlying structure that may be the outcome of large-scale, environmental interactions in ecological dimension. Increased densities of poultry breeders, and increased land coverage by industrial, commercial and transport units were associated with increased heterogeneity in outbreak characteristics. In areas with high breeder densities and with many infrastructures, outbreaks affected mainly industrial turkey/layer flocks. Outbreaks affecting ornamental, commercial and rural multi-species flocks occurred mainly in lowly infrastructured areas of northern Italy. Outbreaks affecting rural layer flocks occurred mainly in areas with low breeder densities in south-central Italy. In savannah-like environments, outbreaks affected mainly commercial flocks of galliformes. Suggestive evidence that ecological ordination makes sense genetically was also provided, as virus strains showing high genetic similarity clustered into ecologically similar outbreaks. Findings were informed by hypotheses about how ecological interactions among poultry populations, viruses and their environments can be related to the observed patterns of H5N2 LPAI occurrence. This may prove useful in enhancing future interventions by developing site-specific, ecologically-grounded strategies.

Editor: Yi Guan, The University of Hong Kong, China

Funding: The authors have no support or funding to report.

Competing Interests: The authors have declared that no competing interests exist.

* E-mail: lapo.mughini.gras@rivm.nl

¤ Current address: National Institute of Public Health and the Environment (RIVM), Centre for Infectious Disease Control (Clb), Epidemiology and Surveillance Unit, Bilthoven, The Netherlands

Introduction

Low pathogenicity avian influenza (LPAI) viruses of H5 and H7 subtypes have the potential for mutation into highly pathogenic avian influenza (HPAI) virus strains [1], causing huge economic losses due to the high bird mortality rates and costs of control measures in domestic poultry [2], not to mention the potentially serious implications for human health [3]. This is the case of H5N2 LPAI viruses, for which there are several documented instances of increased virulence in domestic poultry [4–6], as well as human infection [7,8].

Italy has experienced several incursions of avian influenza (AI) viruses of H5N2 subtype in domestic poultry. In 1997–1998, eight H5N2 HPAI outbreaks were detected in multi-species backyard and small commercial poultry flocks in the north-eastern regions of Veneto and Friuli Venetia Giulia [6]. Seven years later in 2005, 15 outbreaks of H5N2 LPAI were detected in industrial meat turkey flocks, clustered within a radius of 14 km in the northern region of Lombardy [9], causing more than € 5 million losses in both direct and consequential costs [2]. At that time, prophylactic vaccination of long-living poultry species (mainly turkeys and layers) with a bivalent H5/H7 vaccine based on the DIVA (Differentiating Infected from Vaccinated Animals) strategy was in force, according to European Commission Decision 2004/666/EC [10]. The increased resistance to the field virus challenge of vaccinated birds and the reduction of virus shedding, combined with on-farm active surveillance and strict control measures, resulted in a rapid eradication of the epidemic [9]. In 2007, a H5N2 LPAI outbreak was detected in a free-range duck and goose breeder flock in the northern region of Emilia-Romagna, but phylogenetic analyses revealed a poor relationship to the previous one isolated in Lombardy [11]. On the contrary, the HA and NA genes resulted to be highly similar (>99.8%) to a H5N2 LPAI virus isolated from mallards in the same time period and geographical area, suggesting a possible introduction from the wild reservoir [11]. Further outbreaks caused by H5N2 LPAI virus in domestic poultry in Italy were detected in 2010–2012 all over the country, indicating recurrent viral introductions and/or increased viral circulation in various sectors of the Italian poultry industry.

The reoccurrence and spread of H5N2 AI viruses in geographically distant areas and different poultry production sectors, despite significant control efforts implemented, is likely to be driven by dynamic interactions between poultry and the environment where these viruses circulate, and these interactions may vary along ecological gradients in a more or less structured way. Several environmental variables have been found to be related to the observed spatio-temporal patterns and molecular evolution of AI viruses, particularly HPAI virus strains, in many parts of the world [12–15]. It is therefore increasingly apparent that a disease ecology framework, which posits that disease emergence is the result of multifactorial interactions with the environment as a whole [16], may well provide insights about whether and how the observed spatio-temporal pattern of H5N2 AI outbreaks in varying sectors of the Italian poultry industry is driven by certain ecological pressures. This means to identify environmental variables associated with outbreak heterogeneity, which is defined here in epidemiological terms as the dissimilarity in outbreak characteristics such as time and place of outbreak occurrence, type of affected flocks, number and species of birds therein. Focussing on domestic poultry in Italy, this study aims at identifying potential environmental drivers of H5N2 LPAI outbreak heterogeneity, with an account of molecular and phylogenetic characteristics of those virus strains for which genetic sequence data were available. Although inferring any relationship of causation was beyond the objectives of this study, results were expected to help us in disentangling the pattern of (dis)similarities among the different H5N2 LPAI outbreaks as the outcome of possible large-scale, environmental interactions in ecological dimension.

Materials and Methods

Outbreak Description

A total of 27 outbreaks of H5N2 LPAI virus, detected in domestic poultry in Italy between January 2010 and October 2012, formed the basis of this study (Table 1 and Figure 1). The first outbreak (10/1) was detected on 12 January 2010 during mandatory pre-movement testing in an multi-species ornamental poultry flock in the Veneto region. Further outbreaks were detected in backyard, ornamental, grower, dealer and industrial poultry flocks through official controls within the framework of the Italian national AI monitoring plan based on serological (hemagglutination inhibition [HI] assay in blood samples) and/or virological (real-time reverse transcriptase PCR in swab samples) testing according to European Commission Decisions 2006/437/EC and 2010/367/EC. In the seven outbreaks affecting industrial turkey flocks (11/6, 12/5, 12/6, 12/11, 12/12, 12/13 and 12/14), respiratory symptoms, reduced feed consumption and/or increased mortality were observed. All diagnoses of AI were confirmed by the National Reference Laboratory for Avian Influenza and Newcastle Disease in Legnaro (Padua), Italy.

In 2012, following the identification of outbreak 12/5 in the Lombardy region, the Italian Ministry of Health strengthen monitoring activities in the northern regions of Veneto, Lombardy, Piedmont and Emilia-Romagna on industrial and commercial poultry flocks. In addition, fairs, markets, shows or other kinds of gathering of poultry were prohibited. Control measures were implemented by veterinary authorities according to European Council Directive 2005/94/EC, including stamping out of affected premises, cleansing and disinfection, quarantine measures and movement restrictions. In two occasions (outbreaks 10/2 and 12/9), derogation for killing ornamental bird species kept for non-commercial purposes was granted. All control measures were lifted on 21 November 2012.

Epidemiological investigations allowed veterinary authorities to detect outbreak 12/6, which affected an industrial turkey flock in the Lombardy region. Epidemiological investigation also revealed a direct connection (the same poultry company holder, workers and veterinarians, as well as very close geographical proximity) between outbreaks 12/14 and 12/12, and between outbreaks 12/12 and 12/13.

Information on coordinate locations of the affected flocks, date of first sampling, flock type (backyard, ornamental, grower, dealer or industrial), number and species of birds therein, were available for all the outbreaks.

Environmental Variables

Table 2 shows the 34 environmental variables hypothesized to be related to the characteristics of the 27 H5N2 LPAI outbreaks under the disease ecology framework proposed by Carrel *et al.* [15]. Each of these variables was computed within a buffer radius of 10 km around each of the outbreaks. Data on the environmental variables were obtained from different sources (Table 2). Human population density (people/km^2) was computed based on the figures provided by the Italian National Institute of Statistics (ISTAT) at the level of municipal census block for the year 2010. Densities of domestic poultry farms (farms/km^2) and their respective populations (birds/km^2), divided into turkey breeders, fattening turkeys, broiler breeders, fattening broilers, layers, ducks and geese, and other poultry, were calculated based on the georeferenced (lat./long.) official poultry census data for the year 2012 obtained from the Italian National Poultry Registry. Percent land coverage was calculated for each of the 15 land cover classes defined by the second classification level of the Corine Land Cover (CLC) 2006 dataset, version 13, provided by the European Environment Agency at a resolution of 100 m. Median, minimum and maximum elevation (m above sea level) were computed using the the Shuttle Radar Topography Mission (SRTM) Digital Elevation Data provided by the Consortium for Spatial Information of the Consultative Group for International Agricultural Research (CGIAR–CSI) at a resolution of 90 m. Wild waterfowl density (birds/km^2) was obtained from the Italian Institute for Environmental Protection and Research (ISPRA) and associated with the related digitized wetlands at a scale 1:10000, obtained from satellite (Land Sat) imagery.

Ordination Analysis

Non-metric multidimensional scaling (NMDS) [17] was used as ordination technique to explore the underlying pattern of (dis)similarities among the 27 H5N2 LPAI outbreaks, and to further relate this pattern to the 34 environmental variables, as proposed by Carrel *et al.* [15] for scaling H5N1 AI virus isolates in Vietnam.

NMDS was used to find a configuration for the set of n points represented by the 27 H5N2 LPAI outbreaks in multidimensional space such that the inter-point distances corresponded as closely as possible to the observed (dis)similarities measured in p elements represented by the sampling date and geographical coordinates of the outbreaks, the type of affected flocks and the avian species therein (Table 1), according to a goodness-of-fit criterion called 'stress', which measures the degree of distortion of the ordination with respect to the original input data [17]. Conventionally, stress values $<5\%$ indicate a good fit for the data, whereas stress values $>20\%$ give unreliable results [15,17]. After data standardization, a

Table 1. H5N2 LPAI outbreaks in Italy between 2010 and 2012.

Outbreak ID	Region	Province	Flock type	Turkeys	Broilers	Layers	Guinea fowls	Pigeons	Quails	Pheasants	Ducks	Geese	Ornamental species	First sampling	HI[1]	PCR[2]
10/1	Veneto	Rovigo	Ornamental	69	534		26	622			262		142	12/01/2010	X	X
10/2	Emilia Romagna	Forlì-Cesena	Ornamental	20	188				22	18	388	21	2870	18/11/2010	X	X
10/3	Friuli Venetia Giulia	Udine	Backyard			45								02/12/2010	X	
11/1	Veneto	Verona	Backyard	8	15			15			83	38	3	14/02/2011	X	
11/2	Veneto	Venetia	Dealer		24	40	5	10			12			21/03/2011	X	X
11/3	Campania	Salerno	Grower		250									25/11/2011	X	
11/4	Campania	Avellino	Backyard			47								29/11/2011	X	
11/5	Emilia Romagna	Parma	Grower	21	271		2	45	28	18	46	9	294	30/11/2011	X	X
11/6	Latium	Viterbo	Industrial	5600										12/12/2011	X	
11/7	Calabria	Catanzaro	Backyard			30								21/12/2011	X	
11/8	Latium	Rome	Ornamental											21/12/2011	X	
11/9	Campania	Napoli	Backyard			104								28/12/2011		X
12/1	Emilia Romagna	Forlì-Cesena	Ornamental		110			13			777	36		26/01/2012		X
12/2	Campania	Napoli	Industrial			4800								03/05/2012	X	
12/3	Calabria	Cosenza	Industrial			1300								15/06/2012	X	
12/4	Campania	Napoli	Backyard			70								28/06/2012	X	
12/5	Lombardy	Brescia	Industrial	15000										31/08/2012		X
12/6	Lombardy	Brescia	Industrial	11700										03/09/2012		X
12/7	Tuscany	Prato	Backyard	1		100	21				12	34		04/09/2012		X
12/8	Lombardy	Mantua	Grower	2	5653	6			200		25			18/09/2012		X
12/9	Veneto	Treviso	Ornamental		186			100			240	128	96	24/09/2012	X	
12/10	Veneto	Verona	Grower		1950	3900	2750							28/09/2012		X
12/11	Lombardy	Mantua	Industrial	37962										05/10/2012	X	X
12/12	Lombardy	Mantua	Industrial	28268										05/10/2012	X	
12/13	Lombardy	Mantua	Industrial	14456										05/10/2012		X
12/14	Lombardy	Mantua	Industrial	16533										08/10/2012		X
12/15	Latium	Rome	Dealer			750								18/10/2012	X	

1. Diagnosis by hemagglutination inhibition assay in blood samples.
2. Diagnosis by real-time reverse transcriptase PCR in swab samples.

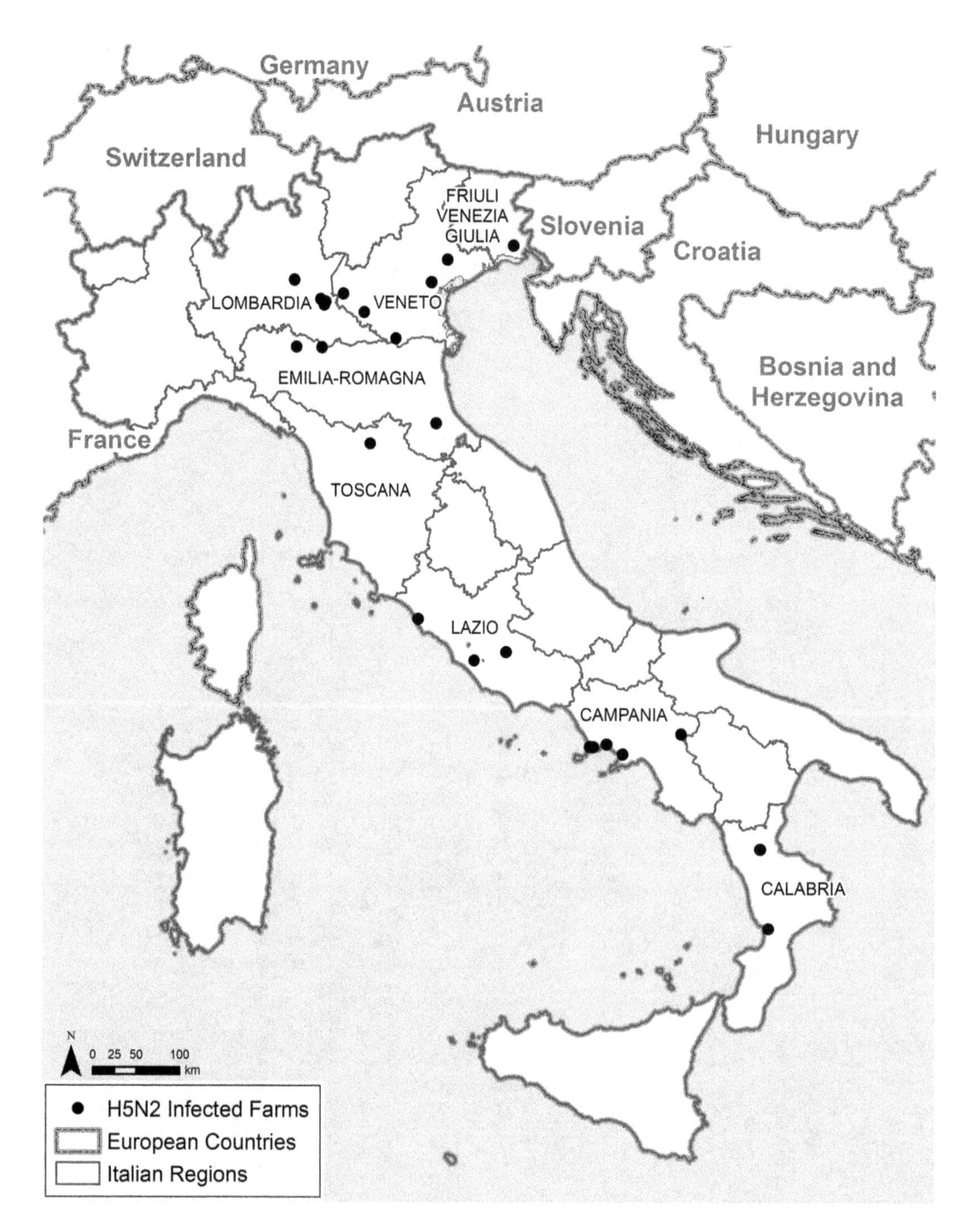

Figure 1. Map of the 27 Italian H5N2 LPAI outbreaks in domestic poultry, 2010–2012.

n by *p* input matrix of all pair-wise distances was calculated using the Gower's similarity measure for mixed data types [18]. The configuration of the points in the final ordination therefore summarized the differences in spatio-temporal and poultry population (flock-related) characteristics of the 27 H5N2 LPAI outbreaks. The optimum number of dimensions used in the NMDS was chosen as to minimize stress without compromising the utility of the ordination.

Once the 27 H5N2 LPAI outbreaks were ordinated according their spatio-temporal and flock-related characteristics, each of the 34 environmental variables was aligned in the ordination space where its correlation was maximal. The coefficient of determination (R^2 statistic) was used to identify the most important environmental variables associated with the ordination configuration.

A three-dimensional NMDS was found to be the optimal configuration for the data. Thus, the obtained three sets of NMDS dimension scores, which reflected the amount of difference in the spatio-temporal and flock-related characteristics of the outbreaks, were regressed on the environmental variables using segmented (piecewise) multiple linear regression analysis as to identify the independent environmental correlates of outbreak heterogeneity. Segmented linear regression allows for the independent variables to be partitioned into intervals in order to fit a separate regression

Table 2. Environmental variables hypothesized to be related to differentiation among the H5N2 LPAI outbreaks.

Variable ID	Environmental variable	Measure computed (within the buffer)	Resolution of raw data	Reference year	Source
A	Human population density	People/km^2	Municipal census blocks	2010	*
	Poultry population density				
B1	Turkey breeders	Birds/km^2	Cartesian coordinates	2012	**
B2	Fattening turkeys	Birds/km^2	Cartesian coordinates	2012	**
B3	Broiler breeders	Birds/km^2	Cartesian coordinates	2012	**
B4	Fattening broilers	Birds/km^2	Cartesian coordinates	2012	**
B5	Laying hens	Birds/km^2	Cartesian coordinates	2012	**
B6	Ducks and geese	Birds/km^2	Cartesian coordinates	2012	**
B7	Other poultry	Birds/km^2	Cartesian coordinates	2012	**
	Poultry farm density				
C1	Turkey breeders	Farms/km^2	Cartesian coordinates	2012	**
C2	Fattening turkeys	Farms/km^2	Cartesian coordinates	2012	**
C3	Broiler breeders	Farms/km^2	Cartesian coordinates	2012	**
C4	Fattening broilers	Farms/km^2	Cartesian coordinates	2012	**
C5	Laying hens	Farms/km^2	Cartesian coordinates	2012	**
C6	Ducks and geese	Farms/km^2	Cartesian coordinates	2012	**
C7	Other poultry	Farms/km^2	Cartesian coordinates	2012	**
	Land surface area				
D1	Urban fabric	Percent	100 m grid raster	2006	†
D2	Industrial, commercial and transport units	Percent	100 m grid raster	2006	†
D3	Mine, dump and construction sites	Percent	100 m grid raster	2006	†
D4	Artificial, non-agricultural vegetated areas	Percent	100 m grid raster	2006	†
D5	Arable land	Percent	100 m grid raster	2006	†
D6	Permanent crops	Percent	100 m grid raster	2006	†
D7	Pastures	Percent	100 m grid raster	2006	†
D8	Heterogeneous agricultural areas	Percent	100 m grid raster	2006	†
D9	Forests	Percent	100 m grid raster	2006	†
D10	Scrub and/or herbaceous vegetation associations	Percent	100 m grid raster	2006	†
D11	Open spaces with little or no vegetation	Percent	100 m grid raster	2006	†
D12	Inland wetlands	Percent	100 m grid raster	2006	†
D13	Maritime wetlands	Percent	100 m grid raster	2006	†
D14	Inland waters	Percent	100 m grid raster	2006	†
D15	Marine waters	Percent	100 m grid raster	2006	†
	Elevation				
E1	Median elevation	m above sea level	90 m grid raster	2008	‡
E2	Minimum elevation	m above sea level	90 m grid raster	2008	‡
E3	Maximum elevation	m above sea level	90 m grid raster	2008	‡
F	Wild waterfowl population density	Birds/km^2	1:10000 digitized shapefile	2010	§

*Italian National Institute of Statistics (ISTAT); **National Poultry Registry (BDN); †Corine Land Cover (CLC) 2006 dataset, version 13; ‡Shuttle Radar Topography Mission (SRTM), CGIAR-CSI, 2008, version 4; §Italian Institute for Environmental Protection and Research (ISPRA).

line segment to each interval. This allowed us to jointly examine the relationships of the environmental variables with outbreak heterogeneity in the three different NMDS dimensions using the same regression model. The segment breakpoints were therefore those between the first and second dimensions, and between the second and third dimensions. A positive association indicated that the environmental variable was associated with an increased outbreak heterogeneity, whereas a negative association indicated that the environmental variable was associated with a decreased outbreak heterogeneity (i.e. increased outbreak homogeneity). Phrased differently, where a positively associated variable is more represented, outbreaks would tend to occur more unpredictably in time, space and affected poultry populations, vice versa for a negatively associated variable.

Following Carrel *et al.* [15], to limit the number of variables that the analysis will focus on, only those variables with R^2 values

>0.10 were included in the initial model. Variables were then dropped if they exhibited high collinearity with other variables, as indicated by a Variance Inflate Factor (VIF) >5, with the choice of which among the collinear variables to retain in the model being made based upon improved R^2. A stepwise variable selection approach was then applied to fit a parsimonious multivariable model in which variables were significantly associated with the outcome ($p<0.05$) and the Akaike Information Criterion (AIC) was minimized. Also the effect of removing variables on the other variables included in the model was monitored.

Cluster Analysis

The ordinated 27 H5N2 LPAI outbreaks were assigned into clusters using Ward's minimum variance method [19]. The Ward's method is an iterative clustering procedure where all clusters are initially considered as singletons and then merged stepwise based on groups that lead to the minimum increase in the total within-cluster variance. The variance of the differences in attributes within a cluster is minimized using a distance algorithm based on the sum of squares of differences in the attributes, thereby combining clusters whose merge minimizes the increase in the within-group total sum of squares error [19].

Differences in environmental variables with R^2 values >0.10 were tested among the identified clusters using Kruskal-Wallis' test (KW) and post-hoc pair-wise comparisons based on Mann-Whitney U test (MW) with Bonferroni's adjustment of the p-value. Environmental variables significantly associated with the subdivision of outbreaks into clusters indicated how differentiation of environmental variables corresponded to differentiation in cluster assignment.

Statistical analysis was performed using STATA 11.2 (StataCorp LP, College Station, USA) and when applicable statistical significance was set to $p<0.05$.

Viral Sequence Data and Phylogenetic Analysis

The complete genome sequences of five H5N2 LPAI viruses representative of the outbreaks 10/1, 11/9, 12/5, 12/10 and 12/14 were generated according to methods described previously [20]. In addition, HA and NA genes were obtained from a virus collected from the outbreak 12/6, and the HA gene only was sequenced for viruses identified from the outbreaks 12/8. The nucleotide sequences obtained in this study are made available in the GISAID database under the following accession numbers: EPI464915–EPI464957.

Nucleotide sequence alignments were manually constructed for each gene segment and for the concatenated whole genome using the Se-Al program [21]. To infer the evolutionary relationships for each gene segment, we employed the maximum likelihood (ML) method available in the PhyML program, incorporating a GTR model of nucleotide substitution with gamma-distributed rate variation among sites (with four rate categories, $\Gamma4$) and a heuristic SPR branch-swapping search procedure [22]. A bootstrap resampling process (1000 replications) using the neighbour-joining (NJ) method and incorporating the ML substitution model defined above, was employed to assess the robustness of individual nodes of the phylogeny using PAUP* [23]. Parameter values for the GTR substitution matrix, base composition, gamma distribution of the rate variation among sites, and proportion of invariant sites (I) were estimated directly from the data using MODELTEST [24].

Results

Outbreak Ordination

At three dimensions, stress was minimized (3.8%), indicating that the final three-dimensional NMDS ordination represented well the scaled spatio-temporal and flock-related differences among the 27 H5N2 LPAI outbreaks. The first NMDS dimension explained 69% of the variance in the data, while the second and third dimensions explained 29% and 2% of the variance, respectively.

The first two NMDS dimensions are plotted in Figure 2 together with the 34 environmental variables hypothesized to be related to outbreak heterogeneity. Most of these variables had their axes of differentiation with the same alignment in the ordination space, mainly directed towards the higher scores of the first and second dimensions. Plotting the environmental variables onto the ordination space also indicated their differing strengths of association, with longer axes reflecting larger R^2 values.

Sixteen variables had R^2 values >0.10 (Figure 3), these were variables specifying the amounts of arable lands and permanent crops, broiler breeder farm density, broiler and turkey breeder population densities, duck and goose population density, maximum elevation, amounts of land occupied by mine, dump and construction sites, urban fabric, marine waters, scrub and/or herbaceous vegetation associations, density of poultry populations other than turkeys, layers, broilers, ducks and geese (TLBDG), duck and goose farm density, human density, industrial, commercial and transport units, and median elevation.

Of the 16 variables with R^2 values >0.10 considered for inclusion in the regression model, 10 were iteratively dropped, leading to a final model consisting of six significant predictors of outbreak heterogeneity (Table 3). Outbreak heterogeneity was positively associated with densities of broiler and turkey breeders, and with the amount of land covered by industrial, commercial and transport units. On the other hand, outbreak heterogeneity was negatively associated with the amount of arable land, and with densities of duck and goose farms, and of poultry other than TLBDG.

Cluster Assignment

The ordinated outbreaks were assigned into six clusters based upon the Ward's method (Figure 2). Cluster 1 included mainly early outbreaks affecting ornamental, commercial and rural multi-species flocks in northern Italy; cluster 2 included mainly later outbreaks affecting rural layer flocks in south-central Italy (with the only exception of outbreak 10/3 in the north-eastern part of the country); cluster 3 included predominantly late outbreaks affecting commercial flocks with galliformes birds located throughout the country; cluster 4 included all the outbreaks affecting industrial flocks with longer-living poultry species (turkeys and layers) located throughout the country; clusters 5 and 6 consisted of only one outbreak apiece (outbreaks 11/5 and 12/9), which therefore appeared to possess the greatest ecological distances from the other outbreaks in ordination space. Among the considered outbreak characteristics, flock type was the one that determined cluster assignment more incisively, i.e. for which the within-cluster variance was minimized the most, followed by place, time and affected birds.

Examining the variation in the 16 environmental variables with $R^2>0.10$ over the four clusters containing >1 outbreak revealed that outbreaks in each of the clusters had different associations with the environmental variables. In Figure 4, box plots of the variables significantly associated with clustering are displayed, these were: 1) amount of land occupied by industrial, commercial

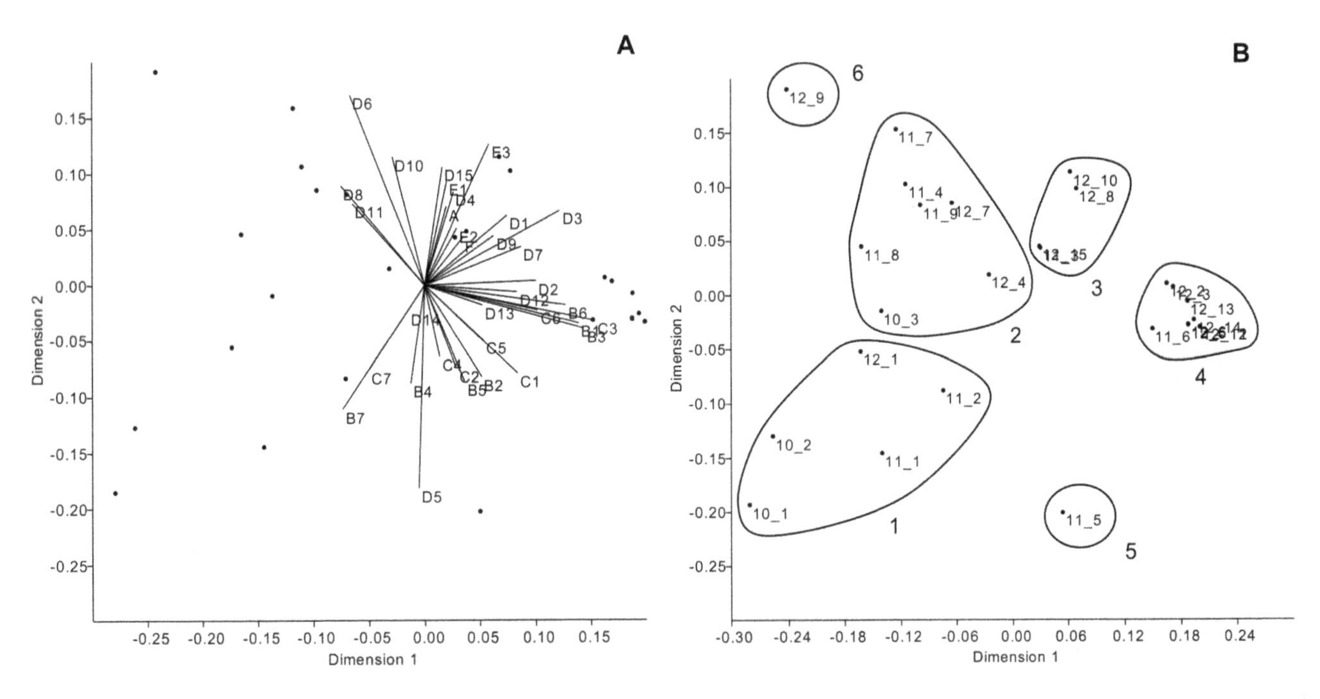

Figure 2. Plots of the Italian 2010–2012 H5N2 LPAI outbreaks in the first two NMDS dimensions. Plot A shows each environmental variable's axis of differentiation, aligned in the ordination space, over the scaled spatio-temporal and flock type-related characteristics of the outbreaks. Plot B shows the final configuration of the outbreaks in the first two NMDS dimensions charted according to cluster assignment. See Tables 1 and 2 for details about the outbreaks and the environmental variables.

and transport units (KW, $p = 0.017$), significantly higher in cluster 4 than cluster 1 (MW, adjusted $p = 0.019$); 2) broiler breeder farm density (KW, $p = 0.027$), significantly higher in cluster 4 compared to cluster 2 (MW, adjusted $p = 0.030$); 3) broiler breeder population density (KW, $p = 0.027$), significantly higher in cluster 4 compared to cluster 2 (MW, adjusted $p = 0.030$); 4) turkey breeder population density (KW, $p = 0.018$), significantly higher in cluster 4 compared to cluster 2 (MW, adjusted $p = 0.048$); and 5) amount of land covered by scrub and/or herbaceous vegetation associations (KW, $p = 0.022$), significantly higher in cluster 3 compared to cluster 4 (MW, adjusted $p = 0.012$).

Molecular Characterization and Phylogenetic Analysis

To explore the whole-genome evolution of viruses obtained from the Italian H5N2 LPAI outbreaks, ML trees were inferred using virus sequences available from seven different outbreaks. The topology of the HA phylogeny indicated that the virus strains from outbreak 12/10 identified in Veneto in 2012 and those collected in the same year in Lombardy (12/5, 12/6, 12/8 and 12/14) clustered in the same branch of the phylogenetic tree showing a high percentage of similarity (99.7%–99.9%) with each other and with the strain isolated from outbreak 11/9 (99.0%–99.1% similarity), which occurred in Campania in 2011. An identity of 97.6%–98% was found with the strain isolated from outbreak 10/1 which occurred in Veneto in 2010. This strain had a higher similarity with the strain A/spur-winged goose/Nigeria/2/2008 (H5N2) (99%) (Figure 5). All the viruses analyzed had typical low pathogenicity motifs at the HA cleavage site. In this portion of the HA gene, all 2012 virus strains presented a substitution of a non-basic amino acid (PQRETR*G) with a basic amino acid (PQRKTR*G). Phylogenies inferred for the other seven genome segments showed the same pattern as that of the HA gene for the viruses collected from the 2012 outbreaks (12/5, 12/10 and 12/14): the three 2012 viruses clustered together showing high similarity (99.6%–100%) with one another and with the strain isolated from outbreak 11/9 (98.7%–99.4%). Differently, the strain isolated from outbreak 10/1 resulted to be genetically distant (92.3%–97.8%) from the 2011–2012 strains for the NA gene (Figure 6) and for all the internal genes, with the only exception of the PB1 gene.

Amino acid analysis of the 2011–2012 strains demonstrated that all these viruses were characterized by an additional glycosylation site (NDA to NDT) in the position 253 of the HA gene, a stalk deletion in the NA gene (position 65–83) and by an elongation of the NS gene due to a mutation on the first base of the codon that usually encodes a stop signal in position 231. This mutation is a LVT (lenght variation type) reported in the literature as LVT (+7), which so far has been found only in human and swine viruses, with the exceptions of avian H7N3, H4N8, and H13N2 virus strains [20].

Discussion

This study aimed at identifying potential environmental correlates of H5N2 LPAI outbreak heterogeneity as to determine environmental conditions where control efforts may be targeted to specific poultry populations. In parallel, phylogenetic characteristics of some of the virus strains were also investigated.

Despite general awareness that the observed patterns of AI viruses may entail some underlying ecological structure, there is a notable paucity of studies that have investigated the environmental drivers of H5N2 AI virus emergence in domestic poultry, e.g. [25,26]. It is clear that ecological variation and disease occurrence are often mediated by complex, large-scale processes that are not immediately amenable to traditional approaches to causal inference [27]. Yet, a comprehensive approach assembling methodologies from diverse disciplines, including ecology, epidemiology, genetics and biogeography has been proposed to explore the spatio-temporal distribution of AI viruses through the

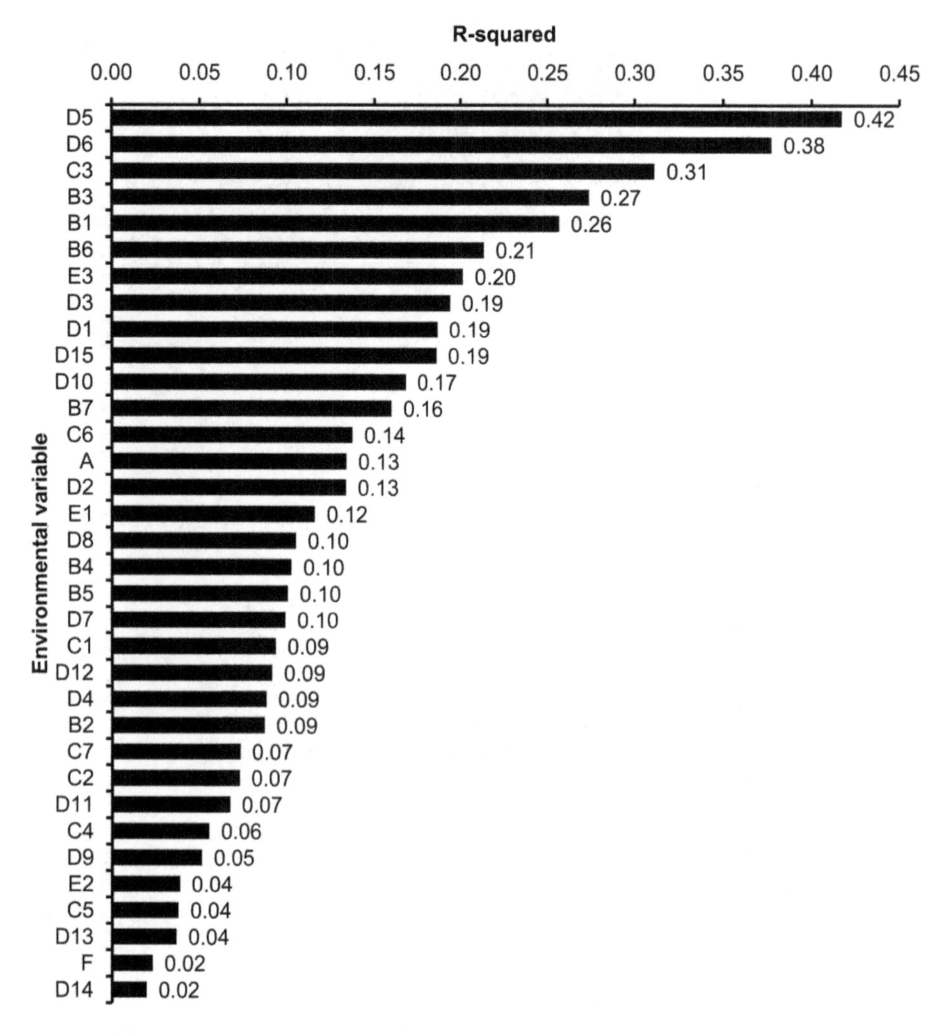

Figure 3. Goodness-of-fit scores (R^2) for the 34 environmental variables aligned in the ordination space. See Table 2 for details about the environmental variables.

combined application of statistical techniques from multivariate data analysis (ordination and clustering methods) and classical regression modelling [15]. These techniques are well suited to frame the disease ecology perspective using, for instance, data obtained from ongoing surveillance activities, allowing the findings to be informed by hypotheses about how ecological interactions among poultry populations, viruses and their environments can be related to the observed pattern of disease emergence [15].

Several environmental variables were dropped to build the final regression model because of their "weak" association with the outbreak ordination configuration. This allowed us to discern which of the considered variables could be deemed as robust and significant predictors of outbreak heterogeneity. The hypotheses

Table 3. Results from the final piecewise multiple linear regression model.

Environmental variable (associated dimension)	Coefficient (95% confidence interval)	p value
Turkey breeder population density (1st)	0.137 (0.058 to 0.216)	0.001
Broiler breeder population density (1st)	0.065 (0.028 to 0.102)	0.001
Industrial, commercial and transport units (1st)	0.041 (0.001 to 0.080)	0.042
Arable land (2nd)	−0.061 (−0.094 to −0.028)	<0.0001
Other poultry population density (2nd)	−0.037 (−0.070 to −0.003)	0.032
Duck and goose farm density (1st)	−0.113 (−0.195 to 0.031)	0.007

Coefficients, 95% confidence intervals and p values for the associations between the selected environmental variables with the three sets of NMDS dimension scores for the 27 H5N2 LPAI outbreaks.

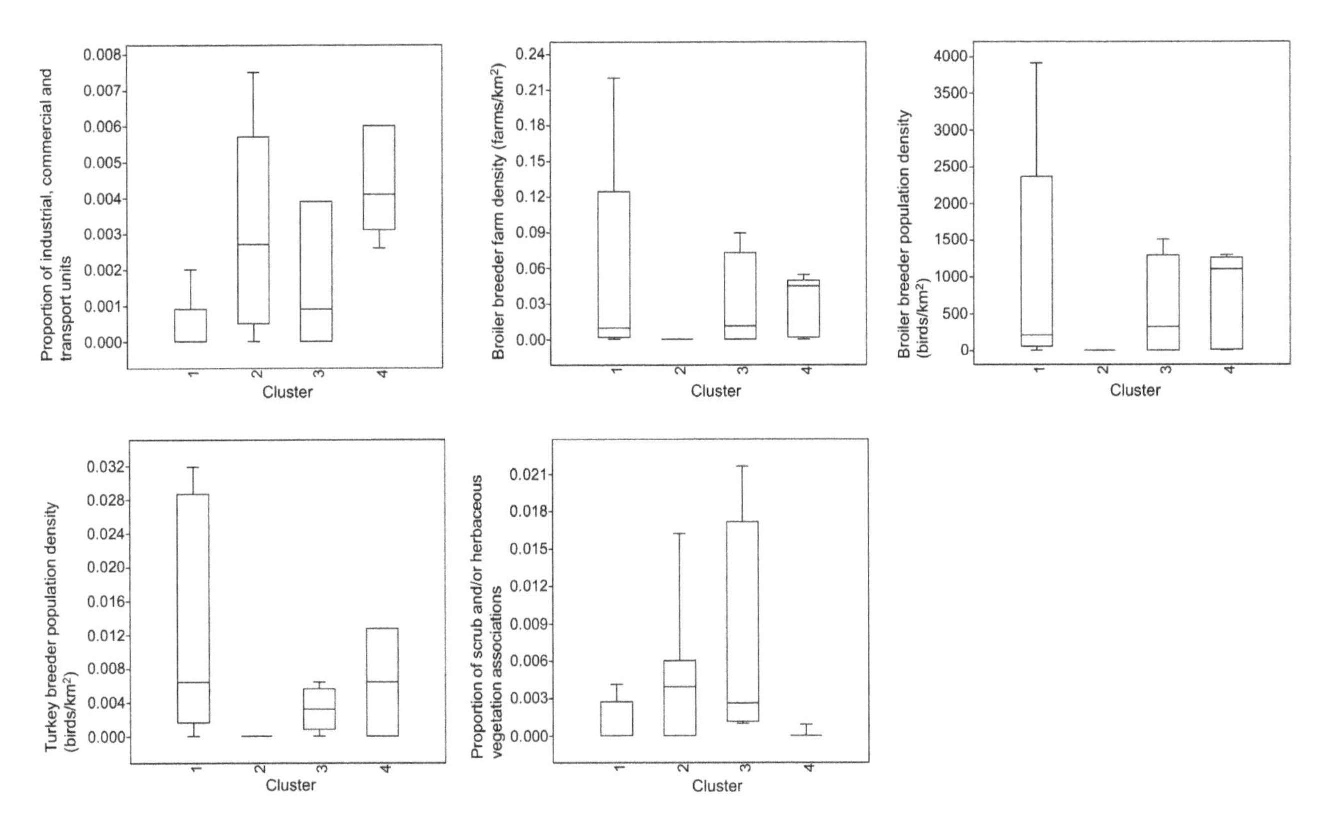

Figure 4. Box-plots of the five environmental variables significantly associated with the subdivision of outbreaks into clusters.

behind the possible association of the considered environmental variables with outbreak heterogeneity varied according to the variable in question. For instance, it was hypothesized that larger human population densities around the infected premises would increase the odds of interactions between poultry and people, as well as the chance of viruses being transferred elsewhere, so that outbreak occurrence would be more heterogeneous (and so less predictable) over time, space and susceptible poultry populations. This also applies to those land cover classes specifying urbanized or infrastructured environments, as these would promote increased movements of human beings and commodities, including poultry and poultry products, which could favor long-distance viral spread to varying poultry populations. It was therefore not surprising to find that increased amount of land covered by industrial, commercial and transport units was significantly associated with increased outbreak heterogeneity. Conversely, the amount of arable land was negatively associated with outbreak differentiation. Indeed increased rural or agricultural areas may plausibly reduce outbreak heterogeneity, as both the low human density and the less developed infrastructural system usually present in the countryside would create an unfavorable condition for extensive viral circulation outside a given timeframe, geographical area or poultry production sector. Consequently, in such circumstances outbreaks would tend to be more (ecologically) similar to each other, involving only specific poultry populations and being temporally and spatially closer to one another.

Similarly to the hypothesized effect of human density, the density of susceptible hosts would influence outbreak differentiation via increased chance of viral circulation outside a given ecological context. Interestingly, outbreak heterogeneity was positively associated with the density of turkey and broiler breeder populations, but as the density of duck and goose farms and the density of poultry populations other than TLBDG increased, then outbreak heterogeneity decreased. While there is no apparent explanation for these two latter negative associations, it is true that poultry breeders are usually characterized by a longer productive lifespan than their fattening counterparts. Therefore, their likely exposure to, and amplification of, AI viruses might be more prolonged, possibly resulting in more extensive viral circulation and so increased unpredictability of outbreak occurrence. This speculation is supported, to a certain extent, by similar findings reported from Japan, where a high density of layer flocks containing "end-of-lay" hens (which usually have a relatively long productive lifespan of 1–2 years) was associated with the occurrence of 41 H5N2 LPAI outbreaks in 2005 [26,25]. Yet, while the Japanese outbreaks also affected layers, breeder flocks were not affected in any of the Italian outbreaks. Thus it seems that outbreak heterogeneity is somehow only related to the poultry circuits at the top of the production chain and not to their longer production cycles per se.

Besides high poultry and human density areas, other important areas where increased viral circulation may take place are wetlands and water surfaces because of their usually high densities of (migratory) waterfowls, as well as the increased chance for surface waters in such circumstances to act as a vehicle for fecal-oral transmission of AI viruses [28]. Also the elevation can influence the risk for a poultry farm to become infected with AI viruses, with farms located at more than 150 m above sea level in Italy being at significantly lower risk than those located at lower altitudes [29], possibly because of increased geographical isolation and the presence of unrecognized local environmental conditions that make such areas less prone to experiencing AI outbreaks. Despite this hypothesized effect, our variables specifying surface water sites, altitude and waterfowl population around the infected flocks were not significant in the final regression model, although

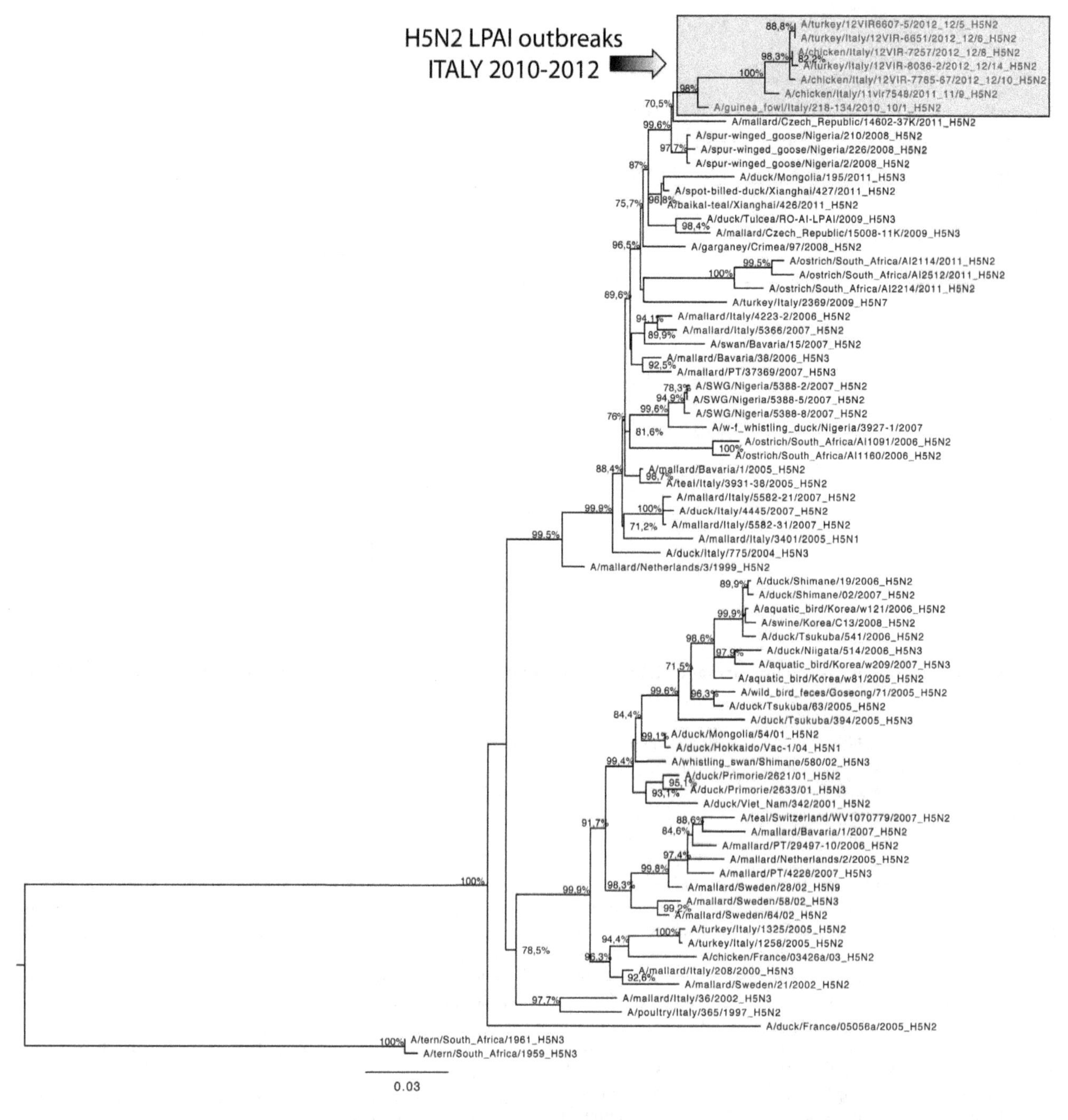

Figure 5. Maximum-likelihood phylogenetic tree for the hemagglutinin gene segment of Italian 2010–2012 H5N2 LPAI viruses. Viruses sequenced and characterized in this study are in red. Numbers at the nodes represent bootstrap values.

some of them exhibited a certain degree of correlation with the outbreak ordination configuration.

Flock type was the main determining factor of cluster assignment, suggesting that commonalities of flocks in the same poultry production circuits may overcome geography, time and type/amount of susceptible hosts in modulating the epidemiological similarities of the outbreaks. Studying associations between the environmental variables and the subdivision of outbreaks into clusters indicated that in areas where industrial, commercial and transport units were widespread, outbreaks were significantly more likely to affect industrial flocks of turkeys and layers nationwide. This also applied to areas with high densities of broiler and turkey breeders. In contrast, lowly infrastructured areas were significantly more likely to experience outbreaks in ornamental, commercial and rural multi-species flocks in the northern part of the country, while rural layer flocks in south-central Italy were significantly more prone to infection in areas with low densities of broiler and turkey breeders. Finally, areas with increased amounts of land covered by scrub and/or herbaceous vegetation (savannah-like environments) were significantly more likely to experience outbreaks in commercial flocks of galliformes, and less likely in industrial flocks, all over the country. Although it is not entirely clear as to how these environmental variables may be associated with the occurrence of H5N2 LPAI outbreaks in these specific

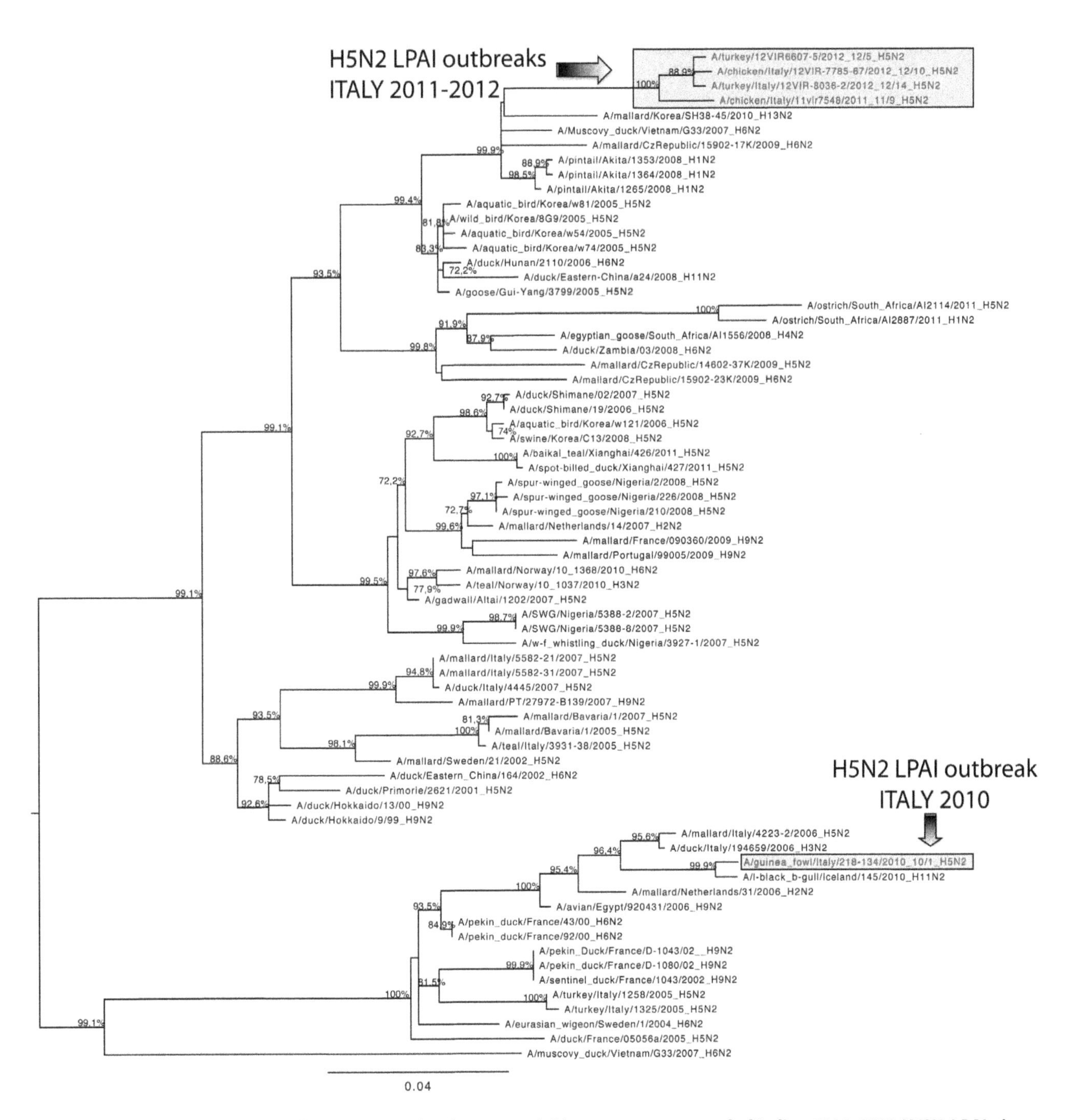

Figure 6. Maximum-likelihood phylogenetic tree for the neuraminidase gene segment of of Italian 2010–2012 H5N2 LPAI viruses. Viruses sequenced and characterized in this study are in red. Numbers at the nodes represent bootstrap values.

ecological contexts, they may prove useful in categorizing several site-specific conditions where AI control efforts may be targeted, enhancing any intervention by matching a specific strategy rather than applying across-the-board methods. This seems to be applicable as in Italy most of the poultry industry has been concentrated in specific areas, where the high density of varying poultry holdings, hatcheries, abattoirs, feed mills, litter processing plants and other establishments is convenient from an organizational point of view, but it may pose a series of drawbacks in terms of increased risk of AI viruses spread [30]. Moreover, certain locations in Italy possess a more pronounced vocation to specific poultry productions than others. For instance, the Veneto region (particularly the province of Verona) is characterized by a high density of turkeys, while laying hens for table eggs are far more concentrated in Lombardy region. These are characteristics which may be framed in the specific environmental background to identify potential targets for control activities.

Despite the low number of available sequences, the results of the phylogenetic analyses substantiate, to some extent, those of the clustering based on the ecological characteristics of the outbreaks. For instance, the virus strains from outbreaks 12/10, 12/5, 12/6, 12/8 and 12/14 showed very high genetic similarity with one another and common molecular signatures were also identified in their HA, NS and NA genes. This was also reflected in the ecological clustering, as all these outbreaks clustered in two clusters (3 and 4) close to each other in ordination space (Figure 2). While this is suggestive of similar ecological niches for these virus strains, the limited number of sequences made it difficult to determine whether or not common environmental drivers were shaping the genetic commonalities identified among the viruses. Moreover, other likely determinants, such as the epidemiological connections among the outbreaks, may have played a major role as well. A larger array of sequences is needed to compare the evolutionary patterns (including rates of nucleotide substitution and selection pressures) of the viruses involved in different outbreaks in order to link the virus evolutionary properties to ecological factors.

Besides the lack of detailed genetic data for all the outbreaks as to examine to which extent genetic differentiation corresponds to ecological structuring, this study has some evident limitations due to the relatively low number of outbreaks which the analysis is based upon and the risk of falling into the trap of "ecological fallacy" in pursuing such a correlative analytical approach. Nevertheless, this study was based on a comprehensive set of fine-scale spatial data that allowed us to identify several environmental correlates of H5N2 LPAI outbreak heterogeneity. To the authors' best knowledge, this is the first time that such analysis is applied to this kind of data.

As a final point, it is worth mentioning that the buffer radius of 10 km was chosen because it appears to be the average threshold for geographical spread of AI viruses between domestic poultry farms, i.e. where most of the viral transmission takes place as revealed by the clustering of infected farms [9] and the risk of transmission from infected to uninfected farms as a function of the inter-farm distance [31]. However, sensitivity analysis of the effect of the choice of the buffer radius on the results was also performed (data not shown). This was done by repeating the analysis with buffers at 1, 5, 15 and 20 km. As the results did not change significantly from one buffer to another, no further results were presented.

Conclusions

Although it was not possible to infer any direct relationship of causation, results indicated that the pattern of (dis)similarities among the Italian H5N2 LPAI outbreaks entailed an underlying structure that may be the outcome of possible large-scale, environmental interactions in ecological dimension. Among the numerous environmental variables tested we identified increased population densities of broiler and turkey breeders, and increased amount of land devoted to industrial, commercial and transport units as possible ecological determinants of outbreak heterogeneity in time, space and susceptible poultry populations. Several environmental conditions associated with increased proneness of specific poultry populations to experiencing H5N2 LPAI outbreaks were also identified. These were conditions related to infrastructured and savannah-like environments and areas densely populated by poultry breeders. Although analytic integration of ecology and genetics was not possible due to the limited number of available sequences, suggestive evidence that ecological ordination makes sense genetically was provided, as virus strains showing high genetic similarity and common molecular signatures clustered into ecologically similar outbreaks. These results are in good agreement with current (eco)epidemiological knowledge on LPAI infection in Italy as derived from more traditional analytical approaches [9,11,32–34].

Considering outbreak heterogeneity rather than treating all LPAI outbreaks equally is important given the potential of these viruses to spread, evolve and increase pathogenicity. A better understanding of how ecological interactions among poultry populations, viruses and their environments can be related to the observed patterns of AI outbreaks may eventually enhance future interventions by implementing site-specific, ecologically-grounded strategies.

Author Contributions

Conceived and designed the experiments: LMG LB SM. Performed the experiments: LMG PM IM. Analyzed the data: LMG PM IM. Contributed reagents/materials/analysis tools: LMG VG PC IM. Wrote the paper: LMG LB PM IM VG PC SM.

References

1. Alexander DJ (2003) Should we change the definition of avian influenza for eradication purposes? Avian Dis 47: 976–981.
2. Sartore S, Bonfanti L, Lorenzetto M, Cecchinato M, Marangon S (2010) The effects of control measures on the economic burden associated with epidemics of avian influenza in Italy. Poult Sci 89: 1115–1121.
3. Horimoto T, Kawaoka Y (2001) Pandemic threat posed by avian influenza A viruses. Clin Microbiol Rev 14: 129–149.
4. Kawaoka Y, Naeve CW, Webster RG (1984) Is virulence of H5N2 influenza viruses in chickens associated with loss of carbohydrate from the hemagglutinin? Virology 139: 303–316.
5. Horimoto T, Rivera E, Pearson J, Senne D, Krauss S, et al. (1995) Origin and molecular changes associated with emergence of a highly pathogenic H5N2 influenza virus in Mexico. Virology 213: 223–230.
6. Capua I, Marangon S, Selli L, Alexander DJ, Swayne DE, et al. (1999) Outbreaks of highly pathogenic avian influenza (H5N2) in Italy during October 1997 to January 1998. Avian Pathol 28: 455–460.
7. Ogata T, Yamazaki Y, Okabe N, Nakamura Y, Tashiro M, et al. (2008) Human H5N2 avian influenza infection in Japan and the factors associated with high H5N2-neutralizing antibody titer. J Epidemiol 18: 160–166.
8. Yamazaki Y, Doy M, Okabe N, Yasui Y, Nakashima K, et al. (2009) Serological survey of avian H5N2-subtype influenza virus infections in human populations. Arch Virol 154: 421–427.
9. Mulatti P, Bos MEH, Busani L, Nielen M, Marangon S (2010) Evaluation of interventions and vaccination strategies for low pathogenicity avian influenza: spatial and space-time analyses and quantification of the spread of infection. Epidemiol Infect 138: 813–824.
10. Capua I, Marangon S (2007) Control and prevention of avian influenza in an evolving scenario. Vaccine 25: 5645–5652.
11. Cecchinato M, Ceolin C, Busani L, Dalla Pozza M, Terregino C, et al. (2010) Low pathogenicity avian influenza in Italy during 2007 and 2008: epidemiology and control. Avian Dis 54: 323–328.
12. Martin V, Pfeiffer DU, Zhou X, Xiao X, Prosser DJ, et al. (2011) Spatial distribution and risk factors of highly pathogenic avian influenza (HPAI) H5N1 in China. PLoS Pathog 7: e1001308.
13. Gilbert M, Pfeiffer DU (2012) Risk factor modelling of the spatio-temporal patterns of highly pathogenic avian influenza (HPAIV) H5N1: a review. Spat Spatiotemporal Epidemiol 3: 173–183.
14. Pfeiffer DU, Minh PQ, Martin V, Epprecht M, Otte MJ (2007) An analysis of the spatial and temporal patterns of highly pathogenic avian influenza occurrence in Vietnam using national surveillance data. Vet J 174: 302–309.
15. Carrel MA, Emch M, Nguyen T, Todd Jobe R, Wan X-F (2012) Population-environment drivers of H5N1 avian influenza molecular change in Vietnam. Health Place 18: 1122–1131.
16. Mayer JD (2000) Geography, ecology and emerging infectious diseases. Soc Sci Med 50: 937–952.
17. Cox TF, Cox MAA (2000) Multidimensional Scaling. London: Chapman and Hall/CRC. 328 p.
18. Gower JC (1971) A general coefficient of similarity and some of its properties. Biometrics 27: 857.
19. Ward JH (1963) Hierarchical grouping to optimize an objective function. J Am Stat Ass 58: 236.
20. Cattoli G, Monne I, Fusaro A, Joannis TM, Lombin LH, et al. (2009) Highly pathogenic avian influenza virus subtype H5N1 in Africa: a comprehensive phylogenetic analysis and molecular characterization of isolates. PLoS ONE 4: e4842.

21. Rambaut A (2007) Sequence alignment editor, version 2.0. Available from: http://tree.bio.ed.ac.uk/software/seal/.
22. Guindon S, Gascuel O (2003) A simple, fast, and accurate algorithm to estimate large phylogenies by maximum likelihood. Syst Biol 52: 696–704.
23. Wilgenbusch JC, Swofford D (2003) Inferring evolutionary trees with PAUP*. Curr Protoc Bioinformatics 6: 6.4.
24. Posada D, Crandall KA (1998) MODELTEST: testing the model of DNA substitution. Bioinformatics 14: 817–818.
25. Nishiguchi A, Kobayashi S, Yamamoto T, Ouchi Y, Sugizaki T, et al (2007) Risk factors for the introduction of avian influenza virus into commercial layer chicken farms during the outbreaks caused by a low-pathogenic H5N2 virus in Japan in 2005. Zoonoses Public Health 54: 337–343.
26. Nishiguchi A, Kobayashi S, Ouchi Y, Yamamoto T, Hayama Y, et al. (2009) Spatial analysis of low pathogenic H5N2 avian influenza outbreaks in Japan in 2005. J Vet Med Sci 71: 979–982.
27. Plowright RK, Sokolow SH, Gorman ME, Daszak P, Foley JE (2008) Causal inference in disease ecology: investigating ecological drivers of disease emergence. Front Ecol Environ 6: 420–429.
28. Brown JD, Swayne DE, Cooper RJ, Burns RE, Stallknecht DE (2007) Persistence of H5 and H7 avian influenza viruses in water. Avian Dis 51: 285–289.
29. Busani L, Valsecchi MG, Rossi E, Toson M, Ferrè N, et al. (2009) Risk factors for highly pathogenic H7N1 avian influenza virus infection in poultry during the 1999–2000 epidemic in Italy. Vet J 181: 171–177.
30. Capua I, Marangon S (2000) The avian influenza epidemic in Italy, 1999–2000: a review. Avian Pathol 29: 289–294.
31. Boender GJ, Meester R, Gies E, De Jong MCM (2007) The local threshold for geographical spread of infectious diseases between farms. Prev Vet Med 82: 90–101.
32. Marangon S, Bortolotti L, Capua I, Bettio M, Dalla Pozza M (2003) Low-pathogenicity avian influenza (LPAI) in Italy (2000–01): epidemiology and control. Avian Dis 47: 1006–1009.
33. Cecchinato M, Comin A, Bonfanti L, Terregino C, Monne I, et al. (2011) Epidemiology and control of low pathogenicity avian influenza infections in rural poultry in Italy. Avian Dis 55: 13–20.
34. Busani L, Toson M, Stegeman A, Pozza MD, Comin A, et al. (2009) Vaccination reduced the incidence of outbreaks of low pathogenicity avian influenza in northern Italy. Vaccine 27: 3655–3661.

Permissions

All chapters in this book were first published in PLOS ONE, by The Public Library of Science; hereby published with permission under the Creative Commons Attribution License or equivalent. Every chapter published in this book has been scrutinized by our experts. Their significance has been extensively debated. The topics covered herein carry significant findings which will fuel the growth of the discipline. They may even be implemented as practical applications or may be referred to as a beginning point for another development.

The contributors of this book come from diverse backgrounds, making this book a truly international effort. This book will bring forth new frontiers with its revolutionizing research information and detailed analysis of the nascent developments around the world.

We would like to thank all the contributing authors for lending their expertise to make the book truly unique. They have played a crucial role in the development of this book. Without their invaluable contributions this book wouldn't have been possible. They have made vital efforts to compile up to date information on the varied aspects of this subject to make this book a valuable addition to the collection of many professionals and students.

This book was conceptualized with the vision of imparting up-to-date information and advanced data in this field. To ensure the same, a matchless editorial board was set up. Every individual on the board went through rigorous rounds of assessment to prove their worth. After which they invested a large part of their time researching and compiling the most relevant data for our readers.

The editorial board has been involved in producing this book since its inception. They have spent rigorous hours researching and exploring the diverse topics which have resulted in the successful publishing of this book. They have passed on their knowledge of decades through this book. To expedite this challenging task, the publisher supported the team at every step. A small team of assistant editors was also appointed to further simplify the editing procedure and attain best results for the readers.

Apart from the editorial board, the designing team has also invested a significant amount of their time in understanding the subject and creating the most relevant covers. They scrutinized every image to scout for the most suitable representation of the subject and create an appropriate cover for the book.

The publishing team has been an ardent support to the editorial, designing and production team. Their endless efforts to recruit the best for this project, has resulted in the accomplishment of this book. They are a veteran in the field of academics and their pool of knowledge is as vast as their experience in printing. Their expertise and guidance has proved useful at every step. Their uncompromising quality standards have made this book an exceptional effort. Their encouragement from time to time has been an inspiration for everyone.

The publisher and the editorial board hope that this book will prove to be a valuable piece of knowledge for researchers, students, practitioners and scholars across the globe.

List of Contributors

Aurelio Briones
Department of Plant, Soil and Entomological Sciences, University of Idaho, Moscow, Idaho, United States of America

Erik Coats and Cynthia Brinkman
Department of Civil Engineering, University of Idaho, Moscow, Idaho, United States of America

Dylan P. Smith and Kabir G. Peay
Department of Biology, Stanford University, Stanford, California, United States of America

Xavier Raynaud
Sorbonne Universités, UPMC Univ Paris 06, Institute of Ecology and Environmental Sciences - Paris, Paris, France

Naoise Nunan
CNRS, Institute of Ecology and Environmental Sciences - Paris, Campus AgroParisTech, Thiverval-Grignon, France

Julianne L. Baron
Department of Infectious Diseases and Microbiology, University of Pittsburgh, Graduate School of Public Health, Pittsburgh, Pennsylvania, United States of America
Special Pathogens Laboratory, Pittsburgh, Pennsylvania, United States of America

Scott Duda
Special Pathogens Laboratory, Pittsburgh, Pennsylvania, United States of America

Janet E. Stout
Special Pathogens Laboratory, Pittsburgh, Pennsylvania, United States of America
Department of Civil and Environmental Engineering, University of Pittsburgh, Swanson School of Engineering, Pittsburgh, Pennsylvania, United States of America

Amit Vikram
Department of Civil and Environmental Engineering, University of Pittsburgh, Swanson School of Engineering, Pittsburgh, Pennsylvania, United States of America

Kyle Bibby
Department of Civil and Environmental Engineering, University of Pittsburgh, Swanson School of Engineering, Pittsburgh, Pennsylvania, United States of America
Department of Computational and Systems Biology, University of Pittsburgh Medical School, Pittsburgh, Pennsylvania, United States of America

Jay Siddharth and Scott J. Parkinson
Host Commensal Hub, Developmental and Molecular Pathways, Novartis Institutes for Biomedical Research, Basel, Switzerland

Nicholas Holway
Scientific Computing, NIBR IT, Novartis Institutes Biomedical Research, Basel, Switzerland

Ricardo Luis Louro Berbara
Soil Science Department, Agronomy Institute, Federal Rural University of Rio de Janeiro, Seropédica-RJ, Brazil

Francy Junio Gonçalves Lisboa
Soil Science Department, Agronomy Institute, Federal Rural University of Rio de Janeiro, Seropédica-RJ, Brazil

Pedro R. Peres-Neto
Canada Research Chair in Spatial Modelling and Biodiversity; Universitédu Québec á Montréal, Département des sciences biologiques, Queébec, Canada

Guilherme Montandon Chaer and Ederson da Conceição Jesus
Embrapa Agrobiologia, Seropédica-RJ, Brazil

Ruth Joy Mitchell and Stephen James Chapman
The James Hutton Institute, Craigiebuckler, Aberdeen, United Kingdom

Utpal Bose, Amitha K. Hewavitharana and P. Nicholas Shaw
School of Pharmacy, The University of Queensland, Brisbane, Queensland, Australia

Miranda E. Vidgen, Yi Kai Ng and John A. Fuerst
School of Chemistry and Molecular Biosciences, The University of Queensland, Brisbane, Queensland, Australia

Mark P. Hodson
Metabolomics Australia, Australian Institute for Bioengineering and Nanotechnology, The University of Queensland, Brisbane, Queensland, Australia

Roberto Murgas Torrazza and Josef Neu
Department of Pediatrics, College of Medicine University of Florida, Gainesville, Florida, United States of America

Maria Ukhanova, Xiaoyu Wang and Volker Mai
Department of Epidemiology, College of Public Health and Health Professions and College of Medicine and Emerging Pathogens Institute, University of Florida, Gainesville, Florida, United States of America

Renu Sharma and Mark Lawrence Hudak
Department of Pediatrics University of Florida College of Medicine, Jacksonville, Florida, United States of America

Yifei Zhang, Jianwei Hu, Ning Du and Feng Chen
Central Laboratory, School of Stomatology, Peking University, Beijing, P. R. China

Yunfei Zheng
Department of Periodontology, School of Stomatology, Peking University, Beijing, P. R. China

Fábio P. Dornas, Lorena C. F. Silva, Gabriel M. de Almeida, Rafael K. Campos, Paulo V. M. Boratto, Ana P. M. Franco-Luiz, Paulo C. P. Ferreira, Erna G. Kroon and Jônatas S. Abrahão
Universidade Federal de Minas Gerais, Instituto de Ciências Biológicas, Laboratório de Viírus, Belo Horizonte, Minas Gerais, Brazil

Bernard La Scola
URMITE CNRS UMR 6236- IRD 3R198, Aix Marseille Universite, Marseille, France

Kristijn R. R. Swinnen, Jonas Reijniers, Matteo Breno and Herwig Leirs
Evolutionary Ecology Group, Biology Department, University of Antwerp, Antwerpen, Belgium

Naraporn Somboonna, Duangjai Sangsrakru, Sithichoke Tangphatsornruang and Sissades Tongsima
Department of Microbiology, Faculty of Science, Chulalongkorn University, Bangkok, Thailand

Alisa Wilantho
Genome Institute, National Center for Genetic Engineering and Biotechnology, Pathumthani, Thailand

Kruawun Jankaew
Department of Geology, Faculty of Science, Chulalongkorn University, Bangkok, Thailand

Anunchai Assawamakin
Department of Pharmacology, Faculty of Pharmacy, Mahidol University, Bangkok, Thailand

Reti Hai, Yulin Wang, Xiaohui Wang and Yuan Li, Zhize Du
Department of Environmental Science and Engineering, Beijing University of Chemical Technology, Beijing, China

Giuliana D. Noratto and Hercia S. Martino
Department of Nutrition and Food Science, Texas A&M University, College Station, Texas, United States of America

Susanne U. Mertens-Talcott
Department of Nutrition and Food Science, Texas A&M University, College Station, Texas, United States of America
Veterinary Physiology and Pharmacology, Texas A&M University, College Station, Texas, United States of America

Jose F. Garcia-Mazcorro
Facultad de Medicina Veterinaria y Zootecnia, Universidad Autónoma de Nuevo León, General Escobedo, Nuevo León, México

Melissa Markel, Yasushi Minamoto, Jörg M. Steiner and Jan S. Suchodolski
Gastrointestinal Laboratory, Texas A&M University, College Station, Texas, United States of America

David Byrne
Department of Horticultural Sciences, Texas A&M University, College Station, Texas, United States of America

Kevin J. Purdy
School of Life Sciences, University of Warwick, Coventry, United Kingdom

Franck Carbonero
School of Life Sciences, University of Warwick, Coventry, United Kingdom
Department of Food Science, University of Arkansas, Fayetteville, Arkansas, United States of America

Brian B. Oakley
United States Department of Agriculture, Agricultural Research Service, Richard B. Russell Research Center, Athens, Georgia, United States of America

Denise Küng and Leyla R. Davis
Institute of Evolutionary Biology and Environmental Studies, University of Zurich, Zurich, Switzerland

Douglas C. Woodhams
Institute of Evolutionary Biology and Environmental Studies, University of Zurich, Zurich, Switzerland

Smithsonian Tropical Research Institute, Balboa, Ancón, Panamá, República de Panama

Laurent Bigler
Institute of Organic Chemistry, University of Zurich, Zurich, Switzerland

Brian Gratwicke
Center for Species Survival, Conservation and Science, National Zoological Park, Smithsonian Institution, Washington DC, United States of America

Edgardo Griffith
El Valle Amphibian Conservation Center, El Valle, República de Panamá

Nina Adam
The Wallenberg Laboratory and Sahlgrenska Center for Cardiovascular and Metabolic Research, Department of Molecular and Clinical Medicine, University of Gothenburg, Gothenburg, Sweden

Felix Sommer
The Wallenberg Laboratory and Sahlgrenska Center for Cardiovascular and Metabolic Research, Department of Molecular and Clinical Medicine, University of Gothenburg, Gothenburg, Sweden
Mucin Biology Group, Department of Medical Biochemistry, University of Gothenburg, Gothenburg, Sweden

Fredrik Bäckhed
The Wallenberg Laboratory and Sahlgrenska Center for Cardiovascular and Metabolic Research, Department of Molecular and Clinical Medicine, University of Gothenburg, Gothenburg, Sweden
Novo Nordisk Foundation Center for Basic Metabolic Research, Section for Metabolic Receptology and Enteroendocrinology, Faculty of Health Sciences, University of Copenhagen, Copenhagen, Denmark

Malin E. V. Johansson and Gunnar C. Hansson
Mucin Biology Group, Department of Medical Biochemistry, University of Gothenburg, Gothenburg, Sweden

Lijun Xia
Cardiovascular Biology Research Program, Oklahoma Medical Research Foundation, Oklahoma City, Oklahoma, United States of America

David A. Head
School of Computing, University of Leeds, Leeds, United Kingdom

Phil D. Marsh
Microbiology Services, PHE Porton, Salisbury, United Kingdom
Department of Oral Biology, School of Dentistry, University of Leeds, United Kingdom

Deirdre A. Devine
Department of Oral Biology, School of Dentistry, University of Leeds, United Kingdom

Steven W. Kembel
Département des sciences biologiques, Universitédu Québec áMontréal, Montréal, Québec, Canada
Biology and the Built Environment Center, University of Oregon, Eugene, Oregon, United States of America
Institute of Ecology and Evolution, University of Oregon, Eugene, Oregon, United States of America

James F. Meadow and Brendan J. M. Bohannan
Biology and the Built Environment Center, University of Oregon, Eugene, Oregon, United States of America
Institute of Ecology and Evolution, University of Oregon, Eugene, Oregon, United States of America

Timothy K. O'Connor
Biology and the Built Environment Center, University of Oregon, Eugene, Oregon, United States of America
Institute of Ecology and Evolution, University of Oregon, Eugene, Oregon, United States of America
Department of Ecology and Evolutionary Biology, University of Arizona, Tucson, Arizona, United States of America

Gwynne Mhuireach, Dale Northcutt, Jeff Kline and Maxwell Moriyama
Biology and the Built Environment Center, University of Oregon, Eugene, Oregon, United States of America
Energy Studies in Buildings Laboratory, University of Oregon, Eugene, Oregon, United States of America

G. Z. Brown
Biology and the Built Environment Center, University of Oregon, Eugene, Oregon, United States of America
Energy Studies in Buildings Laboratory, University of Oregon, Eugene, Oregon, United States of America
Department of Architecture, University of Oregon, Eugene, Oregon, United States of America

Jessica L. Green
Biology and the Built Environment Center, University of Oregon, Eugene, Oregon, United States of America
Institute of Ecology and Evolution, University of Oregon, Eugene, Oregon, United States of America
Santa Fe Institute, Santa Fe, New Mexico, United States of America

Shane Golden, Reuven Dukas
Animal Behaviour Group, Department of Psychology, Neuroscience & Behaviour, McMaster University, Hamilton, Ontario, Canada

Charles K. Fisher, Pankaj Mehta
Department of Physics, Boston University, Boston, Massachusetts, United States of America

Lapo Mughini-Gras, Lebana Bonfanti, Paolo Mulatti, Isabella Monne and Stefano Marangon
Istituto Zooprofilattico Sperimentale delle Venezie (IZSVe), Legnaro, Padua, Italy

Vittorio Guberti
Institute for Environmental Protection and Research (ISPRA), Ozzano dell'Emilia, Bologna, Italy

Paolo Cordioli
Istituto Zooprofilattico Sperimentale della Lombardia e dell'Emilia Romagna (IZLER), Brescia, Italy

Index

P

R

S

T

V

W

Y